Y0-AAX-536

NATO ASI Series

Advanced Science Institutes Series

A series presenting the results of activities sponsored by the NATO Science Committee, which aims at the dissemation of advanced scientific and technological knowledge, with a view to strengthening links between scientific communities.

The series is published by an international board of publishers in conjunction with the NATO Scientific Affairs Division

A	Life Sciences	Plenum Publishing Corporation
B	Physics	London and New York
C	Mathematical and Physical Sciences	D. Reidel Publishing Company Dordrecht, Boston and Lancaster
D	Behavioural and Social Sciences	Martinus Nijhoff Publishers
E	Engineering and Materials Sciences	The Hague, Boston and Lancaster
F	Computer and Systems Sciences	Springer Verlag
G	Ecological Sciences	Heidelberg

Mathematical and Physical Sciences No. 108

Air-Sea Exchange
of Gases and Particles

Series C: N

Air-Sea Exchange of Gases and Particles

edited by

Peter S. Liss

School of Environmental Sciences,
University of East Anglia, Norwich, U.K.

and

W. George N. Slinn

Pacific Northwest Laboratory,
Battelle Memorial Institute, Richland, Washington, U.S.A.

D. Reidel Publishing Company

Dordrecht / Boston / Lancaster

Published in cooperation with NATO Scientific Affairs Division

Proceedings of the NATO Advanced Study Institute on
Air-Sea Exchange of Gases and Particles
University of New Hampshire, Durham, New Hampshire, U.S.A.
19-30 July, 1982

Library of Congress Cataloging in Publication Data

Main entry under title:

Air-sea exchange of gases and particles.

(NATO ASI series. Series C, Mathematical and physical sciences; no. 108)
"Advanced Study Institute on Air-Sea Exchange of Gases and Particles
held at the University of New Hampshire from 19 to 30 July 1982"—P.
Bibliography : p.
Includes index.
1. Ocean—atmosphere interaction—Congresses. 2. Gases—Congresses.
3. Particles—Congresses. I. Liss, P. S. II. Slinn, W. George N., 1938-
III. Advanced Study Institute on Air-Sea Exchange of Gases and Particles
(1982 : University of New Hampshire) IV. Series: NATO ASI series.
Series C, Mathematical and physical sciences ; 108.
GC190.A35   1983   551.57'09162   83-9521
ISBN 90-277-1610-2

Published by D. Reidel Publishing Company
P.O. Box 17, 3300 AA Dordrecht, Holland

Sold and distributed in the U.S.A. and Canada
by Kluwer Academic Publishers,
190 Old Derby Street, Hingham, MA 02043, U.S.A.

In all other countries, sold and distributed
by Kluwer Academic Publishers Group,
P.O. Box 322, 3300 AH Dordrecht, Holland

D. Reidel Publishing Company is a member of the Kluwer Academic Publishers Group

Printed in The Netherlands

CONTENTS

INTERFACE - P. S. Liss and W. G. N. Slinn   xi

ACKNOWLEDGMENTS   xiii

INTRODUCTORY METEOROLOGY AND FLUID DYNAMICS - L. Hasse   1

    1. Some Deductions from Basic Physics   1
  2 2. Large-Scale Atmospheric Circulations   11
    3. Synoptic Scale Circulations   23
    4. Mesoscale Circulations   27
    5. Convection and Stability   30
    6. Turbulence and the Planetary Boundary Layer   35
    7. Interface Dynamics   39
        7.1 Flow at a solid, level surface   39
        7.2 Flow at a fluid interface   42
        7.3 Stability effects in the surface layer   42
        7.4 Parameterization   43
    8. Practical Aspects of Atmospheric Transports   45
        8.1 Vertical transports   45
        8.2 Horizontal transports   46

INTRODUCTORY PHYSICAL OCEANOGRAPHY - F. Dobson   53

    Large-scale Oceanography   53
        Geostrophic, baroclinic and barotropic
        motions   54
        "Ekman" transport   56
        "Sverdrup" transport   58
        Westward intensifcation   59
        Physical properties of sea water   62
        Water property distributions and tracers   63
        Thermohaline circulation   65
        Diffusion and mixing   68
        Turbulent mixing   69
    Mesoscale Oceanography   71
        Ocean eddies   71
        Gulf stream rings   72
        Frontal processes   76
        Tides and large-scale waves   80
        Internal waves   83
    The Oceanic Mixed Layer   84
        Time and space scales   84
        Introduction   85
        Modelling the mixed layer   85
        Forcing variables   87

Solar radiation   87
Wind forcing   90
Buoyancy   91
Conservation relations   92
Shear instabilities   92
Entrainment   93
Advection   93
Langmuir circulations   94
Role of the mixed layer in ocean dynamics
and climate   97
Sea Surface Waves   98
Introduction   98
Wave breaking   103
Wave generation and growth   106
Appendix : Definitions   111

MICROBIOLOGICAL AND ORGANIC-CHEMICAL PROCESSES IN THE
SURFACE AND MIXED LAYERS - J. McN. Sieburth   121

Introduction   121
1. Planktonic Microorganisms and Their Trophic
    Structure   123
    1.1 Evolution of decomposing and synthesizing
         microorganisms and their processes   123
    1.2 Trophic structure and transfer   126
    1.3 Primary producers (the Phototrophs and
         Chemolithotrophs)   129
    1.4 Saprotrophs (the Heterotrophic Bacteria)   130
    1.5 Biotrophs (the Protozoa)   131
2. Uptake and Release of Gases During the Synthesis
    and Decay of Organic Matter and its Control by
    the Solar Cycle   135
    2.1 Cycling of gases and organic matter by
         aerobic processes   135
    2.2 Cycling of gases and organic matter by
         anaerobic processes   138
    2.3 Diel cycle of chemical activity   140
    2.4 Size fractions of plankton and organic
         matter   143
    2.5 Physical factors that control microbiological
         processes   146
3. Organic-Microbial Films that Form the Sea's Skin   148
    3.1 Physical factors holding molecules and
         microbes at the sea-air interface   148
    3.2 Nature of the organic films   149
    3.3 Nature of the microorganisms colonizing
         and aggregating at the surface film   154
    3.4 Spatial and temporal variation in the film   158
    3.5 Crucial function of the surface film   162

GASES AND THEIR PRECIPITATION SCAVENGING IN THE MARINE
ATMOSPHERE - L. K. Peters   173

  1. Introduction   173
  2. Trace Gas Concentrations in Marine Air   175
      2.1 Concentrations of long-lived species   176
      2.2 Concentrations of short-lived species   177
      2.3 Concentrations of free radicals   178
  3. Homogeneous Marine Air Chemistry   179
      3.1 Photochemical considerations   179
      3.2 Pseudo-steady state   181
      3.3 The $NO-NO_2-O_3$ subsystem   182
      3.4 The CH_4- CO oxidation sequence   184
      3.5 The HO_x and RO_x radical chemistry   186
      3.6 The chemistry of SO_2 188
  4. Exchange of Simple Soluble Gases with Hydrometeors   189
      4.1 Basic principles   189
      4.2 The overall mass transfer coefficient   196
      4.3 Individual phase-controlled processes   198
  5. Exchange of Reactive and Ionizable Gases   199
      5.1 Equilibrium considerations   200
      5.2 Enhancement of mass transfer rate   201
      5.3 SO_2 absorption by water   205
  6. The Role of Circulation Within Drops   211
      6.1 The work of Kronig and Brink   211
      6.2 Circulation with chemical reaction   214
      6.3 Comparison of theory with experimental
          results   216
  7. Heterogeneous Processes   217
      7.1 General transport equations   218
      7.2 Particle-gas interaction   219
      7.3 Hydrometeor-gas interaction   225
  8. Scavenging in Complex Systems   229
      8.1 Spatial effects and equilibrium and rever-
          sible scavenging   229
      8.2 Simultaneous absorption of several gases   230

GAS TRANSFER: EXPERIMENTS AND GEOCHEMICAL IMPLICATIONS -
P. S. Liss   241

  1. Introduction   241
  2. Laboratory Sutdies of Air-Water Gas Transfer   243
      2.1 Physical transfer processes   244
      2.2 Modelling approaches   251
      2.3 Gases for which $r_a \gg r_w$ 253
      2.4 Chemical transfer processes   254
  3. Air-Sea Gas Transfer in the Field   260
      3.1 Gases for which $r_w \gg r_a$ 260
      3.2 The role of bubbles   268

3.3 Gases for which $r_a \gg r_w$ 269
4. Possible Role of Natural and Pollutant Films
    in Air-Sea Gas Transfer   270
    4.1 Laboratory studies   270
    4.2 The role of films in gas transfer at the
         sea surface.   274
5. The Role of Air-Sea Gas Exchange in Geochemical
    Cycling   276
    5.1 Introduction   276
    5.2 Hydrogen   277
    5.3 Carbon   280
    5.4 Nitrogen   283
    5.5 Oxygen   285
    5.6 Sulphur   285
    5.7 Chlorine   287
    5.8 Iodine   289
    5.9 Conclusions   289

AIR-TO-SEA TRANSFER OF PARTICLES - *W. G. N. Slinn*   299

1. Introduction   299
2. Residence Times   300
    2.1 Aerosol particles   301
    2.2 Wet and dry deposition velocities   303
    2.3 A simple model for τ_w and τ_d 308
    2.4 Scavenging ratios and τ_w 309
    2.5 Scavening rates and τ_w 310
    2.6 Scavenging efficiencies and τ_w 312
    2.7 Scavenging as a stochastic process   313
    2.8 Summary   317
3. Scavenging by Precipitation   317
    3.1 Rain scavenging by bugs   318
    3.2 Scavenging rates and the collection
         efficiency   319
    3.3 Collection by diffusion   321
    3.4 Impaction, interception, etc.   325
    3.5 Collision efficiency summary   330
    3.6 Approximate scavenging rates   333
4. Scavenging by Storms   338
    4.1 Bomb debris data and Junge's model   339
    4.2 Particle-size dependence of scavenging
         ratios   344
    4.3 Scavenging and precipitation efficiencies   349
    4.4 Storm efficiencies   354
    4.5 Summary for τ_w 357
5. Dry Deposition of Particles   361
    5.1 The dry flux   361
    5.2 Turbulence   363
    5.3 A simple two-layer model   368

5.4 Some small-scale problems   376
6. More Problems with Dry Deposition   382
    6.1 The continuity equation and boundary
         conditions   382
    6.2 Resuspension and spray scavenging   385
    6.3 The constant-flux condition   388
    6.4 Resistance models   391
    6.5 Horizontal and vertical concentration
         gradients   393
    6.6 Actual deposition and mass-mean values   394
    6.7 Residence times, revisited   396

THE PRODUCTION, DISTRIBUTION AND BACTERIAL ENRICHMENT
OF THE SEA-SALT AEROSOL - D. C. Blanchard   407

1. Introduction   407
2. Production of the Sea-Salt Aerosol   408
    2.1 Bubble-size distributions   408
    2.2 Bubble dissolution   412
    2.3 Percentage of sea surface producing
         sea-salt particles   413
    2.4 Ejection of sea-salt particles from the
         sea   414
    2.5 Seawater in the atmosphere   420
3. Distribtuion of the Sea-Salt Aerosol   421
    3.1 Sea-salt concentration as a function of
         wind speed   422
    3.2 Size distribution of sea-salt particles   426
    3.3 The sea-salt inversion   428
    3.4 An attempt to find the salt inversion   432
4. Enrichment of Bacteria in Jet and Film Drops   433
    4.1 Jet drop enrichment   434
    4.2 Film drop enrichment   441
    4.3 Health hazards   442
    4.4 Bacterial as ice nuclei   443
5. Conclusions   444

PARTICLE GEOCHEMISTRY IN THE ATMOSPHERE AND OCEANS -
P. Buat-Ménard   455

1. Introduction   455
2. Geochemistry of Marine Aerosols   455
    2.1 Introduction   455
    2.2 Rationale   456
    2.3 The data bank   458
    2.4 Basic criteria for source identification   458
    2.5 General considerations of the composition
         of crustal and oceanic source aerosols   460

2.6   Two examples of source differentation
       for marine aerosols   463
2.7   Tracers and source markers   466
2.8   Interaction between trace gases and
       particles in the marine atmosphere   476
2.9   Conclusions   478
3.   Geochemistry of Particle Transfer Between the
     Atmosphere and the Oceans   479
3.1   Introduction   479
3.2   Bubbles and the chemical composition of
       sea-source aerosols   479
3.3   Enrichment processes at the air-sea and
       bubble-seawater interfaces   484
3.4   The fate of reactive elements entering the
       ocean   491
4.   Assessment of Particulate Fluxes from and to the
     Air-Sea Interface : Geochemical Implications   496
4.1   Introduction   496
4.2   Basis for the calculation of atmospheric
       particulate fluxes   496
4.3   General considerations of the role of air-
       sea particulate exchange in geochemical
       cycling   503
4.4   A case example: The air-sea exchange of
       trace metals   507
5.   Conclusions   520

APPENDIX A - List of Symbols   533

APPENDIX B - Annotated Bibliography of Sampling   543

SUBJECT INDEX   551

INTERFACE

This book arises from a NATO-sponsored Advanced Study
Institute on 'Air-Sea Exchange of Gases and Particles' held at
the University of New Hampshire (UNH) from 19 to 30 July 1982.
The chapters of the book are the written versions of the lectures
given at the Institute. The authors' combined aim is to present
an up-to-date, pedagogical account of the subject. Although the
chapters contain much current research material, the main objec-
tive is to give the reader a basic understanding of how material
exchange takes place at the air-sea interface.

Just as this introduction forms an interface between authors
and readers (hence its title), so the sea surface is the channel
of communication for material transfer between the atmosphere and
the oceans. Such exchange can be in either direction and the
transferring material may be in gas, liquid or solid phases.
When the exchange is purely gaseous or by deposition of particles,
a clear identification of the phases involved can sometimes be
made. Even here the case is not always clear-cut since, for
example, 'dry' particles in a humid atmosphere (close to the sea
surface) will in reality be substantially 'wet'. The situation
for transfer of liquid drops in either direction is even more
complex since, as well as the water itself and its dissolved ions,
the drops may contain dissolved gases and inorganic and organic
(both living and dead) particulate material. Added to all this,
in order to treat interfacial transfer, there must be some understanding of basic meteorology and oceanography (including biologically mediated processes).

In this book fundamental atmospheric and oceanographic knowledge is described in the first three chapters (Hasse, Dobson,
Sieburth). They are followed by chapters on atmospheric gases
and their removal by rain (Peters) and gas exchange across the
sea surface (Liss). Finally there is a group of three chapters
mainly concerned with transfer in solid and liquid phases: wet
and dry deposition of particles (Slinn), production of bubbles
and the input of micro-organisms and sea-salt particles to the
atmosphere (Blanchard), and the geochemistry of particles in the

P. S. Liss and W. G. N. Slinn (eds.), Air-Sea Exchange of Gases and Particles, xi–xii.
Copyright © 1983 by D. Reidel Publishing Company.

marine environment (Buat-Menard).

After the chapters are two appendices, the first of which contains a list of symbols, together with a brief evaluation of the physical significance of most of the dimensionless groups referred to in the text. The second appendix is a short, annotated bibliography of sampling techniques used in air-sea exchange studies. Since authors in their chapters had insufficient space to deal with sampling, it is hoped that the bibliography will give readers an entry into this very important aspect of the subject.

Clearly the book could have been improved by more strenuous editing. For example, although some overlap (and underlap) between chapters has been eliminated following the presentations at the Institute, it still exists. However, to have removed it all would have involved substantial iteration of texts between authors and editors, with consequent delay to the publication. Furthermore, despite the hope that readers partake of the book as a whole, some are likely to examine only parts. Elimination of all overlap would have meant that individual chapters would not have stood by themselves, thus, possibly, decreasing their usefulness. Also, some deviations from the list of symbols exist, and may even be desirable in particular cases. The readers' indulgence to these faults is requested – with a subject advancing as rapidly as the present one, speedy transmission of knowledge is probably more important than final polishing of the text.

Understanding material transfer across the sea surface requires input of knowledge from many disciplines. This diversity is at one and the same time a large part of the fascination of the subject, and a prime difficulty. The eighty participants in the Institute (including the lecturers) acted as both teachers of what they were expert in and learners of other aspects. Doubtless, our readers will find themselves in both roles as they read what follows.

In the present state of knowledge in this area, it is inevitable that the chapters are individually authored and run largely along disciplinary lines. If, by gathering the separate contributions in one volume, progress can be made in raising the level of understanding and communication between the disciplines involved, then a major task will have been initiated. The subject will have come of age when it is possible to synthesise knowledge in a truly multidisciplinary framework. In this task there is still much work to be done.

December 1982.   Peter Liss, Norwich, UK.
                                       George Slinn, Richland, USA.

ACKNOWLEDGMENTS

The idea of holding the Institute originated from the NATO
Science Committee's Special Programme Panel on Air-Sea Inter-
action. Financial support came from the NATO Advanced Study
Institute Programme.

During the Institute the following case studies were presented: Prof. W.S. Broecker, 'The Role of the Oceans in the CO_2/
Climate Issue'; Dr. J.N. Galloway, 'The Acidity of Rain Over the
Oceans'; Dr. K.A. Rahn, 'Long-Range Transport of Pollutants to
the Arctic'. We much appreciate the efforts made by the three
presenters in addressing, in a short space of time, some of the
more applied aspects of the subject.

We thank Janet Doty, Jane Parsons and the staff of the New
England Center (UNH) for running the domestic side of the
Institute. Jim Love and Ted Loder of the Department of Earth
Sciences (UNH) provided innumerable services and organised a
marvellous social programme which provided invaluable breaks in
a hard-working meeting. The friendship, help and consideration
they showed was greatly appreciated by the participants.

Finally, the authors would like to thank the participants in
the Institute for the active and constructive way in which they
discussed the lecture material, both inside and outside the
sessions, and for giving of their ideas so freely. Written and
verbal comments from participants have greatly aided conversion
of the preprint material handed out at the Institute into the
chapters of the book. The enthusiasm showed by all those who took
part was a vital ingredient in making the Institute both successful and enjoyable.

INTRODUCTORY METEOROLOGY AND FLUID DYNAMICS

Lutz Hasse

Institut für Meereskunde an der
Universität Kiel
D 2300 Kiel, B.R. Deutschland

1. SOME DEDUCTIONS FROM BASIC PHYSICS

Atmosphere and ocean are probably the most complex system
which is treated in physics, since the scales of motions involved
span such a wide range:some 10,000 km for long planetary waves,
through a few millimeters for the scale of decaying turbulent
eddies. The motions at different scales interact. An overview
of such a complex system is necessarily incomplete. Emphasis is
on such processes which relate more or less directly to the transport of gases and particles. The treatment in the introductory
chapters on meteorology and physical oceanography is descriptive
rather than deductive.

Meteorologists and oceanographers describe the motion and
thermodynamics of the atmosphere and the ocean by use of the
basic laws of physics. Hence, the understanding is based not only
on observations, but also on physical reasoning and theoretical
deductions from first principles. There is conservation of momentum, conservation of mass (for air, water vapor, water and salt),
conservation of energy and equations of state for ideal gases and
a thermohaline sea.

The balance of momentum yields the equations of motion, also
called Navier-Stokes equations. According to Newton's second law;
mass times acceleration is equal to the sum of forces acting on
the mass. The balance is given in (1) for a volume of fluid,
written as an equation for accelerations, that is Newton's law is
divided by the mass per volume, the density ρ.

P. S. Liss and W. G. N. Slinn (eds.), Air-Sea Exchange of Gases and Particles, 1—51.
Copyright © 1983 by D. Reidel Publishing Company.

$$\frac{du}{dt} = -\frac{1}{\rho}\frac{\partial p}{\partial x} + fv - \frac{1}{\rho}\frac{\partial \tau_{xz}}{\partial z} \tag{1.1}$$

$$\frac{dv}{dt} = -\frac{1}{\rho}\frac{\partial p}{\partial y} - fu - \frac{1}{\rho}\frac{\partial \tau_{yz}}{\partial z} \tag{1.2}$$

$$\frac{dw}{dt} = -\frac{1}{\rho}\frac{\partial p}{\partial z} - g \tag{1.3}$$

Velocity components u,v,w are given in a cartesian coordinate system with z vertical so that the acceleration due to gravity, g, is in the negative z direction. The other terms on the right hand side are the acceleration due to the pressure gradient, the Coriolis acceleration and the acceleration due to friction. Newton's law is valid only in an unaccelerated coordinate system. Since we are measuring in a coordinate system rotating with the earth, a fictious acceleration appears, called Coriolis acceleration or acceleration due to the rotation of the earth. A motion which is straight and unaccelerated in an inertial (unaccelerated) system will appear as curved to an observer rotating with the earth. Looking with the flow, the deflection is to the right on the Northern Hemisphere and to the left on the Southern. The factor

$$f = 2\Omega\sin\phi, \tag{2}$$

where ϕ is latitude and Ω rotation rate of the earth, is called the Coriolis parameter. Hence the Coriolis acceleration vanishes at the equator where f changes its sign.

The third term on the right hand side is the friction term. τ denotes a stress that is a force per area. The index xz denotes that a force in x direction acts on a plane with the normal in z direction. The difference of the stresses at the sides of any volume is felt as a frictional force by the volume. The stress may also be considered as the x-momentum going through the plane with normal z per unit time and area. This momentum transport is brought about by molecular movements (Newtonian friction). The equations of motion are derived for the forces at a volume in a given moment. They are also used for the 10 minute means of velocity as usually determined in routine measurements. If used

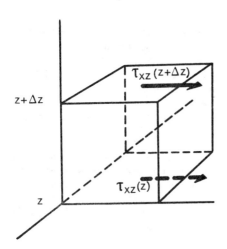

for mean velocity components, the shorter scale motions also transfer momentum and, therefore, additional stresses, the turbulent or
Reynolds stresses, appear. The stress components are then the sum
of turbulent and molecular stress, and so is the friction term.
Except very near to the interface (see the interface section below) the turbulent stresses are much larger than the molecular
and the friction term is essentially due to turbulence.

The left hand side of (1) gives the acceleration of a fluid
parcel. Since the forces operate on the mass of the parcel, this
is called a "substantive" or "material" derivative. This is the
derivative following a moving parcel. Since measurements are more
often done at one point over a time or at one time over a spatial
distance, Euler's relation (3) is used heavily,

$$\frac{dp}{dt} = \frac{\partial p}{\partial t} + u \frac{\partial p}{\partial x} + v \frac{\partial p}{\partial y} + w \frac{\partial p}{\partial z} \tag{3}$$

where p stands for any meteorological or oceanographical variable
$p(x,y,z,t)$ like a temperature or velocity component.

The notations $\partial p/\partial t$, $\partial p/\partial x$, $\partial p/\partial y$, $\partial p/\partial z$ denote partial derivatives with respect to time and the three spatial coordinates.
The notation implies that although p is a function of four variables, the remaining three are held constant when differentiating.
The partial derivatives are also called local derivatives in contrast to the substantive derivative. The last three terms on the

Scale	Inertial terms horiz.	vert.	Pressure term	Coriolis term	Friction term
OCEAN Large-scale	10^{-8}	10^{-8}	10^{-5}	10^{-5}	10^{-8}
Meso-scale	10^{-4}	10^{-5}	10^{-4}	10^{-4}	10^{-5}
Small-scale	10^{-2}	10^{-2}	10^{-2}	10^{-4}	10^{-2}
ATMOSPHERE Synoptic-scale	10^{-4}	10^{-4}	10^{-3}	10^{-3}	10^{-5}
Meso-scale	$3 \cdot 10^{-4}$	10^{-4}	$3 \cdot 10^{-4}$	$3 \cdot 10^{-4}$	10^{-5}
Cumulus-scale	10^{-3}	10^{-3}	10^{-3}	$3 \cdot 10^{-4}$	10^{-4}
Small-scale	10^{-4}	10^{-4}	10^{-3}	10^{-3}	10^{-3}

Table 1. Magnitude of terms (in m/s^2) in the horizontal
equations of motion for various scales of motion in
the ocean and the atmosphere.

right are called the advective terms, since if there is a gradient

$$\frac{\partial p}{\partial x}, \frac{\partial p}{\partial y}, \frac{\partial p}{\partial z}$$

of the property. p, this will be advected (transported) with the
flow, given by the velocity vector (u,v,w). The local rate of
change at a point is determined by

$$\frac{\partial p}{\partial t} = \frac{dp}{dt} - \left(u \frac{\partial p}{\partial x} + v \frac{\partial p}{\partial x} + w \frac{\partial p}{\partial z} \right) \qquad (3)$$

that is, by the rate of change occurring in the parcel which happens to be at the point, minus the rate of change due to the fact
that parcels with different values of p are carried over the point
by the flow of the fluid.

If (3) is used to rewrite the terms on the left side of (1),
it is seen that the Navier-Stokes equations are non-linear, that
is, the variables u,v,w appear as products and squares. This indicates that the different scales of motion will interact. Hence,
we have a very complex interactive system. With the presence of
small scales and the interaction between scales, the atmosphere and
ocean possess a certain degree of unpredictability. (Small scale
in this sense means smaller than we could ever resolve in experimental determination of an initial state of the atmosphere and
ocean or in their numerical modelling. This includes scales as
large as cumulus clouds.)

The equations of motion contain the density ρ. In meteorology,
it is customary to measure air temperature and pressure, while the
density or specific volume is not. The mixture of gases that constitute the dry atmosphere behave like an ideal gas. Hence, the
equation of state

$$p/\rho = RT \qquad (4)$$

(where R is the individual gas constant of dry air) can be used to
calculate density from temperature and pressure. The effects of
humidity are taken into account separately. In the case of sea
water the density depends, in a weakly nonlinear way, on both its
temperature and its concentration of dissolved salts, referred to
as its "salinity" (See Dobson, this volume.)

In large-scale flow the vertical accelerations are small in
the sense that dw/dt can be neglected against g. Hence, the vertical component of the Navier-Stokes equations reduce to the hydrostatic equation:

$$\frac{\partial p}{\partial z} = - g\rho \qquad (5)$$

In static equilibrium the pressure is the weight of the fluid in the
vertical column over any area. Hence, the pressure field is given
by the density field. The equation of state for air tells us that

for a given pressure the density depends only on the temperature. Hence, the wind field is known from the temperature field, except near the surface where a deviation is forced by friction.

The situation is similar in the ocean in the sense that temperature and salinity determine the density field and hence the pressure field. This dual dependence means that flows exist in the ocean which are driven by the influence of both temperature and salinity; they are called "thermohaline" circulations. Yet, determination of flow by aid of the Navier-Stokes equations is more difficult in the ocean than in the atmosphere. The hydrostatic equation is a differential equation and must be integrated with height to obtain a relationship between pressure and height. In the atmosphere the integration is done from the mean sea surface level. In the ocean such reference level is not available and oceanographers must infer the absolute velocity by other means.

The equations of motion, as given in (1), are slightly simplified, but perfectly sufficient for the purpose of this book. An analysis of the order of the terms is given in Table 1. In many cases even simpler forms can be used, see for example the hydrostatic equation (5) as an approximation of the vertical component of the equations of motion. This is a good approximation for all except the smallest scales of motion (that is for most oceanic and atmospheric flows except for convection, surface and internal waves, and turbulence).

For the horizontal components (1.1, 1.2) the simplest type of equilibrium flow is a balance of the pressure gradient and the Coriolis accelerations:

$$u_g = -\frac{1}{\rho f} \frac{\partial p}{\partial y}$$
$$v_g = +\frac{1}{\rho f} \frac{\partial p}{\partial x}$$

(6)

This flow is called the geostrophic wind or current. The geostrophic equilibrium is a good approximation for the velocity field in the free atmosphere (i.e. above the influence of surface friction) and in the interior of the ocean, away from boundaries. It provides for the flow to follow the direction of the isobars (lines of constant pressure). Hence, in the atmosphere we have the typical flow direction around lows and highs as indicated in Figure 1. Geostrophic flow is a helpful device, since it relates the wind or current in a simple way to the pressure field.

Warm fluid is less dense than cold fluid. As the Figure 2 shows, a horizontal temperature gradient yields different horizontal pressure gradients at different heights. Consequently the geo-

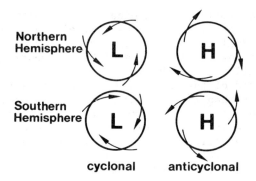

Figure 1. Circulation pattern around highs and lows at
the Northern and Southern Hemisphere. In this idealised
sketch the circles are isobars. The wind arrows would
be parallel to the isobars in the free atmosphere. Convergence into the lows and divergence out of the highs
is found at the surface due to friction.

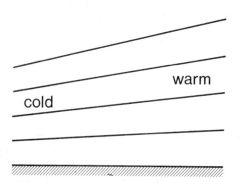

Figure 2. Sketch showing the effect of horizontal temperature gradients on the height of isobaric surfaces
(full lines) in a cross section. The increasing slope
of the isobaric surfaces with increasing height yields,
in geostrophic equilibrium, an increasing wind component
in the direction into the paper (in the Northern Hemisphere, reversed direction in the Southern Hemisphere).

strophic velocity varies with height. This change of geostrophic
flow with height is called the "thermal wind", since it is related to the horizontal temperature gradient. It is also called the
"geostrophic shear". In the atmosphere the most prominent example
of the thermal wind is at the strong temperature gradients of
frontal zones. The polar and subtropical jet streams are found at
the top of these frontal zones. Similar fields are caused in the
ocean by salinity and temperature (i.e. density) gradients.

Though in the free atmosphere the geostrophic equilibrium is
a good approximation, the small deviations from this equilibrium
are responsible for the change of weather. Also, the friction at
the surface has considerable influence on weather. Near the surface, there is an approximate equilibrium between Coriolis acceleration, pressure gradient and friction. Friction yields a reduced wind speed compared to geostrophic flow and a deviation of the
wind direction from the isobars towards low pressure. Hence, near
the surface there is an outflow from a high and an inflow into a
low (see Figure 1). This results in ascending motion at low pressure and descending at highs. Much of the motion in the ocean is
in geostrophic balance but, like the atmosphere, small deviations
from the balance have important effects when superimposed on the
geostrophic flow. In the surface waters, where friction is important, the wind drives geostrophic "Ekman" flows which diverge and
converge in response to large-scale variations in the strength
and direction of the wind; such effects in turn induce the large-
scale gyral motions in the ocean.

Conservation of mass is used to derive the continuity equation

$$\frac{\partial u}{\partial x} + \frac{\partial v}{\partial y} + \frac{\partial w}{\partial z} = -\frac{1}{\rho}\frac{d\rho}{dt} \tag{7}$$

This simply states that if the volume occupied by a given mass is
changed, its density must change in reverse. The continuity equation is mostly used in the incompressible form, that is for constant density $d\rho/dt = 0$. This is correct in the sense that the effect of density variations is smaller than that of velocity variations. Under this condition, the continuity equation can be used
to calculate vertical velocities from horizontal convergences or
divergences.

To understand why ascending and descending motions in the
atmosphere produce quite different effects we have to mention a
few other processes.

Conservation of energy is used to derive the first law of
thermodynamics. This says in the simple form

$$dU = dQ - pdV \tag{8}$$

that for a closed thermodynamic system the rate of change of internal energy dU is proportional to the rate at which thermal energy
dQ is applied to the system minus the rate at which mechanical
work is done. The mechanical work is typically the work done by
compression or expansion of fluid under the pressure of the fluid
around it. Since, in meteorology, density or specific volume is
usually not measured, the first law is rewritten in terms of temperature and pressure as

$$\frac{dQ}{T} = c_p \cdot \frac{dT}{T} - (c_p - c_v) \frac{dp}{p} \qquad (9)$$

An important application of this equation is for vertical
movements of fluid parcels. Since there are always strong vertical
pressure gradients in the air and sea water, a vertical movement
will lead to a pressure change, which, by virtue of the ideal gas
equation for air and the equation of state for water, will result
in a change of volume and/or temperature. The most simple case is
the one where no thermal energy is supplied, called an "adiabatic
process". In this case in the atmosphere there exists a unique
relationship between pressure and temperature, called Poisson's
equation. A parcel, which moves vertically under adiabatic conditions, has a "dry adiabatic lapse rate" (temperature drop with
height) of about 1 K/100 m in the atmosphere. Since processes –
aside from condensation/evaporation processes – that provide heat
to an atmospheric parcel typically have a time scale of order one
day, while many vertical movements are of shorter duration, the
assumption of adiabatic vertical movement is often a good approximation. The same applies to the deep ocean, except that mixing
rates are much slower, so the equivalent time scale comparison is
months to days instead of days to hours. In the ocean, the adiabatic lapse rate is about 0.15 K/km at a depth of 5 km.

One very important constituent of the atmosphere is water
vapor. The conservation of mass is also used to obtain a balance
equation for the water vapor content. If $\hat{W}$ is the rate of condensation, and q, specific humidity, then

$$\frac{\partial q}{\partial t} + u \frac{\partial q}{\partial x} + v \frac{\partial q}{\partial y} + w \frac{\partial q}{\partial z} = - \hat{W}/\rho \qquad (10)$$

A similar equation holds for the salt content in sea water. (The
equation is derived for a parcel, and the velocity components are
momentary components. The change of property q of the parcel by
molecular diffusion across the fluid boundaries of the parcel is
not considered; if important, that must be included in $\hat{W}$.)

Basic thermodynamics are also used to describe the phase
change of water vapor, fluid water and ice in the atmosphere. The
importance of water vapor in the atmosphere stems from the role of

evaporation-condensation-precipitation in the hydrological cycle, from the radiational properties of water vapor and of liquid water (in clouds), and from the energy transfer with the evaporated water. The heat of vaporisation is 2.5 kJ/g, the heat of fusion 0.33 kJ/g (both slightly temperature dependent). In the absence of liquid water, the water vapor follows the equation of an ideal gas with sufficient accuracy. In the presence of liquid water, there is an equilibrium between the gaseous and the liquid phase. This equilibrium is temperature dependent, as described by the saturation water vapor curve (Figure 3). This Figure is the graphical presentation of the Magnus equation which is derived from the Clausius-Clapeyron equation. The temperature dependence of the saturation

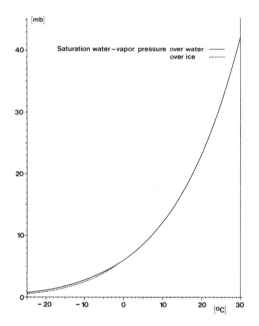

Figure 3. Equilibrium saturation water vapor curve. At vapor pressure below the curve, the relative humidity is less than 100%. At vapor pressure above the curve, the vapor would condense to either liquid water (mist or cloud droplets) or ice crystals (at low temperatures).

water vapor pressure is the reason for the higher water vapor content of tropical air compared to air of the temperate zones and the polar regions. In a similar way, since the temperature decreases on average by about 0.6 K/100 m in the troposphere, the water vapor content is mainly concentrated in the lower 1 km to 3 km. In the ocean there are no phase changes of comparable effect to that of water vapor in the air. Freezing and melting occur in

higher latitudes, but they are highly localized in time and space.
Evaporation at the sea surface increases water density both by
cooling and by increase in salinity, causing convective overturning of the surface waters. Precipitation causes stable stratification by a layer of fresh water on top of the salt water,noticable
at low wind speeds.

The cooling by expansion, together with the water vapor content of the atmosphere, is responsible for most of the "weather".
Consider rising air; by expansion, the temperature of the air is
lowered, but the air takes its moisture with it. With the cooling,
the humidity of the air may reach saturation and condensation will
take place with a release of latent heat. Cloud droplets and eventually rain drops will be formed. This is the typical case of
cumulus convection and of the ascending air in a depression. The
opposite case is found in the free air between cumulus clouds and
in high pressure centres. The sinking air is heated by adiabatic
(or nearly adiabatic) expansion, clouds evaporate and the air becomes dry. There is no analogous process in the ocean.

The role of the Coriolis acceleration in atmospheric and
oceanic flow is quite interesting. Consider a container filled with
fluid of different temperatures (densities) in an unaccelerated
system as in Figure 4a. If the divider is removed, the pressure
difference at the interface would produce a motion and in the final
stage (Figure 4c) the warm fluid would be stratified above the
cold fluid. Some of the system's potential energy has been converted into kinetic energy. The initial situation (without divider)
is unstable.

a.   b.   c.

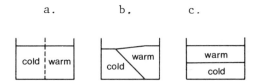

Figure 4. Neighbouring cold and warm fluid is unstably
stratified. In a non-accelerated system, the resulting
pressure gradient would drive the fluid until warm is
above cold (a→b→c). At the rotating earth a sloping
front, like (b), can be stable between moving cold and
warm fluid, since the Coriolis force can balance the
pressure gradient force.

On the rotating earth, in large-scale flow, if we neglect
friction, pressure gradients are balanced by the Coriolis acceleration. Hence,the flow is along the isobars and parallel to the
interface between cold and warm. A situation like Figure 4b can
be in equilibrium (that is, no flow perpendicular to the interface)
with geostrophic flow on both sides parallel to the interface.

This flow must fulfil a boundary condition similar to the thermal
wind condition depicted in Figure 2 (on the Northern Hemisphere,
the wind component into the paper increases with height). This is
remarkable, since in this way potential energy can be accumulated
when temperature differences due to differential heating build up.

If the horizontal temperature gradients become too large, the
equilibrium becomes unstable. That means, if a small disturbance
is superimposed on the geostrophic flow, part of the potential
energy is used and the disturbance is amplified, its kinetic energy is increased. The typical depressions of the midlatitudes (see
below) develop. This instability is called baroclinic instability.
'Baroclinic' indicates, that the surfaces of equal temperature
are inclined against the pressure surfaces. This is the case in
the center of Figure 2 and in Figure 4a and b. The situation with
temperature and pressure surfaces parallel, as in Figure 4c, is
called barotropic. Baroclinic instability is an important feature
of the large-scale flow on the rotating earth.

2. LARGE-SCALE ATMOSPHERIC CIRCULATIONS

The original source of energy for all atmospheric motion is
the radiation from the sun. Of the solar radiation impinging on
the outer edge of the atmosphere, about 30% (the so-called planetary albedo) is returned to space by reflection and scattering in
the atmosphere and reflection at the ground. The averaged daily
flux of solar energy on a horizontal plane at the top of the atmosphere is given in Figure 5. Due to the solar declination, the
annual amount of energy received at the equator is only larger by
a factor of 2.4 than that at the pole.

The energy received as solar (or short-wave) energy from the
sun by the atmosphere-plus-earth system must be returned to space
as terrestrial (or long-wave) radiation (Figure 6). The surface
of the earth emits according to Stefan-Boltzmann law, that is,
the terrestrial radiation increases with the fourth power of the
surface temperature. Since some atmospheric gases (especially water vapor and carbon dioxide) and aerosols (airborne particles and
water droplets of clouds) absorb and emit long-wave radiation, the
earth's surface also receives energy from the atmosphere, the back-
radiation. At the same time, the atmosphere radiates towards space.

In total, the atmosphere emits more energy than it absorbs
short- and long-wave radiation. The heat balance is closed by the
transfer of sensible heat and latent heat (via evaporation) from
the earth's surface to the atmosphere. That is, the solar radiation
received at the ground is used partially to heat the air and to
evaporate water. The heat used for evaporation is released to the

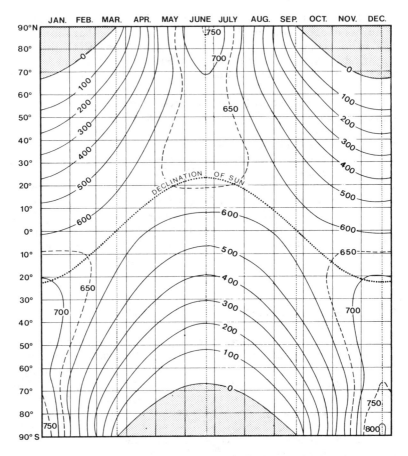

Figure 5. Daily average of solar radiation impinging
on a horizontal surface at the outer edge of the atmosphere, in W/m^2. Solar constant is 1360 W/m^2. Adapted
from List, 1956.)

atmosphere when the water vapor condenses in clouds or fog. It is
thus carried with the air as so-called latent heat. Transport of
sensible and latent heat from the surface by turbulence and order-
ed motions closes the energy balance of the atmosphere, as shown
schematically in Figure 7. Note, however, that such graphs are
meant to show the relative importance of the processes: they give
a global picture, and are not meant to be applicable to any specific place and time. The latitudinal and seasonal variations in
this energy balance lead to corresponding temperature gradients
which drive the atmospheric circulations.

For an understanding of atmospheric circulations it is important to recognize that water and land surfaces have different

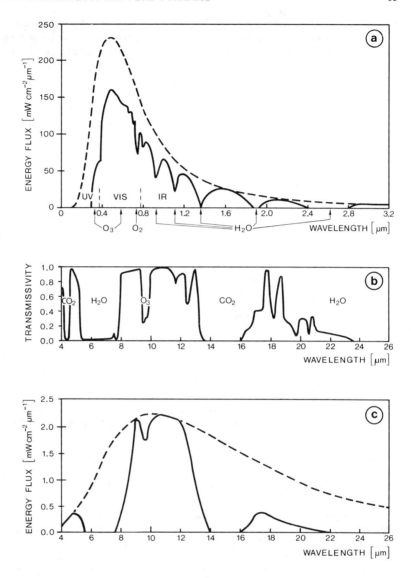

Figure 6. Spectral distribution of solar and terrestrial
radiation. The figure is simplified, the very detailed
structure of absorption as function of frequency is not
considered. (a) Dashed line gives the solar radiation
(assumed to be a black body at 6000 K) as received at
the outer edge of the atmosphere. Full line is the
solar radiation received at the ground after absorption
and scattering in the atmosphere. The main absorbers are
indicated below the abszissa. (b) Transmissivity of the
atmosphere in the infrared. (c) Terrestrial net radiation

at the ground (full line). Net radiation is the balance
between upwelling radiation from the surface minus downwelling radiation from atmospheric gases and particles.
The dashed line gives the black body radiation of the
surface.

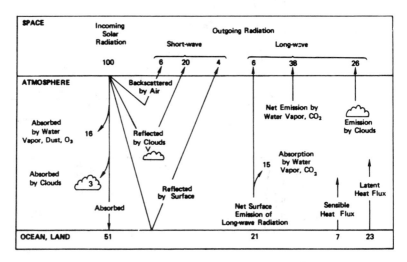

Figure 7.   Energy redistribution in the earth-plus-
atmosphere system. The energies are given as fractions
of the solar radiation impinging on a horizontal surface at the outer edge of the atmosphere. (After NAS,
1975.)

properties. This concerns radiation, heat and moisture storage,
and roughness. The albedo of natural water surfaces, about 6%, is
the lowest of any natural surface (Figure 8). The heat storage is
also quite different for water versus solid surfaces (Tables 2, 3).
In water, part of the solar energy penetrates to several meters
depth. Also, in water turbulent motions distribute the heat absorbed in the uppermost few meters throughout a deeper layer (the
"mixed layer"). For land, the main way to transport absorbed solar
energy is by ineffective molecular conduction. Hence, the land surface temperature responds very fast to any changes in the net surface energy balance, while the sea surface responds only very slowly. The third property, which is different over land and sea,
is the friction (Table 4). The sea surface - even though waves
may look fairly rough - is comparatively smooth. The land is much
rougher for most types of surfaces.

      So far, we have mentioned only land and sea. Ice and snow are
somewhat different. Snow has the highest albedo of any surface.
Also, for snow, an energy gain is first used to thaw the snow be-

fore the temperature may rise above zero degrees Celsius. The roughness of level snow (and also of sand, e.g. of tidal flats) is nearly the same as of the sea surface. About 25% of the land and water
surface are covered with snow and ice (where again the ice may be
covered with snow).

ALBEDO

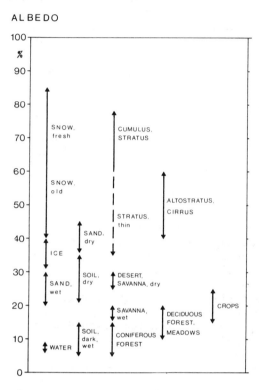

Figure 8. Albedo of different natural surfaces. Albedo
is the shortwave energy returned to space by reflection
and scattering, given as a fraction of incoming solar
radiation.

	Solar radiation albedo	penetration		IR radiation emission coefficient
Water	.06	∿ 50	cm	.92 – .96
Natural surfaces	.10–.30	∿ 1	mm	.90
Snow	.85–.40	∿ 1	cm	.99 – .80
Ice	.30–.40	∿ 10	cm	.96

Table 2. Radiational properties

	ρc Ws cm^{-3}K^{-1}	κ or K cm^{2}s^{-1}	$\rho c\sqrt{\kappa}$ or $\rho c\sqrt{K}$
Snow, new	0.13	0.006	0.01
old	0.9	0.003	0.05
Ice	1.9	0.012	0.21
Sand, dry	1.25	0.0013	0.05
wet	1.7	0.01	0.17
Soil	~ 2.6	~ 0.04	0.5
Ocean, very stable	4.2	0.1	1.3
moderately stable	4.2	50	30
well mixed	4.2	300	70
Atmosphere, very stable	$1.2 \cdot 10^{-3}$	10^3	0.04
near neutral	$1.2 \cdot 10^{-3}$	10^5	0.4
very unstable	$1.2 \cdot 10^{-3}$	10^7	4

Table 3. Physical properties of different media. The second column contains either molecular κ or eddy K conductivity as applicable. The third column gives the 'conductive capacity', which takes into account that not only the heat capacity ρc is different for different media, but also that the penetration of a temperature change with depth scales with the square root of conductivity. The larger the conductive capacity, the more energy is needed to produce a given temperature change. (After Priestley, 1959).

	z_o cm	C_D
sea surface, moderate winds	.015	.0013
high winds	.13	.0020
snow over open natural surfaces	.1	.0019
short grass	1	.0034
long grass	5	.0057
natural surface, open vegetation	10	.0075
forest	20	.010

Table 4. Typical roughness parameters of various surfaces. For definition of the roughness height z_o and the drag coefficient C_D see section on interface dynamics. The values are given for illustration rather than application. For vegetation there is a tendency to decreasing values with increasing wind. The roughness of the sea surface does not depend directly on wind speed, but rather on sea state. C_D is for 10 m reference height.

The large-scale circulations in the lower atmosphere, the Troposphere, can be classified into three main regimes (Figure 9): i) The Hadley Cell, a meridional circulation like a vertically standing wheel, is driven by the heating at the sea surface. In its lower branch, the trade winds move over the sea surface, where part of the sun's energy is used for evaporation and water vapor is carried towards the equator. The trade winds from the Southern and Northern Hemisphere converge near the equator in the Inter-Tropical Convergence Zone, ITCZ. The convergent flow leads to large cumulus and cumulonimbus clouds. With the rise of the air the water vapor condenses and releases heat. Hence in the ITCZ the latent heat collected in the trade wind belt is released and drives the Hadley circulation. The descending branch of this circulation is found in the subtropical high pressure areas. While the trade winds are quasi-permanent winds with high directional stability, the upper poleward branch is less well evident in the observations (but must have the same mass flow).

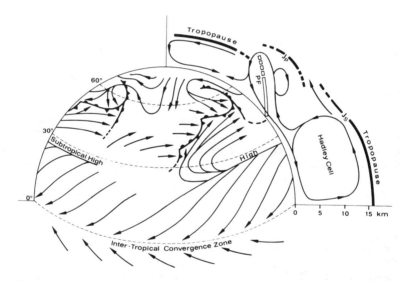

Figure 9. Scheme of the general circulation of the atmosphere at the surface and in a meridional cross section. Jp and Js indicate the average positions of the Polar Front and Subtropical Jet Streams, PF is the Polar Front. (Adapted from Defant & Defant, 1958.)

ii) The second main regime is the west wind belt of the temperate zones. The latitudinal gradient in heating leads to different temperatures of the surface and the troposphere. The strongest temperature gradients of the lower troposphere occur in the temperate zones. Such temperature contrasts are found locally over small distances: the so-called fronts. In principle, from meridio-

nal temperature and, hence, density and pressure gradients one would
expect, in geostrophic equilibrium, a zonal flow. A pure westerly
flow, with meridional temperature gradients exceeding a certain
limit becomes unstable (baroclinic instability). A small disturbance of the flow will be amplified and the zonal flow becomes
wavy. In the middle layers of the troposphere, e.g. at the 500 mb
level, troughs and ridges of low and high pressure are formed
(Figure 10), and in the lower layers the typical disturbances of

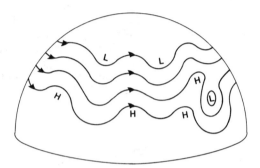

Figure 10.   Diagram of flow patterns at the 500 mb
level (about 5.5 km height).

the temperate zones develop. The result is, that the warm air, originally bordering the cold air, is lifted above it. Potential energy is thus converted into mechanical energy of storms (depressions),
and at the same time, latent heat is released by condensation of
water vapor in the clouds at the fronts. Thus, the two regimes are
quite different, both in their way of converting energy and in
their large-scale transport properties.

iii) A third regime is usually identified: the polar regions are
covered (mostly in winter time) with the so-called Polar High. Since
there is descending air in a high, which must be replenished in the
upper levels, this forms a third, but considerably less energetic
circulation cell.

So far we have emphasized the circulations in the lower atmosphere, the Troposphere. This name denotes the part of the atmosphere which is turned over by weather. Above the Troposphere is
the Stratosphere, which is more stably stratified. But the lower
part of the Stratosphere passively takes part in weather: it is
known from ozone observations that stratospheric air enters the
Troposphere. The Stratosphere shows quite different flow patterns
from the Troposphere: the winter pole has low pressure with westerly winds, and the summer pole has high pressure with easterly
winds above 20 km height. Wind directions reverse in spring and

fall,often rather abruptly.

    A summary of the three regimes is given in Figure 11, which
gives a meridional cross section of the tropospheric and lower
stratospheric circulation. The area of the tropical meridional Hadley cell is seen to have easterly winds at the equator at all
heights. The trade winds have an easterly component, while at higher
levels westerly winds are found, which at some times or at some
places circle the globe with rather high speeds. These are known
as the Subtropical Jet Streams. A similar jet stream is found meandering (and with varying intensity) at the polar frontal zone,
which divides cold polar air from the warmer subtropical air.
Winds are westerly in the midlatitudes, at least on the average.
Near the ground, the disturbances may also produce easterly components. At the pole the Polar High in the lower troposphere produces easterly winds, while above a few kilometers the wind is
westerly.

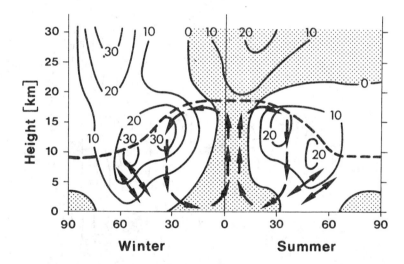

Figure 11. Meridional cross section of the atmosphere
showing schematically the zonal winds. Wind speeds are
indicated in m/s, blank areas are west winds, shaded
areas east winds. The arrows indicate the meridional
Hadley cell in the tropics, and double arrows represent
the sloping motions of the midlatitude disturbances.

    If we look at such a zonally averaged picture, the subtropic-
al highs and the Intertropical Convergence Zone (ITCZ) seem to
confine the air to their circulation cells, thus hindering interhemispheric transports and even exchanges between tropical and
midlatitudes. This is not true. The zonally averaged view does not
allow for the uneven distribution of land and sea over the globe.

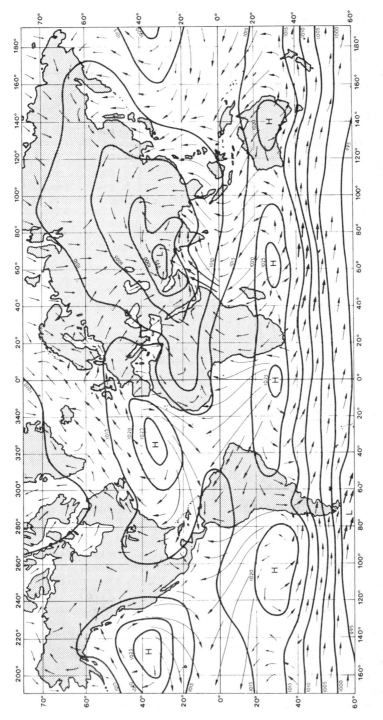

Figure 12a. Mean pressure and winds at the surface in July. Arrows fly with the wind, longer arrows indicate persistent winds. Wind speed code: → 0 to 6 m/s, → 6 to 12 m/s, → >12 m/s. (Adapted from Köppen, 1899.)

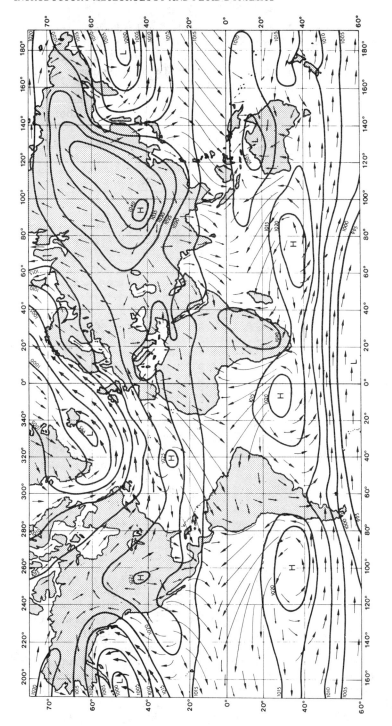

Figure 12b, as Figure 12a, but for January.

Since land and sea have different thermal, radiational, and frictional properties, their distribution modifies the air flow pattern considerably. Consider the wind and pressure field at the
surface in summer and winter (Figure 12): In north-summer (July)
and south-summer (January) we find high pressure over the oceans
and low pressure over the continents. On the Northern Hemisphere
the highs are circled clockwise by the wind, on the Southern Hemisphere, counterclockwise (and vice versa for the lows). Low pressure persists at the equator, and the southern midlatitudes have
persistently strong westerly winds. Due to the vast masses of the
Asian continent, the picture of the Northern Hemisphere is more
complicated: there is the winter high pressure center and the summer monsoon low over Asia.

Let me describe some monsoonal effects. The Indian monsoon
and similar circulations are brought about by the strong summertime heating. In a more general meaning, the term 'monsoon' is used
for any wind system which seasonally reverses wind direction. The
monsoon inflow into the Indian sub-continent in summer completely
cancels the expected northeasterly trade wind. A similar monsoonal
inflow is found into all Southeast Asia. It is also interesting
to see that in north-summer the ITCZ is found considerably north
of the equator at the Atlantic, while in the area of the Indian
Ocean the south-east trade winds turn to a south-west flow and
join the monsoonal inflow into India without any ITCZ. In winter
the northeast trade winds of the North Atlantic cross the equator
and continue as a monsoonal inflow into South America.

It is evident that the transport properties of the lowest 2
or 3 km are not adequately described by a zonally averaged picture.
Similarly, in the middle and higher troposphere, the midlatitude
westerlies show meandering and considerable deviation from a
smooth, confined flow; therefore there is substantial meridional
(north-south) transport (Newell et al., 1972). An interhemispheric
flow in the middle and higher troposphere and in the lower stratosphere is less evident from the observations, but there is evidence from atmospheric tracers that interhemispheric transport
exists.

With respect to atmospheric transport, it is also interesting
to look at the global water balance. In units of 1000 cubic kilometers of water per annum (approximately 2 mm precipitation per
annum), evaporation at the ocean amounts to 448 and precipitation
to 411. The difference (37 units, which is of order 10% of the
total evaporation) is carried over land. Over land 65 units are
evaporated, and 102 are precipitated. With a water content of the
atmosphere of 12400 km^3 and a total precipitation rate of 513
units, the average residence time of water vapor in the atmosphere
is about nine days.

3. SYNOPTIC SCALE CIRCULATIONS

The trade winds of the tropical seas are the most persistent
wind systems of the atmosphere. Yet there are variations of the
wind speed with variations of the positions of the Subtropical
High and of the ITCZ. The ITCZ also shows variations of intensity.
Cloud clusters travel westward, with clear areas following, so
that heavy precipitation is followed by suppressed convection
with a period of three days or so.

The disturbances of the tropics are usually classed according to their wind strength as "easterly waves", "tropical depressions", "tropical storms", and "tropical cyclones" (which locally have other names as hurricane, typhoon, etc.). The easterly
waves are difficult to identify in the pressure field. A slight
trough extends poleward in the isobars of the trade winds. There
is convergence before and divergence behind an easterly wave
(Figure 13). This produces heavy rain in showers and thunder showers. Tropical depressions are found frequently near the ITCZ

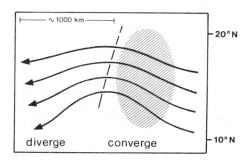

Figure 13.   Diagram of the flow pattern of an Easterly
Wave. The convergent flow east of the trough line
(dashed) leads to precipitation (shaded area).

and less frequently in the trade-wind belt. Some of them develop
to storm speeds, especially in the warm humid season. The tropical
depressions have a low pressure center, but still are mostly poorly developed. They are important since they produce heavy rain.
The weather in the monsoon areas of India and Southeast Asia is
modulated by these depressions. They are also found in the West
Indies and drift into Mexico or the Atlantic coast of North
America.

Some of the tropical storms develop into tropical cyclones
(hurricanes). The tropical cyclones are almost perfectly circular.
They start out as unsuspicious easterly waves, squall lines or
tropical depressions. In the growing stage, the radius of the

band of hurricane strength winds (say above 35 m/s, i.e.Bft 12
and higher) is only of order 50 km, while in mature stage, storm
strength may extend to 300 km radius. A cross section is given in
Figure 14. Tropical cyclones develop only over warm water and
more than 5 degrees latitude off the equator. Their energy is
taken from the release of latent heat with strong convection.

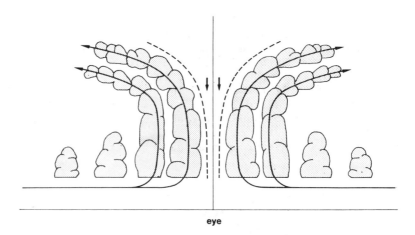

eye

Figure 14.   Schematic cross section of a tropical
cyclone. Hurricane wind strength is found within a
radius of about 80 km, except in the 'eye'. The eye
is cloudless. The storm area has a radius of 300 km
and is filled with convective clouds in spiraling
bands.

Thereby, the storm lives off energy collected in a larger area than
is covered by itself. In the storm area, evaporation is increased
and spray is formed by heavy waves and swell. Tropical cyclones
are seen in satellite pictures as characteristic spiraling cloud
bands and with a cloudless "eye" in the center.

    Weather in the midlatitudes is mainly brought about by travel-
ling depressions. This can be traced back to the latitudinal variation of heating by the sun and the corresponding temperature distribution. The resulting temperature gradients outside the
tropics are often found as strong, localized temperature contrasts,
named fronts. Near the ground, the main temperature contrasts are
found in midlatitudes. This front is taken to divide polar and
subtropical air and is called the Polar Front. Because of the
earth's rotation, neighbouring cold and warm air masses can be in
equilibrium, if certain conditions for the relative velocity components are fulfilled (similar to the thermal wind). Hence potential energy can be built up, which is afterwards released as
energy of storms.

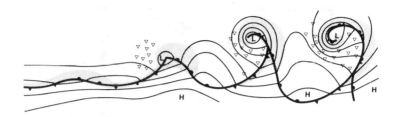

Figure 15. Northern Hemisphere cyclone family according
to Bjerkness and the Norwegian school. Cyclones often
follow each other as disturbances of the Polar Front.
The spatial sequence is also a time sequence with the
youngest developing stage at the left, and the decaying
stage at the right.

The Polar Front can become unstable and disturbances develop.
The life cycle of such a disturbance is depicted in Figure 15. A
cross section of a well developed depression is shown in Figure
16. In the beginning, the cold air was bordering the warm air. In
the developing stage, the warm air glides above the cold air at
the sloping front. This gives rise to condensation and (in the
idealized case) some hours of persistent rain. At the rear, the
cold air advances against the warm air. The scheme shows the cold
air nicely fitted below the warm air. This would mean a fairly
stable situation. The amount of rain would depend only on the velocity with which the warm air is lifted. More often the cold air
advances more rapidly aloft (due to less friction). If cold air
arrives above warm air, the situation is unstable and showers and
thunder showers are formed. The development of a depression stops
when the cold air from the back has reached the cold air in front.
The depression is then filled with cold air near the bottom and the
warm air is lifted above the cold air. Part of the available potential energy and also some of the heat released by condensation
have been converted into the kinetic energy of the depression.
When the energy is spent, the depression is weakened, and the
fronts become inactive and disappear. In the cold air at the rear
of a developed depression, the isobars often converge somewhat
and a line of showers is found. Note that the pictures are exaggerated: the typical slope of a front is 1 in 100. Warm fronts are
inversions in the sense that the normal vertical temperature decrease of 0.6 K/100 m is replaced by a temperature increase with
height.

The high pressure cells provide quite a different weather
type. Friction at the ground forces flow into the low and out from
the high and,by continuity,ascending and descending motions. Ascending motion is destabilizing; descending motion is stabilizing.
The latter provides not only for suppresion of convection in

L. HASSE

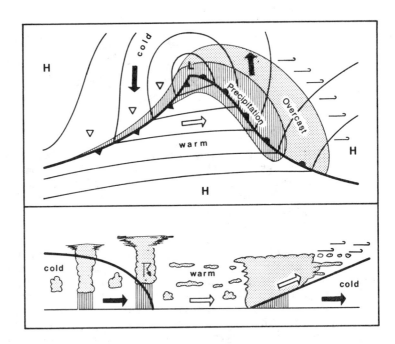

Figure 16. Cross section (below) and plan view (above)
of a well developed midlatitude depression (Northern
Hemisphere) according to Bjerknes and the Norwegian
school. Triangles indicate showers.

highs, but also for formation of inversions as a typical feature
in the extended high pressure cells. In general, the movements,
development and decay of highs are much slower than those of depressions (except for the intermediate highs in a series of lows). The
highs are rather passive, whereas there is conversion from potential to kinetic energy in the lows.

   From a genetic point of view, dynamic and thermal highs are
usually distinguished. Cold air is heavy, and so one would expect
high pressure at the ground in cold air. This is true of the continental (e.g. Eurasian) high in wintertime and the polar high. In
this case the cold air is at the surface (up to 2 or 3 km) only,
and a low is above. The outflow at the ground produces a sinking
motion aloft. The sinking air is warmed by compression and cooled
by radiation, but radiational cooling is more effective at the
ground. The sinking air in the polar highs is dry and naturally
outstandingly clean (both low numbers of nuclei and low humidity,
that is, dry dust only). Recently, however, pollution due to advection of sulfates has increased. On the other hand, high pressure
areas may exist for dynamic reasons, as part of a circulation system. These are warm at the ground due to adiabatic heating from

the sinking motion and high insolation under cloudless conditions.
This is the case for the subtropical highs, which form the poleward,
sinking branch of the Hadley circulation.

4. MESOSCALE CIRCULATIONS

Proceeding to smaller scales of atmospheric motion, we now
come to those circulation systems, which are sometimes called thermal or secondary circulations, or local wind systems. These terms
are used to indicate that temperature differences are responsible
for the onset and existence of local circulations, which are superimposed on the synoptic scale mean flow. There are two well-known
representatives of these local wind systems: the land and sea
breeze systems and the mountain-slope and valley winds. Both are
found mainly with high pressure systems during summer, when the
synoptic scale wind is low and insolation is strong.

The development of a sea breeze due to the temperature difference between land and sea is shown in Figure 17. The reasons
for the different heating of land and sea have been discussed already. The daytime sea-breeze is usually stronger than the nighttime land breeze. The height of the lower branch of the circulation is up to 1.5 km, the height of the returning branch is up to
3.5 km. Typical speeds are 4 to 7 m/s. In the tropics the sea
breeze may penetrate 100 or 200 km inland. These local circula-

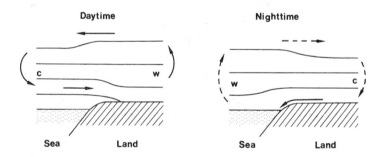

Figure 17.   Scheme of the land-sea breeze system. C and
W stand for cooler and warmer. The thin lines indicate
the height of pressure surfaces (vertically exaggerated).
During daytime, when the land is strongly heated by
insolation, the pressure gradients produce an acceleration
in the lower layers towards land and in the higher layers
towards the sea. At night, the land surface is cooled
by outgoing radiation and the pressure difference and
flow direction are reversed.

tions are also found over larger lakes. They are a distinct feature of the weather in the lower latitudes. Here the rising branch
of the secondary circulations may enhance convection and lead to
strong precipitation. A full understanding of such local circulations would also include consideration of roughness differences
and effects of terrain irregularities.

Slope winds are caused by a similar effect. Daytime heating
of a mountain slope produces uphill motion, while nighttime cooling produces downhill flow of cold air. In a similar way the
mean circulation in a mountain valley is up-valley during daytime,
while the cold air flows down the valley during the night (Figure
18). The latter are called the drainage or katabatic winds. They
are often produced on a larger scale by cold air flowing (in winter time) from high lands towards the sea, following the valley
structure. These winds in certain areas attain high speeds and
strong gustiness. They are known by local names (Borha, Mistral,
Santa Ana). Katabatic winds are the predominant surface winds of
the Antarctic continent. Because of the cold surface, the stratification is extremely stable. The cold, katabatic flow has its
maximum already at about 50 m height and may be stronger than the
synoptic-scale wind aloft. A similar wind field is found at Green-
land. Mountain-slope and valley winds are in principle easily explained. In reality, things seem to be somewhat more difficult.
The gain of solar energy by a slope is very much dependent on its
steepness and orientation and on solar height and azimuth, as well

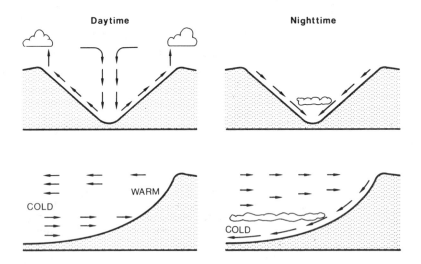

Figure 18.   Schematic section of mountain slope and
valley winds.

as on the interaction of the local and mesoscale field with the synoptic field.

There appear to be very few genuine mesoscale atmospheric circulations on the open ocean(except convection, see below). Coupling between atmosphere and ocean is weak due to the different time and velocity scales, and therefore mesoscale features of the sea surface do not significantly influence processes in the atmosphere.

The land-sea transition can influence mesoscale flow in a variety of ways (in addition to the land-sea breeze systems, mentioned above, and perhaps coastal mountain effects). The so-called coastal convergence is produced by the different roughness of land and sea: with an onshore wind, the smaller roughness of the sea permits higher winds than over land. Hence, on crossing the shore line, the flow is retarded. By continuity, a convergence some 10 km inland is found, leading to enhanced convection at some distance from the shore inland and less convection offshore. This is presumably so even over flat terrain, but it is also true that small orographic features (of 50 m height, say) may induce or reduce convective activity through induced upward or downward motions.

The different thermal properties of land and sea produce a horizontal temperature gradient. Even if this does not result in a land-sea breeze system, a thermal wind is imposed that influences the wind profile in the Planetary Boundary Layer. Since the temperature gradient is more or less perpendicular to the shore, the thermal wind will be parallel to the shore. The relationship between the surface wind vector and the large-scale geostrophic wind is accordingly modified. There is evidence that this effect is felt some 30 km offshore.

Turbulence and hence turbulent vertical transports in the Planetary Boundary Layer are also influenced by the advection of turbulence. Consider flow from sea to land; with higher roughness on land, turbulence will be increased. An internal boundary will develop, below which higher turbulence intensity is established, while aloft lesser turbulence levels characteristic of the sea still exist. With increased production of turbulence, the internal boundary will rise within some 10 km to its equilibrium height. In the opposite case (flow from land over the comparatively smooth sea) the advected turbulent energy is larger than in the equilibrium case. The excess energy will mainly be used up by dissipation, which is a relatively slow process. It is estimated that the higher turbulence level is felt even 50 km offshore. This would produce higher diffusion, a larger cross isobar angle and a smaller surface-to-geostrophic wind speed ratio at sea with offshore winds.

Additional effects at a somewhat smaller scale are expected

from bluff shore lines. There is a shadowing effect in the wind
field caused by bluff shores. Channeling effects are also known:
in a valley or an ocean sound between mountains, the flow is parallel to the sound even when the wind aloft and the geostrophic
wind are across the sound. Also, where the sound narrows, higher
wind speeds are attained, which are sometimes in excess of the
geostrophic speed.

Another mesoscale flow is the island or lake effect. Again,
because of different thermal properties of land and sea, an island in the ocean or an inland lake may cause a local circulation.
This is mainly felt when the island or lake is warmer than the
surroundings and induces convection. This may lead to enhanced
precipitation downstream. Also, satellite pictures have given
spectacular evidence that mountainous islands or capes produce
eddies of von Karman vortex street type in the air and water
downstream of the disturbances.

5. CONVECTION AND STABILITY

We now consider the motions of air volumes of smaller extent,
which we may call parcels or blobs, and which may have typical
diameters from say 100 m to a few kilometers. We have described
the larger scale circulations as coherent flows of air masses.
For example at a warm front of a depression the warm air is gli-
ding slowly upwards above the cold air. On the much smaller scale
which we now consider, we must introduce the notion that a parcel
has individual motions, where its characteristic variables may
change. We need to distinguish these from the variables of the
surrounding air, which we conceive as being at rest. The latter
may have a vertical variation of temperature and humidity. These
are called geometric derivatives in order to distinguish them
from the so-called individual changes that a parcel will experience during vertical movements. The motions of a parcel are strongly influenced by the stratification of the ambient air.

Consider a parcel of air. If it is warmer, it is lighter
than its surroundings and will rise, if it is cooler, it is denser
than the surroundings and will be accelerated downwards. Vertical
movements of isolated parcels are accompanied by compression
(downward) or expansion (upward) and an "adiabatic" heating or
cooling of 1 K/100 m results. Hence it is difficult to compare
temperatures from different levels. In meteorology, therefore, a
hypothetical temperature is introduced, which remains constant
with adiabatic movements: the so-called potential temperature.
This is the temperature a parcel of air would have if it were
brought adiabatically to the pressure level of 1000 mb, the approximate pressure at sea level (1 mb = 100 Pa).

Suppose, the vertical temperature profile of the air is measured e.g. by radiosondes. The stratification is stable, when the
potential temperature increases with height (this is the typical
case in the free atmosphere). A parcel originally in equilibrium
with its surroundings will keep its potential temperature when removed adiabatically from its place. At a higher level, it will be
cooler than the surroundings (and at a lower level, warmer)and
hence it will be accelerated to return to its equilibrium level
(Figure 19). In the opposite case, with the potential temperature
of the surroundings decreasing with height, any small adiabatic
move of a parcel from its equilibrium position would render its
temperature such that it would be accelerated away from its original position. This state is called unstable, the vertical stratification is called lapse (the actual temperature decrease with
height is stronger than 1 K/100 m).

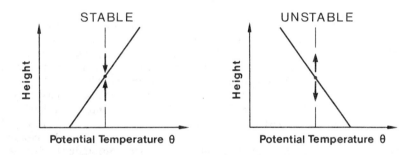

Figure 19. Stable and unstable stratification. The full
line gives the mean temperature profile of the surrounding air. A parcel of air which is originally at the
same temperature as the ambient air keeps its potential
temperature when displaced vertically and hence will be
accelerated either back to its starting point (inversion
condition, stable stratification, left part of the figure)
or will be accelerated away from its equilibrium position
(lapse condition, unstable stratification, right part
of the figure).

An unstable situation is unlikely to persist in the free atmosphere, except near the surface, where the vertical movements (due
to continuity) are hindered by the surface. Here, unstable stratification may persist in a kind of dynamic equilibrium: the stratification is adjusted such that the vertical mixing from the overturning by thermal instability and by mechanically generated turbulence transports heat from the surface at the same rate as it is
supplied. With strong insolation during daytime and infrared radia-
tive cooling during nighttime, there is a strong diurnal cycle of
the net radiation at the surface. The imposed diurnal cycle of surface

temperature produces a typical diurnal cycle of stability in the surface layer over land (unstable in daytime, stable during night).

Over the sea, for most areas of the oceans,we find the air one or two degrees colder than the water, at least on the average. Of course, there are larger variations of the air temperature caused by advection of cooler or warmer air masses with synoptic or meso-scale systems. In certain parts of the oceans, the surface water is cool, caused by upwelling of cooler water from deeper layers to the surface. But since more solar energy is absorbed at the surface than is radiated back by IR radiation (see Figure 7), commonly the sea surface is warmer than the air. When the air is warmer than the water, the air is cooled at the surface and a stable stratification results. If the air is cooler than the water, the air is warmed from below und unstable stratification results.

In an unstable situation,because a parcel removed from its equilibrium condition is accelerated, we must consider the stability of density stratification over a greater height range. The rising parcel is cooled adiabatically. It will continue to rise as long as it remains warmer than its surroundings. With the cooling, the saturation water vapor pressure decreases, and the relative humidity may eventually reach 100%. Even below 100%,part of the water vapor is used to increase the water content of aerosol particles. At 100% humidity further cooling leads to condensation of the excess water vapor into cloud droplets (or ice particles). The latent heat is given to the air and yields a less rapid decrease of air temperature during further ascent. This individual change of temperature of the ascending air is called the wet adiabatic lapse rate. Consequently, ambient air which is stably stratified for dry adiabatic ascent, may be unstably stratified for wet adiabatic ascent, that is, of cloud air. This may give rise to fair weather cumulus clouds or congested forms of cumuli, and in extreme cases to thunder showers from deep cumulonimbi.The situation is depicted in Figure 20. The process of vertical motion in congested clouds, brought about by unstable stratification (both dry and wet), is commonly called convection.

Different types of convection are found in fair weather situations, at cold fronts of synoptic systems, and at squall lines. Also satellite photographs show convection to be a very common feature over the world oceans, especially in the cold air at the rear of depressions. They also show that convection often possesses horizontal patterns, for example, in the form of hexagonal cells. A pattern, which typically has a smaller space scale and can be observed from below, is formed by the so-called "cloud streets". These are rows of low cumulus clouds. It is assumed that cloud streets indicate roll motions in the boundary layer of the atmosphere. The motion is presumably in the form of helical rolls orientated along (or nearly along) the wind direction, and reaching from the surface to a capping inversion (Figure 21).

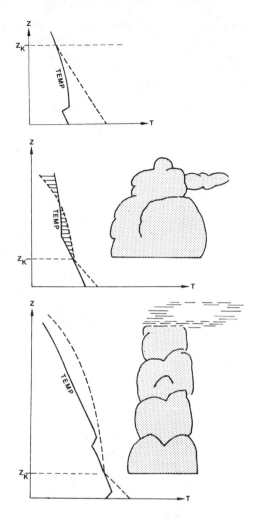

Figure 20. Different conditions of convective cloud
formation. z_K is the 'lifting condensation level'. Full
line is the temperature-height curve of the ambient air
as measured by a radiosonde ascent. Dashed line is the
individual temperature change along dry and wet adiabats.
In the upper graph, a parcel lifted from the ground with
the indicated temperature would not reach condensation.
In the middle graph, a lifted parcel would change its
temperature following the dry adiabat up to the condensation level and then following the wet adiabat until
it is stopped by the inversion. A shallow cumulus layer
would result. In the lower graph, a lifted parcel, which
has reached the condensation level, will (on its wet
adiabatic ascent) remain warmer than the ambient air

and be accelerated upwards. Deep cumulus convection
results. In the latter two cases, the stratification
above the condensation level is called conditionally
unstable, because the situation becomes unstable only
if condensation is reached.

The upward branch may or may not be made visible by cloud streets.
These helical rolls are difficult to measure (because of their
scale), but where they exist, their vertical motions are efficient
in carrying admixtures from the surface to the top of the boundary
layer, or vice versa.

In convective clouds, the vertical motions within the clouds
are concentrated in funnels of higher speed. Updrafts reach 10 or
20 m/s in well developed cumuli. The rising air is diluted by entrainment of dryer air from the sides. With raining clouds, evaporation of rain leads to cooling of the air, so that cold, wet
air is accelerated downwards and spreads at the ground. Also
downdrafts are produced by the friction of falling rain. Between the clouds we also find slow sinking motion. This is often
evidenced by the clear sky between cumulus clouds. This downward
motion is small and difficult to measure, but must be there for
continuity reasons. In a field of convection the area fraction
covered with clouds in a radar picture is 5% or 10%. Hence the
sinking motion is by a factor of 10 or 20 smaller than the upward
motion. The sign of the net mass flux is determined by the large-
scale convergence or divergence.

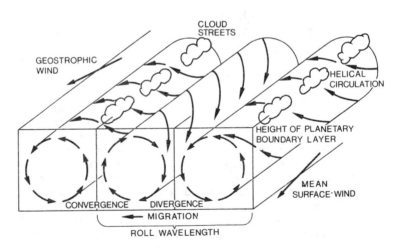

Figure 21. Perspective schematic presentation of helical vortices and cloud streets. (Adapted from LeMone,
1972.)

6. TURBULENCE AND THE PLANETARY BOUNDARY LAYER

The different properties of the surface (land, sea), compared
to the air, have a strong influence on atmospheric flow and other
properties near the surface. The influence of the surface decreases with height. The layer where either in the motion field or in
the temperature structure the direct influence of the surface is
felt, is called the Planetary Boundary Layer (PBL) of the atmosphere. The Planetary Boundary Layer is most often defined in
terms of the influence of friction. In the free atmosphere, to a
good approximation, geostrophic equilibrium prevails; that is,
there is equilibrium between pressure gradient and Coriolis acceleration, and the flow is parallel to the isobars. Near the surface friction is effective as a third acceleration, and the flow
deviates from the isobars towards low pressure. The frictional influence and hence the angular deviation (cross-isobar angle) decreases with height until geostrophic equilibrium is reached
(Figure 22).

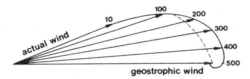

Figure 22. Wind veering in the planetary boundary layer of the atmosphere. The arrows represent wind vectors
at the indicated heights. The tips of the arrows, as a
function of height, form a spiral curve. Under simplifying assumptions, this is a logarithmic spiral, called
the Ekman spiral. The dashed line is more typical for
observations. The upper graph is for conditions over
land, the lower one for sea. Actual spirals show a wide
range of variation from the given ones.

This level is taken as the height of the Planetary Boundary Layer.
It is usually 1000 m to 1500 m over the land (daytime) and 300
to 600 m over the sea.

In order to understand the processes in the Planetary Boundary
Layer, we discuss turbulent motion first. Any fluid (this includes

liquids and gases) may flow either in a laminar or turbulent state
of motion. These two states are best demonstrated with an experiment by Reynolds. In his study of flow through a pipe he made the
movements of the fluid visible by introducing a fine filament of
colour. He found two distinct types of flow: in "laminar" flow,
the filament formed a straight line, indicating that the flow was
layered. With higher speed, the filament became wavy and broke up,
the dye being mixed over the entire fluid. This second type of
fluid motion is called turbulent. It is a characteristic of turbulent flow that irregular motions in all three directions are
superimposed on the mean flow. These fluctuating velocity components mix the fluid. If there exists a gradient of a certain property in the fluid, the mixing can transport the property down its
gradient.

Turbulence can be characterized by the kinetic energy of the
turbulent velocity components. The energy of turbulence is taken
from the mean flow (which is driven by the horizontal pressure
gradients). The friction at the surface, for example, produces
a shear flow, that is, flow with a vertically changing wind vector.
This shear is instrumental in converting mean flow energy into
turbulent kinetic energy. The turbulence intensity is strongly
influenced by the ambient density stratification. In unstable
stratification, vertical motions are accelerated and receive energy from the density stratification. With stable conditions, on the
other hand, vertical motions induced by friction at the surface
must work against the density stratification. The work is taken
from the energy of turbulence and hence, under strongly stable
stratification, turbulence may decay or can not exist. The effects
of stability are often measured in terms of a Richardson-Number:

$$Ri = \frac{g}{T} \cdot \frac{\frac{\partial \theta}{\partial z}}{(\frac{\partial u}{\partial z})^2} \tag{11}$$

Instead of the Ri-number, other dimensionless numbers are used
with slightly different definitions, e.g. z/L, where L is the
Monin-Obukhov length (see e.g. Haugen, 1973). The Ri-number and
similar numbers measure the rate of work done by or against the
buoyant forces versus the rate of shear-production of turbulent
energy (unstable, Ri < 0; stable, Ri > 0).

Except for a very thin layer (a few millimeters deep) at the
surface, where the influence of the molecular viscosity is felt,
the motions in the atmosphere are always turbulent. Turbulence is
very active as a transporting agent. This is seen when by analogy
to the molecular transport, the turbulent transport is described
by aid of an eddy diffusivity K

$$F = - \rho K \frac{\partial m_r}{\partial z} \qquad\qquad\qquad (12)$$

where F is the flux of a certain property and m_r is its mixing
ratio $|kg/kg|$. A typical value of K (except near the surface) is
5 m^2/s. This is by a factor of 10^3 or 10^4 higher than the molecular diffusivities in air (see e.g. Table 3). Equation (12) is
applicable only for such properties which remain unchanged with
vertical motions. It can be derived in various ways, e.g. by dimensional analysis. It is usually a good description in fully turbulent flows such as found in the atmospheric surface layer.

Aside from the formal analogy of molecular and eddy diffusivity,
the physical processes of molecular and turbulent transport are
quite different. Molecular diffusivities depend on the property
exchanged, on the fluid in which the transport takes place, and
on the temperature of the fluid. The eddy diffusivity is independent
of temperature and more or less the same for all exchanged properties, but is highly dependent on the intensity of turbulence. This
means that the eddy diffusivity is proportional to wind speed
(more exactly, to the friction velocity, see below), modified by
stability, and is dependent on the distance from the surface (the
eddy size is limited near the surface for continuity reasons, and
hence the turbulent transport is less efficient). The typical
picture (see Figure 23) is a linear increase of the diffusivity
with height near the surface until 50 to 150 m, followed by a
decrease to a small (but non zero) value in the free atmosphere.

The description of eddy motion and transport processes in the
Planetary Boundary Layer is quite difficult. From the defining
equation of eddy diffusivity (12) it would seem, that only the
mean vertical gradient and the eddy diffusivity are needed in order to calculate the flux. But the eddy diffusivity is highly
variable, since it depends on the physics of the processes in the
Planetary Boundary Layer. This is true of other descriptions of
the Planetary Boundary Layer too. Consider the mechanical generation of turbulence. This depends on the friction at the surface
and on the variation of the wind speed and direction with height,
the wind shear. Surface friction can be characterized by a roughness length or a drag coefficient (which to a first approximation
depend only on surface geometry). The wind shear is brought about
by turbulent mixing and is variable; it is also influenced by the
shear of the geostrophic wind, that is the so-called thermal wind.

The most important parameter to influence the turbulent transport and hence the eddy diffusivity in the Planetary Boundary
Layer is the stability of density stratification, as measured e.g.
by the Richardson-Number. Even with small deviations from neutral
stability there is a distinct difference in the turbulent processes between stable and unstable stratification. On the stable side
the destruction of kinetic energy, by work against buoyancy, damps

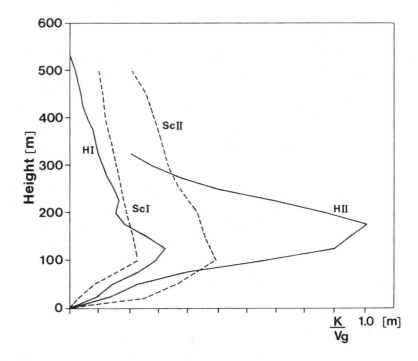

Figure 23.   Variations of eddy diffusivity with height.
Since the eddy diffusivity increases with wind speed,
it is here normalised with the speed of the geostrophic
wind. Examples are from pilot-balloon-derived wind profiles at the Scilly and Helgoland islands, so the diffusivities are for momentum (eddy viscosities) in the
example. (Adapted from Hasse, 1968.)

turbulence. Hence transports are inefficient under stable conditions. Inversions (that are layers of temperature increase with
height), therefore, are barriers for turbulent (or diffusive) transport of atmospheric admixtures. On the other side, in unstable
conditions, the buoyancy induced vertical motions enhance turbulence. In the extreme, in strongly unstable conditions, buoyancy-
driven convection dominates the mechanically generated turbulence.
Near the surface, turbulence(in the sense of irregular, random
motions) will still exist. But at some distance from the surface,
the vertical motions become organized, and the concepts of eddy
diffusivity and transport by turbulence fails. One finds constant
potential temperature in the ascending and in the descending air
so that the vertical temperature gradient vanishes. Hence from
(12) zero vertical flux would be calculated, although there is
strong transport by convective motions. In convective conditions
we, therefore, usually find a surface layer with non-zero gradients

of temperature and wind speed, up to a few tens of meters, with a
well-mixed layer above (constant potential temperature, no variation of wind speed and direction).

In the simplest case of atmospheric boundary layer theory, in
neutral density stratification, the height of the boundary layer
is determined by the friction at the surface and is proportional
to the wind velocity (e.g., in the Ekman spiral). In the stable
situation, the boundary layer is compressed to the height range,
where frictionally induced turbulence can overcome buoyant damping. In the unstable case, the height of the boundary layer is the
height of the unstable or well-mixed layer. Unstable conditions
exist only near the ground. In the free atmosphere the stratification is stable, the potential temperature increases with height.

Over land, the height of the temperature inversion capping
the unstable boundary layer is mainly determined by the sensible
heat flux at the surface and can be estimated fairly well (Driedonks, 1982). Over the sea, the height of the unstable boundary
layer is determined not only by the heat flux at the surface, but
also by large-scale divergence or convergence and radiational cooling.

7. INTERFACE DYNAMICS

Within the Planetary Boundary Layer, the layer adjacent to
the surface is usually referred to as the surface layer, or constant flux layer, or Prandtl layer. This layer is typically of
order 20 m in height. The fluxes of heat and momentum are not
really constant within this layer, but a good description may be
obtained by assuming constant fluxes. Due to its moderate height
the time constant of this layer is of order a few minutes, and
horizontal homogeneity generally is a good assumption. Processes
in this layer are essentially influenced by the turbulent character of the flow. Since turbulent energy is produced by friction,
the roughness of the surface is one important parameter. Again,
turbulent energy is influenced by stability. We will discuss the
case of neutral stability first and consider the influence of
stability separately. We will discuss flow at a solid, level surface before we proceed to a fluid interface like the sea surface.

7.1 Flow at a solid, level surface

The model envisages a decreasing diameter of eddies as the
surface is approached. Hence an eddy scale-length or "mixing
length" is defined, which is proportional to the distance z from
the wall, say κz. Assuming a constant flux of momentum, given by
the shear stress τ at the surface, the vertical profile of the

modulus of the mean wind vector (the wind speed) u becomes

$$\frac{\partial u}{\partial z} = \frac{u_*}{\kappa z} \tag{13}$$

where u_* is called the "friction velocity" and is defined by

$$\tau = \rho_a\, u_{*a}^2 \tag{14}$$

τ is the modulus of the horizontal shear stress vector, the direction of which is in the mean wind direction. Integrating along z, which we assume to be perpendicular to the wall, we obtain

$$u(z) - u(z_o) = \frac{u_*}{\kappa} \ln \frac{z}{z_o} \tag{15}$$

This is called the logarithmic wind profile, applicable for neutral stability. Here the integration is extended to a variable height z. Usually it is assumed that there is no slip at the surface, $u(z_o) = 0$ and the integration constant z_o so-defined is called the 'roughness length'. z_o is a measure of the surface roughness. When Prandtl introduced this concept, he .specifically called it a trick, which allows the logarithmic profile to be extended right to the surface. This implies that the eddy motions decrease in size in proportion to z right down to the wall. This is a good description as long as z is large compared to z_o, say by a factor of 100.

At the "wall" (which for the present we take to be a fixed, flat, but rough wall) the eddies are hindered in their movements perpendicular to the wall by its presence. Due to continuity constraints, the transport by eddies decreases more rapidly than linearly as the wall is approached. The decreasing energy of turbulence close to the wall has led to the definition of a "laminar" or better "viscous sublayer", where the fluid viscosity becomes important in the dynamics and the flow is dominated by viscous forces.

Experimental evidence (Figure 24) shows the typical profile of laminar flow near the wall, a transition region, and at larger heights the logarithmic profile of turbulent flow. The conceptional model is of a layer with decreasing turbulent transport as the wall is approached. Molecular transport adds to turbulent transport, since movement of the molecules takes place anyway, independent of additional eddy motions. In the transition region, molecular and eddy transport are of the same order, while with increasing distance from the wall eddy transport takes over.

Such continuity and boundary condition type arguments give only a gross picture. It is not really known how the flow behaves

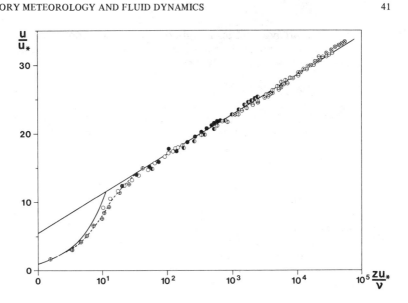

Figure 24. The velocity profile near the wall. Abscissa
(distance from the wall in dimensionless units) is logarithmic, hence the logarithmic profile is given by a
straight line. The linear velocity profile of laminar
flow on the left hand side is distorted according to
the logarithmic scale. (Adapted from Schlichting, 1965.)

near the wall. Does the scale and energy of the eddies decrease as
they approach the wall such that no eddy may really remove fluid
from the vicinity of the wall? Or is there just a decrease in
probability of an eddy reaching a given distance from the wall?
In principle, the power laws for parallel and perpendicular eddy
velocity components can be different, although there is a cross-
coupling between eddy velocity components effected by pressure
fluctuations. The question of how the eddy velocity components
and turbulent exchange decreases on approaching the wall is of
considerable interest, since molecular transport is so much more
ineffective than turbulent transport, especially for heat and gases with small molecular diffusivities. A review of power laws near
the wall is given by Monin and Yaglom (1965, Vol. I, Sec. 5.3 and
5.7).

The roughness in wind tunnel work is customarily produced by
gluing sand of a certain grain size to the wall or by determining
an "equivalent sand-grain size". The thickness of the viscous sublayer depends on the kinematic viscosity ν and is proportional to
ν/u_*. The transition from laminar to turbulent flow occurs at a
height of about 11 ν/u_*.

If the roughness elements are smaller than the thickness of

the viscous sublayer, the roughness elements do not influence the
flow and the integration constant z_o may be replaced by a length
proportional to ν/u_*:

$$u(z) = \frac{u_*}{\kappa} \ln \left(\frac{9u_*z}{\nu}\right) \tag{16}$$

This flow is called hydrodynamically smooth flow, while in the alternative case the flow is called hydrodynamically rough.

7.2 Flow at a fluid interface

In the case of a moving interface such as the sea surface,
the definitions of smooth and rough flow are not directly applicable. In the sense used in physics, both gas and water are fluids,
and the atmosphere-ocean interface can be seen as an internal interface, with waves propagating coherently on both sides of the
interface (see e.g. Kraus, 1972). Waves and surface currents
take up and transfer momentum and hence the momentum transfer
is certainly different from the case of a solid wall.

Despite this, it can in practice be assumed that many of the
concepts from laboratory work on smooth and rough surfaces can be
applied at the mobile air-sea interface. The arguments are as follows. For a turbulent eddy approaching the interface, movements
perpendicular to the interface are restricted as they are on approach to a solid wall. This is so because to deform the interface,
work against gravity and surface tension would be necessary, and
this work is much larger than the kinetic energy of turbulent
motion of a small fluid parcel. Although the boundary conditions
at a liquid/gas interface are different, the same power law is expected to hold as at a solid wall (Hasse and Liss, 1980).

A complication exists, however, if waves are present. The
above mentioned derivation is for a flat, fluid interface. As long
as the wave scales are large compared to the typical thickness of
the viscous sublayer (a few millimeters or fractions thereof), the
wave induced curvature of the sea surface is unimportant. This is not
so in the case of capillary waves, whose radius of curvature may
have the same order as the thickness of the viscous sublayer. Unfortunately, both the life cycle of capillary waves and wave-in-
duced-near-interface-flows and their influence in gas exchange at
the sea surface are not well understood.

7.3 Stability effects in the surface layer

Micrometeorologists have expended considerable efforts to determine the influence of stability on the fluxes of momentum, heat
and water vapor in the surface layer. The stability dependence of
turbulence-controlled fluxes is dealt with by similarity theory,
often called "Monin-Obukhov similarity". In short, similarity theory

provides a relationship between fluxes and gradients of the transported property in the surface layer. This relationship depends
on a dimensionless stability parameter like the Richardson-number
or z/L, where L is the Monin-Obukhov length. A review is given in
Haugen (1973). Since the stability effects become important only
above the region of wave influence, stability functions determined
over land are applicable also at sea. It can be assumed that surface layer stability is not important for gas or particle exchange.
Stability modifies the turbulent energy and hence the turbulent
transport in the surface layer (or higher layers) only. Transport in the viscous sublayer or near to it is not influenced
directly. Near the surface, turbulence is mainly generated mechanically; the influence of stability becomes increasingly important
only above a height of a meter or so. There is an indirect influence of stability on the transports through the viscous sublayer,
since the sublayer thickness is inversely proportional to the
friction velocity. The latter depends on surface layer stability,
but this variation is of order 10% only if parameterization (see
below) is made in terms of surface layer variables.

7.4 Parameterization

The momentum flux at the surface can be parameterized reasonably well (with a scatter of 20% and a probable uncertainty of
50% at higher wind speeds) with aid of a drag coefficient C_D
(Figure 25) and the mean wind speed u:

$$\tau = C_D \, \rho_a \, u^2 \tag{17}$$

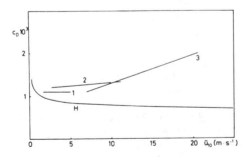

Figure 25. Roughness of the sea surface. Comparison between measured drag coefficients C_D (numbered lines)
and hydrodynamical smooth flow (curve marked H). The implication is that the increase of momentum flux, compared to smooth flow, is brought about by form drag of
gravity waves and ripples. The average value at moderate
wind speeds is fairly well known both from direct and
profile measurements, while the value at high wind
speeds is based on few direct measurements. (After Hasse
and Liss, 1980.)

With (14) the thickness of the viscous sublayer in air can be determined. (The equation is written for the modulus of the horizontal stress component only, which enters in u_*.)

It is also possible to relate the surface stress to the geostrophic wind speed V_g, that is, to the surface pressure field.
This is done either by boundary layer modelling or by a similarity
theory known as "resistance law". A geostrophic drag coefficient
C_g is defined by

$$\tau = C_g \, \rho_a \, V_g^2 \tag{18}$$

The geostrophic drag coefficient depends on all processes which influence turbulence and convection in the Planetary Boundary Layer,
and is consequently quite variable. With the same neglect of detailed description of processes it is also possible to use a surface to geostrophic wind speed relationship (Figure 26) together
with the surface layer parameterization (17).

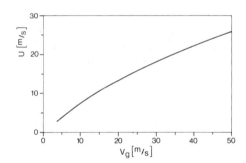

Figure 26. Surface to geostrophic wind relationship
from observations in the German Bight. (Adapted from
Luthardt and Hasse, 1982.)

The viscous sublayer thickness is determined by aid of the
friction velocity u_* in the respective media. The parameterizations
(17,18) are for the atmosphere. The friction velocity in the sea
is obtained by assuming constant flux of momentum across the interface:

$$\rho_a u_{*a}^2 = \rho_s u_{*s}^2 \tag{19}$$

In papers on gas and particle exchange usually a deposition
(or transfer or piston) velocity is used, which has dimension of a
velocity and hence will probably be proportional to some characteristic velocity of the problem (e.g. friction velocity). In micrometeorology, however, use of a dimensionless transfer coefficient

is preferred, like the drag coefficient (17). Take for example
the water vapor flux (evaporation) E. With q specific humidity,
and Δq the difference of specific humidity between air and sea
surface,

$$E = k_a \, \rho_a \, \Delta q \qquad\qquad\qquad (20)$$

$$E = C_E \, \rho_a \, u\Delta q \qquad\qquad\qquad (21)$$

Hence, from $k_a = C_E u$, micrometeorological results (see e.g. Kraus,
1972) can be used to obtain deposition velocities (C_E is about
1.3×10^{-3}). The use of dimensionless transfer coefficients like
C_D , C_H , C_E has the advantage that the rough proportionality of
turbulent transports to mean wind speed is taken care of and the
transfer coefficients are less variable.

8. PRACTICAL ASPECTS OF ATMOSPHERIC TRANSPORTS

In the preceding sections atmospheric motions of different
scales have been discussed. Since the atmospheric processes are
very complex, for an understanding of meteorological conditions
for air/sea exchange of gases and particles it will often be
necessary to simplify and mainly consider processes at the scale
of interest. There is a large disparity between the horizontal
and vertical extent of the atmosphere and, similarly, in the horizontal and vertical components of atmospheric circulations, caused
by the different magnitudes of Coriolis acceleration ($\sim 10^{-3} \text{m/s}^2$)
and acceleration of gravity (10 m/s^2). Hence, it simplifies discussion if we deal with horizontal and vertical transports separately, although de facto horizontal and vertical motions are always coupled.

8.1 Vertical transports

We have already discussed some processes, which provide vertical upward transports: The mixing by turbulence causes vertical
transports if there is a gradient of a certain property. This
transport is effective only in the surface layer and the Planetary
Boundary Layer. If the source of a property is at the surface,
turbulence helps to transfer this property from the surface into
the lower troposhere, where other mechanisms may take over. Convection can be quite effective in transporting properties into
higher layers of the atmosphere. Note that with the diurnal variation of insolation over land there is enhanced turbulence and convection during daytime and suppressed turbulence and convection
during nighttime, with decoupled larger scale flow above the
nighttime inversion. A third mechanism producing upward transports

is the rising motion in midlatitude and tropical disturbances.
Additionally, there are mesoscale circulations, which may be important in given circumstances and areas.

Downward transports are different from upward transports,
due to different energetics of upward and downward motions. Turbulence mixes and provides downward transports if there exists a
gradient. Convection provides for downward transports both by
downdrafts and by the slow sinking motion in between clouds. Sinking motion in the high pressure cells is slow. If it were not
for gravitational settling and wet removal, the downward transport
of any admixtures would be slower than the upward transport.

In our picture vertical fluxes of gases or particles pass
through a series of layers: free atmosphere, planetary boundary
layer, surface layer, viscous sublayer in the air, interface, viscous sublayer in the ocean, mixed layer, deep ocean. These layers
have different physical properties and consequently different
transport resistances or conductivities. Depending on the property
being transported, sources and sinks may be in any of these layers,
while in the intermediate layers the fluxes are more or less constant. For a constant flux, it is unimportant in which part of
the path it is determined. In the case of exchange of gases and
particles, the viscous sublayers in the atmosphere and/or ocean
have the highest resistances (except for gravitational settling and
wet removal, which form a bypass). Hence the processes in and
near this layer and its thickness will be decisive.

There are two conclusions, which we can draw for those substances with the highest transport resistance in the viscous sublayer. First, the gradients above the viscous sublayer will be
fairly small and an attempt to determine fluxes by the gradient
method almost surely must fail. Since the admixtures are well
mixed in the surface and Planetary Boundary Layer, the eddy correlation technique also will not work. The correlations between the
fluctuations of the admixture and the vertical velocity component
will be small and side effects become dominant (see Webb, 1982).
Second, the turbulent exchange coefficient K is probably nearly
the same for all passive admixtures. The surface layer has a time
constant of a few minutes. This is the time needed to reestablish
an equilibrium profile after a disturbance of the profile occurred.
The time constant of the Planetary Boundary Layer is of order a
few hours. Hence, with an average advection speed it can be estimated that a fetch of order 100 km is sufficient to establish an
equilibrium profile of admixtures in the Planetary Boundary Layer.

8.2 Horizontal transports

The atmosphere possesses a large spectrum of motions from
planetary waves, synoptic scale disturbances, mesoscale circula-

Residence time	Mechanism	Methods
Years	Stratospheric transport. Interhemispheric exchange.	Analysis of nuclear test products and atmospheric contaminants.
Months	Tropospheric transport.	Statistics of trajectories, e.g. from numerical models possible.
Days	Advection with synoptic disturbances and larger scale tropospheric flow.	Trajectory analysis from weather maps or models.
24 hours	Advection by synoptic and local wind.	Mesoscale models available, but wind systems depend strongly on orography.
3 hours	Advection by local winds, diffusion by turbulence.	Stability depdndent diffusion models.

Table 5.   Transport classification.

tions, to turbulent fluctuations. Which of these scales are important will depend on the atmospheric residence time of the gas/
particle in question. A coarse classification is given in Table
5. The columns "residence time" and "mechanism" are not really
independent. For example, the troposphere is turned over by
weather systems. Thus, contaminants removed by contact at the sea
(or land) surface are depleted from the entire troposphere. In
contrast, the stratosphere is stably stratified, with less vertical motion and ineffective vertical diffusion. Hence, long-lived
contaminants follow the stratospheric motions for a longer time.
Some few explanatory remarks may be added:

(i) Trajectory analysis is most easily performed from
weather-service synoptic maps, which commonly are available as
contour lines (topography) of height of selected pressures, say
500 mb pressure. Winds are then given by the slope of the heights
of pressure levels, and trajectories are calculated as if the
identified air parcels would remain at these pressure levels. But
if we assume adiabatic behaviour of an air parcel, we have seen
from the first law of thermodynamics that in a stable atmosphere
a parcel would remain at the level of its potential temperature.
If we move the parcel in predominantly horizontal direction, it
would tend to follow the surface of its equilibrium potential
temperature. Such surfaces are called isentropic surfaces. They

may be inclined to the pressure surfaces, which usually are taken
as calculation levels in numerical models. After some time, the
parcel will reach some other pressure level and have a different
wind direction from the one determined at the original pressure
level. In principle, since the potential temperature (above the
Planetary Boundary Layer) increases with height, the vertical coordinate (i.e. pressure levels) could be replaced by isentropic
surfaces (see Danielson, 1974).

Complications arise since an air parcel will remain on its
isentropic surface only under adiabatic conditions. On the time
scales of a day, diabatic processes become effective, e.g. temperature changes caused by radiation and evaporation/condensation.
The change of potential temperature could in principle be calculated, and the parcel moved to a revised isentropic level. But
whether trajectories are determined on pressure levels or isentropic surfaces, the difficulty remains: either one needs vertical
movements explicitly (to select the appropriate pressure level),
or one needs diabatic heating by condensation, which is modelled
from vertical movements. Also, radiational heating and cooling
depend very much on cloud cover, type, and height distribution,
and these depend on vertical motions. At present, vertical motions
are poorly determined in numerical modelling.

Additionally, trajectory analysis has limitations resulting
from the large range of scales of atmospheric motions. A cluster
of parcels released together may be separated by diffusion and
the parcels can then be included in different convective or mesoscale motions. After some time, the parcels will have quite
different pathways. This would be so even if numerical models
were sufficiently detailed and exact. In fact, additional uncertainties are added through the coarse resolution either of numerical models or observational networks and observing frequencies
(Sykes and Hatton, 1976; Pack et al., 1978).

(ii) Mesoscale transport would take place with land-sea
breeze systems. Since the system reverses between day and night,
one might suspect that it would bring no net transport. In general,
this is not true even if sea and land breezes are of equal strength.
Consider a source over land near the surface so that the transport
would be in the lower branch of the land/sea breeze system. During
night-time, contaminants would be carried towards the sea. The same
air might return to the land in daytime. If the residence time is
of order 12 or 24 hours, some of the contaminants may be deposited
at the sea surface. Hence there will be a net transport of contaminants from the land to sea. In general, because of the considerable variability of shoreline orography, there is probably no mesoscale model suitable to consider all actual cases. Hence,if it is
necessary to consider actual fluxes at a given place, it is advis-
able to investigate locally for the typical flow pattern.

(iii) Diffusion-models for local dispersion of gases or suspended matter are in wide use for different types of souces(point,
area) and operations (continuous, burst). Note that the commonly
used Pasquill-classes or similar stability- or diffusion-categories
have been developed for use over land(e.g. Pasquill, 1962, p.209).
The input parameters, which are used to characterize the radiation
balance and hence (implicitly) the atmospheric stability over land,
are not appropriate over sea. Because of different thermal properties, the stability over land is determined by the radiation
balance, while over sea, stability is mainly influenced by advection of cold or warm air or water, as represented by the air-sea
temperature difference. In order to translate the Pasquill classes
from land to sea, the following scheme seems feasible (Weber and
Hasse, 1982, unpublished) : The stability classification given in
terms of radiation balance and hence vertical heat flux over land
could be replaced by a parameterization of the sensible heat flux
H at sea, e.g. by

$$H = C_H \cdot C_p \rho u \Delta T \qquad\qquad\qquad (22)$$

where ΔT is the air minus sea potential temperature difference and
the bulk transfer coefficeint C_H is about 1.3×10^{-3}. In order to
account for the different roughness of land and sea, the windspeeds
given for land should be increased by a factor of 2 for use over
sea.

GENERAL REFERENCES

Petterssen, S., 1969: Introduction to meteorology. Third Edition,
    McGraw-Hill, New York, 333 pp.
Wallace, J.M. and P.V. Hobbs, 1977: Atmospheric science. An intro-
    ductory survey. Academic Press, New York, 467 pp.
Kraus, E.B., 1972: Atmosphere-ocean interaction. Clarendon Press,
    Oxford, 275 pp.
Haugen D.A.,(Editor)1973: Workshop on micrometeorology. American
    Met. Soc., Boston, 392 pp.
Nieuwstadt, F.T.M. and H. van Dop (Editors), 1982: Atmospheric
    turbulence and air pollution modelling. D. Reidel, Dordrecht,
    358 pp.
Danielsen, E.F. and J.W. Deardorff, 1978: Modeling the atmospheric
    transport of pollutants and other substances from sources to
    the oceans. In: U.S. Nat. Acad. Sci., The tropospheric trans-
    port of pollutants and other substances to the oceans, pp.
    25-52.
Eliassen, A., 1980: A review of long-range transport modeling.
    J. Appl. Meteorol. 19 , 231-240.
Frenkiel, F.N. and R.E. Munn (Editors), 1974: Turbulent diffusion
    in environmental pollution. Adv. Geophys. 18 B, Academic
    Press, New York, 389 pp.

REFERENCES

Danielsen, E.F., 1974: Review of trajectory methods. In:Frenkiel
    and Munn (Editors), Turbulent diffusion and environmental
    pollution. Adv. Geophys. 18 B, Academic Press,New York,
    73-94.
Defant, A. and Fr. Defant, 1958: Physikalische Dynamik der Atmo-
    sphäre. Akad. Verlagsges., Frankfurt, 527 pp.
Driedonks, A.G.M., 1982: Models and observations of the growth of
    the atmospheric boundary layer. Boundary-Layer Meteorol.
    23, 283-306.
Hasse,L., 1968: Zur Bestimmung der vertikalen Transporte von Im-
    puls und fühlbarer Wärme in der wassernahen Luftschicht
    über See. Hamburger Geophys. Einzelschriften, 11, Cram de
    Gruyter Verlag, Hamburg, 70 pp.
Hasse, L. and P.S. Liss, 1980: Gas exchange across the air-sea
    interface. Tellus 32, 470-481.
Köppen, W., 1899: Grundlinien der maritimen Meteorologie.
    Niemeyer Verlag, Hamburg, 88 pp.
LeMone, M.A., 1972: The structure and dynamics of the horizontal
    roll vortices in the planetary boundary layer. Ph.D. thesis,
    Univ. Wash., Seattle, 128 pp.
List, R.J., 1958: Smithsonian meteorological tables. Sixth re-
    ised edition.Smithsonian Inst.,Washington D.C., 527 pp.
Luthardt, H. and L. Hasse, 1982: The relationship between pressure
    field and surface wind in the German Bight area at high wind
    speeds. In: Sundermann and Lenz (Editors), North Sea
    Dynamics. Springer Verlag, Berlin, 340-348.
Monin, A.S. and A.M. Yaglom, 1965: Statistical fluid mechanics.
    Transl. from Russian, Cambridge, MIT Press, Vol. I, 1971,
    769 pp., Vol. II, 1975, 874 pp.
NAS, 1975: Understanding climate change. National Academy Press,
    Washington, D.C., 239 pp.
Newell, R.E., J.W. Kidson, D.G. Vincent and G.J. Boer, 1972: The
    general circulation of the tropical atmosphere and inter-
    actions with extratropical latitudes. MIT Press, Cambridge,
    Mass., Vol. 1, 258 pp., Vol. 2, 371 pp.
Pack, D.H., G.F. Ferber, J.L. Heffter, K. Telegadas, J.K. Angell,
    W.H. Hoecker and L. Machta, 1978: Meteorology of long-range
    transport. Atmos. Environ. 12, 425-444.
Pasquill, F., 1962: Atmospheric diffusion. Van Nostrand, London,
    297 pp.
Priestley, C.H.B., 1959: Turbulent transfer in the lower atmosphere.
    Univ. Chicago Press, Chicago, 130 pp.
Schlichting, H., 1965: Grenzschicht-Theorie. 5. Aufl., Braun,
    Karlsruhe, 736 pp., Engl. ed.: Boundary-layer theory, McGraw-
    Hill, New York.
Sykes, R.I. and L. Hatton, 1976: Computation of horizontal traj-
    ectories based on the surface geostrophic wind. Atmos.
    Environ. 10, 925-934.

Webb, E.K., 1982: On the correction of flux measurements for the effects of heat and water vapor transfer. Boundary-Layer Meteorol. 23, 251-254.

INTRODUCTORY PHYSICAL OCEANOGRAPHY

Fred Dobson

Bedford Institute of Oceanography
Dartmouth, N.S. B2Y 4A2
Canada

LARGE-SCALE OCEANOGRAPHY

The oceans are typically one thousand times broader than their depth. One can infer from this that their dynamics ought to be primarily two-dimensional, and this is in fact so, but important deviations occur. Both horizontal and vertical circulations have vital roles to play, as we shall see. The principal driving forces for the ocean are solar heating, the winds, and the attraction of the moon.

Both ocean and atmosphere obey the same physical laws, including thermodynamics, but there are important differences between them, which lead to differences in response of the two systems to external forcing and in our techniques for observing them.

The large-scale features of the tropospheric circulations in the atmosphere - the Aleutian low, the Iceland low, the Azores high, the subtropical convergence, the Hadley cells, and other analagous features - set up wind stress (i.e. wind drag) patterns on the surface of the sea. These stresses drive the large ocean gyres by causing surfaces of constant density to tilt, which in turn causes ocean currents to flow.

Whereas the atmosphere is free to move everywhere on the earth's surface, being constrained only in the sense that the mountain ranges exert more friction and to some extent direct the air flow at low levels ("orographic" flow), the oceans are constrained by the land to move within fixed basins. The resulting flow patterns in the atmosphere are roughly zonal (that is, along

53

P. S. Liss and W. G. N. Slinn (eds.), Air-Sea Exchange of Gases and Particles, 53–119.
Copyright © 1983 by D. Reidel Publishing Company.

latitude lines) and gyral (circular) in the oceans. Figure 1,
from Sverdrup et al. (1942), displays the major ocean currents in
February/March.

     Whereas meteorologists can accurately measure the surface
wind speed and the surface pressure differences which cause geostrophic flows, oceanographers cannot. Even at the edges of the
continents, it is difficult to measure the sea surface elevation
well enough to compute the mean currents. The fundamental
problem is simple: the surface of the sea moves, both vertically
and horizontally, in ways not directly related to the local wind,
and the motions are so large they obscure the signal to be
measured.

     This section will begin with some definitions, and then will
describe the fundamental dynamic and thermodynamic balances which
exist in various parts of the ocean. All are simplifications
from the full set of mathematical relations, describing fluid
flow, which are described in the "Basic Physics" section. The
particular flow configurations they produce have been named after
their discoverers, hence the names like "Ekman" and "Sverdrup".
Taken together, the flows can be used to generate plausible wind-
plus density-driven ocean circulation models. Unfortunately,
although the models so far developed produce qualitatively correct ocean circulation maps, none are completely quantitative,
and therein lies a lot of the excitement of theoretical ocean
dynamics.

Geostrophic, Baroclinic and Barotropic Motions

     In the atmosphere, two important considerations make it
possible to measure or infer the three-dimensional velocity
field. First, the density field, the pressure field and the
velocity field of the atmosphere can be monitored routinely at
fixed locations. Second, the geostrophic relations (see "Basic
Physics" section) specify that flow velocities must be (horizontal pressure gradient) ÷ (Coriolis parameter x density). In
the case of the atmosphere at mid-latitudes a 100 Pa/m pressure
difference over 1000 km produces a flow of 10 ms^{-1}, which is
relatively easy to measure to an accuracy of, say, 10%. In large
areas of the ocean the corresponding flow rate is 0.1 ms^{-1}, and
such currents can typically be measured to ±10% only after
averaging over periods of several months to more than a year,
either with floats which follow water parcels or fixed current
meters. The data from both floats and current meters, averaged
over a year, have typical precisions of ±0.01 ms^{-1}. As a result,
oceanographers only very rarely know the absolute current at a
given location.

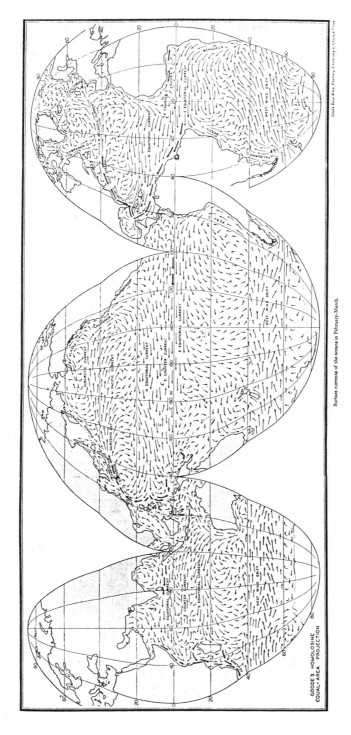

Figure 1. Chart of the ocean currents of the world in February-March. (The only major currents which change a significant amount with season are those of the Indian Ocean.) Much of the information shown was reconstructed from accumulated measurements of ship drift. (After Sverdrup, Johnson and Fleming, 1942.)

Setting direct current measurements aside, oceanographers determine relative currents using the geostrophic relations, the hydrostatic relation, and depth profiles of seawater density. Typically a "depth of no motion" is assumed, usually in very deep water where vertical pressure gradients are very uniform, although if it is available (as it might become if absolute satellite altimetry ever comes to pass) a "depth of known motion" is of course superior. Horizontal gradients of density, vertically integrated with hydrostatics to give pressure, then give the geostrophic currents relative to that level, or the so-called "geostrophic shear". A typical calculation is given in Pond and Pickard (1978), p. 60 ff.

Such calculations, supplemented by routine measurements of ship drift and inferences from property distributions (see later),have been the principal tools by which oceanographers have mapped the currents of the world's oceans over the last 50 years. The principal drawback is that the calculations give relative currents only, (i.e. "baroclinic" motions resulting from non-parallel pressure and density surfaces, that is, "isobars" and "isopycnals"). As a result,oceanographers have had to rely on inferential techniques for estimating the reference, or "barotropic" motions, which are due to a uniform tilt of both isobars and isopycnals.

It should be noted that the definitions of "baroclinic" and "barotropic" flow given here are not always held to by oceanographers. Sometimes "barotropic" is taken to mean "depth mean" and "baroclinic" to mean "deviation from the depth mean". The reference velocity used for baroclinic flows is generally that of the deep water, but sometimes use is made of the surface velocity or a velocity where measurements exist.

"Ekman" Transport

V.W. Ekman (1905) considered an infinitely deep and infinitely broad ocean (i.e. no bottom friction and no horizontal boundaries), being acted on by a steady wind, with no pressure gradients or density inhomogeneities. He allowed for friction at the sea surface and within the system by vertical shear of the current and used a shear friction coefficient independent of depth. His solution (Figure 2) gave a spiral current, which at the surface travelled at 45° to the right of the wind (cum sole) in the Northern Hemisphere, in which the Coriolis force was everywhere balanced by friction. The total mass transport (i.e. integrated over the depth to which the wind-driven currents penetrate-generally taken to be the depth of the mixed layer) is found to be

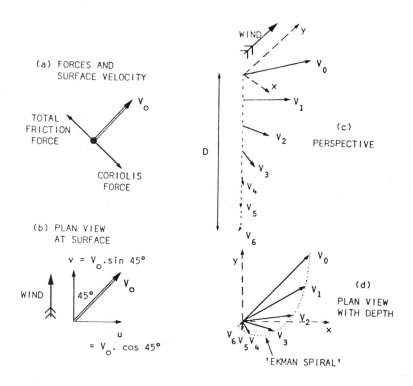

Figure 2. Ekman's solution (in the Northern hemisphere) for
wind-driven currents, in which friction in the fluid is balanced
by the Coriolis force. The surface current $\vec{V}_0$, is caused by
the wind; the current is seen to decrease and turn clockwise with
increasing depth. (After Pond and Pickard, 1978.)

$$M_{xE} = \tau_y/f \tag{1}$$

$$M_{yE} = -\tau_x/f \tag{2}$$

where $f = 2\Omega \sin \phi$ is the Coriolis parameter; Ω is the earth's angular rotation rate and ϕ is the latitude. Here τ_y and τ_x are the meridional (north-south) and zonal (east-west) components of the "wind stress", or the rate per unit area at which the wind imparts its horizontally-directed momentum to the sea surface. In the case of a steady current flowing over the sea bottom with friction, the "Ekman spiral" is also expected to apply, and has in fact been observed, although it is often obscured by other effects. Although (1) and (2) involve many simplifications, and probably do not represent physical reality, they produce quantatively useful estimates of wind-driven transports in the oceans.

"Sverdrup" Transport

Harald Sverdrup (1947) started with the linearized (i.e. no "advective" terms) basic equations, as did Ekman. He included pressure gradients but ignored horizontal friction. As did Ekman, he assumed his steady-state "ocean" was driven by a steady wind stress. He deviated from Ekman in allowing the wind stress to vary with latitude and longitude. Only the depth-integrated currents – the mass transports – were considered, and no lateral oceanic boundaries were permitted.

Sverdrup's equation was

$$M_y \, \partial f/\partial y = (\partial \tau_y/\partial x - \partial \tau_x/\partial y) \tag{3}$$

The derivative $\partial f/\partial y$ on the left hand side results from the commonly-used approximation

$$f = f_o + y\partial f/\partial y \equiv f_o + \beta y \tag{4}$$

where f_0 is the value of f at the latitude of y. The right hand side of the equation is the vertical component of the "rotational" derivative of the wind stress, or curl. Thus (3) can be written

$$\beta M_y = \text{curl}_z \, \vec{\tau} \tag{5}$$

This is known as the "Sverdrup equation" for the meridional oceanic transport. The "Sverdrup balance" is widely assumed to hold in the open ocean, away from western boundary currents (see

later). The zonal and meridional components of the mass transport, M_x and M_y, can be separated unambiguously into depth-
integrated "Ekman" and "geostrophic" components

$$\int_{-D}^{0} \rho U dz = M_x = M_{xE} + M_{xg} = (1/f) \ (\tau_y + \int_{-D}^{0} (\partial p / \partial y) \ dz) \quad (6)$$

$$\int_{-D}^{0} \rho V dz = M_y = M_{yE} + M_{yg} = (1/f) \ (-\tau_x + \int_{-D}^{0} (\partial p / \partial x) \ dz) \quad (7)$$

$$= \text{curl}_z \ \vec{\tau} / \beta$$

Westward Intensification

The next step in understanding the ocean circulation came
with Henry Stommel's (1948) explanation of why the oceanic gyres
are stronger on the westward sides of the oceans ("westward
intensification"). This represented a large step forward in
dynamic oceanography, and understanding Stommel's explanation is
central to understanding ocean circulations in general.

The mean global wind patterns (Figure 3) show a reversal
in direction at mid-latitudes, from the high-latitude westerlies
to the low-latitude trades. In the Northern Hemisphere this
combination induces surface (Ekman) flows in the ocean at 90° to
the right of the winds, which converge at mid-latitudes. The
ocean, on a long time-scale, responds baroclinically, that is, by
changing its vertical density structure. The water does not
"pile up", but instead the lighter surface water pushes downwards
on the heavier water below the mid-depth permanent temperature
discontinuity known as the "main thermocline" (see the Mixed
Layer Section), forcing the deep water outwards. The response of
the ocean to the outwards acceleration is to form a geostrophic
gyre, clockwise in the Northern Hemisphere (see Figure 4). The
waters of the gyre, and in fact all fluids on a rotating plat-
form, such as the earth, are also subject to the constraint that
they must keep their total angular rotation rate, or spin, a constant (the actual rule is that they must keep total spin divided
by depth constant, but we ignore depth variations in this simplified argument). This means that on the eastern side of the gyre,
the equatorwards-travelling water, forced to gain clockwise spin
relative to its surroundings as it increases its distance from
the earth's axis of rotation, must acquire a counterclockwise
spin to keep its total spin constant. Similarly, polewards-
travelling water on the western side of the gyre has a tendency
to spin clockwise. The wind stress pattern at the same time

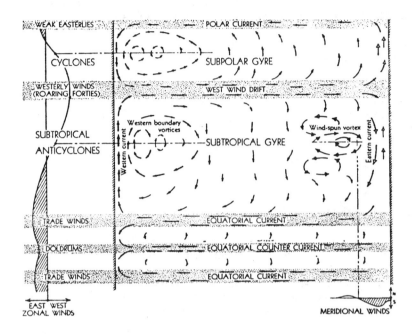

Figure 3. Schematic representation of the circulation in a
rectangular basin resulting from the principally east-west
(zonal) winds over the oceans (north-south (meridional) profile
of zonal winds at left). The nomenclature applies to either
hemisphere, but in the Southern Hemisphere the subpolar gyre is
replaced by the Antarctic Circumpolar Current (the "west wind
drift" shown in Figure 1). (After Munk, 1950.)

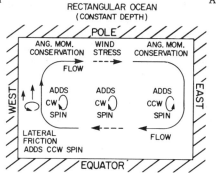

Intrinsic CCW (Counter
ClockWise) angular momentum
of fluid on earth's surface
is zero at poles, maximum
at equator: distance r
from spin axis increases,
equatorward.

Angular momentum

Polar-
moving water
conserves
total angular
momentum by
spinning CW to
offset CCW
tendency
caused by
decreasing dis-
tance from
spin axis.

Angular momentum

Equatorward-
moving water
conserves
total angular
momentum by
spinning CCW
to offset CW
tendency
caused by in-
creasing dis-
tance from
spin axis.

Spins do not balance.
Flow speeds up until
lateral friction with
boundary provides bal-
ancing CCW spin.

Spins balance,
flow in equil-
brium.

Figure 4: Westward Intensification.   Diagram of the flow in a
rectangular, flat-bottomed "ocean" on a rotating earth, induced
by a wind stress pattern (dotted line) which is eastwards near
the pole and westwards near the equator. To conserve total
spin (vorticity) in the gyral flow, the polewards-moving water
on the western side of the "ocean" must speed up until its CW
spin is balanced by lateral friction with the boundary. The
result is a circulation which is "Westward intensified". The
Gulf Stream and the Kuroshio are examples of the effect.

induces everywhere a clockwise spin (in the Northern Hemisphere)
to the surface waters. On the western side of the basin, the
water gains clockwise spin from the wind and from its polewards
motion, and, this leaves an imbalance. The flow on the western
side of the ocean is forced to speed up; it intensifies until a
balance exists between the clockwise spin gained from the wind
and by polewards motion and lost by lateral friction against the
western boundary. On the eastern side of the basin the spins
induced by wind and equatorwards motion are opposite and roughly
equal in magnitude, so a balance exists and no speed-up of the
flow occurs.

   The two most prominent examples of westward-intensified
ocean currents are the Gulf Stream and the Kuroshio. Both have
been extensively modelled; N.P. Fofonoff gives an up-to-date
review of the present status of Gulf Stream models in Warren and
Wunsch (1981). An article by G. Veronis in the same book provides an excellent theoretical account of large-scale ocean circulation dynamics. For a physically more complete, mathematically less complicated treatment, the book by Stommel (1960)
provides first-class reading.

Physical Properties of Sea Water

   Sea water density is defined by a rather complicated
equation of state:

$$\rho(s,T,p) = a_1(s,T,p)s + a_2(s,T,p)T + a_3(s,T,p)p \qquad (8)$$

where a_1, a_2 and a_3 are empirically determined constants, which
has been extensively tabulated. The present standard is presented as a set of algorithms giving the full equation of state
and relating salinity to electrical conductivity, temperature and
depth, in UNESCO (1981a,b). Before the advent of reliable
conductivity-measuring techniques, the salinity was determined
chemically by titration with silver nitrate. Density was obtained from temperature and salinity using the tables of Knudsen
(1901).

   The compressibility of water, although small, is not negligible in the ocean. The water in the deep ocean (below 4 km) is
under a sufficiently heavy load that its in situ temperature
actually increases with depth due to compression. To avoid
problems in estimating the stability of the deep waters and in
measuring the density of water moving over large ranges in depth,
it is customary to use "potential" temperature, usually referred
to as θ, and potential density, or sigma-theta. These are the
temperature and density the water would have if decompressed from
in situ to sea surface pressure without thermal contact with the
surrounding water (that is, "adiabatically").

The formula to convert in situ temperature T to potential temperature θ is (see e.g. Phillips, 1977)

$$\theta = T - \int_{p_o}^{p} (\partial T/\partial p)_{S,s} \, dp \qquad (9)$$

where p_o is the surface pressure and p that in situ, S is entropy and s is salinity (the subscripts mean the quantities mentioned are held constant). For potential density ρ_{pot}

$$\rho_{pot} = \rho - \int_{p_o}^{p} (\partial \rho/\partial p)_{S,s} \, dp \qquad (10)$$

Water Property Distributions and Tracers

Oceanographers have for many years used distributions of various water properties and solutes to infer large-scale flows in the ocean (see, for example, the charts at the end of "The Oceans", by Sverdrup et al., 1942). The original technique was to map the large-scale distribution of a given "tracer", make some assumptions about how the tracer might change its concentration in situ with the passage of time, and then attempt to reconstruct the path from its source. Water "masses" are defined which retain some property, such as a particular shape on a plot of temperature versus salinity (T-s diagram), as they spread from the region where they are formed. Plots of one property versus another (T-s, s-O_2, O_2-Si, etc.) are much used by descriptive oceanographers as a means of characterizing water masses. Figure 5, from Gordon (1982) is a typical example of such a plot. With such diagrams mixing of two water masses with conservative characteristics (i.e. no processes exist which create or destroy the relevant properties) occurs on straight lines joining the two points defined by the two types. A "water type" is defined, and used here, as water with uniform temperature and salinity; that is, it forms a point on a T-s or θ-s diagram. "Thermostad", "halostad" and "pycnostad" refer to regions in the ocean with uniform temperature, salinity, or density.

Since the period of extensive nuclear weapons testing in the 1960's, many new radioactive tracers have become available. Since they were injected at the sea surface over a relatively short period and their half-lives are well-known, they can be used to delineate pathways by which water parcels travel in the ocean. The principal uncertainty is whether the water carrying the tracer has been carried there by currents (i.e. "advected") or the tracer has diffused into the region. The GEOSECS program

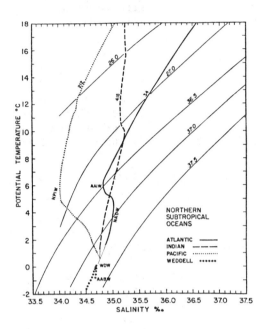

Figure 5. Temperature-salinity (T-s) diagram for the waters of
the northern subtropical oceans. (For a definition of potential
temperature see text, page 10.) The lines shown for the various
water masses are averaged over many observations. The thin
diagonal curves are constant-density lines, and the densities
given are 1000 (potential density in kg m^{-3} - 1.000), commonly
referred to as "sigma-theta". The numbers given beside each
water mass line are specific GEOSECS station numbers. NPIW is
"North Pacific Intermediate" water; AAI is "Antarctic Intermediate"; NAD is "North Atlantic Deep"; WD is "Weddell (Sea)
Deep"; and AAB is "Antarctic Bottom". (After Gordon, 1982.)

(Geochemical Ocean Sections Study) provided excellent baseline
surveys in the 1970's, and the TTO program (Transient Tracers in
the Ocean) is now providing additional data on the short halflife
tracers, to study the mechanisms by which large-scale oceanic
mixing occurs on climatic time scales. The reader interested in
pursuing the details should consult the article by W.S. Broecker
in Warren and Wunsch (1981), and Roether (1982).

Thermohaline Circulation

     Flows which are caused by the sinking of newly-formed, dense
water near the poles and subsequent upwards diffusion elsewhere
are called "thermohaline". Over most of the ocean the upper
mixed layer, with its relatively stable, almost frictionless
lower boundary (the so-called" permanent thermocline"), isolates
the deep waters from the direct influence of the driving forces
of solar radiation and wind. Therefore the mechanisms by which
the deep circulations are driven are quite different from those
which drive the surface flows. Only in a few places on the
globe, notably near the poles, are the deep waters not thus isolated. In the polar regions relatively warm, salty surface water
from lower latitudes encounters intensely cold, dry arctic air.
Three processes occur: the warm, saline water loses heat rapidly
to the air through evaporation and direct conduction, and large
volumes of freshwater runoff enter the system, freezing, and
releasing salt. All three cause the density of the surface
waters to increase; these waters then sink, usually in small
(<10 km diameter) regions. Lateral mixing with surrounding water
(often the surrounding water has its origins at higher latitudes,
and is relatively cold and fresh) produces large amounts of water
with closely-defined temperature and salinity characteristics,
which then spreads out from the polar regions along the bottom of
the worlds' oceans. Examples of these deep waters are to be seen
in Figure 5, at the bottom of the θ-S curves. A clear example of
the way they flow out of the arctic regions is shown in Figure 6.

     To complete the thermohaline circulation, return "flows" are
needed of magnitude equal to the global-average rate of deep
water formation. Stommel and Aarons (1960) proposed the first
real model; how quantitatively correct it is remains unknown.
The sources of the oceans' deep waters are taken to be in small
regions near the poles, and the return flow is by a general slow
upward motion of the deep water. The latter accounts for the
presence of the permanent thermocline (a relatively sharp change
in temperature at intermediate depths - 300 to 1200 m - which is
found everywhere), since it balances the downwards diffusion of
heat from the warm upper layer, which tends to eliminate the
thermocline. The mean horizontal flow in the deep ocean is
assumed to be geostrophic, and, as Stommel and Aarons point out,

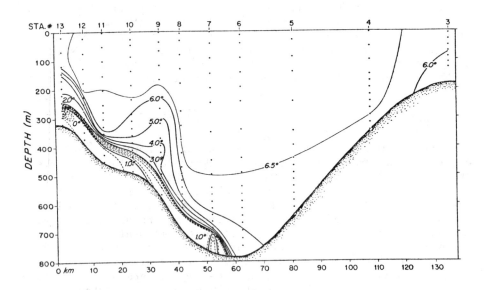

Figure 6. Temperature section across the Denmark Strait, looking polewards (Lat. 65-66N). The cold water is flowing southwards from the Norwegian Sea, hugging the western side of the Strait. From CSS Hudson Cruise BI-02-67, January 1967. (After Worthington, 1969).)

this implies, contrary to intuition, that the thermohaline flows over most of the ocean must be polewards, towards their source. The only way to get the dense polar water south is to posit deep equatorward western boundary currents, which is what Stommel and Aarons did. The arguments leading to westward intensification of the thermohaline circulation are identical in spirit to those given in the preceding section; they are well summarized in the article on the deep circulation of the ocean by B.A. Warren in Warren and Wunsch (1981). Stommel's (1960) schematic of how things might work in the Atlantic is given in Figure 7.

It should be realized that the deep, density-driven flows are not understood quantitatively. Only rarely has the strength of the deep flows been measured well: the difficulties are enormous. The flows typically have large temporal and spatial variability (due partly to "barotropic" - i.e. depth-independent-components of mesoscale eddies, and partly to the deep waters' proclivity for following bottom contours), so that long-term (several-year) measurement programs are required to obtain useful statistics of the deep flows. And because the deep waters are so voluminous, very small mean flows can account for enormous transports of water. (In water 5 km deep, a 1000 km wide mean current of 0.5 cm s^{-1} would transport 25x10^6 m^3 s^{-1}, about the same as the mean Gulf Stream transport through the Florida Straits.)

The thermohaline circulation, as poorly known as it is, has recently become the subject of intense interest, partly because the deep waters are an enormous sink for anthropogenic increments in airborne CO_2, and partly because man wants to store nasty things with long half-lives on or in the sea bottom (If they are out of sight, can they really be put out of mind?). The GEOSECS and TTO programs were both designed to define the mean state of the deep waters and to estimate lifetimes of tracers injected in polar regions. The next phase of exploration, now being planned as part of the World Climate Research Program, will be to study the response of the deep circulations to perturbations in their driving processes. The reader with curiosity about the details would do well to read Stommel's (1960) book and the Stommel memorial volume by Warren and Wunsch (1981). The CO_2 problem is lucidly addressed by Broecker et al. (1979) and the WMO White Paper on impact of CO_2 on climate variations (1981).

Future studies will involve not only investigations of the deep flows themselves. Their driving mechanisms are at the surface, and they must be understood, too. High on the priority list are high-latitude deep convection in the oceans, ice formation, and shallow-shelf runoff processes, all of which occur only sporadically in time and space, and with many required preconditions on the state of the ocean and the atmosphere (see, for example, the article by Gascard, in Kraus and Fieux, 1981).

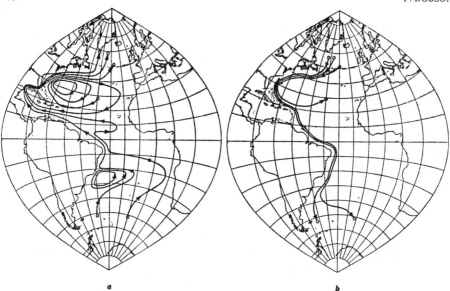

a   b

Figure 7. Schematic charts of the Atlantic Ocean, showing
idealized upper ocean (a, at left) and lower ocean (b, at right)
transport streamlines. Devised by Stommel (1960).

Variations in the rate at which the deep waters are formed by
these processes must surely affect the deep circulation, but no
one knows by how much. The subantarctic fronts have a role to
play, as do low-latitude and equatorial upwelling and distributed
diffusion processes (see the following section). Once again,
little is known quantitatively about the dynamics of the oceanic
response to forcing via such mechanisms.

Diffusion and Mixing

     There are many processes by which ocean water can be mixed
and thereby merge its original properties (T,s, momentum, vorticity, O_2, CO_2 and nutrient concentrations, etc.) with those of
nearby water. The most straightforward mixing process is molecular diffusion, in which a water property is exchanged at a rate
proportional to the spatial rate of change of the property. The
resulting, so-called "Fickian", diffusion equation (see e.g.
Batchelor, 1967) is

$$\partial C/\partial t = K_D\left[\partial^2 C/\partial x^2 + \partial^2 C/\partial y^2 + \partial^2 C/\partial z^2\right] \qquad (11)$$

where K_D is the "diffusion coefficient" of the water properties
with concentration C (K_D has units of length^2 time^{-1}). For

molecular diffusion, K_D depends only on the property diffusing
(e.g. salt); it is not time or space-dependent.

Because the diffusivity of heat (or temperature, T) and salt
(or salinity, s) differ in the ocean ($K_{DT} \approx 100 \times K_{Ds}$),
fluids that are stably stratified in the sense that their density
increases continuously with depth, may become unstable and mix.
Two examples of such "double diffusive" processes, referred to as
"salt fingering" and "cabbeling" are thought to be of some
importance in the ocean.

Salt fingering occurs when hot, salty water lies above cold,
fresher water (see Figure 8). As soon as the interface is perturbed, the fluid which bulges down loses heat to its surroundings faster than it loses salt, becomes more dense, and sinks;
the upward bulges gain heat and rise, and "fingers" of the two
water types interleave vertically. The process is thought to
occur on a widespread scale in the ocean, where the bottom waters
(formed at high latitudes by deep convection and overflow from
polar basins) are typically fresher than the "intermediate"
waters immediately above them. Salt fingering is also considered
to be an important mixing mechanism in fronts (e.g. see Bowman
and Esaias, 1978).

Cabbeling results because the relation between the density
of seawater and its s and T is nonlinear, particularly at low
temperatures. Thus if adjacent water types mix, the mixture may
be denser than either, resulting in further mixing by convective
overturning and turbulence as the mixture sinks. Such processes
are important in the formation of Antarctic Bottom water. For a
recent review of double diffusion see the article by S. Turner in
Warren and Wunsch (1981). For more details, see Turner's (1979)
book.

Turbulent Mixing

Turbulence is by far the most efficient mixing process in
the oceans. When a fluid becomes turbulent, it is continually
being deformed by random fluid motions. These deformations-
-spinning, stretching, interleaving -- cause an originally contiguous parcel of water to be formed into fine sheets or filaments. Thereby gradients in water properties between neighboring
parcels are continually sharpened, since the process of filament-
ation brings them into ever-closer proximity, so that molecular
diffusion can occur with great efficiency.

Turbulence can be generated in a number of ways, either by
direct mechanical "stirring", or through the action of buoyancy
forces. One source of turbulence in the ocean is the breaking of
sea surface waves. Such turbulence carries with it momentum from

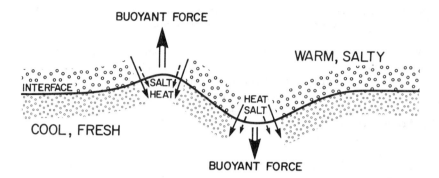

Figure 8. Salt fingering: a double-diffusive process known to
be important in the ocean, particularly in the vicinity of
fronts. Because heat diffuses 100 times faster than salt across
the (initially sharp and horizontal) interface, any vertical perturbations result in buoyant forces which further perturb the
initial state. Upward extensions of fresh (i.e., low-salinity)
water are heated and rise; downward extensions of salty water are
cooled and sink.

the wave field, and as well as mixing the surface water to a
depth of one or two wave amplitudes, it transfers the wave
momentum as a current to the upper mixed layer (see the section
on the Oceanic Mixed Layer). The action of a velocity shear can
also cause turbulence. Buoyancy forces can act to stabilize the
water column, if heating from above or sea surface precipitation
are dominant in determining the vertical variation of water
density, or to destabilize the water column, if cooling from
above or sea surface evaporation are dominant. In the presence
of an otherwise stable density gradient, a laminar flow will
become unstable when it reaches a critical ratio of buoyancy-
induced stability, to shear-induced "mixing capability". For
vertical shears the ratio (one of many such "numbers" having to
do with fluid stability) has been named after L.F. Richardson,
and it is defined as

$$Ri = (-g \, \partial\rho/\partial z)/\rho(\partial u/\partial z)^2 \qquad\qquad (12)$$

This particular version is called the "gradient" Richardson
number, since it is defined in terms of gradients. If the shear
is large ($\partial u/\partial z \gg 1$), then Ri becomes smaller, and hence a fluid
is more likely to be turbulent if the Richardson number is small.
Large density gradients ($-\partial\rho/\partial z \gg 1$, i.e. increasing density
with depth) cause Ri to be larger, and hence turbulence is
suppressed for large Richardson number.

Whenever density gradients exist in a fluid, internal waves can exist on the gradient (precisely as sea surface waves exist on the enormous air-water density gradient). They oscillate with displacements of the isopycnals from their mean depth with a frequency

$$N = \{-(g/\rho)\ (\partial\rho/\partial z)\}^{1/2} \tag{13}$$

N is called the Brunt-Väisala frequency, and

$$Ri = N^2/(\partial u/\partial z)^2 \tag{14}$$

Engineers often work with the inverse square root of Ri; it is called the "Internal Froude Number" and is usually designated Fr.

The criterion for stability is not simply determined, either theoretically or experimentally, in the laboratory or for field work: there are many types of shear-buoyancy instabilities. In general, flows are stable for $Ri > 1$ and can be unstable for $Ri \leq 1/4$.

MESOSCALE OCEANOGRAPHY
Ocean Eddies

Eddies in the ocean are the dynamical equivalent of "synoptic-scale" disturbances, or storms, in the atmosphere. They differ in important ways from their atmospheric analogs. There are no "fronts" in the interior of the eddies, as there are in typical high-latitude cyclones; in the ocean the frontal processes are confined to the regions of large current shear at the edges of the eddies. In the ocean, there is no analog to the evaporative/condensative processes which play such a central role in the dynamics of atmospheric storms. Oceanic eddies, once formed (often by instabilities in the western boundary currents), simply rotate, slowly converting their initial store of potential energy to kinetic energy. The scale of such disturbances has been defined by Rossby (1938) to be the distance over which a disturbance will be directly transmitted by pressure gradients, before the effects of the earth's rotation become appreciable. It is defined as the ratio of the speed of the wave in the medium to the Coriolis parameter,-which for a stratified fluid is

$$L_R \equiv \text{Rossby deformation radius} = (gh\Delta\rho/\rho)^{1/2}/f \tag{15}$$

where h is the vertical extent of the disturbance and $\Delta\rho/\rho$ is the degree of density stratification. L_R is about 800 km for the atmosphere and 50 km for the ocean.

Oceanic eddies have been a central concern of oceanographers
over the 1970-1980 decade (see e.g. Oceanus, 1976). The MODE and
POLYMODE experiments were designed to elucidate the role of mesoscale processes in ocean dynamics, and have greatly increased our
understanding of such processes. Ocean dynamic models are now in
existence (Holland and Rhines, 1980) that resolve the oceanic
eddies and incorporate their dynamics into the general oceanic
circulation. Such models, although they cannot yet be said to be
predictive, do reproduce the general circulation remarkably well.

Gulf Stream Rings

Figure 9a describes the birth of a cold-core ring from the
Gulf Stream; figure 9b shows a typical cross-section for a cold-
core ring. Similar (but not identical) rings are spawned by the
Kuroshio system (White and McCreary, 1976). The rings form as
the result of instabilities in the strong currents, which begin
as "meanders" and end as pinched-off "rings", which can be either
cold-core and rotating counterclockwise (in the Northern Hemisphere) when pinched off equatorward of the stream, or warmcore
and rotating clockwise when pinched off to poleward. Both types
have now been extensively observed, most effectively with satel-
lites (Figure 10) and with drifters that report their positions
regularly to a satellite (Richardson et al., 1977).

The dynamics of such rings are understood in general, but
their interactions with the general circulation, and hence with
the global climate system,are not well-understood at all. A
cold-core ring, for example, consists of a well-defined area of
water 50 to 100 km in diameter, typically cooler by about 5°C
than its surroundings, rotating with tangential velocities up to
1 ms^{-1} in the core. In the Gulf Stream the typical vertical
extent of the rings is from the surface to 1500 m. The density
distribution within the rings shows a general raising of the main
pycnocline by as much as 500-600 m, and a sea surface depression
of 0.4 to 0.5 m. Both contribute to a geostrophically balanced
counterclockwise flow in the Northern Hemisphere. Warm-core
eddies rotate clockwise in the Northern Hemisphere, and are
characterized by a sea surface elevation and depression of the
main thermocline, both of the same magnitude as for the cold-core
eddies.

The eddies are spawned from the Gulf Stream at the rate of 5
to 8 per year on each side of the Stream. Their life time,
before decaying or being reabsorbed, is up to two years for cold-
core, and less than one year for warm-core rings. They move at
rates of 2 to 10 km/day, and, in the generally southerly Gulf
Stream recirculation region tend to move in a southwesterly
direction.

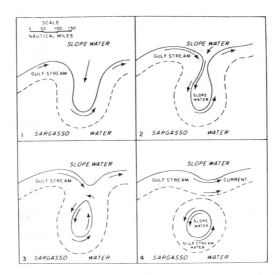

Figure 9a.   Diagram of the birth of a cold core ring from the
Gulf Stream.   (After Parker, 1971.)

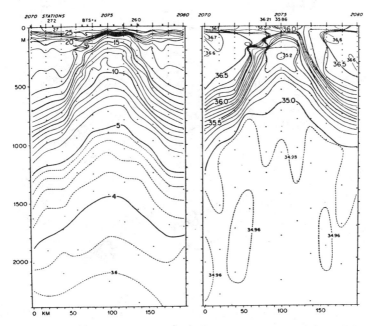

Figure 9b.   Temperature and salinity sections through a Gulf
Stream "cold core" ring, or eddy.   Location is 36.5°N, 64°W, on
31 July 1967; data from R.V. "Crawford" cruise 158.   (After
Fuglister, in Angel, 1977.)

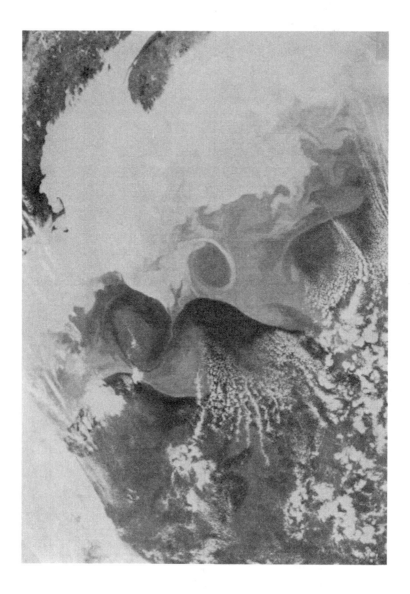

Figure 10. NOAA-5 infrared image of the Gulf Stream off Cape
Cod, MA. The lighter-coloured waters are cool; darker-coloured
regions are warm, Gulf Stream water (the white areas in the lower
right are clouds). Several warm-core eddies are evident; the one
at the center is particularly well-formed. (Photo courtesy of
Atlantic Environment Group, U.S. National Fisheries Service,
Narragansett, RI).

To see the reason for the southwesterly drift, consider a warm-core, counterclockwise-rotating eddy which is in geostrophic equilibrium, so its pressure gradients are balanced exactly by the mean Coriolis force at its central latitude. Since water equatorwards of the central latitude has higher counterclockwise angular momentum than the water polewards of the central latitude, the equatorwards water will turn poleward with a slightly smaller radius of curvature than that of the polewards water. This effect, due to the latitude variation of the Coriolis parameter, causes all eddies, warm-core and cold-core, to propagate westwards.

Rings cause significant disturbances in their environment, particularly when they approach the shore, as warm-core Gulf Stream rings do. They interfere with frontal and long-wave generation processes at the shelf break, and with surface layer circulation and mixing, and hence with mass exchange over the continental slope and shelf (Csanady, 1979). Dissipation of such rings and the lateral mixing they cause may be an effective way of transporting heat, salt and angular momentum in the ocean. Eddies decay by using up their store of potential energy by converting it to kinetic energy of spinning (in a young cyclonic ring the main thermocline is raised 500-600 m, giving it an "available potential energy" of 10^{17} joules; using this up in 2 years, averaged over a 50 km diameter eddy, gives an energy dissipation rate of 0.8 W m^{-2}). Eddies are known to be an integral part of ocean dynamics, but their overall significance is not yet fully understood.

Eddy motions are important for another reason: they make sampling very difficult. N. Fofonoff, in Warren and Wunsch (1981), states "The Gulf Stream, if it exists near the bottom above 70°W ..... is nearly completely masked by the strong deep eddy field". P. Rhines, in Oceanus (1976) writes, referring to the entire ocean, "The attempt at understanding the balance of forces and flow of energy in the ocean is made difficult by the variability of the currents: that is, by the eddies. The currents are in fact far too capricious to be mapped once and for all, so that the procedure of the classical oceanographer – the gradual filling in of a jigsaw picture of the ocean – may simply not work". He then goes on to suggest that overcoming such a fundamental difficulty will involve the formulation of theories that describe the eddy fields statistically rather than causally. Meanwhile, existing eddy-resolving models (e.g. Holland, 1978) strain the largest available computers to their limits, and even remote sensing techniques such as satellite altimetry (Wunsch, 1981) have a difficult time resolving eddy-scale motions in time and space except in a statistical sense.

The eddies have been found to be ubiquitous, at least in the Northern oceans, and oceanographers, fully aware of the sampling problems eddies create, are presently reassessing their "error bars" and their experimental strategy. More than any other single phenomenon, eddies have changed the face of theoretical and experimental oceanography. The new age belongs to the statistically-oriented numerical simulation, the drifting float, and the satellite. Deep soundings from ships, although far from outliving their usefulness, must now be interpreted from an entirely new perspective: is the resulting field really a time (or space) average, or is it the result of sampling a field of eddies?

Frontal Processes

Oceanic fronts are best defined as boundaries between water masses with dissimilar properties. The fronts are often marked on the surface by disturbances of the waves there or by lines of foam or debris. They generally have associated with them regions, of convergence and relatively strong vertical motions. Such circulations, as well as the various diffusive processes and the large gradients of water properties at such boundaries, form effective mixing agents. Consequently, fronts are considered to be major contributors to the formation of new water masses.

Estimates of the volumes of water mixed within the various types of fronts are, unfortunately, not yet available. The small-scale processes by which mixing occurs within fronts are the subject of intense study (for an excellent recent review see Garrett, 1979), but only crude estimates, such as that given for the decay rate of eddies in the section on Gulf Stream rings, are available now. Fronts occur not only at the sea surface, but also at any depth in the water column, including the vicinity of the sea bottom. They occur on all spatial scales, from less than a meter to thousands of kilometers. In general they are caused and maintained in the oceans by processes similar to those in the atmosphere, but fronts in the ocean take on a greater diversity of form and strength than they do in the atmosphere.

Following Bowman and Esaias (1978), fronts may be classified into six types, according to their scale and dynamical origin.

Planetary fronts are those associated with large-scale con-vergence of surface Ekman transports, and are typically found in mid-ocean. A good example is the Antarctic convergence zone, or Polar Front (see Figure 11a and b), which is at about 60°S and forms the northern limit of the so-called Antarctic Surface Water formed by ice-melt in summer and surface cooling in winter.

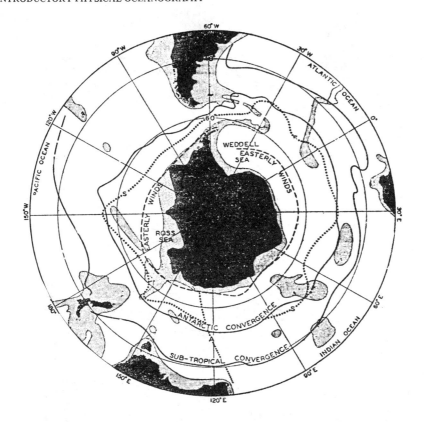

Figure 11a. Map of Antarctica, showing (solid lines) the position of the Sub-tropical and (winter) Antarctic Convergence
Zones, and (dotted line) the summer Antarctic Convergence zone;
the dashed line shows the outer limits of the Antarctic easter-
lies. The roughly meridional dotted line marked "A" is the position of the line of stations from which Figure 11b is derived.

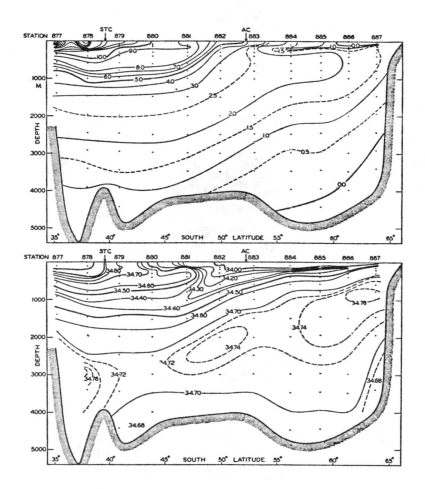

Figure 11b. Temperature (top) and salinity (bottom) sections
from Australia (on the left) to Antarctica (section marked "A" on
Figure 11a). Positions of Sub-tropical and Antarctic Convergences marked "STC" and "AC". (After Sverdrup, Johnson and
Fleming, 1942.)

Western boundary current edges often behave as fronts.
Warm, salty tropical water is brought into relatively close
proximity with cold, fresh high-latitude water by the western
boundary currents. Sharp boundaries and mixing on all but the
planetary scales is the inevitable result. For a beautiful
example of such a front, see the frontespiece to Stommel's (1960)
book; see also Figure 10. That such fronts are important contributors to oceanic dynamics and mixing is discussed further in
the preceding sections on "Ocean Eddies" and "Gulf Stream Rings".

Shelf break fronts are formed at the edges of continental
shelves, where waters characteristic of the shallow shelf regions
meet and mingle with water types typical of the continental
slopes and the deep ocean. All types of diffusion and mixing
occur, and the flows can be either baroclinic (surfaces of constant pressure and density tilted) or barotropic (isobars and
isopycnals parallel) depending on whether or not the salinity and
temperature fronts coincide, and which controls the water
density.

Upwelling fronts (see Figure 12) are generally associated
with offshore Ekman transports generated by longshore winds.
They are characterized by a pycnocline that reaches the surface
with surface water offshore and deep water at the surface inshore. Such fronts commonly occur off the West coasts of continents (California, Peru, Africa). They interact strongly with the
local meteorology, and are one of the few locations on the globe
where waters beneath the permanent pycnocline come into direct
contact with the atmosphere. They thus act as an important pathway by which airborne gases and particles can be exchanged with
the deep ocean.

Plume fronts occur on the boundaries of the fresh-water discharge plumes produced by large rivers, such as the Amazon and
Columbia Rivers. They act to stir the fresh runoff waters, with
their load of terrestrial solutes,into the ocean.

Shallow-sea fronts are formed in shallow seas and estuaries
and around islands, shoals, capes, etc. They form the boundary
between well-stratified offshore waters and waters in the shallow
seas, which are well-mixed by winds and tides. Simpson and
Hunter (1974) have defined a useful "stratification ratio", R, as
the rate of production of potential energy by surface heat flux Q
to the rate of tidal energy dissipation:

$$R = (gh\beta_T \; Q/2c_p\rho)/(C_d u^3) \; \approx \; h/u^3 \tag{16}$$

providing Q is constant. Here β_T is the volume expansion
coefficient of the water, h its depth, c_p its specific heat at

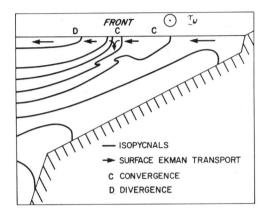

Figure 12.   Diagram of an "upwelling front" in the Northern
Hemisphere. A wind stress τ_w, out of the paper, drives surface
water offshore ("surface Ekman transport" arrows). The dynamics
by which the front is maintained between the inshore upwelled
deep water and the offshore surface water causes convergence and
sinking at the front ("C") and divergence offshore ("D"). (After
Bowman and Esaias, 1978.)

constant pressure, ρ its density, and C_d a "drag coefficient"
relating the drag exerted by the bottom on the water above it to
the square of the water speed u. (The reader should note that R
is really a type of Richardson Number: see the section on
Turbulent Mixing, equation 12.) Plots of the logarithm of h/u^3
are commonly used to distinguish well-stratified (large R) from
well-mixed (small R) areas, and the fronts that lie between them.
Figure 13, from Bowman and Esaias (1978) is an example of such a
"h/u^3" plot; fronts are visible on the eastern sides of the
Celtic and Irish Seas, and are also present in the nearshore
areas on the south side of the English Channel.

Tides and Large-Scale Waves

     Three excellent reviews of tidal motions are "Ebb and Flow"
by Defant (1958), "The Tides" by Darwin (1968 edition), and the
more technical review by Cartwright (1977). The forces which
generate the tides can be described from the dynamics of the
earth-moon system.

     The earth and the moon are two spheres rotating around a
centre of mass that is within the earth but is not at the earth's
centre, being shifted towards the moon. The earth itself (which
also spins, a fact that we ignore for the moment) then rotates <u>as
a solid body</u> (like one end of a "dumbell") about the centre of
mass of the earth-moon system, and each point on the earth's
surface experiences a centrifugal force inversely proportional to

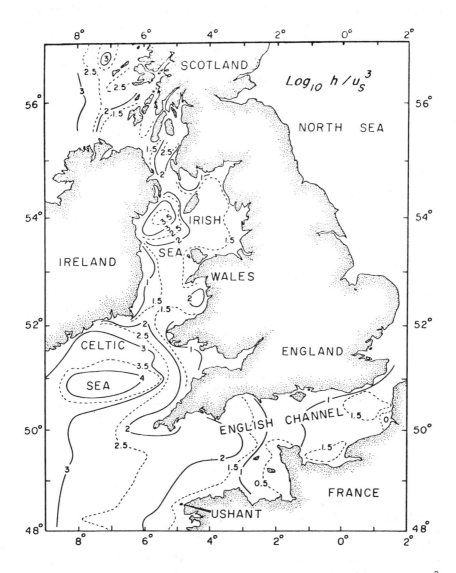

Figure 13. Plot of the log₁₀ of the tidal mixing parameter h/u^3, where h is the water depth and u_s is the surface tidal speed of the water, for the seas of southwest England and Ireland. Tidal mixing fronts are found between regions of little mixing (Celtic Sea, western Irish Sea) and strong mixing (the coast of South Wales, eastern Irish Sea), and in the vicinity of major headlands (Land's End and Ouessant or "Ushant"). (After Bowman and Esaias, 1978.)

its distance from the centre of mass and directed away from the
moon. Each point experiences the moon's gravitational pull as
well as the earth's. The only asymmetrical forces are the moon's
pull (the side of the earth away from the moon is about sixty-one
earth radii from the moon, and that facing the moon is fifty-
nine, so the gravitational pull is stronger on the side nearest
the moon) and the centrifugal force caused by the earth's solid-
body rotation (since the centre of mass of the earth-moon system
is shifted toward the moon from the earth's centre, the centrifugal force is stronger on the side of the earth furthest from
the moon). When the tidal forces are added, there is a net out-
ward force on the earth's surface at the points nearest and most
distant from the moon, and net inward forces on points on the
earth at right angles to them. As the earth spins, these forces
act on the oceans, causing two tides per day. The tidal forces
are not just those from the moon but also from the sun (the sun's
force is about half that of the moon, because it is so much
further away from the earth).

The tidal forces have various distinct frequencies, caused
by perturbations in the orbits of the earth-moon-sun system. The
actual magnitude of the tide at a given location is determined by
interactions of the tide-generated waves with local topography.
Since they are "body" forces, the tidal forces act everywhere,
not just at the sea surface; there are measurable tides in the
earth itself, as well as in the atmosphere and on pycnoclines in
the ocean. The latter are called "internal tides", and are
clearly evident in most oceanic current meter records. They are
generated by direct forcing and also indirectly, as the sea surface tide (the "barotropic" tide) impinges on continental
shelves, causing the pycnocline to oscillate and radiate "baroclinic" tidal-frequency waves seaward.

The wavelength of tidally-generated waves is half the circumference of the globe; since the depth of the ocean is only a
small fraction of a wavelength, they obey the equations for
shallow-water waves. There are various other waves, also
shallow-water by nature, which occur in the ocean: "Poincare"
waves, coastally-trapped "Kelvin" waves, and others. Their
exposition lies beyond the scope of this text, and the interested
reader is referred to texts on physical oceanography or geophysical fluid dynamics (e.g. Pedlosky, 1979).

Oceanic tides are important in transport only in the sense
that they induce mixing and generate internal waves, which in
turn break down to produce mixing. Most tidally-induced mixing
occurs at the edge of or on the continental shelves. The so-
called "h/u^3" criterion for determining where tidal mixing is
likely to be important is discussed in the section on Frontal
Processes. Such mixing in shallow seas can form new water

masses, particularly at high latitudes, where warm, salty waters
delivered from low latitudes by western boundary currents can be
mixed tidally with cooler, fresher northern waters. The eastern
continental shelves of North America are breeding grounds for
several at least partly tidally-generated water types, as are the
shallow waters of the Barents and East Siberian Seas in the
Arctic.

Internal Waves

    Gravity waves occur on all density gradients in the atmos-
phere and the ocean; in the latter the "internal waves" on
various density gradients are as ubiquitous as are surface
gravity waves. They have wavelengths from hundreds of meters to
tens of kilometers and periods from tens of minutes to tens of
hours. Their amplitudes can be tens of meters. The currents and
current shears associated with them often interact with short-
wavelength (≈ 10 cm) surface waves, producing surface patterns of
alternately ruffled and unruffled water that are highly visible
to the (elevated) eye or to obliquely incident radars (see e.g.
Fu and Holt, 1982). The large displacements of isopycnals (and
hence isohalines, isotherms, and other isolines) are an extremely
important source of sampling error to oceanographers. Classical
geostrophic transports are calculated from salinity and temperature information collected from specific depths at various locations. It is easy to see how an internal wave, by displacing
isolines vertically by distances of 10 m or more over distances
of one or more km, could change the outcome of such calculations,
particularly since internal waves can propagate with speeds of
0.1 m s^{-1} or more.

    Classical internal waves (that is, those occurring on a
relatively sharp density discontinuity) have their largest vertical displacements, and hence their largest horizontal velocity
differences, or shears, at the discontinuity. Such velocity
shears are known to become unstable (see the section on Mixed
Layers) when the ratio of the stabilizing influence of the
density difference to the destabilizing influence of the shear
(i.e. the Richardson Number Ri) becomes small. Below Ri $\approx 1/4$,
internal waves begin to break and to mix the waters above and
below the interface. This mixing mechanism is important in the
ocean, since internal waves and shears are so common. Internal
waves have been extensively studied in the ocean and a number of
useful (although often somewhat mathematical) reviews are available. Phillips (1977) gives a relatively complete and authorata-
tive coverage. Perhaps the most useful reviews for estimating
quantitatively the effects of internal-wave mixing in the ocean
are those of Munk (1966) and Garrett (1979).

THE OCEANIC MIXED LAYER
Introduction

     The oceanic mixed layer is defined as the region of the
upper ocean which is influenced by mixing processes, with predominant seasonal and daily cycles, which originate at the sea
surface. It has uniform properties and is bounded below by a
large density gradient, called the "pycnocline" (gradients in
temperature are "thermoclines" and gradients in salinity are
"haloclines". For dynamical reasons discussed further in the
section on Thermohaline Circulation, the ocean, the density of
which is controlled primarily by temperature, has a "permanent
thermocline", delineating the lower limit of the annual cooling
cycle, at depths ranging from less than 100 m in some equatorial
regions to more than 1000 m in the centre of the large oceanic
gyres. Above this a series of seasonal thermoclines are formed,
which have the permanent thermocline as their limiting winter
depth, and are reformed at the sea surface every spring. The
various density discontinuities of the mixed layer, and in particular the permanent thermocline, act as barriers to the downwards mixing of substances in the mixed layer because they are
very efficient at damping out buoyant convection. The "rate
constants" of the various pycnoclines for various mixing processes can at the moment only be guessed at, or inferred from
large-scale distributions of the various tracers.

     The atmosphere, like the ocean, has a mixed layer, but the
dynamics of the phenomena in the two media are dominated by
different processes. In the atmosphere, mechanically-generated
turbulence has its source mostly near the sea surface, from
instabilities in the strongly sheared flow there. In the ocean,
the waves generate some turbulence, but equally important sources
are convection and shear instabilities at the base of the mixed
layer. In the atmosphere, the clouds have a strong influence on
absorption and transmission of solar energy, both incoming short-
wave and outgoing longwave. In the ocean, most of the incoming
solar radiation is absorbed within one meter of the surface, and
infrared absorption and emission is in the top millimeter. Solar
heating causes instability in the air and stability in the water;
evaporation causes convective instability in both media.

     The oceanic mixed layer is of intense interest because its
upper surface is the region of first interaction of the various
gases and particles with the ocean. Heat and salt are stored;
CO_2 is chemically altered and stored; the various atmospheric
daters and tracers start to be mixed into the ocean here. The
time scales are seconds to years, the vertical scales are centimeters to decimeters, and these scales overlap with those of the
surface (on the short end) and the deep ocean (on the long end).

Dynamically the mixed layer is a terra incognita. Velocity
measurements (or good density measurements) are very difficult to
make there. The principal measurement difficulty, both in the
fluid and as an influence on measurement platforms, results from
surface wave motions. The mixed layer is a region of strong air-
sea interactions, and hence is governed by strongly nonlinear
dynamics. (Surface and internal wave breaking, mixing of heat
and salt across isentropic surfaces ("diabatic" mixing), particularly in coastal regions, advection, absorption and re-
emission of solar radiation, and strong biological activity all
contribute to the dynamics). This makes analytical solutions
unlikely. Consequently, the approach has been to parameterize
the strong effects in terms of measurable quantities and do
simple budgets.

Time and Space Scales

Figure 14, from Tabata and Giovando (1963), shows the time
variation of the depth of the oceanic mixed layer at Ocean
Weather Station "P" (50°N, 145°W, in the North Pacific) averaged
over four years. Note that, in contrast to the atmospheric mixed
layer, the oceanic mixed layer is deeper in winter than in
summer. Figure 15, after Woods, 1982, shows schematically the
yearly cycle of heating/cooling, including the range of diurnal
variation. There are also large spatial variations in the
observed depth of the mixed layer, as indicated in the Robinson,
Bauer and Schroeder (1979) atlas (Fig. 16a).

Since the convectively mixed layer contains, distributed
within it, all gases and particles transferred from the atmosphere to the ocean over a time interval short compared with the
equilibration time of the deep ocean, it will pay to understand
the dynamics of the mixed layer well.

Modelling the Mixed Layer

The modelling of mixed layers is actively evolving, after a
big push in the mid-70's by a whole series of "one-dimensional"
models, which are detailed in Kraus (1977). These one-
dimensional models use as their basic premise the observed fact
that vertical gradients are much stronger than horizontal ones at
most places in the ocean. Hence the mixed layer can be thought
to have a vertical but no horizontal structure, and the forces at
work can be considerably simplified. Time variation is, of
course, allowed. Models allowing for both time and three-
dimensional space variation are being developed and tested now.
They must allow for advection: the transport of water and its
properties from place to place by oceanic currents, either driven
by large-scale forcing external to the problem (Ekman flux
divergence, for example) or by local forcing due to wind and/or

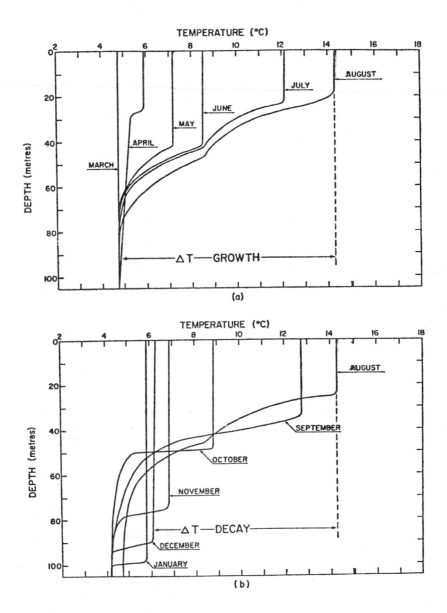

Figure 14. Time variation of mixed layer thermal structure at
Ocean Weather Station "P". (a) March-August, (b) August-
January. (After Tabata and Giovando, 1963.)

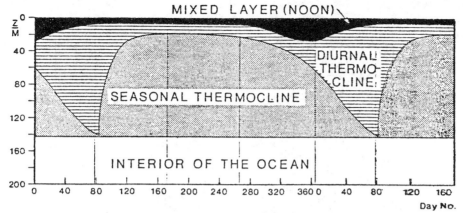

Figure 15. Idealized yearly cycle of mixed layer depth, including depth variations of the extent of the noon thermocline (dark),
the extent of the diurnal thermocline (hatched), and the seasonal
thermocline (stippled). (After Woods, in Kraus and Fieux, 1981.)

buoyancy effects. An illustration of the results of one of the
earlier models (Zubov, N.N. in Gorshkov, 1978) is given in Figure
16b. Comparison with observations (Fig. 16a) shows that the
models are far from perfect.

Forcing Variables

The mixed layer is forced primarily by solar heating by
shortwave radiation, the wind stress, and the vertical fluxes of
latent and sensible heat through the sea surface. Smaller but in
some circumstances important driving forces are precipitation and
the mixing effects of internal waves. The presence of slicks
affects both the heat fluxes and the influence of surface waves.
The response of the mixed layer to such forcing is radiation back
to the atmosphere at long wavelengths, downwards turbulent mixing, including entrainment at the bottom of the mixed layer by
mechanical and buoyant forces, and the generation of currents on
scales ranging from oceanic (Ekman flux divergence) to microscale
(turbulence).

Solar Radiation

Solar heating is a complex process. Once the spectrum of
radiation reaching the sea surface after transmission through the
atmosphere is known, and once the reflectivity (or albedo) of the
sea surface is known, then an attempt can be made to estimate the
scattering and the penetration depths (and thus the heating
capabilities) of the various wavelengths of the incoming radiation.   The necessary procedures, and the difficulties involved,
are succinctly summarized in the article by A. Ivanoff in Kraus
(1977), and given in more detail in the books by Budyko (1974)
and Jerlov (1976). Figure 17, from Augstein, 1981, shows the

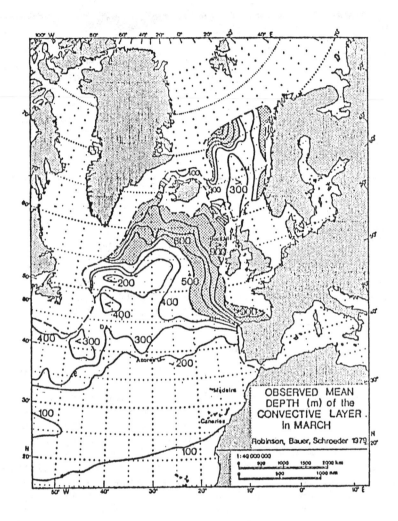

Figure 16a. Variation of the annual mean depth reached by the
mixed layer for the Northeast Atlantic Ocean. The region of
total ice cover is crosshatched. (After Robinson et al., 1979.)

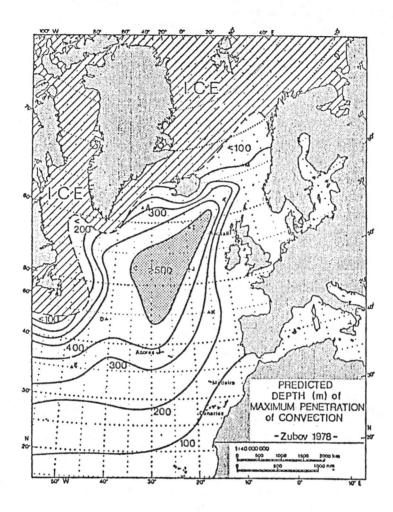

Figure 16b. Prediction for the annual mean depth reached by the
mixed layer for the Northeast Atlantic Ocean, using a mixed-layer
model attributed to N.N. Zubov. (After Gorshkov, 1978, Plate
141.)

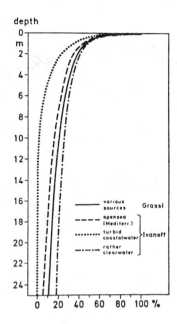

Figure 17. Percentage of incident solar energy (allowing for tha
reflected) absorbed per meter of water as a function of depth.
The solid curve is from H. Grassl (pers. comm.); the other curves
are from the article by A. Ivanoff in Kraus, 1977. (After
Augstein, 1981.)

fraction of incident solar energy (allowing for that reflected)
absorbed per meter of water as a function of depth, for different
types of seawater. It is plain to see that a large fraction of
the incident solar energy is absorbed in the first meter. Also
the "turbidity" of the water, which is strongly influenced by
biological and nearshore processes, is important in determining
the vertical distribution of solar heating in the mixed layer.
How to model these effects remains one of the central questions
in mixed-layer theory.

Wind Forcing

    Wind forcing introduces a number of effects. First, the
wind produces surface gravity waves, and although their effects
diminish exponentially with depth, they produce orbital currents
which typically exceed other currents by factors of 2 to 10 in
magnitude. Of no small consequence are the motions induced by
the waves on measurement platforms. The reader is referred to
articles by J. McCullough, R. Davis and R. Weller, W.
Blendermann, and H. Berteaux in Dobson, Hasse and Davis (1980)
for discussions on wave-induced errors in current meters and the
wave-induced motions of buoys and moorings.

It remains unclear how the momentum and energy of the wind
is transferred to mean currents in the open ocean. One possibility (Dobson 1971) is that it is first transferred to the waves
and then, through wave dissipation, "handed over" to the mixed
layer below. In any case, the "hand-over" process is a complex
one with many possibilities for fluid motion. The small-scale
wave-generation and dissipation processes are dealt with later;
we will concentrate here on the larger-scale phenomena. Pollard
(1970) has investigated the generation of inertial oscillations
(horizontal circular motions of the entire mixed layer, which
result whenever the water is forced to move impulsively, followed
by a removal of the forcing) in the ocean. He finds that the
energy of the oscillations is confined almost entirely to the
surface mixed layer (the oscillations, as we shall see, help to
create the mixed layer); the amplitude of the oscillations is
strongly affected by the initial depth of the mixed layer; it is
nearly independent of the horizontal scale of the wind field or
of the vertical stratification in the mixed layer. What matters
is the time history of the wind stress vector. If the wind speed
and direction vary on time scales much less than one inertial
period (f^{-1} sec; about 17 h at 45° latitude), then inertial
oscillations result. If, for instance, a cold front embedded in
a low pressure area passes a given point in the ocean, then the
sharp clockwise (in the Northern Hemisphere) shift in wind direction associated with the front will efficiently generate clockwise inertial oscillations, which will then persist until they
disperse out of the region (their rate of travel, however, is
very slow), decay, or are damped by an anticlockwise wind shift.
Pollard found the latter mechanism was necessary to explain the
transient nature of the observations, and was hence important in
the ocean.

Pollard, Rhines and Thompson (1973) invoked inertial oscillations to explain part of the mixing in the surface layers of
the ocean. In a one-dimensional model they assumed the entire
mixed layer moved in inertial oscillations under the action of a
time-varying wind stress. Since the oscillations are strongly
confined to the surface layer, large vertical shears (variations
of current speed and direction over small vertical distances)
occur at the bottom of the mixed layer. These shears in turn
cause instabilities in the flow, which generate turbulence and
mix the fluid from below. The authors were successful in
explaining the sudden deepening of the mixed layer on the passage
of storms.

Buoyancy

The mixed layer is forced by density fluctuations as well as
by mechanical stirring, and evaporation is perhaps the most
important buoyancy-driving mechanism. Evaporation acts on both

the temperature and the salinity to increase the density of the
surface water, giving it a tendency to sink convectively. The
opposite process--rainfall-- decreases the surface water salinity
but may change the sea surface temperature in either direction.

Conservation Relations

One-dimensional mixed-layer models solve simultaneous conservation equations for momentum, buoyancy, heat and salt. For
example, P. Niiler and E. Kraus, in Kraus (1981) use, for the
buoyancy b,

$$b = -g(\rho-\rho_r)/\rho_r = g\{\alpha(T-T_r) - \beta(s-s_r)\} \qquad (17)$$

where ρ is density, T is temperature, and s salinity; the "r"
subscript denotes constant "reference" values of quantities in
the mixed layer. α and β are the coefficients which describe the
(linearized) effect of temperature and salinity on density.
Given suitable boundary conditions for the fluxes at the surface
and at the base of the mixed layer, all the equations can be
solved for the vertical motion, temperature, salinity, and depth
of the mixed layer. At the moment the solutions depend on a
great deal of mathematical and physical simplification of the
turbulent transport processes in the layer. Such use of formulae
relating easily-measured quantities to represent complex physical
processes is called "parameterization".

Shear Instabilities

In the presence of a stable vertical density gradient, such
as that at the base of the mixed layer, instabilities can be produced in the (otherwise laminar) flow by velocity shear, i.e.
vertical gradients in the horizontal velocity. Such instabilities are common in nature; they manifest themselves as wavelike
"billows" in clouds at the top of the atmospheric mixed layer,
for example. That such "billows" exist in the thermocline has
been elegantly demonstrated by Woods (1968).

The shear-induced billows generate turbulence, which then
enhances the mixing process in the vicinity of the base of the
mixed layer, and allows the "mixing down" of fluid to occur much
more efficiently than would happen if the mixing was purely convective. Whether or not the billows form is determined by the
local ratio of hydrostatic stability (as measured by Δb, the
buoyancy difference across the base of the mixed layer), to the
shear instability (as estimated by $(\Delta V)^2/h$, where h is an estimate of the thickness of the sheared layer). The ratio is a
Richardson Number

$$Ri = \overline{\Delta bh/(\Delta V)}^2 \tag{18}$$

Pollard, Rhines and Thompson (1973) compute $\overline{(\Delta V)}^2$ from their inertial oscillation solutions, thereby assuming that it is storm-induced oscillations which create shear at the base of the mixed layer and produce the observed rapid storm-induced deepening of the oceanic mixed layer. The theory is still considered valid for the initial deepening; other processes are assumed to dominate at later times. The Richardson number so defined needs to be less than about unity, for shear instability to be an important source of mixing energy.

Entrainment

An excellent discussion of entrainment will be found in O.M. Phillips' contribution to the Kraus (1977) volume. The process of migration of the boundary between turbulent and laminar regions, with the turbulence continually incorporating more and more of the laminar fluid, or "eroding" it, is called entrainment. It is obviously central to our understanding of downward mixing in the ocean, but it is in general a highly complex, nonlinear process. For this reason, Phillips and others have performed a number of elegant laboratory and field studies that more or less empirically relate the vertical movement of the bottom of the mixed layer, or "entrainment velocity" w_e, to various measurable properties of the fluid and its surroundings, such as the rms (root-mean-square) turbulent velocity shear $(u^2)^{1/2}/\ell$ where ℓ is a local scale length, and the buoyancy $g\Delta\rho/\rho$ (ie the Richardson number $g\Delta\rho\ell/u^2$). The entrainment velocity is related to the rate of increase in depth of the mixed layer, which is usually a prediction of the mixed-layer theories, available for comparison with experiments.

Advection

Large-scale ocean currents (and spatial inhomogeneities in such currents) move water horizontally through a given region, and produce convergences and divergences of heat and salt which are quite independent of local forcing. Such movement of water and its properties are accounted for in the equations of motion by the so-called "advective" terms $\vec{V}.\nabla s$, $\vec{V}.\nabla T$ and $\vec{V}.\nabla \vec{V}$.

These terms introduce nonlinearities into the equations, and the basic equations are typically intractable analytically. Advection is, unfortunately, crucial to any attempt to relate observations to theory. Without careful investigation, it is very difficult to discover if an observed time rate of change in a given quantity has a local cause, or is simply the result of some large-scale gradient in the quantity being carried through the observation area. In the study of mixed-layer development in the ocean advection must always be taken into account. In practice it is generally allowed for in the observational program, by large-scale surveys of the quantities of interest and the currents which advect them.

Langmuir Circulations

Irving Langmuir (1938) analyzed the wind-aligned streaks, or "windrows", often found on lakes, giving penetrating insights into the causative mechanisms. Since then such "Langmuir" circulations (Figure 18) have been observed on all bodies of water. Because they can cause downwelling velocities as high as 1% of the wind speed, Langmuir circulations represent a potentially important mechanism for the downwards mixing of surface waters, and so have been extensively studied. For a recent review see the article by R. Pollard in Angel (1977).

The mechanism for their formation (Leibovich and Paolucci, 1980) is now thought to be that described in Figure 19. A small, initially crosswind perturbation on the downwind surface current causes crosswind velocity shears, and hence water vorticity as shown in (a). Coupled with the vertical shear in the water caused by the Stokes' drift in the wave field (for a description of the Stokes' drift see the section on Sea Surface Waves), shown in (b), a vorticity field like that in (c) is set up. This leads to convergence of surface water (long downwind-oriented lines where the surface currents are larger than normal), and divergences between. (In the figure, the convergences would coincide with the maximum of the forward (downwind) moving section of the initial perturbation (i.e., between 2 & 3 in a) and the divergences, with the upwind-moving perturbation.) The surface currents are acted upon by the wind stress, and since low-momentum water reaching the surface is accelerated downwind by the surface stress, the downwind velocities are greatest near the convergences and the forcing is amplified (there is a positive feedback). A final equilibrium state is presumably reached in which the wind stress and the wave field/Stokes' drift it produces and amplifies are balanced by friction, including turbulent dissipation. The Craik-Leibovich (1976) theory predicts most of the well-known features of these circulations. It predicts a transfer of energy from small scales (the vertical distances over which the Stokes' drift shears occur) to large scales, with a transport efficiency that increases strongly with scale size. Eventually the Coriolis force must come into play, at which point Ekman (1905) dynamics must become important, too.

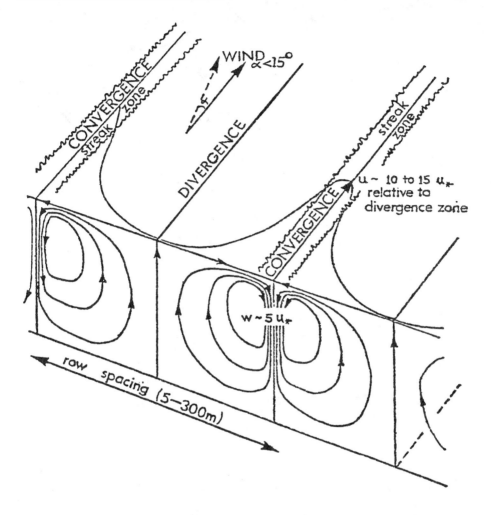

Figure 18. Schematic diagram of the observed structure of
Langmuir circulations. Here u_{*_w} = (wind stress/water
density)$^{1/2}$ is the friction velocity in the water. (After
Pollard in Angel, 1977.)

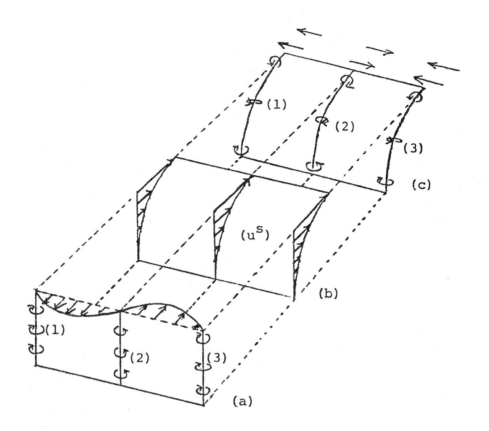

Figure 19. Schematic diagram of the distribution of vorticity in
a wave field caused by a perturbation in surface current. (a) is
the current perturbation; (b) is the Stokes drift u^S associated
with the waves; (c) is the resulting deformation of the vorticity
field, showing how Langmuir "cells" can be set up. The mechanism
is that proposed by Craik and Leibovich (1976); the figure is
from Smith (1980).

Langmuir circulations act as an efficient heat pump, stirring relatively warm surface waters downwards and cool deeper waters towards the surface, where they may gain heat from the air and the sun. Leibovich and Paolucci (1980) define a "mixing efficiency" m_ℓ as the time rate of increase of potential energy of the mixed layer from work done by the Langmuir circulations, divided by the total rate of input of kinetic energy averaged over one Langmuir wavelength in the crosswind and downwind directions (that is, one circuit of a water parcel in the Langmuir "cell"):

$$m_\ell = (d(pe)/dt)/\rho_a \, u_*^3 \, L_c L_d$$

where the "c" and "d" subscripts on the length scales L mean crosswind and downwind. The numerator is in turn given by

$$d(pe)/dt = \int \rho_w \, g\beta_T \, \overline{\hat{w}\hat{\theta}} \, dV$$

where the integral is over the volume dV of the "cell", β_T is the thermal expansion coefficient of·water and $w\theta$ is the rate of vertical transport of heat per unit area caused by the Langmuir circulations. They find values of m_ℓ of 5 to 15 in typical oceanographic situations, as compared to values near 1 for most mixed-layer models. The reader should note that the Leibovich-Paolucci estimate of m_ℓ implicitly assumes that the surface of the entire ocean is covered with Langmuir cells, which of course it is not. To be truly comparable with the values of m_ℓ from the mixed-layer models, the Leibovich-Paolucci m_ℓ estimates should be multiplied by the probability of finding Langmuir circulations at a given place at a given time.

There has not,as yet,been published an attempt to merge the Leibovich-Paolucci model with a generalized mixed-layer model that includes the effects of solar heating, inertial current mixing, and the larger-scale Ekman circulations. Some such model will be necessary for serious modeling of the dynamics and thermodynamics of the world oceans such as, for example, numerical weather or climate forecasting.

Role of the Mixed Layer in Ocean Dynamics and Climate

The oceanic mixed layer responds to forcing in a manner which has far-reaching dynamic effects. It absorbs three-quarters of the solar radiation reaching the bottom of the atmosphere, and stores it for release at a later time and another

place. The latitudinal variation of rate of storage of heat in
the ocean, according to Oort and Vonder Haar (1976), is compared
in Figure 20 with other terms in the global heat budget. Almost
all the oceanic heat storage occurs above the seasonal thermocline, and so heat storage in the mixed layer, can be seen to be a
crucial part of the earth's climate system. On the relatively
sharp density change at the bottom of the mixed layer, internal
waves can form and propagate; such waves are found everywhere in
the ocean and serve to make observations difficult. They provide
a widespread mechanism by which energy and momentum can be fed
from the relatively active mixed layer to the less active water
beneath. The mixed layer effectively decouples the deep ocean
from all but the largest-scale, longest-term forcing by the
atmosphere. A good example of this is the well-confined behavior
of storm-forced inertial oscillations (Pollard, 1970). The only
places where this isolation breaks down is at high latitudes
(where strong forced convection sometimes mixes the ocean to
great depths), in upwelling regions near coastlines, and in major
frontal zones, such as at the Antarctic Convergence Zone (see the
Mesoscale Oceanography section).

SEA SURFACE WAVES
Introduction

    Surface waves are perhaps the most obvious and ubiquitous
feature of the oceans. They interact with the air above them and
with the near-surface layer of the ocean beneath, and with each
other. The dynamics of all the interactions have not been worked
out, in spite of more than a century of intensive study.

    Much can be learned about ocean waves by thoughtful inspec-
tion. In the open sea in a strong wind (Beaufort Force 6, for
example), one can find wavelengths from less than a centimeter
(ripples, or capillaries, which make "catspaws" visible on the
sea surface as gusts pass) to hundreds of meters (the largest
waves at Beaufort Force 6 would be about 4 m high from peak to
trough and have wavelengths of about 150 m). The longest wavelengths travel at the greatest speeds and have the largest
heights: the great sea waves can travel at 30 m s^{-1} (60 knots)
and be as high as 30 m. In an actively growing sea the waves
with wavelengths intermediate between the capillaries and the
dominant waves are normally the steepest, and show the greatest
tendency to break. "Breaking" takes many forms, from the turbulent overturning of the crest of a large, long-wavelength wave to
the production of trains of capillary waves moving down the leeward face of steep, short-wavelength waves. The shape of the
larger waves is not quite symmetrical: the wave crests are
sharper than the troughs.

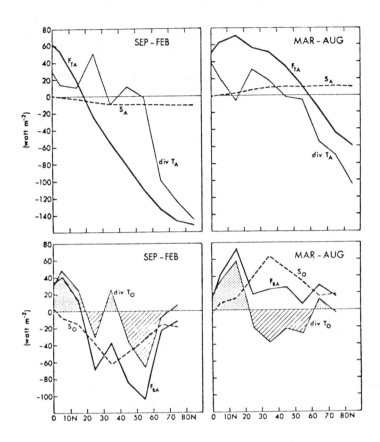

Figure 20. Latitudinal variation of various components of the
earth's heat transport budget for the atmosphere (top) and the
ocean (bottom). The subscript code is T=top, B=bottom;
A=atmosphere, O=ocean. S is "rate of heat storage", F is "vertical flux", and divT is "divergence of transport". Thus S_O
means "rate of storage of heat in the ocean". (After Oort and
Vonder Haar, 1976.)

The dominant waves have a distinct tendency to travel in
"groups"; that is, the waves which pass a given point vary their
height, three or four large ones normally following after a
period of relative calm. If observed carefully, individual waves
can be seen "passing through" the groups, indicating that the
speed of passage of a given surface undulation is faster than
that of the groups. [It turns out that the energy contained in
the waves travels at the speed of the wave groups: see the
Appendix, relation (A6)]. Also to be seen "passing through" the
waves generated by the local wind (the "sea") are trains of relatively long-crested waves not generated locally (known as
"swell"). Such waves, after their initial generation by a storm,
propagate enormous distances with little change in form.

Storm seas are among the most spectacular of natural phenomena: gigantic mountains of water, travelling at high speed,
occasionally breaking at their crests, producing enormous quantities of spray and spume, filling the air with wind-blown spray
and the water with spume. To fully appreciate their effect on
air-sea exchange of gases and particles, one must experience,
them.

Waves result when any density discontinuity (e.g., the discontinuity of 1000:1 which is the sea surface) is perturbed. Two
forces act to return the sea surface to static equilibrium:
gravity, and surface tension. Although they operate in different
ways (gravity is a vertical attractive "body" force, while surface tension is an elastic force acting to minimize the surface
area of the water), both forces can be included in the equations
of motion or boundary conditions of the problem, and mathematical
solutions can be found at least for simple cases.

The principal source of difficulty is in specifying the surface boundary conditions. Since the surface is free to move, one
must know the solution to the problem (i.e. the shape of the surface) before the boundary condition can be set. A further difficulty arises because the boundary conditions make the equations
nonlinear. We will give only some "classical" solutions to
linearized problems below and in the Appendix; nonlinear cases
are being extensively studied, but are well beyond the scope of
this book. In any case, the solutions given will provide an
adequate background for understanding most wave phenomena.

Ocean waves occur over a large range of scales, or wave
lengths (Figure 21). Their wavelengths vary from millimeters for
capillaries to thousands of kilometers for tsunamis (so-called
"tidal waves") and the lunar/solar tides (which are forced
shallow water waves with lengths of half the circumference of the
earth). Ocean waves occur with periods of 0.1 second to 1 day.
In this discussion, however, we will limit out treatment to waves

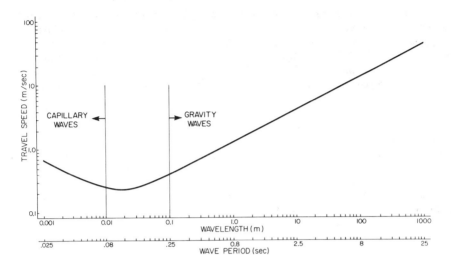

Figure 21. Dispersion relation, or relation between wavelength
or period and travel speed (phase speed), for sea surface waves.
The wave components enclosed by the thin vertical lines are
referred to as gravity/capillary waves. Note that the travel
speed of gravity waves increases with wavelength, while for
capillary waves the travel speed decreases with wavelength.
(After Dobson, 1974.)

generated directly by the wind, i.e. an upper wavelength limit
of about 0.5 kilometers and a longest period to about 20 seconds.
A discussion on tides can be found in the section on Mesoscale
Oceanography.

The amplitude of ocean waves is a strong function of their
wavelength, because water waves become unstable and break if
their steepness (that is, the ratio of wave height to length)
becomes too large. Therefore the longest waves are the largest.
Short wavelength waves are more easily seen than measured (their
visibility is a function of their steepness); typical heights for
10 cm wavelength waves are fractions of a millimeter. The influence of sea waves on the above- and below- water environment
extends only a fraction of a wavelength (wave-induced motions
typically vary as exp $(-k|z|)$, where k is $(2\pi \div$ wave-length) and
$|z|$ is the distance from the interface). This means that in the
air above the sea, where there is usually a mean wind many times
stronger than the wave-induced flows, the wave-induced motions
are hard to detect. In the ocean, on the other hand, where the
mean currents are typically very small, the wave-induced flows
dominate the surface layers.

Under a deep-water gravity wave of amplitude a, individual
parcels of water move in circles with radii a exp kd_o, where
d_o is the mean depth of the particle. Thus a fixed current
meter, located below the troughs, would record a zero mean current in the absence of outside influences: the water in contact
with it would move to an fro with an amplitude proportional to
exp kd_o. On the other hand, a neutrally-buoyant float which
maintained the same mean depth as the current meter, would drift
slowly in the direction of the waves. The effect is caused by
the reduction in the amplitude of wave orbital motions with
depth. Under the crests, at the top of the float's excursion,
its speed in the wave travel direction would be slightly greater
than its return speed under the wave troughs. The phenomenon is
known as the "Stokes drift", after its discoverer. For deep
water gravity waves the Stokes drift speed is given by

$$U_{Stokes} = \omega k a^2 \exp 2kd_o$$

Analysis of the wave equations becomes more complex (and the
solutions closer to reality) for water with nonzero viscosity and
nonzero inherent spin, or vorticity (note that water parcels can
move in circular orbits, as they do beneath waves, and still have
zero vorticity). The effect of viscosity is only evident in very
thin layers near the surface and bottom boundaries, but there,
considerable vorticity can be generated. The thickness δ of
these viscous boundary layers is, approximately,

$$\delta \simeq (2\nu/\omega)^{1/2} \tag{19}$$

where ν is the so-called "kinematic viscosity" and is equal to
about 10^{-6} m^2 s^{-1} for pure water at 20°C.

In the surface layer, the motion induced by the presence of
viscosity is (naturally enough) a strong function of the surface
boundary condition, that is of the type of stresses applied and
of the surface tension. With no surface stress the (rotational)
perturbation velocities tangential to the wavy surface (u') and
perpendicular to it (w') have amplitudes, in deep water in a
coordinate system moving with the waves (see Phillips, 1977,
pp. 48-49):

$$u' = \delta ak\omega \exp (x'/\delta) \qquad \text{(in deep water,} \quad (20)$$
$$w' = -2\nu ak^2 \qquad\qquad\qquad \text{clean surface)}$$

where the prime is for the moving coordinate system. In the
presence of a slick that has essentially no tangential compressibility, the perturbation amplitudes are

$$u' = -\omega a \exp (x'/\delta) \tag{21}$$
$$w' = 2\omega a/k\delta.$$

The attenuation coefficient $\beta = -\partial E/\partial t/2E$, where E is the wave energy density, also varies with the condition of the surface. For a clean surface in deep water (Phillips, 1977)

$$\beta_\nu = 2\nu k^2 \qquad \text{(deep water)} \tag{22}$$

In the presence of a tangentially incompressible film

$$\beta_f = \nu k/2\delta \qquad \text{(deep water)} \tag{23}$$

When waves are attenuated by surface slicks, account must be taken of the residual wave momentum, since in general, momentum must be conserved. The excess momentum is in fact taken up by an excess velocity (or "streaming") in the surface boundary layer of thickness δ where viscosity is important in the dynamics. In the presence of a tangentially incompressible slick, for instance, the velocity difference ΔU which exists across the boundary layer is (Phillips, 1977)

$$\Delta U = 3/4 \ \omega k a^2 \tag{24}$$

Wave Breaking

As waves of a given wavelength grow by the action of the wind, they eventually reach height-to-wavelength ratios (i.e. slopes) where they become unstable (see e.g. the article by M.S. Longuet-Higgins in Favre and Hasselmann, 1978). With the addition of more energy, they break in a variety of ways. The larger gravity waves tend to spill over at their crests, creating turbulent water in a wake which may extend over a number of wavelengths. A plot of the observed fractional ocean coverage with white caps vs wind speed (Figure 22) indicates the large variability involved. Wave breaking injects bubbles into the water down to a depth of about one wave height (M. Donelan in Favre and Hasselmann, 1978), and as the bubbles burst, they inject spray droplets to heights of one or two wave amplitudes into the turbulent air flow above. The smaller, gravity-capillary waves break in a very different way. They create few bubbles, but the longer-wavelength components among them create small areas of turbulence at their crests; their most characteristic feature is a group of forced capillaries that runs down the front face of the wave, with the shortest-wavelength components leading the longest-wavelength ones.

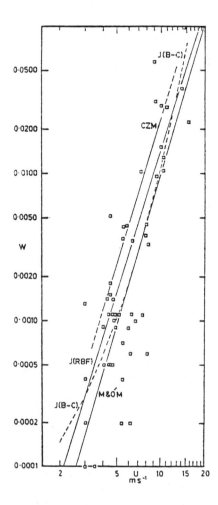

Figure 22. The fraction w of the ocean surface area covered with
whitecaps as a function of wind speed u, from observations made
during JASIN 1978 (Pollard, 1978) by Monahan and O'Muircheartaigh
(JASIN Newsletter No. 25, March 1982). The open squares are the
observations; the lines are various earlier empirical curves.
The most recent data are fitted best by the central solid line.

Breaking of the short waves is ubiquitous in the real ocean, since short-wavelength gravity waves and long-wavelength capillary waves travel slowly and hence gain energy rapidly from the wind. Very soon after the onset of the wind they reach amplitudes beyond which they cannot grow without breaking (such waves are termed to have reached their "saturation" phase). The phenomenon is easily observed. As the wind begins to steepen the shorter waves, their steepness makes them more visible: the water surface darkens as less and less of its surface becomes available to reflect light from the sun. Patches of short waves formed by wind gusts are referred to as "catspaws", because of their shape. Within a typical catspaw, every wavelet with less than a 10 cm wavelength will be breaking, continuously.

Breaking of the sort described above has been studied theoretically and experimentally by Banner and Phillips (1974). They note that the breaking of small-wavelength waves is strongly influenced by the presence of a shallow, surface-wind-induced layer of highly sheared mean water speed. This layer is caused by the direct frictional influence of the air on the viscous surface layer in the water, and is typically of thickness

$$\delta \simeq 0.5 \quad \rho_w \nu / \rho_a u_* \tag{25}$$

where the subscripts "a" and "w" refer to air and water and where u_* (the so-called "friction velocity" in the air: see the section in Interface Dynamics), is defined by

$$u_* = (\tau/\rho_a)^{1/2} \tag{26}$$

where τ is the drag stress per unit area exerted by the air on the water. At sea δ is about 2 mm for a 6 ms^{-1} wind speed. The magnitude of the frictionally induced surface water motion varies along the wave profile; it can be enhanced at the wave crest to many times that in the trough, with the largest enhancements occurring for the slower-moving waves. The point of incipient breaking is that for which the downwind water speed at the wave crest exceeds the phase speed (i.e. the speed at which the disturbance moves through the water). Stokes (1880) calculated the maximum crest elevation before a wave breaks to be

$$\zeta_{max} = c^2/2g \tag{27}$$

while Banner and Phillips (1974) found it to be

$$\zeta_{max} = (c-q_o)^2/2g \tag{28}$$

where q_0 is the downwind drift rate of the surface water at the
point of the wave where $\zeta = 0$.

Wave Generation and Growth

The wave generation process begins, visually, with the
appearance of short-wavelength ripples on calm water, either as
one proceeds away from the beach in an offshore wind or in the
open sea immediately after a wind has sprung up. The initial
waves are composed of a wide spectrum of wavelengths and directions, generated by random fluctuations of air pressure
(Phillips, 1957). Of these initial waves, only those that have a
component of their velocity in the direction of the wind continue
to grow, and their energy (proportional to the square of their
amplitude, a) grows linearly with time, t:

$$a^2(t) \; \alpha \; \Pi(\omega) \; t \; \cos\theta/\rho_w C$$

where θ is the angle   the waves make with the wind, Π is the
spectrum of turbulent air pressure fluctuations, ρ is water
density, and C is the phase speed of the waves.

Shortly after the wind has begun, and in parallel with the
mechanism just described, a second mechanism becomes active,
which produces short-wavelength waves of a more regular (long-
crested) form. They grow exponentially, because they result from
an instability in the strongly sheared viscous flows within 1-2
mm of the surface in both air and water. Such flows have been
extensively investigated; the most elegant and most convincing
demonstration of their existence and importance in the wave
generation process has been given by the late S. Kawai (1979).
He found that the initial wavelets grew very fast for about 100
wave periods, and then became random in both frequency and direction, losing their long-crested character. The frequency of the
initial wavelets generated was 10 Hz for $u* = 0.1$ ms^{-1} in the
air, and 25 Hz for $u* = 0.3$ ms^{-1}. The process whereby the
waves become random was not elucidated by Kawai. Higher har-
monics of the initial frequencies are clearly visible in the
observed spectra, and in all likelihood the wavelets reach limiting slopes and "break", radiating their surplus energy and
momentum to other wavelengths and directions.

One of the most important aspects of Kawai's work is his
detailed study of the form of the shear flow in the water, which
turned out in his case not to be as predicted by standard viscous
sublayer "theory": the shear was "neither steady nor logarithmic", but rather the water friction velocity varied, starting
from zero and reaching equilibrium with the air friction velocity
(i.e. $\rho_a u^2_{*a} = \rho_w u^2_{*w}$) just prior to the initial

appearance of the waves. An empirical relation was used to
describe the velocity profile in the water, and with it the
author was able to make excellent predictions of the generation
and growth of the wavelets.

Once the waves have been initiated by one of the two mechanisms described, then other effects take over. At high frequencies, where the (capillary/gravity) waves travel slowly relative
to the wind, growth occurs rapidly. In fact, waves of a given
frequency grow so rapidly that they "overshoot" their equilibrium
energy (Figure 23), i.e. they temporarily have slopes larger than
those they can support when in their final equilibrium, or
"saturated" state. In that state, the wave spectrum has the form
(Phillips, 1958)

$$\Phi(f) = \alpha \, g^2 f^{-5}$$

where the "constant" α is in fact a weak function of the dimensionless fetch $x = xg/u^2_*$, varying from 5×10^{-2} for $x = 10^5$ to
5×10^{-3} for $x = 10^7$ (Hasselmann et al., 1973).

While the waves are in the "overshot" condition they are in
a nearly-breaking state, and radiate energy and momentum to other
wave spectrum components, both in frequency and directions.
These, so-called, "nonlinear interactions", investigated in detail
by Hasselmann (1967), are a major source of wave dissipation
(loss of energy to the high-frequency part of the wave spectrum,
and of momentum to the mean current) and growth of the large,
faster-moving components. In fact, the lowest-frequency components in any wind-driven sea are generated in this way, their
growth rate being determined by the shape of the wave spectrum at
frequencies from 1.2 to 1.5 times the frequency at the spectral
peak.

The only effective source of new energy for the gravity
waves in a growing sea is wave-coupled fluctuations in air pressure. The air flow over a wavy surface (Figure 24) speeds up
over crests, causing low pressure (the "Bernoulli effect"), and
slows down over troughs, causing high pressure. In the absence
of the logarithmically-sheared wind profile which is commonly
observed at sea, the Bernoulli pressures generated by the waves
would be exactly out of phase with the wave elevation, and hence
would be zero when the water's vertical velocity is a maximum
(upwards ahead of the crests and downwards ahead of the
troughs). It is pressures in phase with the wave vertical
velocity which are necessary for growth (that is, low pressure
over the upwards-moving water on the front face of the wave, and
high pressure on the downwards-moving rear face). Miles (1957)
has shown that in the presence of a logarithmically-sheared air

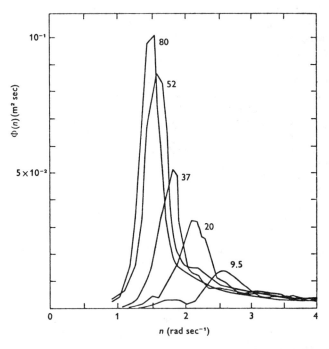

Figure 23. The variation of the wave spectrum with offshore
distance, or fetch. Note the large "overshoot" of waves with
frequencies at or above the local peak frequency, which disappears at the same frequencies at longer fetches as the waves
approach their "saturation" energy. (After Hasselmann et al.,
1973.)

flow typical of that observed over the sea (e.g. Figure 25) the
wave-induced flow is unstable and leads to wave growth.

      To understand how wave-coupled pressures in phase with the
wave vertical velocity are produced, we must view the waves from
a reference frame moving with the wave phase speed. The instability acts as follows (Figure 24). Air particles above the
height where wave speed equals wind speed (the "critical level")
move forward relative to the wave, and those beneath move back-
wards. Air parcels near the critical level move very slowly
relative to the wave, and hence come under the influence of the
wave-induced Bernoulli pressures, mentioned above, which are low
in the region of the crests and high near the troughs. Downwind
from the crests, the adverse pressure gradient in the troughs
causes forward-moving air just above the critical level to be
turned back and downwards, and backward-moving air below to be
turned upwards, forming eddies centered at the crests and
critical level. Because the air flow above the waves is

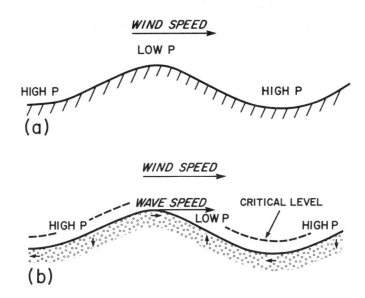

Figure 24. Mean wave-induced air flows and accompanying pressure
fields in the presence of a surface wave, (a) wave speed zero
(i.e. a fixed undulation in a wind tunnel floor; (b) a water wave
propagating downwind with a speed less than that of the wind at a
height of many (>10) wavelengths. The arrows in the water give
the direction of orbital motion of the water as the wave passes
by. Note that the propagation of the wave, by inducing secondary
flows in the air (see the text), causes the pressure field to be
shifted downwind. The pressure field is shown shifted by 1/4 of
a wavelength, at which point the wave growth by pressure forcing
is a maximum.

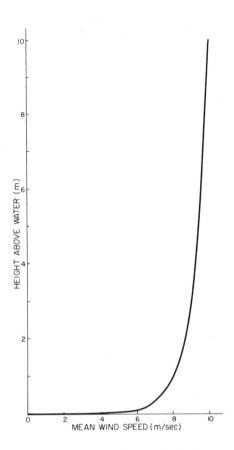

Figure 25. Typical variation of wind speed with height over the ocean. The "mean wind speed" was 10 ms^{-1}, as measured at a height of 10 m. (After Dobson, 1974.)

logarithmically sheared in the vertical, the vorticity, or spin, of the forward-moving air parcels above the critical height is slightly less than that of the rearward-moving air below. Therefore downwards-turning air in the troughs ahead of the crests has a larger radius of curvature than upwards-turning air to the rear of the crests. As a result the eddies are deformed, moving the flow streamlines above them forward relative to the waves. The air flow above encounters wave-coupled pressures which are shifted forwards, i.e. downwind, causing work to be done by the air flow on the water.

Mathematically the wave-coupled pressure is expressed as (Snyder et al., 1981)

$$\hat{p}(t) = \rho_a g \gamma \ \zeta(t)$$

where the hat (^) means "wave-coupled" and γ is a dimensionless complex number

$$\gamma = Re(\gamma) + i \ Im(\gamma)$$

with $i = (-1)^{1/2}$; Re and Im stand for "real and imaginary part of". If the waves grow according to

$$\overline{a^2(t)} = \overline{a_o^2} \exp \left\{ Im(\gamma) \ \rho_a \ (\partial E/\partial t)/\rho_w E \right\}$$

where a_o is the initial wave amplitude, then Im (γ) is in fact (ρ_w/ρ_a) times the fractional increase in wave energy per radian. Snyder et al. (1981) find, empirically,

$$Im(\gamma) \approx 0.25 \ (\vec{k} \cdot \vec{U} - 1)$$

where $\vec{k}$ is the wavenumber, and $\vec{U}$ is the mean wind velocity.

Appendix: Definitions

A travelling wave can be defined (Phillips, 1977) in terms of a sinusoidal disturbance (more complex wave forms are allowed, providing they can be treated as sums of sinusoids, or. Fourier series) travelling in the direction of their vector wave number $\vec{k}$, where

$$\vec{k} = 2\pi\vec{\ell}/\lambda \tag{A1}$$

in which $\vec{\ell} = \vec{k}/k$ and λ is the crest-to-crest length of the wave.

Such a disturbance may be written as

$$\zeta(\vec{r}, \ t) \ = \ a \ \cos \ (\vec{k} \cdot \vec{r} - \omega t) \tag{A2}$$

where ζ is the instantaneous elevation of the sea surface above
the mean height at position r and time t, a is the wave amplitude
(1/2 the crest-trough distance), ω = 2πf is the angular
frequency and f is the frequency in Hz, and $\vec{k} \cdot \vec{r}$ means kr cosθ
where θ is the angle the direction of wave travel makes with the
direction chosen for r. Then, if it is assumed that wave slopes
are small (that is, ka $\ll$ 1), it can be shown that a mathematical
relation, called the "dispersion relation" exists between the
wavelength of a wave and its frequency:

$$\omega^2(k) \ = \ k \ (g \ + \ \gamma k^2/\rho) \ \tanh \ kd \tag{A3}$$

where g is the gravitational acceleration, γ is the surface tension (the force normal to any line on the interface per unit
length of the line), ρ is the water density, and d is the water
depth.

The result (A3) looks complicated, but it simplifies considerably in particular cases. In "deep water", that is, for 2πd
$>$ λ, the hyperbolic tangent is very nearly unity (for d $>$ λ/4,
tanh kd $>$ 0.99). Then

$$\omega^2(k) \ = \ k(g \ + \ \gamma k^2/\rho) \qquad \begin{cases} \text{gravity-capillary} \\ \text{waves in deep water} \end{cases} \tag{A4}$$

The disturbances themselves travel at the "phase speed"

$$C \ = \ \omega/k \ = \ (g \ + \ \gamma k^2/\rho)/\omega \qquad \text{(deep water)} \tag{A5}$$

and transfer their energy at the "group speed" (see the Introduction to the Waves section)

$$C_g \ = \ \partial\omega/\partial k \ = \ (g \ + \ 3\gamma k^2/\rho/2\omega \qquad \text{(deep water)} \tag{A6}$$

On the right side of (A4) through (A6) the first term is the contribution from gravity and the second from surface tension. A
plot of phase speed vs wavelength (Figure 21) shows a minimum in
the gravity-capillary region, for which (in clean water) the
phase velocity is 0.23 ms^{-1} and the wavelength 0.017 m. One has
only to observe how waves "disperse" (that is, separate into
their various component wavelengths) to realize that a minimum
phase velocity exists. Dropping stones in a pool will show that
the capillaries disperse with the short wavelength components
running ahead of the longer wavelengths, while the longer gravity
waves outrun their shorter companions.

Waves have a mean energy density (i.e., energy per unit area)

$$E = \rho \omega^2 a^2 / 2k \qquad \text{(deep water)} \qquad (A7)$$

which is equally divided between potential and kinetic. For gravity waves ($\gamma k^2 / \rho g \ll 1$)

$$E = \rho g a^2 / 2 = \rho g \overline{\zeta^2} \qquad (A8)$$

and for capillary waves ($\gamma k^2 / \rho g \gg 1$)

$$E = \gamma k^2 a^3 / 2. \qquad (A9)$$

Waves also have a momentum density, which is directly related to the energy density by

$$\vec{M} = \vec{\ell} \, \rho \omega a^2 = \vec{\ell} E / c \qquad \text{(deep water)} \qquad (A10)$$

where $\vec{\ell} = \vec{k}/k$ and c is the phase velocity, as before. For small-amplitude pure capillary waves (i.e. $\gamma k^2 / \rho \gg g$) the mean increase in area of surface per unit projected area is,

$$\overline{\Delta A} = \overline{(\partial \zeta / \partial x + \partial \zeta / \partial y + \partial \zeta / \partial z)^2} / 2 \qquad (A11)$$

where terms smaller than $(ka)^2$ have been dropped. For non-small-amplitude waves, ΔA can be large. Cox and Munk (1954) have measured the right side of (A11) in the field; their results are reproduced in Phillips (1977: Fig. 4.23).

GENERAL REFERENCES

Bowman, M.J. and W.E. Esaias (eds.), 1978: Oceanic fronts in coastal processes. Proceedings of a Workshop, SUNY, Stony Brook, N.Y., May 25-27, 1977. Springer-Verlag, 114 pp.

Darwin, G.H., 1962: The tides. W.H. Freeman and Co., San Francisco, 378 pp.

Dobson, F.W., L. Hasse and R. Davis (eds.), 1980: Air-Sea Inter-action: Instruments and Methods. Plenum Press, New York, 801 pp.

Kraus, E.B. (ed.), 1977: Modelling and Prediction of the Upper Layers of the Ocean. Pergamon Press, Oxford, 325 pp.

Oceanus, 1976: Ocean Eddies. Woods Hole Oceanographic Institution 19, 88 pp.

Oceanus, 1981: Oceanography from Space. Woods Hole Oceanographic Institution 24, 76 pp.

Phillips, O.M., 1977: The Dynamics of the Upper Ocean (2nd Ed.), Cambridge University Press, 326 pp.

Royal Society of London, 1981: Circulation and Fronts in Continental Shelf Areas. The Royal Society, London: 177 pp.

Stommel, H., 1960: The Gulf Stream: A Physical and Dynamical Description. University of California Press, 202 pp.

Sverdrup, H.U., M.W. Johnson and R.H. Fleming, 1942: The Oceans: Their Physics, Chemistry, and General Biology. Prentice-Hall, Inc., Englewood Cliffs, N.J. 1087 pp.

Warren, B.A. and C. Wunsch (eds.), 1981: Evolution of Physical Oceanography: Scientific Surveys in Honor of Henry Stommel. The MIT Press, Cambridge, MA, 623 pp.

REFERENCES

Angel, M., ed., 1977: A Voyage of Discovery: George Deacon Memorial Volume. Supplement of Deep-Sea Res., Pergamon Press, 696 pp.

Augstein, E., 1981: Atmosphaerische und ozeanische grenzschichten in den niederen breiten. Hamb. Geosphys. Einzelschr. 53, 148 pp.

Banner, M.L. and O.M. Phillips, 1974: On the incipient breaking of small scale waves. J. Fluid Mech. 65, 647-656.

Batchelor, G.K., 1967: An Introduction to Fluid Mechanics. Cambridge Univ. Press, 615 pp.

Bowman, M.J. and W.E. Esaias, 1978: Oceanic fronts in coastal processes. Proc. Workshop at Marine Sciences Center, SUNY, Stonybrook, NY, May 25-27, 1977. Springer-Verlag, 114 pp.

Broecker, W.S., T. Takahashi, H.J. Simpson, and T.-H. Peng, 1979: Fate of fossil fed carbon dioxide and the global carbon budget. Science, 206, 409-418.

Budyko, M,I., 1974: Climate and Life. Academic Press, 508 pp.

Cartwright, D.E., 1977: Ocean Tides. In Reports on Progress in
    Physics, 40, 665-708.

Cox, C.S. and W.H. Munk, 1954: Statistics of the sea surface
    derived from sun glitter. J. Mar. Res. 13, 198-227.

Craik, A.D.D. and S. Leibovich, 1976: A rational model for
    Langmuir circulations. J. Fluid Mech. 73, 401-426.

Csanady, G.T., 1979: The birth and death of a warm core ring.
    J. Geophys. Res., 84, 777-785.

Darwin, G.H., 1962: The Tides. Reissued by W.H. Freeman and
    Co., 378 pp.

Defant, A., 1958: Ebb and Flow: The Tides of Earth, Air and
    Water. U. Michigan Press, Ann Arbor, 121 pp.

Dobson, F.W., 1971: Measurements of atmospheric pressure on
    wind-generated sea waves. J. Fluid Mech., 48, 91-127.

Dobson, F.W., 1974: The wind blows, the waves come. Oceanus,
    17, 29-36.

Dobson, F.W., L. Hasse, and R. Davis (eds.), 1980: Air-Sea
    Interaction: Instruments and Methods. Plenum Press, 801
    pp.

Ekman, V.W., 1905: On the influence of the earth's rotation on
    ocean currents. Ark. Math. Astr. Fys. (Stockholm), 2,
    1-52.

Favre, A. and K. Hasselmann (eds.), 1978: Turbulent Fluxes
    through the Sea Surface, Wave Dynamics, and Prediction.
    NATO Conference Series V, Plenum Press, 677 pp.

Fu, L.-L., and B. Holt, 1982: Seasat views oceans and sea ice
    with synthetic-aperture radar. NASA, Jet Propulsion Lab.
    Pub. 81-120, 200 pp.

Garrett, C.J.R., 1979: Mixing in the ocean interior. Dyn. Atm.
    and Oceans 3, 239-265.

Gordon, A.L., 1982: World ocean water masses and the saltiness
    of the Atlantic. Presented to Study Conference on Large-
    Scale Oceanographic Experiments in the WCRP, Tokyo, May
    1982. WMO Secretariat, Geneva.

Gorshkov, S.G. (ed.), 1978: World Ocean Atlas, Vol. 2, Atlantic
    and Indian Oceans. Pergamon Press, Oxford.

Hasselmann, K., 1967: Nonlinear interactions treated by the methods of theoretical physics (with application to the generation of waves by wind). Proc. Roy. Soc. A, 299, 77-100.

Hasselmann, K. et XV al., 1973: Measurements of wind wave growth during the Joint North Sea Wave Project (JONSWAP). Deutsches Hydrogr. Zeitschr. A., 12, 95 pp.

Holland, W.R., 1978: The role of mesoscale eddies in the general circulation of the ocean-numerical experiments using a wind-driven quasi-geostrophic model. J. Phys. Oceanogr. 8, 363-392.

Holland, W.R., and P.B. Rhines, 1980: An example of eddy-induced ocean circulation. J. Phys. Oceanogr. 10, 1010-1031.

Jerlov, N.G., 1976: Marine Optics. Elsevier Press, 231 pp.

Kawai, S., 1979: Generation of initial wavelets by instability of a coupled shear flow and their evolution to wind waves. J. Fluid Mech. 93, 661-703.

Knudsen, M., 1901: Hydrographical Tables. G.E.C. Gad, Copenhagen. Tables for the calculation of sigma-t from values of salinity and temperature, p. 63.

Kraus, E.B. (ed.), 1977: Modelling and Prediction of the Upper Layers of the Ocean. Pergamon Press, 325 pp.

Kraus, E.B. and Michèle Fieux (eds.), 1981: Large-scale transport of heat and matter in the ocean. Proceedings of a NATO Advanced Research Institute, Bonas, France, September 1981, 166 pp.

Langmuir, I., 1938: Surface motion of water induced by wind. Science 84, 119-123.

Leibovich, S. and S. Paolucci, 1980: The Langmuir circulation instability as a mixing mechanism in the upper ocean. J. Phys. Oceanogr. 10, 186-207.

Miles, J.W., 1957: On the generation of surface waves by shear flows. J. Fluid Mech. 3, 185-204.

Munk, W.H., 1950: On the wind-driven ocean circulation. J. Meteorol. 7, 79-93.

Munk, W.H., 1966: Abyssal recipes. Deep-Sea Res. 13, 707-730.

Oceanus, 1976: Ocean Eddies. Woods Hole Oceanographic Institu-
    tion 19, 88 pp.

Oort, A.B. and T.H. Vonder Haar, 1976: On the observed annual
    cycle in the ocean-atmosphere heat balance over the Northern
    Hemisphere. J. Phys. Oceanogr. 6, 781-800.

Parker, C.E., 1971: Gulf Stream rings in the Sargasso Sea.
    Deep-Sea Res. 18, 981-993.

Pedlosky, J., 1979: Geophysical Fluid Dynamics. Springer
    Verlag, 624 pp.

Phillips, O.M., 1957: On the generation of waves by turbulent
    wind J. Fluid Mech. 2, 417-445.

Phillips, O.M., 1958: The equilibrium range in the spectrum of
    wind-generated waves. J. Fluid Mech. 4, 426-434.

Phillips, O.M., 1977: The Dynamics of the Upper Ocean.
    Cambridge Univ. Press, Second Ed. 336 pp.

Pollard, R.T., 1970: On the generation by winds of inertial
    waves in the ocean. Deep-Sea Res. 17, 795-812.

Pollard, 1978: The Joint Air-Sea Interaction Experiment - JASIN
    1978. Bull. Amer. Meteorol. Soc. 59, 1310-1318.

Pollard, R.T., P.B. Rhines, and R.O.R.Y. Thompson, 1973: The
    deepening of the wind mixed layer. Geophys. Fluid Dyn. 3,
    381-404.

Pond, S. and G.L. Pickard, 1978: Introductory Dynamic
    Oceanography. Pergamon Press, 241 pp.

Richardson, P.L., R.E. Cheney, and L.A. Martini, 1977: Tracking
    a Gulf Stream ring with a free drifting surface buoy. J.
    Phys. Oceanogr. 7, 581-590.

Robinson, M.K., R.A. Bauer, and E.H. Schroeder, 1979: Atlas of
    North Atlantic - Indian Ocean monthly mean temperatures and
    mean salinities of the surface layer. U.S. Naval Oceano-
    graphic Office Ref. Pub. 18, Washington, D.C.

Roether, W., 1982: Transient Tracers in the Ocean. Contribu-
    tion, Joint CCCO/JSC Study Conference on WCRP Oceanography,
    Tokyo, 1982. WMO WCRP White Paper, WMO Secretariat Geneva.
    To be published.

Rossby, C.-G., 1938: On the mutual adjustment of pressure and
    velocity distributions in certain simple current systems,
    II. J. Mar. Res. 1, 239-263.

Simpson, J.H. and J.R. Hunter, 1974: Fronts in the Irish Sea.
    Nature 250, 404-406.

Smith, J.A., 1980: Waves, currents and Langmuir circulation.
    Ph.D. Dissertation, Inst. Oceanography, Dalhousie University
    242 pp.

Snyder, R.L., F.W. Dobson, J.A. Elliott and R.B. Long, 1981:
    Array measurements of atmospheric pressure above surface
    gravity waves. J. Fluid Mech. 102, 1-59.

Stokes, Sir G.G., 1880: Math. and Physical Papers 1, Cambridge
    University Press, 314-326.

Stommel, H., 1948: The westward intensification of wind-driven
    ocean currents. Trans. Amer. Geophys. Union 29, 202-206.

Stommel, H., 1960: The Gulf Stream, Univ. Calif. Press:
    202 pp.

Stommel, H. and A.B. Aarons, 1960: On the abyssal circulation of
    the world ocean - II. An idealized model of the circulation
    pattern and amplitude in oceanic basins. Deep-Sea Res. 6,
    217-233.

Sverdrup, H.U., 1947: Wind-driven currents in a baroclinic
    ocean; with application to the equatorial current of the
    eastern Pacific. Proc. Nat. Acad. Sci. Wash. 33, 318-326.

Sverdrup, H.U., M.W. Johnson, and R.H. Fleming, 1942: The
    Oceans. Prentice-Hall, Inc., Englewood Cliffs, N.J. 1087
    pp.

Tabata, S. and L. Giovando, L., 1963: The seasonal thermocline
    at Ocean Weather Station "P" during 1956 through 1959.
    Fish. Res. Board Can., MS Rept. Ser. 157, 27 pp.

Turner, J.S., 1979: Buoyancy Effects in Fluids. Cambridge
    Univ. Press, 368 pp.

UNESCO, 1981a: Background Papers and Supporting data on the
    Practical Salinity Scale 1978. UNESCO Technical Papers in
    Marine Science 37, UNESCO, Paris.

UNESCO, 1981b: Background papers and supporting data on the
     International Equation of State of Seawater, 1980. UNESCO
     Technical Papers in Marine Science 38, UNESCO, Paris.

Warren, B.A. and C. Wunsch (eds.), 1981: Evolution of Physical
     Oceanography. MIT Press, 623 pp.

White, W.B. and J.P. McCreary, 1976: On the formation of the
     Kuroshio meander and its relationship to the large-scale
     ocean circulation. Deep-Sea Res. 23, 33-47.

Woods, J.D., 1968: Wave-induced shear instability in the summer
     thermocline. J. Fluid Mech. 32, pp. 791-800.

Woods, J.D., 1980: Diurnal and seasonal variation of convection
     in the wind-mixed layer of the ocean. Quart. J. Roy.
     Meteorol. Soc. 106, 379-394.

Woods, J.D., 1982: Climatology of the upper boundary layer of
     the ocean. In Proceedings of Joint JSC/CCCO Study
     Conference on Oceanographic Experiments in the WCRP. To be
     published, WMO Secretariat, Geneva.

World Meteorological Organization, 1981: On the assessment of
     the role of CO_2 on climate variations and their impact. WMO
     Secretariat WCRP White paper No. 3, 29 pp.

Worthington, V., 1969: An attempt to measure the volume
     transport of Norwegian Sea overflow water through the
     Denmark Strait. F.C. Fuglister 60th Anniv. Vol. Deep-Sea
     Res. 16 (Supplement), 421-432.

Wunsch, C., 1981: The promise of satellite altimetry. Oceanus
     24, 17-26.

MICROBIOLOGICAL AND ORGANIC-CHEMICAL PROCESSES IN THE SURFACE
AND MIXED LAYERS

John McN. Sieburth

Graduate School of Oceanography
University of Rhode Island - Bay Campus
Narragansett, R.I. 02882-1197

INTRODUCTION

The microorganisms, their organic matter and the gases they consume and release are vital components of the hydrosphere which
impinge upon the atmosphere. This chapter considers the microbiological processes, and the organic-chemical transformations
that occur, when microorganisms utilize one gas as a raw material
while releasing another gas as a by-product during the synthesis
and decay of organic matter. The cells and the by-products from
the growth and metabolism of each trophic or feeding type of micro-
organism are in turn used by other trophic types to form an ecosystem, a community of organisms considered as a unit together
with its physical environment. The microorganisms, their organic
matter and the gases that regulate the metabolism within this ecosystem are transformed and balanced in a daily rhythm of production and consumption, controlled directly or indirectly by the
solar cycle. They are also affected by physical processes which
transfer gases, organic matter and microorganisms to and from the
mixed layer and its boundaries at the sea-air interface and at the
thermocline.

   This brief overview is based on twenty five years of observing
the calm lines on rippled waters and similar natural phenomena and
trying to characterize the natural populations of marine microorganisms and their organic-chemical processes which create such phenomena. I have been disappointed when the meteorologists, oceanographers, physicists, and inorganic chemists who study sea-air
interaction and exchanges neglect these aspects in their texts
(Dobson, Hasse and Davis, 1980) and in their research programs
(SEAREX). I was therefore pleased when the organizers of the NATO

P. S. Liss and W. G. N. Slinn (eds.), Air-Sea Exchange of Gases and Particles, 121–172.
Copyright © 1983 by D. Reidel Publishing Company.

Advanced Study Institute, upon which this volume is based, wanted
to include the microbiological processes responsible for the
transformation of gases, generation of particles, and formation
of the sea's skin that play vital roles in air-sea exchange of
gases and particles. As a result of listening to the physicists,
inorganic chemists, engineers, and meteorologists during the two
weeks of the Institute, I gained not only an appreciation of what
they do and why they do it, but an uneasy feeling that we micro-
biologists and organic chemists have not kept pace and have left
a large void. We study the organic matter and microorganisms of
the bulk water but not how they are selectively concentrated on
bubbles and injected into the air along with sea salt to form
aerosols. Distinctive marine microorganisms are potential tracers
of marine aerosols.

Only in the past few years have microscopic and organic chemical procedures been developed that let us characterize the microbial populations and concentrations of organic materials in natural waters. The time is ripe to apply these procedures to determine the fractionation of organic substances and microorganisms
by wind-induced bubbles that form aerosols. The nature and fate
of these materials during desiccation and rehydration, as they
are returned to the earth's or sea's surface by dry and wet deposition, also require microbial characterization. The surface films
which are continually renewed must also be characterized during
their formation, maturation, and removal, by transmission electron
microscopy of thin sections, a procedure familiar to microbiolo-
gists studying the ultrastructure of natural populations of microorganisms. The characterization of the natural surface skin of
the sea is critical, as it not only plays a conspicuous role in
wave dampening, but controls the rates of gas transfer between
the air and the sea.

This chapter attempts to show that microbiological processes
are a central and key part of air-sea exchange. There are large
voids in our understanding. In order to fill in the gaps, while
portraying my view of the emerging picture of open ocean microbiological processes, I have been speculative and conjectural. I
have taken this license in order to re-explore old problems and
perhaps add new insights. Hopefully the scenario of "what may be"
will not be construed as gospel, but as an attempt to explain the
enigmas and paradoxes not accounted for by the classic models and
concepts.

I. PLANKTONIC MICROORGANISMS AND THEIR TROPHIC STRUCTURE

1.1 Evolution of Decomposing and Synthesizing Microorganisms and their Processes.

Almost certainly, bacteria were the first forms of life to develop on this planet. Through eons they developed the structures and functions used to create all other forms of life. The pre-biological seas, before some 4 billion years ago, presumably acted as collection basins for organic substances created through physical-chemical reactions of the gases and condensing waters that were the raw materials provided by the cooling planet. It has been well documented that under a variety of conditions the raw materials of H_2, H_2S, NH_3, CO, CO_2, and H_2O reacted to electrical and photic energy under anaerobic conditions to yield an ever-increasing variety and quantity of organic compounds necessary to build a replicating cell (Margulis, 1981). The orderly assemblage of these lower molecular weight compounds as building blocks for the synthesis of polymeric substances, presumably using the crystalline structure of minerals and ice crystals as templates, apparently yielded the biopolymers still manufactured by today's cells.

In the absence of life, these substances must have accumulated to yield a rich broth, especially at sea-air and sea-bottom interfaces where organic matter (OM) accumulates today (Weyl, 1968). These concentrated organic compounds provided the raw materials not only for cell membranes, cytoplasm, and genetic material needed to construct a replicating cell, but the preformed OM necessary to feed it. The absence of free oxygen and its derivatives (e.g., super oxides and peroxides that degrade OM) allowed the abiotically produced OM to stockpile. The absence of these toxic substances also permitted the ancestral procaryote (cell without distinct organelles) to be built as a simple anaerobe without mechanisms to protect itself from oxygen. But once the pieces fell together to yield spontaneously generating cells, the proliferating ancestral anaerobes started to ferment and consume the stockpiles of OM, and released CO_2, H_2, and N_2, as well as a number of organic acids as by-products. As this consumption exceeded production, the ancestral procaryotes had to learn how to synthesize organic compounds from these end-products, or perish.

A simplified diagram is given in Figure 1 of the hypothetical evolution of bacteria to form a variety of biosynthetic steps, and to evolve cells with increasing diversity. Eventually, this diversity led to the evolution of oxygen and the development of protective mechanisms to survive in this hostile environment. Also shown is how these microorganisms which evolved together,

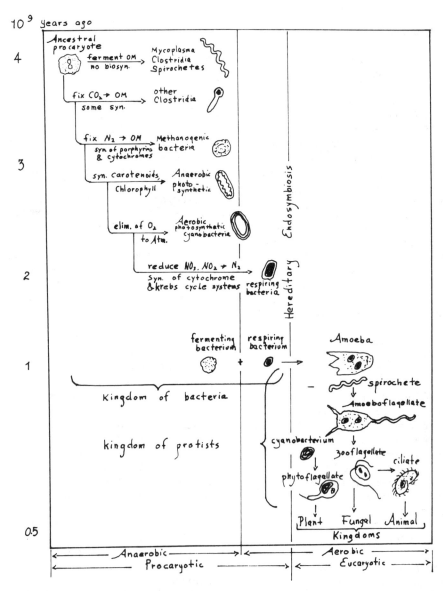

Figure 1. Hypothetical scheme for the evolution of bacteria (procaryotes) and protists (eucaryotes). The bacteria gained diversity
and biosynthetic pathways as they evolved that allowed them to
produce and exist in an oxygenated planet. Through hereditary endo-
symbiosis, certain types of bacteria formed organelles such as the
nucleus and plastids, to form the true (eucaryotic) cells of the
protists that further evolved to fungi, plants and animals.
(Adapted from Whittaker, 1969; Margulis, 1970). OM=organic matter.

adapted to form endosymbiotic associations with one cell living within another cell. When these associations became hereditary, then true cells (eucaryotes) were formed with their mission-specific organelles. These single-celled eucaryotes include synthetic forms reducing CO_2, saprotrophic forms utilizing non-living OM, and biotrophic forms eating a variety of cells. Together these single-celled eucaryotes form the Kingdom of Protists. When the protists learned to colonize and form multicellular organisms with specific tissue types performing specific functions, they evolved into the plant, fungal, and animal kingdoms.

Looking at the major steps in the two billion year evolution to synthesize OM in an oxygenated environment, we see that the early forms which fermented OM to obtain energy had a minimum of synthesizing ability. Each group of more complicated cells that evolved still has living descendants that are habitat-limited to environmental conditions that perpetuate those under which it evolved. The first synthetic ability was to fix atmospheric CO_2 into OM, thereby recycling respired CO_2, a process that occurs in some clostridia which are sporeforming anaerobes. The next biosynthetic processes were the synthesis of porphyrins and cytochromes and an ability to fix atmospheric nitrogen into organic compounds. This yielded more advanced clostridia as well as bacteria that could utilize hydrogen as a source of energy, and those that could reduce CO_2 with hydrogen to form methane. The latter are among the earliest bacteria (archebacteria) to have developed which still exist today, and may be thought of as "living fossils" since their genetic makeup shows their antiquity (Fox et al., 1980). While the atmosphere was still free of oxygen, the biosynthesis of carotenoid compounds and the chlorophylls permitted the evolution of anoxyphotobacteria which photosynthesized OM by reducing CO_2 under anaerobic conditions, using H_2S instead of H_2O as a source of electrons. It was not until the oxyphotobacteria, also with specialized photosynthetic membranes (thylakoids), gained the ability to split water to obtain the reducing power of hydrogen, that O_2 was evolved into the atmosphere. By then the substances needed to combat the toxicity of oxygen, such as catalase and cytochromes, had already evolved. In order for the bacteria to be able to respire the newly synthesized OM being produced, they also had to possess the complete Krebs Cycle and cytochrome systems necessary for respiration. Such microorganisms had developed by about two billion years ago.

It was not until about one billion years ago that the procaryotic microorganisms, which evolved and lived together, also learned to live one within the other in endosymbiotic associations. This yielded the organelles that characterize the true cells of the eucaryotes. A possible scenario, shown in Figure 1, is that the organelles in the simple plastic cells of amoebae were respiring bacteria which became endosymbiotic in simple

membrane-bound fermenting cells. The amoeboflagellates which are
still seen in marine samples and are a necessary life stage in
many more complex organisms may have been formed from amoebae
which obtained the large eucaryote flagellum with its distinctive
microtubules that could have evolved from the periplasmic fibrils
of spirochetes. The incorporation of O_2 evolving cyanobacteria
would have then led to the development of photosynthetic flagellates whose progeny later evolved to produce the metaphyta or Plant
Kingdom. A similar evolution of the amoeboflagellates to the zoo-
flagellates so common in marine waters, may have also led to a line
of saprotrophs that became the Kingdom of Fungi. Similarly, the
formation of multiflagellates to yield the ciliated protozoa would
have formed the basic cells needed to form the Animal Kingdom.
These newcomers to the oxygenated planet still require the present
day representation of the ancient fermentative microorganisms to
duplicate in microenvironments the primordial anaerobic environment
that is necessary to cycle both gases and inorganic nutrients.

1.2 Trophic Structure and Transfer.

     Today, new models or paradigms of the marine food chain are
emerging to challenge old concepts (Pomeroy, 1974). This has resulted from a burgeoning of new information, brought on by new
techniques and approaches for characterizing the microorganisms
and their activities in the sea (Pomeroy, 1979). The classic food
chain, as envisioned by Lindeman (1942), consisted of primary producers and a series of ever. larger consumers, with the waste products from each of these levels going to the decomposers, see Figure 2A. More recently, Wiegert and Owen (1971) recognized that
consumers are also decomposers and developed a model that recognizes a series of organisms feeding on live cells (biophages) as
well as a series of organisms utilizing non-living OM (saprophages),
see Figure 2B. Our recent observations on the distribution and
activities of bacteria and their protozoan predators (Sieburth, in
press) gives us a better idea of how the primary producers, saprotrophs (saprophages) and biotrophs (biophages) are structured and
transfer energy and nutrients. It is clear that the spectrum of
different sized photosynthesizers are not eaten by just one group,
but must be grazed by. a spectrum of different sized biotrophs. It
is also clear that the chemoorganotrophic bacteria and physiologi-
cally similar phycomycetes, which constitute the saprotrophs, are
present as free and attached forms that must act differently in the
transfer of energy and nutrients. Excluding the larger organisms,
a modified food chain model can be constructed to show more precisely the flow of biomass and nutrients, see Figure 2C.

     This model shows that the by-products of chemosynthesis and
photosynthesis supply each others' needs, that photosynthetic

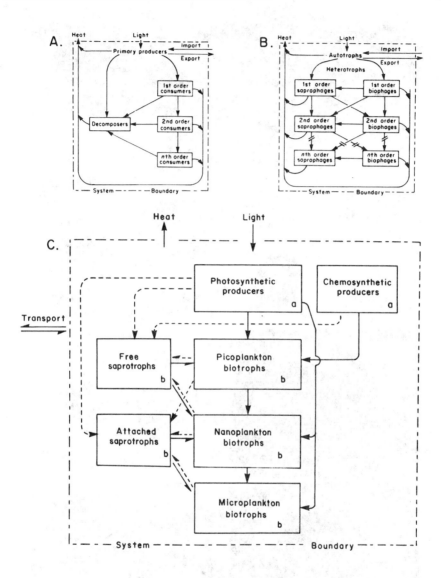

Figure 2.   Models of the food chain showing trophic structure and transfer. A) The classic food chain of Lindeman. B) The more structured model of Wiegert and Owen that distinguishes between organisms utilizing dead and living organic matter. C) A micro-biological version of the above, showing that the saprotrophs and biotrophs (b) release dissolved nutrients required by the auto-trophs (a) and that organic matter released from all trophic forms (---) supports the saprotrophs. (A and B from Wiegert and Owen, 1971; C from Sieburth, in press).

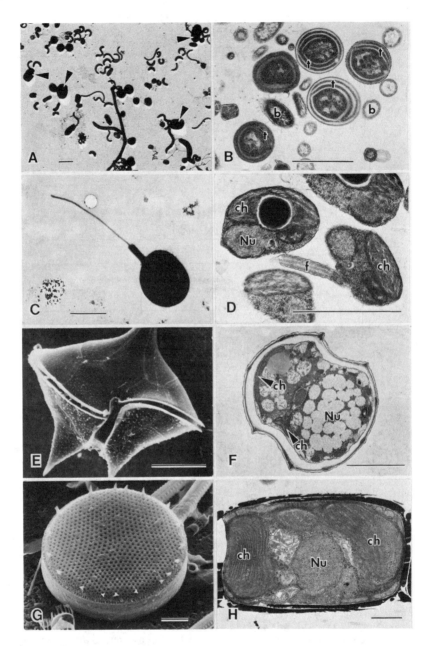

Figure 3. Morphology and ultrastructure of procaryotic (A,B) and
eucaryotic (C-H) cells representative of the photosynthetic plankton of the mixed layer. Chroococcoid cyanobacteria (A, arrows)
can be distinguished from the smaller saprotrophic bacteria (also
b in B) and by the presence in thin sections of intracellular

membranes or thylakoids (t in B). The microflagellate Micromonas pusilla (C, D) has a single flagellum (f), chloroplast (ch) and nucleus (Nu). The dinoflagellates (E, F) have a distinctive cell covering of plates and have multiple chloroplasts (ch) and characteristic nucleus (Nu). Diatoms (G, H) which are distinguished by their silica frustule (house) also have a nucleus (Nu) and multiple chloroplasts (ch). Marker bars A, B, C, D and H =1.0 μm; F =5 μm, G =10 μm and E =50 μm.

---

microorganisms come in a variety of sizes and require different sized biotrophs, that the larger biotrophs also eat the smaller biotrophs, and that the saprotrophs also require different biotrophs, depending on whether they are free in the water or attached to organic debris. Having discussed the apparent game plan for the mixed layer ecosystem, let us discuss the players.

1.3 Primary Producers (The Phototrophs and Chemolithotrophs).

The primary producers or autotrophs that have received the most attention are the microalgae, which are large enough to be retained by plankton nets. These are the larger cells such as the diatoms, more common to coastal waters. These cells are characterized by their siliceous houses (frustules) with their pores in distinctive patterns (see Figure 3G). Although some phytoplanktologists still restrict themselves to net samples, Lohmann (1911) drew attention to the smaller forms of the nanoplankton, which are the dominant cells in the open sea. More recent studies not only confirm this but show that these cells are also dominant in nearshore waters in summer and account for over 50% of year-round productivity (Malone, 1971, 1980a,b; Durbin et al., 1975).

Centric diatoms, which are usually dominant in nearshore plankton, also occur in the open ocean where they range over three orders of magnitude in size, from the giant Ethmodiscus rex, at 2,000 μm, to minute cells of 2 μm (Booth, 1975; Booth et al., 1982). The photosynthetic dinoflagellates (Figures 3E&F) range from small naked forms to large armored cells. The cell walls consist of cellulosic material and appear as plates of varying thickness. Other eucaryotes, with distinctive organelles such as nucleus, chloroplast, mitochondrion, etc., can be as small as 1-2 μm in diameter. An example of this is the ubiquitous coastal microflagellate, Micromonas pusilla, shown in Figures 3C&D. The widespread occurrence of this and other small photosynthetic eucaryotes in estuarine and oceanic waters has recently been described (Johnson and Sieburth, 1982). The ubiquitous existence of unicellular marine cyanobacteria also in concentrations of 10^3 to 10^5 per ml of oceanic water has been recently recognized (Johnson and Sieburth, 1979). The much simpler structure

of these cells is shown in Figure 3B. The common character in
all of these photosynthetic microorganisms is the photosynthetic
membranes or thylakoids, whose lamellar structure can be seen in
the thin sections of Figures 3B,D,F&H. We can see, then, that
the photosynthetic primary producers occur in a spectrum of sizes
including the picoplankton, nanoplankton, and microplankton size
fractions shown in Figure 8.

The smallest of the photosynthetic microorganisms that we
have observed were found by taking glutaraldehyde-fixed seawater
samples, removing the cells larger than 2.0 μm by gentle, reverse-
flow filtration, and concentrating the remaining cells in the fil-
trate by centrifugation to form a pellet, which is then prepared
for thin sectioning and examination by transmission electron microscopy (Johnson and Sieburth, 1979, 1982). Of the various types
of potentially photosynthetic forms, only a very small, nonflag-
ellated prasinophyte, bearing distinctive tiny scales (Johnson and
Sieburth, 1982) and type II and III procaryotic cells (Johnson and
Sieburth, 1979) suggestive of cyanobacteria, have resisted all
attempts at culture. It is possible that the latter may not be
photosynthetic cells at all. One of these ultrastructural types
(type II) could be nitrifying bacteria, possibly responsible for
the early darkperiod uptake in CO_2, Figure 7. The populations
and distribution of these cells are similar to those reported for
nitrifying bacteria in the North Pacific Ocean (Ward et al., 1982),
which are also light sensitive (Olson, 1981a,b; Horrigan et al.,
1981).

1.4 Saprotrophs (the Heterotrophic Bacteria).

Saprotrophy is the utilization or decay of soluble and particulate OM by bacteria or fungi. Microorganisms capable of utilizing this non-living OM must have extracellular enzymes to hydro-
lyze substances such as polysaccharides and proteins, as well as
mechanisms for the diffusion or active transport of the solubilized
substances into the cell. Such microorganisms, depending upon an
intake of soluble substances, are called osmotrophic and include
fungi as well as bacteria.

The fungi are a diverse group and include yeast forms, filamentous forms with the typical mold mycelium, as well as the phycomycetes with their flagellated zoospores. Although these forms
occur in the sea, they are rarely seen during the microscopy of
marine samples because their habitats are so specialized and their
populations are relatively small. The heterotrophic bacteria,
which carry out similar decomposing processes, have greater populations and distribution and are readily apparent in microscopic
preparations.

When particles in seawater are fixed with formaldehyde, stained with a fluorescent dye such as acridine orange (Hobbie et al., 1977) or DAPI (Porter and Feig, 1980), and are concentrated on a 0.2 μm Nuclepore membrane, the fluorescent cells are easily detected with the epifluorescence microscope (Figure 4A&B). Direct counts based on these procedures reveal populations of 5×10^6 cells per ml in estuarine waters to 5×10^5 cells per ml in the mixed layer of oceanic waters. Such procedures, when applied to surface films, will greatly elevate the bacterial enrichment shown in surface films (Sieburth, 1965, 1971a; Sieburth et al., 1976). The biovolume of the average-sized coccobacillary bacteria, which is some 0.9 μm^3 in the estuaries, decreases to as little as 0.04 μm^3 in the open sea. The flakes of marine snow can be heavily colonized (Figure 4B). These cells are much larger than the free planktonic bacteria (Figure 4A) and although encountered less frequently than the free cells, their biomass contribution is greater than their numbers alone would indicate. The thin sections of standing stocks of free (Figure 4C) and attached (Figure 4D) bacteria show this. The two types of saprotrophs also have different nutritional states. The paucity of ribosomes in the small free cells and their relative abundance in the larger attached forms is indicated by the darker staining contents of the cell. The small free cells have more of a "feast and famine" existence than the larger attached cells which are "feasting". This relationship of few ribosomes in "skinny" cells of slow growing cultures and the uniform packing of ribosomes in the "fat" cells of rapidly growing cells was recognized by Maaloe and Kjeldgaard (1966) in their classic monograph on the control of macromolecular synthesis. The free cells shown in Figures 4A&C are responsible for the decay of soluble OM, while those in Figures 4B&D are responsible for the decay of particulate OM. Both forms play a significant role in OM transformation in the sea.

Among the substrates available for heterotrophic bacteria in the mixed aerobic layer are supersaturated concentrations of methane that are apparently produced throughout the mixed layer but peak within the thermocline. The abundant type III cells (Johnson and Sieburth, 1979), at first thought to be oxyphototrophic cyanobacteria, may actually be obligate or facultative methanotrophic bacteria (Patt et al., 1974, 1976), which also have a similar cytomembrane structure (Figure 4E). These cells can be very prevalent in natural populations (Figure 4F) and at times can account for most of the bacterial biomass. These cells apparently divide during the evening (as observed by TEM of thin sections) just prior to the formaldehyde peak (Figure 6), suggesting active methanotrophy.

1.5 Biotrophs (the Protozoa).

The predators of the small primary producers and the saprotrophic bacteria are the marine protozoa, occurring in a size

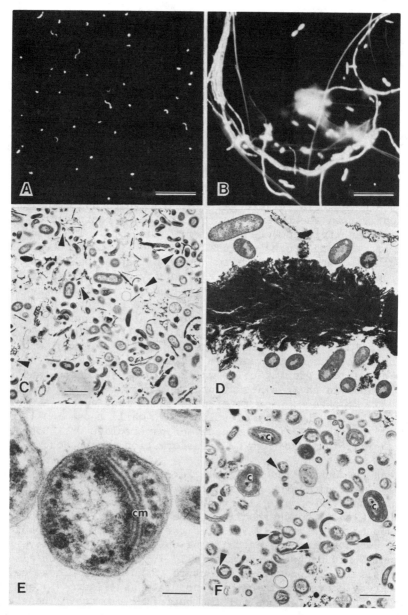

Figure 4. The morphology and ultrastructure of free (A, C, E & F) and particle-associated (B & D) saprotrophic bacteria representative of those in the mixed layer. The cells free in the water (A) are much smaller than the bacteria associated with marine snow (B) as shown by epifluorescence microscopy. The free cells in the pico-plankton (C) include minicells less than 0.15 μm wide (arrows) and appear to be deficient in ribosomes. In contrast, bacteria attached

to a particle of organic detritus (D) are much larger, ribosome-
rich, cells. (C and D are at the same magnification.) Type III
bacteria (E), with distinctive cytomembranes (cm), could possibly
be methanotrophs. This type of saprotroph is prevalent in the
picoplankton (F) of the Caribbean Sea (arrows) along with the
phototrophic cyanobacteria (c). (C-F are transmission electron
micrographs of thin sections; marker bars A & B =10 µm, C, D,
& F =1.0 µm, E = 0.1 µm.)

---

range of roughly 2 to 200 µm (Figure 8). This large size range
accommodates the spectrum of prey available, including themselves,
as indicated in Figure 2C. The term biophage (Wiegert and Owen,
1971) or biotroph is more appropriate than the commonly used term
consumer, since all trophic forms are consumers of some substrate.
Unlike the trophic structure model (Figure 2B) of Wiegert and Owen
(1971), which shows one order of biotrophs eating all the primary
producers, and the second order biotrophs preying upon the first
order, etc., I have broken the biotrophs up into groups feeding
on the distinct planktonic size fractions shown in Figure 8, as
well as upon the next smaller group of biotrophs. What Figure 2C
tells us is that different sized prey have different sized predators. But what are they taxonomically?

The three basic forms of protozoa are shown in Figure 5. The
amoeboid forms have a plastic cytoplasm that extrudes small filaments or larger lobes, which are both feeding and locomotory organelles. The cell in Figure 5A is a naked amoeba in a locomotory
mode while Figure 5B shows a cross section of an amoeba with engulfed bacteria. These forms are adapted to crawling over surfaces colonized by a bacterial film. In addition to the naked
amoebae there are a variety of amoeboid forms with houses (testa-
ceans and foraminifera), rigid filaments like the sun's rays (he-
liozoa), flagella (amoeboflagellates) and translucent skeletons
(radiolaria and acantharia). The foraminifera, whose house (or
lorica) is often calcareous, are mainly benthic, except for the
two well studied oceanic families of planktonic forms (Globiger-
inidae and Globotoralidae) that are used for paleological studies
of the sea (Kennett, 1982). Other oceanic amoeboid forms are the
radiolaria and acantharia, with skeletons of silica and strontium
sulfate, respectively. These forms often have captive algal cells
as endosymbionts, which supply a significant amount of nutrients
to their hosts. These planktonic forms consume the larger microorganisms, as well as prey as large as copepods. The diversity
of the amoeboid forms arises from their specialized habitats such
as surface films, the benthos, and the mixed layers of the sea.
The most poorly understood may be very important: the naked amoebae living in bacterial films developing at all interfaces.

The most numerous protozoans are the 2-5 µm non-pigmented

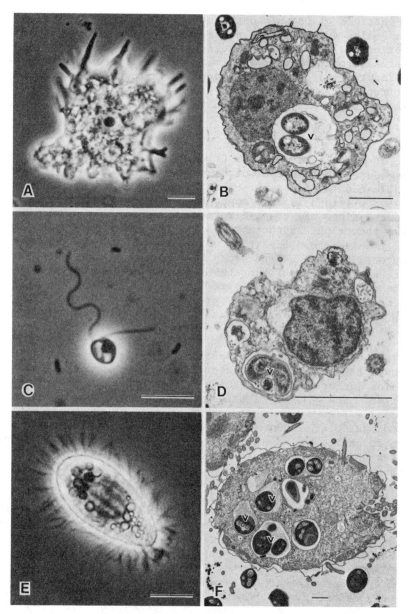

Figure 5. The morphology and ultrastructure of representative
biotrophic miroorganisms with engulfed prey in food vacuoles (v)
as they are found in the mixed layer. These include amoebae (A &
B), flagellates (C & D) and ciliates (E & F) as shown by phase
contrast light microscopy (A, C & E) and transmission electron
microscopy of thin sections (B, D & F). Marker bars A, C & E=10
μm; B, D & F=1.0 μm.

flagellates that primarily graze upon bacteria. The ubiquitous genus Bodo is shown in a light micrograph in Figure 5C. These smaller forms are present in populations of hundreds to thousands per ml of seawater, maintaining a ratio of approximately 1:225 of the larger bacterial cells in nearshore waters and 1:1250 of the much smaller cells in offshore waters. The small flagellates similar to that shown in Figure 5C are apparently the dominant predators on free bacterial cells in the sea. Examples of larger flagellated biotrophs are the non-pigmented dinoflagellates that feed on an assortment of the smaller microorganisms (from bacteria to the nanoplankton). Both the smaller and larger flagellates seem to be primarily of the plankton, but the smaller bacterial grazers also are concentrated on bacteria-rich surfaces, such as surface films, and organic aggregations, e.g. flakes of marine snow (Figure 4B&D).

The ciliates are a very diverse group that have exploited a multitude of highly-specialized microhabitats, from the gills of crustaceans to the large intestine of virtually every marine invertebrate. The most obvious forms free in the water are the tintinnids, which are characterized by their distinctive houses or loricae. Live specimens in net samples, actively feeding on algal flagellates, look like ornate torpedoes streaking through the water. Much less studied are their more ubiquitous, naked cousins that occur at oceanic populations of 1 per 0.2 to 2 ml of seawater. As they do in nearshore waters, these forms in the open sea concentrate around bacterial rich particles such as marine snow (Figure 4B&D), their probable oceanic habitat (Sieburth, in press; Caron et al., in press).

Collectively, the diverse protozooplankters play a vital role in the transfer of OM up the food web and in releasing organic and inorganic nutrients required by the primary producers and saprotrophs. Also included in the biotrophs is the size fraction up to 200 μm which is usually referred to as microzooplankton (which includes the smaller stages of the metazooplankton including copepod nauplii). These would be in the microplankton biotroph group (Figure 2C) along with the larger amoeboid, flagellate, and ciliate species. The nanoplankton biophages would include many organisms from tintinnids to copepods. The picoplankton biotrophs would be mainly the smaller flagellates, but would also include some of the larger forms.

2. UPTAKE AND RELEASE OF GASES DURING THE SYNTHESIS AND DECAY OF ORGANIC MATTER AND ITS CONTROL BY THE SOLAR CYCLE

2.1 Cycling of Gases and Organic Matter by Aerobic Processes.

The depletion of the standing stocks of non-living OM that

resulted from the spontaneous generation and evolution of saprotrophic bacteria discussed in 1.1, then led to the evolution of
photosynthesis that oxygenated the oceans and atmosphere of a
formerly anoxic planet. Photosynthesis is characterized by the
uptake and reduction of carbon dioxide by the hydrogen atoms
split from water through the action of light on chlorophyll-con-
taining membranes with the concomitant release of oxygen as follows:

$$6 \ CO_2 + 6 \ H_2O \longrightarrow C_6H_{12}O_6 + 6 \ O_2 \qquad\qquad (1)$$

Photosynthesis is a good example of a process in which one gas
acts as a raw material or substrate (carbon dioxide) while another
is released as a by-product (oxygen). The photosynthesized carbohydrate provides energy for cell maintenance, and the excess energy is polymerized into polysaccharides that are used as both
structural elements (e.g., cellulose) and reserve food material
(e.g., starch). Some of the carbohydrate is converted to acids
that are aminated and polymerized to form enzymes, and protein
that is used as structural elements for growth and division. Only
a minor fraction of the photosynthesate, usually some 4%, is converted to lipids. The photosynthesate, released by algae directly
through exudation or indirectly through the consumption of living
cells (biotrophy) to form the dissolved organic carbon (DOC) pool
of the sea, is dominated by carbohydrate primarily as polysaccharides, and secondarily by amines mainly as proteins (Melkonian,
1978). The latter are more quickly utilized by bacteria than the
carbohydrates.

The polysaccharide released by the algae and hydrolyzed to glu-
cose by the bacteria are used by the bacteria as energy for cell
maintenance and for cell synthesis. The apparent domination of
the dissolved organic matter (DOM) pool by carbohydrate is probably a protein sparing mechanism whereby carbohydrates are burned
for energy, rather than the more valuable amino acids of protein
whose nitrogen is required for cell growth and division. During
the respiration of the carbohydrates by all trophic forms, oxygen
is consumed as a raw material and carbon dioxide is released as a
by-product:

$$6 \ O_2 + C_6H_{12}O_6 \longrightarrow 6 \ H_2O + 6 \ CO_2 \qquad\qquad (2)$$

The dominance of reaction (1) during the photoperiod with reaction
(2) occurring during the darkperiod yields a daily or diel cycle
in which changes in concentration of oxygen are the inverse of the
changes in carbon dioxide. For a series of diel cycles observed
in the northwestern Caribbean Sea, the diel variation in these two
gases is shown in Figure 6. The photoperiod, with solar radiation
during 0700 to 0900 h, is characterized by an accumulation of oxygen and a depletion of carbon dioxide. During the photoperiod,

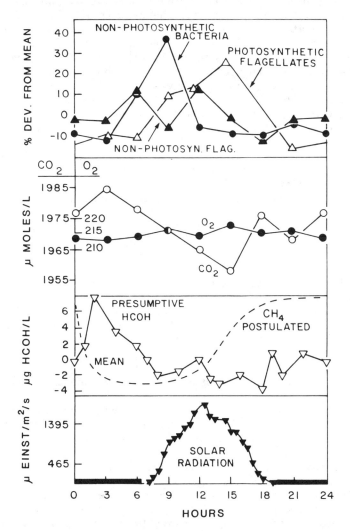

Figure 6. In the mixed layer the aerobic processes of photosynthesis and respiration are shown by inverse changes in CO_2 and O_2
through the solar cycle. The significant daytime peak in O_2 is
minimized by plotting on the same scale as CO_2 to show its 5x
larger changes due in part to chemosynthesis. Such a process is
methanogenesis and the postulated methane curve is indicated by
the inverse presence of formaldehyde, an intermediate in methane
oxidation. Population changes in the photosynthetic flagellates
follow solar radiation while bacterial growth occurs before the
peaking of solar radiation. All data from the northwestern Caribbean (Johnson et al., in prep.) except that for HCOH which is
unpublished Sargasso Sea data from EN-009, May–June 1977.

there was an increase in the dominant biomass of photosynthetic
flagellates in the nanoplankton, the 2-20 μm size fraction (see
Figure 8). What is very striking about these oceanic observations
is that changes in CO_2 exceed O_2 changes by five fold. By plotting these changes on the same scale, the significant cycling of
oxygen is difficult to see in Figure 6. Reasons for this enigmatic difference in CO_2 and O_2 changes occurring during the diel
cycle (Johnson et al., 1981a) are discussed later. Two unusual
features shown in Figure 6 should be noted. One is the marked
uptake of CO_2 starting at 0300, some four hours before there is
sufficient solar radiation for detectable photosynthesis. The
other is that the marked increase in bacteria during the ante
meridian period appears to be markedly inhibited during the post
meridian period, presumably due to photic damage and/or biotrophy
due to the non-photosynthetic flagellate peak at 12 hrs, or both.

In the mixed layer, there are two other gases whose production
is associated with the photoperiod. One is hydrogen, which under-
goes a diurnal change in concentration that parallels those in
solar radiation (Herr, pers. comm.). The other gas that shows
increased concentrations during the photoperiod is carbon monoxide (Swinnerton et al.,1969,1970;Seiler and Schmidt,1974). The
generation of CO as well as H_2 is ascribed to the photolysis of
OM (Wilson et al., 1970; Zika, 1981). These diurnal (daytime)
changes may be significant for the cycling of gases by anaerobic
processes.

2.2 Cycling of Gases and Organic Matter by Anaerobic Processes.

The mixed layer of the sea with oxygen concentrations near
saturation is not an environment in which one would expect anaerobic processes to play a potentially large role. In reviewing
methane cycling in aquatic environments, Rudd and Taylor (1980)
summarized the literature on the production of methane in the
mixed layer of the sea and concluded that methane is 30 to 70%
supersaturated in relation to the atmosphere, and that maxima
associated with the thermocline may be two or threefold greater
than equilibrium stability. Besides transport from nearshore
anoxic sediments, there are strong indications for in-situ production (Lamontagne et al., 1971, 1973; Scranton and Brewer, 1977;
Scranton and Farrington, 1977). Although it is possible that
algae, the gut of animals, and fecal pellets are sources, Rudd
and Taylor (1980) were not confident of a probable source within
the mixed layer.

Methane production in the anoxic sediments of ponds, marshes,
and estuaries is fueled by the accumulation and fermentation of
organic debris. The microsites of methane production in the
mixed layer are probably also the accumulation of organic debris
as discussed in section 2.5. Methanogenic bacteria are obligate

anaerobes which grow in association with fermenting bacteria
that produce small organic acids such as formate and acetate as
well as CO_2 and H_2 as end-products. If allowed to accumulate, H_2
will inhibit the anaerobic decay of OM. Methanogenesis is one
mechanism for preventing this inhibition. Methane is formed by
the reduction of CO_2 using the electrons generated by the oxidation of hydrogen and formate, or the fermentation of acetate or
methanol.

$$CO_2 + 4 H_2 \longrightarrow CH_4 + 2 H_2O \qquad (3)$$

Hydrogen required for CO_2 reduction may arise from OM other than
through fermentation. The photolysis of organic compounds during
the photoperiod releases not only hydrogen but carbon monoxide
(Wilson et al., 1970; Zika, 1981). Both products may be used in
methane formation. Hydrogen by the mechanism discussed above,
while CO can be reduced by methanogens using hydrogen obtained by
the hydrolysis of water (Quayle, 1972).

$$4 CO + 2 H_2O \longrightarrow CH_4 + 3 CO_2 \qquad (4)$$

Hydrogen can also be released from cyanobacteria (Rudd and Taylor,
1980), which are an ubiquitous component of oceanic plankton
(Johnson and Sieburth, 1979).

My colleague Ken Johnson, who is seeking the cause of the
enigmatically larger cycling of CO_2 than O_2, has come to the conclusion that it is due in part to the chemosynthetic fixation of
CO_2 into methane, and the cycling of carbon through the nonconservative gases CH_4, H_2 and CO. In [14]C "primary production" bottle
assays, the production of labeled methane would not be detected.
Diurnal production of H_2 and CO is documented, but Swinnerton and
Linnenbom (1967) failed to detect such a cycle for methane. But
such a cycle is possible. The relative formaldehyde concentrations occurring over a diel cycle obtained in the Sargasso Sea,
shown in Figure 6, was obtained from a series of 200 controls for
the aldehyde specific colorimetric reagent MBTH, used for the
assay of carbohydrates (Johnson and Sieburth, 1977; Burney and
Sieburth, 1977; Johnson et al., 1981b). From this curve of relative formaldehyde concentrations, a postulated inverse methane
curve was obtained. The rationale for this is that formaldehyde
is an intermediate in the utilization of methane by methanotrophic
bacteria (Ribbons et al., 1970).

The oxidation of methane is via methanol (5), formaldehyde (6),
and formate (7) to carbon dioxide (8) according to the following:

$$CH_4 + \tfrac{1}{2} O_2 \longrightarrow CH_3OH \qquad (5)$$

$$CH_3OH + \tfrac{1}{2} O_2 \longrightarrow HCOH + H_2O \qquad (6)$$

$$HCOH + \frac{1}{2} O_2 \longrightarrow HCOO^- + H^+ \tag{7}$$

$$\underline{HCOO^- + H^+ + \frac{1}{2} O_2 \longrightarrow CO_2 + H_2O} \tag{8}$$

$$CH_4 + 2 O_2 \longrightarrow CO_2 + 2 H_2O \tag{9}$$

The marked increase in formaldehyde in the early morning hours is suggestive of methane utilization. The depressed values during the day are suggestive of methane production. A daytime chemosynthetic fixation of CO_2 to form methane would be in addition to the photosynthetic fixation of CO_2 to yield the much larger CO_2 variation than O_2 variation (PQ<1), and is also consistent with the photoperiod accumulation of H_2. The bacteria responsible for methanotrophy are discussed in section 1.4.

2.3 Diel Cycle of Chemical Activity.

The diel cycle of the presence (photoperiod) and absence (darkperiod) of solar radiation and how it apparently controls trophic activity with the uptake and release of specific materials is shown in Figure 7. This figure is from Johnson et al. (in prep.) and is a conceptual model based on published and unpublished data as well as postulated cycles, and summarizes some of the phenomena shown in Figure 2. The photoperiod, which occurs between 0700 and 1800 hours, is paralleled by oxygen release and accumulation until late afternoon, about 1500 hrs. The photosynthetic release of O_2 is usually accompanied by an uptake of CO_2. In the open sea, the uptake as well as release of CO_2 is enigmatically much larger than the accumulation and removal of O_2 (Johnson et al., 1981a). The uptake of CO_2 starts earlier than the photoperiod (during the darkperiod at 0300) and ceases at the time O_2 evolution ceases. What is this mechanism for the dark fixation of CO_2?

Processes other than photosynthesis must be incorporating CO_2 into cell mass, a process controlled in part by the diel cycle. An obvious possibility is chemosynthesis, another mechanism for primary production. Possible chemosynthetic processes include methanogenesis, sulfur oxidation, nitrification (oxidation of NH_3 via NO_2 to NO_3), and nitrogen fixation (incorporation of atmospheric N_2 into amines). The occurrence of methane at concentrations above saturation are definitely associated with the mixed layer and the thermocline (Scranton and Brewer, 1977; Scranton and Farrington, 1977). As already mentioned, there still may be a diel rhythm in CH_4 production and utilization. For CO_2 to be reduced to methane there has to be a source of hydrogen. Herr (pers. comm.) has just completed a cruise in the southern North Atlantic (at a latitude close to that in the

Caribbean where the data were obtained upon which Figure 6 is
based) and found that hydrogen production closely tracked solar
radiation. One should note that heterotrophic processes, such as
the utilization of DOM, are apparently photoinhibited during the
day, permitting total carbohydrates and DOM to accumulate during
the photoperiod (Burney et al., 1982; Mopper and Lindroth, 1982).
Chemosynthetic bacteria, such as nitrifiers, are also apparently
inhibited during the photoperiod (Olson, 1981a,b). We therefore
postulate the uptake of hydrogen and the production of methane
during the late photoperiod, early darkperiod, and suggest that
the methane produced in the day or early evening is utilized in
the early hours before the dark fixation of CO_2. Data to sub-
stantiate this hypothesis are the apparent accumulation of formaldehyde during these early morning hours followed by a decrease in
formaldehyde concentrations (Johnson et al., in prep.). The accumulation of carbon monoxide during peak solar radiation, from
photolysis and possibly biological production, decreases during
the darkperiod and may be reduced and oxidized by methanogenic
bacteria to CH_4 and CO_2, respectively (Daniels et al., 1977) as
shown by equations (3) and (4). The changes in carbonate alkalinity, in the late evening and early morning observed in the
Caribbean, are also suggestive of methane production.

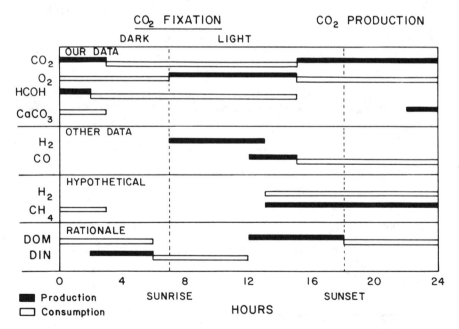

Figure 7. A schematic representation of the coupling of chemoauto-
trophy, photoautotrophy and heterotrophy through the non-conservative gases of seawater during the solar cycle as shown by actual
data and hypothetical events. (From Johnson et al., in prep.)

The dark uptake of CO_2 may also indicate other chemosynthetic processes such as nitrification. The methanotrophs can also fix atmospheric nitrogen and oxidize ammonia (Higgins et al., 1981). But the large populations of nitrifiers (Ward et al., 1982) and possibly methanotrophs, as well as the photosensitivity of nitrifying bacteria (Olson, 1981a,b) would suggest that nitrogen fixation followed by nitrification are active processes in the predawn hours transforming nitrogen into usable forms required by the photosynthetic microorganisms. Likewise, DOM released during the PM period by the photosynthesizers, would not only provide the energy for saprotrophic bacteria, but the growth factors required for autotrophic and saprotrophic bacteria.

The solar cycle therefore orchestrates a continuing sequence of microbiological processes that sustain each other in the mixed layer. The photoperiod appears not only to sustain photosynthesis, but to photoinhibit decomposing bacteria and to produce hydrogen and CO through photolysis. Methanogenesis appears to occur in the late photoperiod and early darkperiod due to different mechanisms. Saprotrophy, which occurs mainly in the darkperiod, not only utilizes DOM to respire CO_2, but also oxidizes the accumulated methane. This scenario is presented in an attempt to explain the temporal changes that can and do occur in the gases which serve as both substrate and by-product in these reactions. An example of the amount of CO_2 being taken up daily in net production and the daily release and uptake of DOC and total carbohydrate occurring in the mixed layer is given in Table 1.

Table 1. Net daily flux of carbon dioxide, total carbohydrate, and dissolved organic carbon in the free water at two drift stations in the Caribbean Sea (adapted from Burney et al., 1982).

Station	4a	5b
Net Flux	μg C/L/d	
ΣCO_2 Production	372	372
DOC Release	149	187
DOC Uptake	104	163
TCHO Release	53	61
TCHO Uptake	30	85

ΣCO_2=sum of CO_2; DOC=dissolved organic carbon; TCHO=total dissolved carbohydrate.

2.4 Size Fractions of Plankton and Organic Matter.

    The most numerous particles in seawater are the smaller
microorganisms, which form the basic fabric of life in the sea.
These smaller microorganisms are also most prone to sea to air
transfer by other mechanisms (e.g., see Blanchard, this volume).
The larger organisms, with sparser populations, require plankton
nets to concentrate them to suitable populations for study. And,
although important to the ecosystem, these larger organisms are
embroidery on the fabric of microbial life in the sea, and have
a minimal effect on the particles and gases in the sea and their
exchange with the air. They cannot be neglected, however, because
they have an effect disproportionate to their low oceanic density,
due to their grazing. The size-fractions of planktonic microorganisms and organic matter that can be quantified in 100 ml portions
of seawater are illustrated in Figure 8. The size of these materials affects not only their transport but their consumption by biotrophs and saprotrophs feeding on living cells and dead organic
matter, respectively (Sections 1.4 and 1.5).

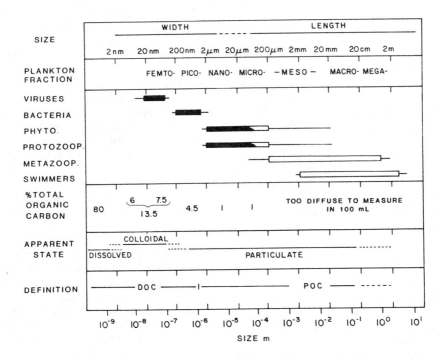

Figure 8. Distribution of planktonic and total organic carbon
size fractions that occur in 100 ml water samples. (Adapted from
Sharp, 1973; Sieburth et al., 1978.)

In order to put all particles in the sea into perspective, the
size range shown in Figure 8 is from 0.2 nm to 20 m, which includes
everything from dissolved organic substances to whales. As noted
on the figure, the particles somewhere below 20 µm are sized by
width (since their size-fractions are determined by water passing
through sieve-like filters), while the particles somewhere above
20 µm are usually measured by length. It is for the particles
within the microplankton size fraction, spanning from about 20 to
200 µm, that a 100 ml portion of seawater becomes too small to
quantitate the larger but sparser particles. However, the particles less than 10 µm account for over 90% of the total biomass in
the plankton (Beers et al., 1975), and the standing stocks above
20 µm can be only 1 or 2 percent of the total organic carbon at
most (Sharp, 1973). Therefore the size fractions and carbon fractions above 20 µm will not be dealt with in detail.

The smallest replicating units are the viruses in the femto-
plankton size fraction. These viruses are observed as free particles as well as associated with the cells in the picoplankton (0.2-
2.0 µm) size fraction (when filtrates less than 2 or 20 µm are
centrifuged at high speeds for several hours and the sediments
are examined by transmission electron microscopy, e.g., see Johnson and Sieburth, 1979, 1982). The viruses occur in the transi-
tional zone between particles and colloidal substances. The smallest microorganisms are planktonic bacteria that mainly occur between 0.1 and 1.0 µm and are discussed in section 1.4 under saprotrophs. The size fraction that includes these microorganisms
accounts for the largest fraction of particulate organic carbon
(4.5% of total in Figure 8). However, as much as half of this can
be microdetritus, which is plainly seen in centrifuged sediments
of the picoplankton size fraction, especially in nearshore waters.
The biovolume within the picoplankton size fraction is apparently
5% photosynthetic and 42% non-photosynthetic accounting for 47%
of the total biovolume less than 20 µm (Sieburth, in press). Much
of the synthetic and decomposing activity in the sea must reside
in this very important fraction as discussed in sections 1.3 and
1.4.

The next larger size fraction is the nanoplankton, which may
account for some 1% of total organic carbon. This size fraction
is relatively free of detrital carbon, and the biovolume apparently consists of approximately 27% photosynthetic biomass and 26%
heterotrophic biomass, dominated by biotrophs to account for some
53% of the total biomass less than 20 µm (Table 2, and Sieburth,
in press). Therefore the picoplankton and nanoplankton size fractions are similar in magnitude, but surprisingly the photosynthesizers barely account for one third of the total biomass. This is
also shown in Table 2, in which the carbon content of phototrophic

and heterotrophic carbon in the <20 µm plankton is shown for the
continental shelf and the open sea. This strongly suggests that
processes other than photosynthesis are contributing to primary
productivity. The microorganisms larger than 10 µm which account
for less than 10% of total particulates are beautiful, highly-
structured organisms, but they are too diffuse in open ocean
waters to be considered here.

Table 2. A comparison of the relative carbon content of the
smaller planktonic compartments over the continental shelf and
in the open sea (Sargasso and Caribbean Seas), showing that
phototrophs account for a constant one third of the total and
that the biomass in the open ocean is 18% that of the shelf.

µg C/L

Oceanic	Phototrophic			Heterotrophic			Total
Province	Pico	Nano	Total	Pico	Nano	Total	
Shelf	1.2	7.4	8.6	11.5	7.2	18.7	27.4
	(5%)	(27%)	(32%)	(42%)	(26%)	(68%)	
Open Ocean	0.1	1.4	1.5	2.0	1.3	3.3	4.8
	(5%)	(29%)	(31%)	(42%)	(27%)	(69%)	

This assumes that the carbon content of microorganisms is 10%
wet weight and that all microorganisms have similar water and
carbon contents. (Data adapted from Sieburth, in press)

The total organic carbon in 100 ml volumes of seawater has
been size-fractioned by Sharp (1973). The values given in Figure
8 were calculated from this data. It appears that some 80% is
truly dissolved, being less than 2 nm in size. Between this size
and 200 nm is organic matter considered to be in the colloidal
size range, which also includes viral particles, and accounts for
some 13.5% of total carbon. It is second only to truly soluble
organic matter in quantity. This amount of carbon is threefold
that in the size fraction containing bacteria, and thirteen and a
half times the carbon in the nanoplankton size fraction, which are
undoubtedly a main contributor to this fraction through the autotrophic and biotrophic processes of the flagellates dominant in
this size fraction. The bottom box of Figure 8 shows that the
conventional definition of dissolved organic carbon (DOC) and particulate organic carbon (POC, less than and greater than 1 µm,
respectively, fails to adequately describe either of these fractions. The apparent twofold-greater concentration of colloidal
substances compared to particulates has been an overlooked phenomenon. The presence of such materials may have a profound effect
on the transport and utilization of OM in the sea, as suggested

in Figure 9 (discussed in section 2.5). An appreciation of the
nature and size of the OM and microorganisms in the mixed layer is
necessary for an understanding of how they contribute not only to
the sea-air interface but to the 0.1-10 µm particles in aerosols
(see Blanchard, Slinn and Buat-Menard, this volume).

2.5 Physical Factors that Control Microbiological Processes.

The schematic drawing in Figure 9 illustrates the environment

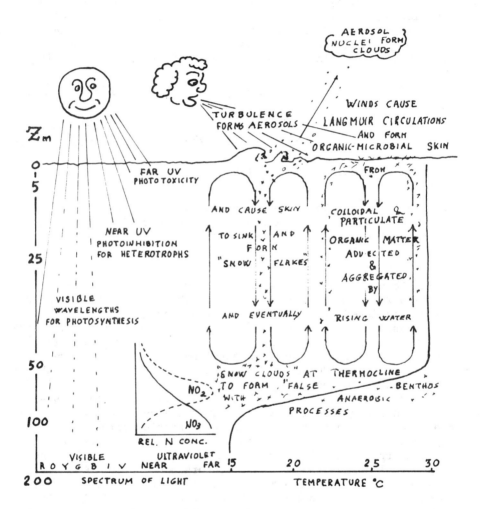

Figure 9. The physical factors that control the microbiological
and organic chemical processes throughout the mixed layer from
the surface layer to the benthic-like layer at the thermocline.

of the mixed layer ecosystem and the physical factors that contain it, and control the microbiological processes within it. The upper boundary is the sea-air interface whose organic-microbial skin depends upon physical, chemical, and microbiological processes for its formation and replenishment. The lower boundary is formed by the density gradient at the thermocline, which appears to effectively create a false bottom to retain low density particles.

Solar radiation, which usually penetrates throughout the mixed layer, has a marked effect on biological processes (Smith, 1977) because it not only provides the wavelengths required for photosynthesis (Jeffrey, 1980), but the UV wavelengths that cause near-surface phototoxicity (Dubinsky, 1980). Photoinhibition affects a number of metabolic processes (Harris, 1980) and is augmented by the higher environmental levels of oxygen, produced during photosynthesis (Pope, 1975). The wavelengths of light causing photoinhibition of all life forms was formerly thought to be restricted to far UV, between 200 and 300 nm, which is severely limited in its penetration of seawater. It is now known that near UV with wavelengths between 300 and 380 nm penetrates seawater well into the mixed layer. This is due mainly to UV-B, with wavelengths of 290-320 nm that reach even to the base of the euphotic zone (Smith and Baker, 1979) where it can affect not only the saprotrophic processes of bacteria which decay OM (Webb and Brown, 1979), but the autotrophic processes of nitrification in which ammonia is oxidized to nitrate. Evidence for the latter is the quite regular occurrence of a nitrite maximum (Wada and Hattori, 1971; see Figure 9), which appears to result from the selective inhibition of the bacteria oxidizing NO_2 to NO_3, but not of those oxidizing NH_3 to NO_3 at the top of the nitricline (Olson, 1981b).

Solar radiation, which provides the energy for photosynthesis, also warms the air to generate winds which form Langmuir circulations (Leibovich and Paolucci, 1980) and internal waves (Pollard, 1970; Pollard et al., 1973) which cause vertical mixing in as little as 3-4 hours; see Dobson, this volume, for a detailed discussion. The smaller planktonic microorganisms are brought through a spectrum of light intensities and wavelengths that have a profound effect on their metabolism (Dubinsky, 1980) within the life span of these rapidly growing cells. The darkperiod free from photoinhibition permits the utilization of OM and the development of a marked heterotrophic microbiota. The dissolved OM, which nourishes the free cells of saprotrophic bacteria under aerobic conditions, is produced and allowed to accumulate during the photoperiod, through the processes of photosynthesis, exudation, and photoinhibition.

In addition to the free bacterial cells in the plankton, there are cells of attached saprotrophic bacteria associated with aggregated OM commonly referred to as marine snow (Riley, 1963; Silver

et al., 1978; Alldredge, 1979; Silver and Alldredge, 1981). Turbulence and mixing at the sea surface apparently causes a slough-
ing of the sea skin to form flakes of colonized OM indistinguishable from those of marine snow. Observations on removal are
sparse (Jannasch, 1973) and the best indication is that of Freedman et al. (1982) shown in Figure 14. These particles, with dense
populations of saprotrophic bacteria and their protozoan grazers,
may be oxygen depleted to a degree that would permit anaerobic
processes such as fermentation and methanogenesis. If such microsites are not effective throughout the mixed layer, they surely
must accumulate to dense clouds (Wangersky, 1978) at the density
discontinuity at the thermocline due to the process of macro-
flocculation (Kranck and Milligan, 1980).

3. ORGANIC-MICROBIAL FILMS THAT FORM THE SEA'S SKIN

3.1 Physical Factors Holding Molecules and Microbes at the
     Sea-Air Interface.

    Buoyancy, electrostatic attraction, physical adsorption,
chemical adsorption, and surface tension are overlapping physical
processes capable of holding a variety of materials at the sea-
air interface. Buoyancy is caused by a density lighter than the
supporting water. Examples of buoyancy are the bladderworts
floating on ponds, pelagic species of the seaweed Sargassum kept
afloat with their bladders, and the purplish Portuguese Man-of-War,
Physalia, and "By-the-Wind Sailor", Velella, with their air bladders. Lesser known are the blue pigmented copepods, water striders,
molluscs, anemones, siphonophores and decapods described by David
(1965b) and Cheng (1975). The term pleuston has been used to describe such organisms that, by buoyancy, reside on top of, in, or
under the surface (see Banse, 1975, for an historical review).
These larger organisms can be sampled by specialized nets held
open across the sea-air interface with floats (Willis, 1963; David,
1965a). Although these larger organisms are important in the
economy of the sea (Zaitsev, 1970), I am more concerned with the
microorganisms.

    In contrast to the pleuston, neuston (a term coined by Nau-
mann, 1917) are microorganisms associated with the surface film,
which are mainly dependent upon surface tension rather than buoyancy for support. In the intriguing text "Life in Moving Fluids,"
Vogel (1981) suggests that in small-scale systems where the Reynolds number is low, water takes on the consistency of asphalt
(Purcell, 1977). Thus at the scale of macromolecules and bacteria,
the sea-air interface is solid enough to "walk on" and the sea is
barely fluid enough to swim in. Because of its consistency, the
interface promotes adherence (Marshall, 1976).

There is some evidence that dissolved organic matter is readily adsorbed at surfaces (Hunter and Liss, 1981). This was beaut-
ifully demonstrated by Neihof and Loeb (1972) for both low and high
molecular weight OM. When the OM was photooxidized, the particles
became charged consistent with their chemical nature. Platinum
metal surfaces and optical polarization analysis were used to confirm the adsorption of OM from particle free natural seawater before but not after the photooxidation of OM (Loeb and Neihof, 1977).
The thickness of the film increased over time. The negatively
charged particles in seawater (Neihof and Loeb, 1972, 1974), from
macromolecules to micromolecules, may be attracted to a positively
charged interface, or to a negatively charged interface, through
the electrical double layer phenomenon, which appears to play an
important part in the interactions between bacteria and solid surfaces (Marshall, 1976).

3.2 Nature of the Organic Film.

Films occur on the surface of all waters, whether there is a
visible slick or not. But it is the smoothing of rippled water
by either natural surface films or applied oil that has attracted
attention. The observations of mariners on the effect of oil on
rough waters aroused the curiosity and ingenuity of Benjamin Franklin, who conducted an experiment recorded in a letter to a scientific acquaintance in 1773 (Goodman, 1931, cited by Garrett, 1972).
After a false start in which he added oil to the leeward side of
a wind-roughened pond, Franklin added a teaspoonful of oil to the
windward side and produced an instant calm spreading to the lee
side "making all that quarter of a pond, perhaps half an acre, as
smooth as a looking glass." But it was Henry David Thoreau who
apparently first recorded the similar action of natural films,in
1858,in which the leading edge of a ripple dampening organic film
is clearly seen as a ridge (cited by McDowell and McCutchen, 1971).
Only four years later, Thompson (1862) wrote a note "on the calm
lines often seen on a rippled sea." The effects of oil on stormy
seas was recorded in the scientific literature by Reynolds (1880)
and Aitken (1884).

The role of organic films in affecting the surface tension
and the dampening of wind induced ripples was studied by Pockels
(1891) and Langmuir (1917). Much of this early literature is
summarized by Gifford Ewing (1950) in his scholarly treatise on
internal wave slicks. These internal wave or calm slicks, which
can be observed during summer conditions along the California
coast, are generated by wind speeds of 1-3 meters per second and
occur at distances of hundreds of meters apart. They are repor-
tedly distinct from the more closely spaced wind-cell slicks generated at speeds above 4 m/sec described by Langmuir (1938). Ewing
concluded that the surfaces of coastal waters and lakes are often
coated with "films of finely particulate or colloidal material

rich in OM." There was no a priori guess as to the chemical nature of these materials.

In a contemporary report, Dietz and Lafond (1950) described
the parallel slicks caused by Langmuir circulations. One of the
authors of this observational-type note admits that, prior to un-
dertaking the study, it was his expectation that natural slicks
possibly were films of oil derived from organisms in the sea. It
is this bias, not supported by analysis of the films, that probably led Dietz and Lafond (1950) to state that "slicks are contam-
inate films of organic oil, probably derived primarily from diatoms which contain droplets of oil in their cells to assist in
flotation and/or as an energy food supply." Despite the fact that
lipids are usually a minor (4%) constituent of diatoms (Miller,
1962), which are a minor constituent of oceanic plankton, and that
lipids are not enriched in surface films, the erroneous notion that
lipids dominate surface films continues to persist (e.g., Norkrans,
1980). Norkrans cites Jarvis (1967) and others in putting forward
her case that lipids are considered to be the major constituents
of the uppermost surface microlayer on all natural waters. Jarvis
et al. (1967) suggested that the more stable components of films
are polar-nonpolar compounds containing long-chain hydrocarbon
groups, such as high-molecular-weight saturated and unsaturated
fatty acids and their esters. Usually overlooked in this major
paper is the statement that the outstanding difference, for natural
film material and for artificial films of oleic acid, in dampening
coefficients versus area occupied, is that the oleic acid films
occupied a much larger area than that of natural film material.
They concluded that the natural film must contain a relatively
large quantity of nonpolar or weakly surface active material which
did not contribute to either dampening or force versus area curve.

The important statement identified in the previous paragraph
was not missed by Williams (1967), who reserved judgment on the
composition of the films. He noted that the films contained high
concentrations of both DOC and POC, and contributed to marine snow,
but that the C:N:P ratios in the films and bulk water were indistinguishable. During the peak of interest in surface slicks and
their lipid nature in the mid-sixties, Sieburth and Conover (1965a)
reported slicks with decreased surface tension associated with
aggregations of the filamentous cyanobacterium Trichodesmium on the
surface of the Sargasso Sea. Homogenates of cell concentrates
lacked antibacterial, lipolytic, and proteolytic activity, but
possessed marked amylolytic activity. This led to the conclusion
that the slicks may result from released carbohydrate, rather than
lipids. Sturdy and Fischer (1966) pursued this observation to show
that surface tension was similarly reduced in the slick patches
associated with the beds of carbohydrate-rich kelp in southern
California.

Baier (1972) developed a technique whereby film material adsorbed onto a germanium slide can be analyzed by infrared absorption. Using this technique to determine the dominant chemical composition of dried sea-surface films, Baier et al. (1974) concluded that all waters, except heavily travelled waterways contaminated with hydrocarbons, contain a film of polysaccharide and polypeptide components in the water at the interface. MacIntyre (1974b) is of the opinion that lipids reported by others were extracted from the microorganisms in the sample. Liss (1975), in reviewing the chemistry of the sea surface, described in detail the work of Garrett (1967) which showed the lipid nature of the film. But Liss also used the DOC data of P. M. Williams (1967) and the lipid data of Garrett (1967) to show that the latter's procedure could analyze at most 25% of the total organic carbon in the sample. Liss (1975) concluded that if polysaccharide and protein are dominant compounds, "it will be necessary to reappraise completely our concepts of the nature of materials present at the sea surface." In studying the surface microlayers of the North Atlantic, Sieburth et al. (1976) determined that carbohydrate alone accounted for 13 to 46% of the dissolved organic carbon, with a mean of 28% compared to 16% for subsurface waters.

In a recent review of organic sea surface films, Hunter and Liss (1981) continue to pursue the lipid versus polysaccharide-protein controversy. They point out that Baier's dehydrated polysaccharide protein film of 20 Angstroms could actually reach dimensions of 1 μm when hydrated. Hunter and Liss (1981) stress that if fatty acids and their lipid derivatives were major components of the film, then they should be enriched by a factor on the order of a thousand over the bulk water, instead of a few percent of this value. They then calculate that fatty lipids can only account for 0.8% of the microlayer, normally, and some 5-10% in the region of slicks. Hunter and Liss conclude that, contrary to intuitive expectation, organic films on the surface of the sea in the absence of petroleum pollution do not consist of classically known, simple surfactants such as fatty acids and their esters, but consist of complex polymerized material, including humic substances, with a high degree of hydroxylation and carboxylation. Thus, it is suggested that simple lipids account for a few percent of the ambient film, at most, except in compressed slicks. They also point out that it would seem unlikely that the dilute concentration of lipids would remain free in a molecular sense to form the classic dry film above a wet film (MacIntyre, 1974a), but would rather become adsorbed by the overwhelmingly greater quantity of macromolecules.

A four year study on the formation, composition, and alteration of sea surface films conducted by Peter M. Williams and Angelo Carlucci of Scripps is apparently the most complete investigation to date on the chemical composition of the surface film

compared to bulk water. Williams (pers. comm.) is now summarizing
the results of their four year study, which includes three cruises.
An analysis of protein, carbohydrate, and lipids accounts for some
30 to 60% of the total dissolved organic carbon that was hydrolyz-
able. The results for surface films and bulk were similar. A
first impression of the overall results is that when protein is
calculated as 100, the mean for carbohydrate is approximately 250,
while that for lipids is approximately 25. Such ratios are consistent with those obtained for materials released from algal culture (Melkonian, 1978). Assuming that this accounts for 50% of
total organic carbon, the rest being presumed to be condensed
humic substances, then protein would account for 13.3% of DOC,
carbohydrate 33.3%, and lipid 3.3%. The carbohydrate value is
similar to the mean of 28% reported by Sieburth et al. (1976),
while that for lipid is within the range (0.8 to 10%) calculated
by Hunter and Liss (1981). There is an indication that the humic
substances may be quite old (Williams, 1971), but there are also
data indicating that these substances are metabolized (de Haan,
1977) or precipitated from solution (Sieburth and Jensen, 1968,
1969).

Now that we have a fairly solid picture of what the components
of a surface film are, it is time to replace the models of 'dry' and
'wet' surfactants (MacIntyre, 1974a; Odham et al., 1978) with a new
one. I submit the sketch in Figure 10 as my conception of dissolved and colloidal material released into the water column. This
material advects through the mixed layer, by Langmuir circulations
and other mixing processes (see Dobson, this volume), to adsorb
onto the "solid" sea-air interface. Amid a few scattered colloidal sized cell fragments are much more numerous colloidal macromolecules, both free and condensed, in an approximately 50:50 ratio.
The smaller, but significant, amount of low molecular weight substances
are left out to avoid clutter. I envision not layers of wet and
dry surfactants, but a hydrated gel which is a matrix of inter-
tangled molecules that might be only 0.1 to 1.0% on a dry weight
basis.

In addition to the naturally derived organic substances,
which act as substrates for the chemoorganotrophic bacteria, there
are natural and man-made chemicals of a toxic nature, which also
concentrate in surface films. Examples are PCB's and heavy metals
(Duce et al., 1972) as well as other organochlorine substances and
phthalic acid esters (Larsson et al., 1974). A phthalate commonly
used as a plasticizer is di-2-ethylhexylphthalate (which is on the
EPA "hit list" of materials finding their way into the sea which
are potential carcinogens). Its fate is being studied at the EPA's
Narragansett Environmental Research Lab. Because of its poor solubility, this phthalate concentrates in the surface film where it
is preferentially decomposed and respired to CO_2 by the bacterial
flora in the film, when water temperatures are above $10°C$ (E. Davey,
pers. comm.).

Natural inhibitors, which apparently also concentrate in the surface film, are phenolic materials (Sieburth, 1968). The antibiotic activity of the pelagic seaweed <u>Sargassum</u> in the Sargasso Sea (Conover and Sieburth, 1964) and the excretion of colored ultraviolet substances by brown marine algae (Craigie and McLaughlin, 1964) were found to be connected and to be caused by polyphenols with tannin-like activity (Sieburth and Conover, 1965b). A comparison of humic substances from terrestrial and marine sources showed that the former were immediately precipitated

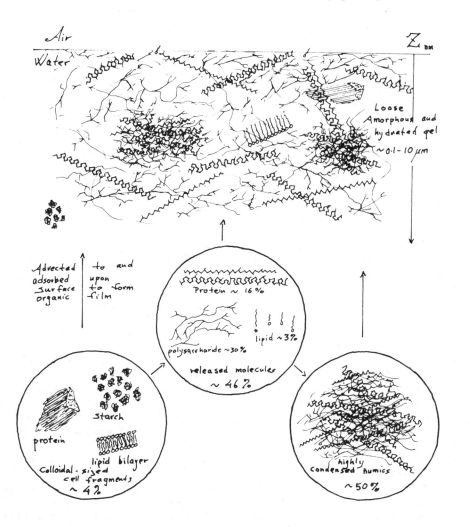

Figure 10. A conceptual diagram of the surface film as a highly hydrated loose gel of tangled macromolecules and colloids held to the surface by surface tension.

in seawater and were distinct from the latter (Sieburth and Jensen, 1968). The humic substances from marine sources seem to originate from brown seaweeds (Sieburth and Jensen, 1969), which can release up to 40% of their photosynthate at certain times of the year (Sieburth, 1969). Marine humics appear to be formed in-situ and proteinaceous and carbohydrate material appear to react with phenols to form complexes that make up a considerable part of marine humic substances (Sieburth and Jensen, 1969). A similar scenario for the in-situ formation of marine humic acids from the decomposition products of plankton was also suggested by Nissenbaum and Kaplan (1972). Recently Carlson and Mayer (1980) found that phenolic-containing material was selectively enriched in coastal waters.

During routine microbiological surveys on surface films made from R/V Trident during 1962 and 1975, instances of detectable and marked bacterial inhibition were observed (Sieburth, 1971b; Sieburth et al., 1976). Such inhibition may result from both man-made as well as naturally toxic substances such as algal polyphenols. These nearshore waters were sampled near Puerto Rico and the New England coast. Suppression of the bacteria in surface films has also been reported in coastal waters of British Columbia (Dietz et al., 1976) and Sweden (Kjelleberg and Hakansson, 1977).

3.3 Nature of the Microorganisms Colonizing and Aggregating at the Surface Film.

Experimental surface films suitable for microscopic examination can be obtained in a bucket of seawater to which a fist-sized intertidal stone or two with its microbiota is added. Many methods have been devised to remove and examine this film from the surface, including Formvar-coated transmission electron microscope grids (Young, 1978), Nuclepore membranes which can be examined directly rather than being eluted and concentrated upon a second membrane (Sewell et al., 1981), or by a wire loop that uses surface tension (Freedman et al., 1982). None seems simpler or more effective, however, than the glass coverslip floated on the surface, as discussed by Charles Renn (1964). Within several days, gel-like surface films in excess of 10 μm are formed and occupied by micro-colonies of bacteria that develop much as they do on the surface of an agar plate. The microcolonies occur as two dominant forms (Figure 11), in which one type consists of cells uniformly separated by extracellular polysaccharide slime (also known as exocellular polymeric substances, EPS, Geesey, 1982). The other type of cells are those which are not equally spaced but can be contiguous. When stained by the fluorochrome, acridine-orange, some cells also have non-staining granules that may be crystals of poly-β-hydroxybutyrate (PHB), which is a storage product.

Both of these polymeric substances, EPS and PHB, are endproducts of overflow mechanisms used by bacteria when the substrate is present in excessive quantities. Marine bacteria, which are apparently eurytrophic, and utilize substrates across the range of micrograms to grams of carbon per liter, show the presence of these polymers when grown in the presence of concentrations of a hundred to thousands of mg C per liter (Baxter, 1982). In addition to a variable staining of cells by acridine orange (some microcolonies fluoresce a faint green while others fluoresce bright orange in epifluorescence microscopy), a gram stain of the neuston produces both gram negative and gram positive microcolonies, indicating a diversity of physiological types of cells. This is of interest since gram positive bacteria will not usually grow in seawater until it is supplemented with OM at gram per liter concentrations. This was first seen during R/V Endeavor cruise 015, during October 1977, when high counts of non-enteric gram positive cocci grew on enteric media inoculated with surface films collected by screen (Sieburth, 1965) and glass plate (Harvey and Burzell,

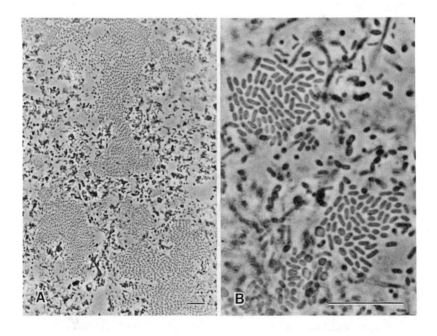

Figure 11. Phase contrast light micrographs of saprotrophic bacteria as they occur in microcolonies in the surface film of seawater in a bucket enrichment as obtained by adsorption onto a glass coverslip. Note that the cells in some microcolonies are evenly separated by polysaccharide, while others are randomly distributed. Marker bars=10 μm.

156   J. McN. SIEBURTH

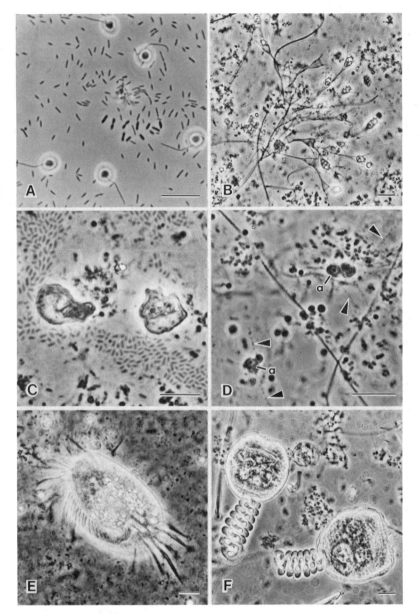

Figure 12. Phase contrast light micrographs of representative biotrophic microorganisms (amoebae, flagellates, and ciliates) observed in the surface film of a seawater enrichment as sampled
by a glass coverslip. Included are examples of motile (A, C & E)
and sessile (B, D & F) forms. A) the monoflagellate Oikomonas;
B) a colony of choanoflagellates; C) crawling forms of amoebae;
D) minute attached amoebae (a) with fine, beaded pseudopodia

(arrows); E) the ciliate, Euplotes; F) the peritrichous ciliate, Vorticella. All marker bars=10 μm.

1972) samplers (Dufour and Sieburth, unpubl. data). Subsequently, Peele et al. (1981) have reported Staphylococcus from the surface waters near the Puerto Rico dump site. Such observations appear to confirm the calculations of Sieburth et al. (1976) that the mean effective concentrations of organic carbon in surface films ranged from 471 to 2,250 mg per liter, with a mean value of 1.427 g carbon per liter of film.

At the very top of the surface film (as observed under a coverslip), the bacteria occur as apparently undisturbed micro-colonies (as seen in Figure 11), but the bacterial cells become sparser as one focuses down through the gel-like surface film. This is apparently due to the biotrophs shown in Figure 12, which graze on the bacteria they can reach. The types of protozoa asso-ciated with the surface film are the amoeboid, flagellated, and ciliated forms, as shown in Figure 5. Representatives occurring in the surface film can be both motile (Figure 12A,C&E) as well as sessile forms (Figure 12B,D&F). The most common protozoans are the unattached flagellates such as Oikomonas, shown in Figure 12A, which cruise through the loose gel and feed upon the cells below the uppermost microcolonial layer. Hanging upside down from the surface film, as described by Norris (1965), are the beautiful choanoflagellates (Thomsen, 1973, 1976), protected within their basket-like houses made from siliceous rods. The flagellates obtain their bacterial prey by trapping them against the outside of their collar by flagellar action (Laval, 1971) or by adhesion to the tentacles that make up the collar.

The amoebae can be observed as they graze, slowly crawling in close proximity to the microcolonies (Figure 12C). In addition to the typical motile forms are the apparently undescribed, attached or sessile forms, an example of which is shown in Figure 12D, where the pseudopods are found to extend out and appear as beaded filaments. Ciliates such as Euplotes (Figure 12E) can be observed as they spin bacteria into their oral cavities with their oral ciliature. Other ciliated forms are the sessile peritrichous ciliates, such as Vorticella, whose stalk can contract into a coil as shown in Figure 12F. These experimental films are dominated by heterotrophic bacteria and their protozoan grazers, as are natural films (Sieburth et al., 1976). Similar enriched films, prepared at the NATO Institute at Durham, were examined by the participants using epifluorescence and phase contrast microscopes, kindly supplied by Galen Jones and Richard Blakemore, in order to show vividly the nature, diversity, and dynamic state of the micro-biota in the surface film.

I have not illustrated the photosynthetic component of the
surface film. It is a highly temporary and selective component.
Marumo et al. (1971) reported that diatoms, especially a species
of Nitzschia, were enriched in the surface layer at an oceanic
station, but with the absence of chromatophores in two of the
minor species, the authors concluded that the algae were killed
by phototoxicity. Sieburth et al. (1976) also noted chlorophyll
enrichment at one offshore station where there was not only a
Nitzschia bloom but a marked reduction of both bacteria and their
amoeban grazers.

In a Georgia salt marsh, as the incoming tide floods over the
soil and cord grass (Spartina alterniflora), a film of pennate
diatoms lifts off to become a part of the surface film, which
becomes enriched several orders of magnitude over that of the
bulk water (Gallagher, 1975). About 37% of the net photosynthesis
occurred in the surface film which provides a mechanism for avoid-
ing shade limited growth. As the tide ebbs, the concentration of
plant pigments in the film drops to levels of the bulk water. The
pennate diatoms of the mud and sand flats, including the species
of Nitzschia found offshore, appear uniquely suited for surviving
the extremes of light at the sea surface. Diversity of the phytoplankton is markedly reduced in the surface microlayer (Manzi et
al., 1977).

Our knowledge of the contribution of phytoplankton to the
surface film is obviously incomplete. Studies on the flagellated
phytoplankton, such as the dinoflagellates, have shown marked diel
migration (Hasle, 1950). Apparently, maximum populations in the
inner Oslo Fjord occurred within the upper 2 m, but not on the
surface. Different species had different phototactic responses,
some rising to the surface by day and descending at night, while
others descended by day and rose towards the surface by night.
Eppley et al. (1968) reported migration rates of dinoflagellates
of 1-2 meters per hour. Freedman et al. (1982) have shown that
in a dystrophic lake, the dinoflagellate Peridinium became a very
temporary but important component of the algal biomass in the surface film (Figure 14). Such diel studies are indicated for the
open sea.

3.4 Spatial and Temporal Variation in the Film.

I have created a picture of the surface film as a loose hydrated gel of macromolecules (Figure 10) colonized by bacteria
(Figure 11) that grow optimally in this rich milieu to form dense
microcolonies, whose cells show evidence of overfeeding by the
presence of EPS and PHB formation. This bacterial flora, in turn
enriches a protozoan community of both motile and attached (sessile) forms (Figure 12). The spatial arrangement of a mature film
is represented in Figure 13. At the interface itself, bacterial

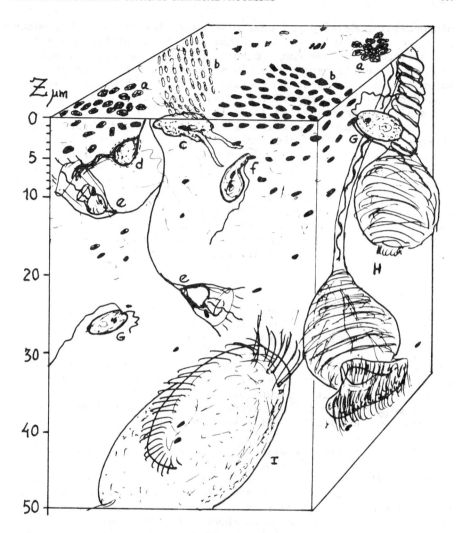

Figure 13. A three dimensional representation of the surface film
microhabitat as reconstructed from the previous observations by
light microscopy. The bacterial microcolonies contain both (a)
touching cells and (b) cells spaced uniformly with extracellular
polysaccharide slime, which are minimally grazed in the top plane
but (as they multiply into the subsurface gel) they are grazed by
(c) an amoeba, (d) the scaled flagellate Paraphysomonas attached
to the film, (e) a colonial choanoflagellate, (f) the flagellate
Rynchomonas feeding through its proboscis, (G) the flagellate
Bodo, (H) the stalked ciliate Vorticella in extended and contracted
states, and (I) the free ciliate Euplotes.

microcolonies develop and remain somewhat intact, probably because
they are unavailable to the grazing protozoans within and below
the surface film. We can see both the evenly spaced (b) and
clumped (a) cells, some possibly with internal granules of PHB.
An amoeba with pseudopods extended (c) works below this layer. The
scaled flagellate Paraphysomonas (d) attaches to the surface with
its stalk as the flagellum draws in free cells. A stalked choanoflagellate (e) moves its cells so that the flagellar-generated
current can suck the bacteria against their cells for absorption.
The motile flagellate Rynchomonas engulfs cells through its proboscis (f), while the biflagellated Bodo (g) engulfs bacteria
randomly distributed within the loose gel of the film. The larger
members of this community are the free ciliate Euplotes (I) and
the attached peritrichous ciliate Vorticella (H). These forms,
characteristic of those occurring on the debris of nearshore waters,
not only are present in experimental films such as the one depicted here, but these or similar species are found on flakes of marine snow in offshore waters (Caron et al., in press.).

The presence of such films at most randomly picked stations,
even in the absence of detectable slicks, indicates that these
films are continually being formed and replenished. This is apparently a characteristic of the surface film (P. M. Williams, pers.
comm., and Dragcevic and Pravdic, 1981). Since surface films slide
over their underlying waters and are quickly adsorbed onto surfaces, observations at frequent intervals on the same film area
over a diel cycle is all but impossible. But observations are
needed to determine if the surface films have a diel cycle similar
to those in the plankton, discussed in section 2.

As we have already discussed, surface films also occur in
lakes. Although the nature of the films on freshwater ponds appear very different from those in the sea (unpubl. observations),
such ponds are ideal for the study of the temporal nature of the
microbiota of the surface film. Danos (1980) undertook an elab-
orate study following seasonal changes in algae, bacteria, and
nutrients in the surface film and at different depths of a pond.
But the real action, that of the diel cycle, was apparently
missed.

Freedman et al. (1982), while studying a small dystrophic
lake in Georgia, noticed the nocturnal accumulation of material
at the water-air interface. This was documented by determining
adenosine triphosphate (ATP) which is essentially present only
in living cells and is an index of total biomass. Algal cell
counts were also taken at four hour intervals over the 24 hr study
period. The mean from four sampling areas is shown for ATP and
cell biovolume in Figure 14. The physical parameters of light,
wind, and temperature for this period are shown in Figure 14A.

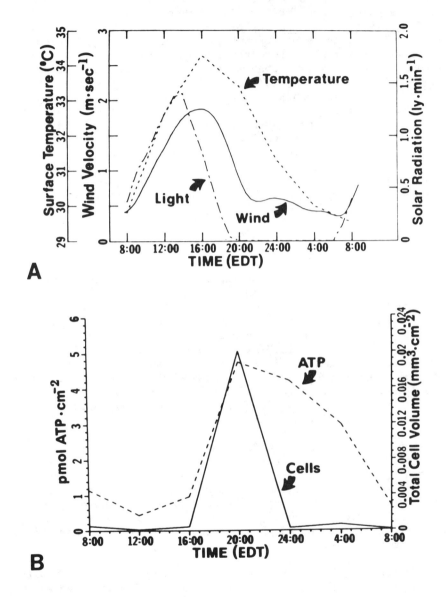

Figure 14. A) The physical parameters of solar radiation, wind, and water temperature on Lake Louise, GA, during a diel cycle when (B) indexes of total cell biomass (adenosine triphosphate= ATP) and algal biomass (biovolume) were observed in the surface film. (From Freedman et al., 1982.)

There is a striking photoperiod depression of total biomass, as
well as algal biomass, followed by an active recruitment of the
dinoflagellate Peridinium for only the first part of the darkperiod. The majority of the biomass remaining during the darkperiod was presumably heterotrophic in nature. The early morning
decrease in biomass is probably caused by mechanical removal
through wind driven compression (Sieburth, unpubl. obs.). Such
diel cycles appear to be a general occurrence on Barbers Pond,
RI, at the author's front door. If such diel cycles also occur
in the sea, this means that most of our samples, obtained during
the day, have missed the real action. Further studies on the
chemistry and microbiology of the surface film must not only
encompass the diel cycle, but document the physical factors such
as solar radiation, wind, surface roughness, and temperature
changes, which will affect the accumulation of OM and the microbiota which degrade it.

3.5   Crucial Function of the Surface Film.

The cyclical nature of the processes in the biosphere of this
planet, controlled directly or indirectly by the solar cycle, may
be thought of as wheels. Then the wheel that concerns the two-
way exchanges between the atmosphere and the water bodies must
have the surface skin of the water as the hub, with the physical,
chemical, and biological processes that connect it to the wheel's
rim being the spokes. I think this analogy is appropriate, since
the processes that take place at the hub radiate their effects
into the atmosphere and into the abysses of the seas. The effects
of the surface film on gas transfer have been discussed by
Liss,   this volume;   while their effect on wave dampening has
been discussed by Dobson, this volume.

Just as the human skin is a major site of exchange between
the inner working of the body and the outer environment, so does
the sea's skin form a locus where processes above and below interact. These are the processes that maintain the biosphere near
steady state. Our most recent techniques and technologies and
their application to the mixed layer of the sea are indicating
how artificial and erroneous have been our notions on how this
ecosystem operates.

We cannot select out specific trophic groups to study without considering how they affect the others. We must have an
overall approach to untangling the intricacies of the mechanisms.
The individuals studying productivity have used shortcut methods,
such as the ^{14}C incubation assays, to obtain rates--while failing
to first determine the parts of the ecosystem and how they might
function.

The purposes of this chapter have been to show you the parts, and to speculate on how the different trophic components may mesh and interact during the solar cycle. What is apparent is that decomposing OM concentrates first at the sea surface, through surface tension, and then, as it sloughs off and sediments, it concentrates at the false bottom of the thermocline. The microbiota that produce, eat, and recycle the primary production in the mixed layer have counterparts in the surface film which work in a very concentrated milieu and probably have among the shortest doubling times of any microorganisms in the sea. Not only are the natural organic substances recycled in the surface film but toxic substances, both synthetic organics and the trace metals generated by our industry, are also concentrated here.

Just as the human skin sloughs off and is renewed, so does the sea's skin, so that the waste products and contaminating substances are removed. If the sea-air interface did not exist, many of the processes would occur at other interfaces, but at very different rates. What makes the sea's skin unique is what makes our skin unique: it is a semi-permeable covering that lets gases and liquids pass, and since it is continually renewed, it rarely plugs up.

ACKNOWLEDGEMENTS

I thank Kenneth M. Johnson for wrestling with the chemistry of the gases and Paul W. Johnson for taking the micrographs and preparing the artwork. Both helped proof the manuscript, which was diligently Apple-printed by Jean Knapp, and meticulously edited by George Slinn. This paper is based upon the research conducted by Ken and Paul and my graduate students cited in the references. It was funded by the Biological Oceanography Program of the National Science Foundation through grants GA-41501X, DES-74-01537, OCE-76-81778, OCE-78-14528, OCE-78-26388, and OCE-81-21881.

REFERENCES

Aitken, J., 1884: On the effect of oil on a stormy sea. Proc.
    Royal Soc. Edinburgh, 12, 56-75.
Alldredge, A.L., 1979: The chemical composition of macroscopic
    aggregates in two neretic seas. Limnol. Oceanogr. 24, 855-866.
Baier, R.E., 1972: Organic films on natural waters: their retrieval,
    identification and modes of elimination. J. Geophys. Res.,
    77, 5062-5075.
Baier, R.E., D.W. Goupil, S. Perlmutter and R. King, 1974: Dominant
    chemical composition of sea-surface films, natural slicks, and
    foams. J. Rech. Atmos., 8, 571-600.
Banse, K., 1975: Pleuston and neuston: on the categories of organ-
    isms in the uppermost pelagial. Int. Revue ges. Hydrobiol.,
    60, 439-447.
Baxter, M., 1982: The response of marine bacteria to carbohydrate.
    M.S. Thesis, Univ. of Rhode Island, 88 pp.
Beers, J.R., F.M.H. Reid and G.L. Stewart, 1975: Microplankton of
    the North Pacific Central Gyre. Population structure and abun-
    dance, June 1973. Int. Revue ges. Hydrobiol., 60, 607-638.
Booth, B.C., 1975: Growth of oceanic phytoplankton in enrichment
    cultures. Limnol. Oceanogr., 20, 865-869.
Booth, B.C., J. Lewin and R.E. Norris, 1982: Nanoplankton species
    predominant in the subarctic Pacific in May and June 1978.
    Deep-Sea Res., 29(2A), 185-200.
Burney, C.M., and J.McN. Sieburth, 1977: Dissolved carbohydrates
    in seawater. II, A spectrophotometric procedure for total car-
    bohydrate analysis and polysaccharide estimation. Mar. Chem.,
    5, 15-28.
Burney, C.M., P.G. Davis, K.M. Johnson and J.McN. Sieburth, 1982:
    The diel relationships of microbial trophic groups and in-situ
    dissolved carbohydrate dynamics in the Caribbean Sea. Mar.
    Biol., 67, 311-322.
Carlson, D.J., and L.M. Mayer, 1980: Enrichment of dissolved
    phenolic material in the surface microlayer of coastal waters.
    Nature, 286, 482-483.
Caron, D.A., P.G. Davis, L.P. Madin and J.McN. Sieburth, in press:
    Heterotrophic bacteria and bacterivorous protozoa in oceanic
    macroaggregates. Science.
Cheng, L., 1975: Marine pleuston--animals at the sea-air interface.
    Oceanogr. Mar. Biol. Ann. Rev., 12, 181-212.
Conover, J.T., and J.McN. Sieburth, 1964: Effect of Sargassum dis-
    tribution on its epibiota and antibacterial activity. Bot.
    Mar., 6, 147-157.
Craigie, J.S., and J. McLachlan, 1964: Excretion of coloured
    ultra-violet substances by marine algae. Can. J. Bot., 42,
    23-33.

Daniels, L., G. Fuchs, R.K. Thauer and J.G. Zeikus, 1977: Carbon
    monoxide oxidation by methanogenic bacteria. J. Bacteriol.,
    132, 118-126.
Danos, S.C., 1980: The seasonal variation and enrichment of nutri-
    ents and chlorophyll in the surface microlayer of two small
    Wisconsin ponds. M.S. Thesis, Univ. of Wisconsin, Milwaukee,
    201 pp.
David, P.M., 1965a: The neuston net. A device for sampling the
    surface fauna of the ocean. J. Mar. Biol. Assoc. U.K., 45,
    313-320.
David, P.M., 1965b: The surface fauna of the ocean. Endeavour,
    24, 95-100.
de Haan, H., 1977: Effect of benzoate on microbial decomposition
    of fulvic acids in Tjeukemeer (the Netherlands). Limnol.
    Oceanogr., 22, 38-44.
Dietz, A.S., L.J. Albright and T. Tuominen, 1976: Heterotrophic
    activities of bacterioneuston and bacterioplankton. Can. J.
    Microbiol., 22, 1699-1709.
Dietz, R.S., and E.C. LaFond, 1950: Natural slicks on the ocean.
    J. Mar. Res., 9, 69-76.
Dobson, F., L. Hasse and R. Davis (eds.), 1980: Air-Sea Interaction.
    Instruments and Methods. Plenum Publ., NY, 814 pp.
Dragcevic, Dj., and V. Pravdic, 1981: Properties of the seawater-
    air interface. 2. Rates of surface film formation under steady
    state conditions. Limnol. Oceanogr., 26, 492-499.
Dubinsky, Z., 1980: Light utilization efficiency in natural phyto-
    plankton communities. In:Primary Productivity in the Sea
    (P.G. Falkowski, ed.), Brookhaven Symposia in Biology No. 31,
    Plenum Press, NY, pp. 83-119.
Duce, R.A., J.G. Quinn, C.E. Olney, S.R. Piotrowicz, B.J. Ray and
    T.L. Wade, 1972: Enrichment of heavy metals and organic com-
    pounds in the surface microlayer of Narragansett Bay, Rhode
    Island. Science, 176, 161-163.
Durbin, E.G., R.W. Krawiec and T.J. Smayda, 1975: Seasonal studies
    on the relative importance of different size fractions of
    phytoplankton in Narragansett Bay (USA). Mar. Biol., 32,
    271-287.
Eppley, R.W., O. Holm-Hansen and J.D.H. Strickland, 1968: Some
    observations on the vertical migration of dinoflagellates.
    J. Phycol., 4, 333-340.
Ewing, G., 1950: Slicks, surface films and internal waves. J.
    Mar. Res., 9, 161-187.
Fox, G.E., E. Stackebrandt, R.B. Hespell, J. Gibson, J. Maniloff,
    T.A. Dyer, R.S. Wolfe, W.E. Balch, R.S. Tanner, L.J. Magrum,
    L.B. Zablen, R. Blakemore, R. Gupta, L. Bonen, B.J. Lewis,
    D.A. Stahl, K.R. Luehrsen, K.N. Chen and C.R. Woese, 1980:
    The phylogeny of prokaryotes. Science, 209, 457-462.
Freedman, M.L., J.J. Hains, Jr., and J.E. Schindler, 1982: Diel
    changes of neuston biomass as measured by ATP and cell counts,
    Lake Louise, Georgia, U.S.A. J. Freshw. Ecol., 1, 373-381.

Gallagher, J.L., 1975: The significance of the surface film in
    salt marsh plankton metabolism. Limnol. Oceanogr., 20, 120-123.
Garrett, W.D., 1967: The organic chemical composition of the ocean
    surface. Deep-Sea Res., 14, 221-227.
Garrett, W.D., 1972: Impact of natural and man-made surface films
    on the properties of the air-sea interface. In: The Changing
    Chemistry of the Oceans, Nobel Symp. 20 (D. Dryssen and D.
    Jagner, eds.), Wiley, NY, pp. 75-91.
Geesey, G.G., 1982: Microbial exopolymers: ecological and economic
    considerations. Amer. Soc. Microbiol. News, 48(1), 9-14.
Goodman, N.G., 1931: The Ingenious Dr. Franklin. Univ. Pennsyl-
    vania Press, Philadelphia.
Harris, G.P., 1980: The measurement of photosynthesis in natural
    populations of phytoplankton. In: The Physiological Ecology
    of Phytoplankton (I. Morris, ed.), Studies in Ecology Vol. 7,
    University of California Press, Berkeley and Los Angeles, pp.
    129-187.
Harvey, G.W., and L.A. Burzell, 1972: A simple microlayer method
    for small samples. Limnol. Oceanogr., 11, 603-618.
Hasle, G.R., 1950: Phototactic vertical migrations in marine
    dinoflagellates. Oikos, 2, 162-175.
Higgins, I.J., D.J. Best, R.C. Hammond and D. Scott, 1981:
    Methane-oxidizing microorganisms. Microbiol. Rev., 45, 556-590.
Hobbie, J.E., R.J. Daley and S. Jasper, 1977: Use of Nuclepore
    filters for counting bacteria by fluorescence microscopy.
    Appl. Environ. Microbiol., 33, 1225-1228.
Horrigan, S.G., A.F. Carlucci and P.M. Williams, 1981: Light inhi-
    bition of nitrification in sea-surface films. J. Mar. Res.,
    39, 557-565.
Hunter, K.A., and P.S. Liss, 1981: Organic sea surface films.
    In: Marine Organic Chemistry (E.K. Duursma and R. Dawson, eds.),
    Elsevier Sci. Publ. Co., Amsterdam, 259-298.
Jannasch, H.W., 1973: Bacterial content of particulate matter in
    offshore surface waters. Limnol. Oceanogr., 18, 340-342.
Jarvis, N.L., 1967: Adsorption of surface-active material at the
    sea-air interface. Limnol. Oceanogr., 12, 213-222.
Jarvis, N.L., W.D. Garrett, M.A. Scheiman and C.O. Timmons, 1967:
    Surface chemical characterization of surface-active material
    in seawater. Limnol. Oceanogr., 12, 88-96.
Jeffrey, S.W., 1980: Algal pigment systems. In: Primary Product-
    ivity in the Sea (P.G. Falkowski, ed.), Brookhaven Symposia
    in Biology No. 31, Plenum Press, NY, pp. 33-58.
Johnson, K.M., and J.McN. Sieburth, 1977: Dissolved carbohydrates
    in sea water. I, A precise spectrophotometric analysis for
    monosaccharides. Mar. Chem., 5, 1-13.
Johnson, K.M., C.M. Burney and J.McN. Sieburth, 1981a: Enigmatic
    marine ecosystem metabolism measured by direct diel ΣCO_2 and
    O_2 flux in conjunction with DOC release and uptake. Mar. Biol.,
    65, 49-60.

Johnson, K.M., C.M. Burney and J.McN. Sieburth, 1981b: Doubling the production and precision of the MBTH spectrophotometric assay for dissolved carbohydrates in seawater. Mar. Chem., 10, 467-473.

Johnson, K.M., P.G. Davis and J.McN. Sieburth: The larger diel variation of CO_2 than O_2 in oceanic waters, and its implications for a significant chemosynthetic production. In prep. for Mar. Biol.

Johnson, P.W., and J.McN. Sieburth, 1979: Chroococcoid cyanobacteria in the sea: A ubiquitous and diverse phototrophic biomass. Limnol. Oceanogr., 24, 928-935.

Johnson, P.W., and J.McN. Sieburth, 1982: In-situ morphology and occurrence of eucaryotic phototrophs of bacterial size in the picoplankton of estuarine and oceanic waters. J. Phycol., 18, 318-327.

Kennett, J.P., 1982: Marine Geology. Prentice-Hall, NY, 813 pp.

Kjelleberg, S., and N. Hakansson, 1977: Distribution of lipolytic, proteolic, and amylolytic marine bacteria between the lipid film and the subsurface water. Mar. Biol., 39, 103-109.

Kranck, K., and T. Milligan, 1980: Macroflocs: Production of marine snow in the laboratory. Mar. Ecol. Prog. Ser., 3, 19-24.

Lamontagne, R.A., J.W. Swinnerton and V.J. Linnenbom, 1971: Nonequilibrium of CO and CH_4 at the air-sea interface. J. Geophys. Res., 76, 5117-5121.

Lamontagne, R.A., J.W. Swinnerton, V.J. Linnenbom and W.D. Smith, 1973: Methane concentrations in various marine environments. J. Geophys. Res. 78, 5317-5323.

Langmuir, I., 1917: The constitution and fundamental properties of solids and liquids. Part 2, Liquids. J. Amer. Chem. Soc., 39, 1848-1906.

Langmuir, I., 1938: Surface motion of water induced by wind. Science, 87, 119-123.

Larsson, K., G. Odham and A. Sodergren, 1974: On lipid surface films on the sea. I, A simple method for sampling and studies of composition. Mar. Chem., 2, 49-57.

Laval, M., 1971: Ultrastructure et mode de nutrition du choanoflagelle Salpingoeca pelagica sp. nov. Comparaison avec les choanocytes des spongiaires. Protistologica, 7, 325-336.

Leibovich, S., and S. Paolucci, 1980: The Langmuir circulation instability as a mixing mechanism in the upper ocean. J. Phys. Oceanogr., 10, 186-207.

Lindeman, R.L., 1942: The trophic-dynamic aspect of ecology. Ecology, 23, 399-417.

Liss, P.S., 1975: Chemistry of the sea surface microlayer. In: Chemical Oceanography (J.P. Riley and G. Skirrow, eds.), Academic Press, NY, pp. 193-243.

Loeb, G.I., and R.A. Neihof, 1977: Adsorption of an organic film at the platinum-seawater interface. J. Mar. Res., 35, 283-291.

Lohmann, H., 1911: Uber das Nannoplankton und die Zentrifugierung
    kleinster Wasserproben zur Gewinnung desselben in lebendem
    Zustande. Int. Revue Ges. Hydrobiol. Hydrogr., 4, 1-38.
Maaloe, O., and N.O. Kjeldgaard, 1966: Control of Macromolecular
    Synthesis. W.A. Benjamin, Inc., NY and Amsterdam, 284 pp.
MacIntyre, F., 1974a: The top millimeter of the ocean. Sci. Amer.,
    230(5), 62-77.
MacIntyre, F., 1974b: Non-lipid-related possibilities for chemical
    fractionation in bubble film caps. J. Rech. Atmos., 8, 515-527.
Malone, T.C., 1971: The relative importance of nannoplankton and
    netplankton as primary producers in tropical oceanic and
    neritic phytoplankton communities. Limnol. Oceanogr., 16,
    633-639.
Malone, T.C., 1980a: Size-fractionated primary productivity of
    marine phytoplankton. In: Primary Productivity in the Sea
    (P.G. Falkowsky, ed.), Brookhaven Symposia in Biology No. 31,
    Plenum Press, NY, pp. 301-319.
Malone, T.C., 1980b: Algal size. In: The Physiological Ecology of
    Phytoplankton (I. Morris, ed.), Studies in Ecology Vol. 7,
    Univ. California Press, Berkeley and Los Angeles, pp. 433-463.
Manzi, J.J., P.E. Stofan and J.L. Dupuy, 1977: Spatial heterogeneity
    of phytoplankton populations in estuarine surface microlayers.
    Mar. Biol., 41, 29-38.
Margulis, L., 1970: Origin of Eukaryotic Cells. Yale University
    Press, New Haven, CT, 349 pp.
Margulis, L., 1981: Symbiosis in Cell Evolution: Life and its En-
    vironment on the Early Earth. W.H. Freeman, San Francisco,
    419 pp.
Marshall, K.C., 1976: Interfaces in Microbial Ecology. Harvard
    Univ. Press, Cambridge, MA, 156 pp.
Marumo, R., N. Taga and T. Nakai, 1971: Neustonic bacteria and
    phytoplankton in surface microlayers of the equatorial waters.
    Bull. Plankton Soc. Japan, 18, 36-41.
McDowell, R.S., and C.W. McCutchen, 1971: The Thoreau-Reynolds
    Ridge, a lost and found phenomenon. Science, 172, 973.
Melkonian, M., 1978: Die Bildung extrazellularer Produkte durch
    die Grunalge Fritschiella tuberosa IYENG (Chaetophorineae) in
    Abhangigkeit vom Wachstum der Alge. Doctoral Dissertation,
    University of Hamburg, 131 pp.
Miller, J.D.A., 1962: Fats and steroids. In: Physiology and Bio-
    chemistry of Algae (R.A. Lewin, ed.), Academic Press, NY and
    London, pp. 357-370.
Mopper, K., and P. Lindroth, 1982: Diel and depth variations in
    dissolved free amino acids and ammonium in the Baltic Sea
    determined by shipboard HPLC analysis. Limnol. Oceanogr. 27,
    336-347.
Naumann, E., 1917: Beitrage zur Kenntnis des Teichnannoplanktons.
    II. Uber das Neuston des Susswassers. Biol. Zentralbl., 37,
    98-106.
Neihof, R.A., and G.I. Loeb, 1972: The surface charge of particu-

late matter in seawater. Limnol. Oceanogr., 17, 7-16.

Neihof, R.A., and G.I. Loeb, 1974: Dissolved organic matter in seawater and the electric charge of immersed surfaces. J. Mar. Res., 32, 5-12.

Nissenbaum, A., and I.R. Kaplan, 1972: Chemical and isotopic evidence for the in situ origin of marine humic substances. Limnol. Oceanogr., 17, 570-582.

Norkrans, B., 1980: Surface microlayers in aquatic environments. Adv. Microbial Ecol., 4, 51-85.

Norris, R.E., 1965: Neustonic marine Craspedomonadales (Choano-flagellates) from Washington and California. J. Protozool., 12, 589-602.

Odham, G., B. Noren, B. Norkrans, A. Sodergren and H. Lofgren, 1978: Biological and chemical aspects of the aquatic lipid surface microlayer. Prog. Chem. Fats Other Lipids, 16, 31-44.

Olson, R.J., 1981a: ^{15}N tracer studies of the primary nitrite maximum. J. Mar. Res., 39, 203-226.

Olson, R.J., 1981b: Differential photoinhibition of marine nitri-fying bacteria: a possible mechanism for the formation of the primary nitrite maximum. J. Mar. Res., 39, 227-238.

Patt, T.E., G.C. Cole, J. Bland and R.S. Hanson, 1974: Isolation and characterization of bacteria that grow on methane and organic compounds as sole sources of carbon and energy. J. Bacteriol., 120, 955-964.

Patt, T.E., G.C. Cole and R.S. Hanson, 1976: Methylobacterium, a new genus of facultatively methylotrophic bacteria. Int. J. Syst. Bacteriol., 26, 226-229.

Peele, E.R., F.L. Singleton, J.W. Deming, B. Cavari and R.R. Col-well, 1981: Effects of pharmaceutical wastes on microbial populations in surface waters at the Puerto Rico dump site in the Atlantic Ocean. Appl. Environ. Microbiol., 41, 873-879.

Pockels, A., 1891: Surface Tension. Nature, 43, 437-439.

Pollard, R.T., 1970: On the generation by winds of inertial waves in the ocean. Deep-Sea Res., 17, 795-812.

Pollard, R.T., P.B. Rhines and R.O.R.Y. Thompson, 1973: The deepening of the wind mixed layer. Geophys. Fluid Dyn., 3, 381-404.

Pomeroy, L.R., 1974: The ocean's food web, a changing paradigm. BioScience, 24, 499-504.

Pomeroy, L.R., 1979: Microbial roles in aquatic food webs. In: Aquatic Microbial Ecology (R.R. Colwell and J. Foster, eds.), Sea Grant Publ., University of Maryland, pp. 85-109.

Pope, D.H., 1975: Effects of light intensity, oxygen concentration, and carbon dioxide concentration on photosynthesis in algae. Microbial Ecol., 2, 1-16.

Porter, K.G., and Y.S. Feig, 1980: The use of DAPI for identifying and counting aquatic microflora. Limnol. Oceanogr., 25, 943-948.

Purcell, E.M., 1977: Life at low Reynolds number. Amer. J. Phys., 45, 3-11.

Quayle, J.R., 1972: The metabolism of one-carbon compounds by
    micro-organisms. Adv. Microbial Physiol., 7, 119-203.
Renn, C.E., 1964: The bacteriology of interfaces. In: Principles
    and Applications in Aquatic Microbiology (H. Heukelekian and
    N.C. Dondero, eds.), John Wiley, NY, pp. 193-201.
Reynolds, O., 1880: On the effect of oil in destroying waves on
    the surface of water. Rep. Brit. Ass., 489-490.
Ribbons, D.W., J.E. Harrison and M. Wadzinski, 1970: Metabolism
    of single carbon compounds. Ann. Rev. Microbiol., 24, 135-158.
Riley, G.A., 1963: Organic aggregates in seawater and the dynamics
    of their formation and utilization. Limnol. Oceanogr., 8,
    372-381.
Rudd, J.W.M., and C.D. Taylor, 1980: Methane cycling in aquatic
    environments. Adv. in Aquatic Microbiol., 2, 77-150.
Scranton, M.I., and P.G. Brewer, 1977: Occurrence of methane in the
    near-surface waters of the western subtropical North Atlantic.
    Deep-Sea Res., 24, 127-138.
Scranton, M.I., and J.W. Farrington, 1977: Methane production in
    the waters off Walvis Bay. J. Geophys. Res., 82, 4947-4953.
Seiler, W., and U. Schmidt, 1974: Dissolved nonconservative gases
    in seawater. In: The Sea, Vol. 5 (E.E. Goldberg, ed.), John
    Wiley & Son, NY, pp. 219-243.
Sewell, L.M., G. Bitton and J.S. Bays, 1981: Evaluation of membrane
    adsorption-epifluorescence microscopy for the enumeration of
    bacteria in coastal surface films. Microb. Ecol., 7, 365-369.
Sharp, J.H., 1973: Size classes of organic carbon in seawater.
    Limnol. Oceanogr., 18, 441-447.
Sieburth, J.McN., 1965: Bacteriological samplers for air-water and
    water-sediment interfaces. Ocean Science and Ocean Engineer-
    ing, Trans. Joint Conf. Mar. Tech. Soc. and ASLO, pp. 1064-1068.
Sieburth, J.McN., 1968: The influence of algal antibiosis on the
    ecology of marine microorganisms. In: Advances in Microbiol-
    ogy of the Sea (M.R. Droop and E.J.F. Wood, eds.), Academic
    Press, London, pp. 63-94.
Sieburth, J.McN., 1969: Studies on algal substances in the sea.
    III. The production of extracellular organic matter by littoral
    marine algae. J. Exp. Mar. Biol. Ecol., 3, 290-309.
Sieburth, J.McN., 1971a: Distribution and activity of oceanic
    bacteria. Deep-Sea Res., 18, 1111-1121.
Sieburth, J.McN., 1971b: An instance of bacterial inhibition in
    oceanic surface water. Mar. Biol., 11, 98-100.
Sieburth, J.McN., 1979: Sea Microbes. Oxford University Press,
    NY, 491 pp.
Sieburth, J.McN., in press: The grazing of bacteria by the proto-
    zooplankton in pelagic marine waters. In: Heterotrophic
    Activity in the Sea (J.E. Hobbie and P.J.LeB. Williams, eds.),
    Plenum Press, NY.
Sieburth, J.McN., and J.T. Conover, 1965a: Slicks associated with
    Trichodesmium blooms in the Sargasso Sea. Nature, 205, 830-831.

Sieburth, J.McN., and J.T. Conover, 1965b: Sargassum tannin, an antibiotic which retards fouling. Nature, 208, 52-53.

Sieburth, J.McN., and A. Jensen, 1968: Studies on algal substances in the sea. I. Gelbstoff (humic material) in terrestrial and marine waters. J. Exp. Mar. Biol. Ecol., 2, 174-189.

Sieburth, J.McN., and A. Jensen, 1969: Studies on algal substances in the sea. II. The formation of Gelbstoff (humic material) by exudates of phaeophyta. J. Exp. Mar. Biol. Ecol., 3, 275-289.

Sieburth, J.McN., P-J. Willis, K.M. Johnson, C.M. Burney, D.M. Lavoie, K.R. Hinga, D.A. Caron, F.W. French III, P.W. Johnson and P.G. Davis, 1976: Dissolved organic matter and heterotrophic microneuston in the surface microlayers of the North Atlantic. Science, 194, 1415-1418.

Sieburth, J.McN., V. Smetacek and J. Lenz, 1978: Pelagic ecosystem structure: heterotrophic compartments of the plankton and their relationship to plankton size fractions. Limnol. Oceanogr. 23:1256-1263.

Silver, M.W., A.L. Shanks and J.D. Trent, 1978: Marine snow: microplankton habitat and source of small-scale patchiness in pelagic populations. Science, 201, 371-373.

Silver, M.W., and A.L. Alldredge, 1981: Bathypelagic marine snow: deep-sea algal and detrital community. J. Mar. Res., 39, 501-530.

Smith, K.C. (ed.), 1977: The Science of Photobiology. Plenum/ Rosetta, NY, 430 pp.

Smith, R.C., and K.S. Baker, 1979: Penetration of UV-B and biologically effective dose-rates in natural waters. Photochem. and Photobiol., 29, 311-323.

Sturdy, G., and W.H. Fischer, 1966: Surface tension of slick patches near kelp beds. Nature, 211, 951-952.

Swinnerton, J.W., and V.J. Linnenbom, 1967: Gaseous hydrocarbons in seawater: determination. Science, 156, 1119-1120.

Swinnerton, J.W., V.J. Linnenbom and C.H. Cheek, 1969: Distribution of methane and carbon monoxide between the atmosphere and natural waters. Environ. Sci. Technol., 3, 836-838.

Swinnerton, J.W., V.J. Linnenbom and R.A. Lamontagne, 1970: The ocean: a natural source of carbon monoxide. Science, 167, 984-986.

Thomsen, H.A., 1973: Studies on marine choanoflagellates. I. Silicified choanoflagellates of the Isefjord (Denmark). Ophelia, 12, 1-26.

Thomsen, H.A., 1976: Studies on marine choanoflagellates. II. Fine-structural observations on some silicified choanoflagellates from the Isefjord (Denmark), including the description of two new species. Norwegian J. Bot., 23, 33-51.

Thompson, J., 1862: On the calm lines often seen on a rippled sea. Philos. Mag., 4, 247-248.

Vogel, S., 1981: Life in Moving Fluids. Willard Grant Press, Boston. 352 pp.

Wada, E., and A. Hattori, 1971: Nitrite metabolism in the euphotic layer of the central North Pacific Ocean. Limnol. Oceanogr., 16, 766-772.

Wangersky, P.J., 1978: The distribution of particulate organic carbon in the oceans: ecological implications. Int. Revue ges. Hydrobiol., 63, 567-574.

Ward, B.B., R.J. Olson and M.J. Perry, 1982: Microbial nitrification rates in the primary nitrite maximum off southern California. Deep-Sea Res., 29, 247-255.

Webb, R.B., and M.S. Brown, 1979: Action spectra for oxygen-dependent and independent inactivation of Escherichia coli WP2s from 254 to 460 nm. Photochem. and Photobiol., 29, 407-409.

Weyl, P.K., 1968: Precambrian marine environment and the development of life. Science, 161, 158-160.

Whittaker, R.H., 1969: New concepts of Kingdoms of organisms. Science, 163, 150-160.

Wiegert, R.G., and D.F. Owen, 1971: Trophic structure, available resources, and population density in terrestrial vs. aquatic ecosystems. J. Theor. Biol., 30, 69-81.

Williams, P.M., 1967: Sea surface chemistry: organic carbon and organic and inorganic nitrogen and phosphorus in surface films and subsurface waters. Deep-Sea Res., 14, 791-800.

Williams, P.M., 1971: The distribution and cycling of organic matter in the ocean. In: Organic Compounds in Aquatic Environments (S.J. Faust and J.V. Hunter, eds.). Marcel Dekker, NY, pp. 145-163.

Willis, R.P., 1963: A small towed net for ocean surface sampling. New Zealand J. Sci., 6, 120-126.

Wilson, D.F., J.W. Swinnerton and R.A. Lamontagne, 1970: Production of carbon monoxide and gaseous hydrocarbons in seawater: relation to dissolved organic carbon. Science, 168, 1577-1579.

Young, L.Y., 1978: Bacterioneuston examined with critical point drying and transmission electron microscopy. Microbial Ecol., 4, 267-277.

Zaitsev, Y.P., 1970: Marine Neustonology (K.A. Vinogradov, ed., Transl. by Israel Prog. Sci. Transl., Jerusalem, 1971), 207 pp.

Zika, R.G., 1981: Marine organic photochemistry. In: Marine Organic Chemistry (E.K. Duursma and R. Dawson, eds.), Elsevier Oceanogr. Ser., 31, pp. 299-325.

GASES AND THEIR PRECIPITATION SCAVENGING IN THE MARINE ATMOSPHERE

Leonard K. Peters

Department of Chemical Engineering
University of Kentucky
Lexington, Kentucky 40506-0046 U.S.A.

1. INTRODUCTION

    The marine atmosphere contains numerous trace gaseous
species that can potentially interact with water droplets in
clouds or during precipitation events. These gases, when dissolved in water droplets, modify the chemistry of the droplet
phase and, then during precipitation, can be transferred to the
oceans. This process is in addition to the direct transfer of
the trace gases from the air to the ocean, described in this
volume by Broecker and Liss.

    There is a complex sequence of events that takes place
during the transfer of gases to the ocean phase by precipitation.
During cloud formation, gases can be swept up from the boundary
layer into the cloud, and the trace gases, depending on their
solubility, dissolve to a lesser or greater degree in the cloud
drops that are being formed. Since the cloud droplets form
around condensation nuclei that are generally soluble in liquid
water (e.g., sea-salt particles), the gas transfer is into an
aqueous solution which can initially be quite concentrated until
many water molecules have deposited on individual nuclei. As a
result, the scavenging of gases in cloud droplets can be greatly
affected by the chemical nature of the condensation nuclei.

    Once the cloud droplets have grown sufficiently large and/or
coagulated with other cloud droplets, their fall velocity becomes
significant compared to the cloud's updraft velocity. During the
fall to the surface, the rain drops can continue to absorb gases.
If the droplets do not evaporate prior to reaching the surface,
then these gases will be deposited during the precipitation event.

P. S. Liss and W. G. N. Slinn (eds.), Air-Sea Exchange of Gases and Particles, 173–240.
Copyright © 1983 by D. Reidel Publishing Company.

On the other hand, raindrops may evaporate before reaching the
surface. If this occurs, at least two important effects have resulted. First, the gases and nuclei have been transported very
rapidly from the cloud to lower portions of the atmosphere.
Secondly, the nuclei and gases may have changed their chemical
nature. For example, gaseous SO_2 absorbed during the formation
of clouds may be released nearer the Earth's surface as a sulfate
particle.

     Figure 1 is a schematic representation of the scavenging
processes already mentioned. The rates at which these processes
occur depend on numerous physical and chemical characteristics of
both the trace gases and the droplets. Since the species present
exhibit a wide range of physicochemical behavior, each species
behaves differently and must be evaluated keeping its individual
properties in mind. In addition, there are numerous pathways
that can lead to physical and chemical interactions which should
be considered.

     In this chapter, we will present the basic principles of
some of these processes. The first part of the chapter is
devoted to an overview of the homogeneous chemistry of the marine
atmosphere. This will detail which species are relevant and the
important chemical reactions based on current knowledge. These
discussions are followed by. consideration of the exchange of
gases with droplets. Simple soluble gases are considered first
followed by increasingly complex systems, where the gases can

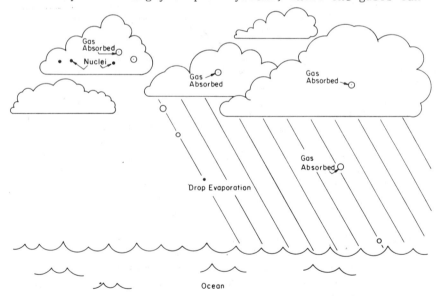

Figure 1. Schematic representation of the precipitation
scavenging processes affecting trace gases.

ionize or react in the liquid phase. Finally, complex interacting gas absorption systems will be discussed in light of our
current knowledge. In this chapter, scavenging of gases by snow
will not be considered because its importance is not yet clear.

2. TRACE GAS CONCENTRATIONS IN MARINE AIR

The numerous trace gas concentrations present in the atmosphere occur from both natural and anthropogenic sources. Many
of the species in the marine atmosphere tend to be dominated by
natural sources, although it is virtually impossible to find a
region in the Earth's troposphere that is not impacted to some
extent by the anthropogenic sources of these species. Nevertheless, it is widely held that the marine atmosphere is more representative of background or normal clean air, and the concentrations of the trace gas species are usually lower over the ocean.

Trace gases can interact chemically with one another, and
since there are many such species, it is difficult (and almost
impossible) to discuss one species without considering one or
more others. These gaseous compounds show varying degrees of
chemical reactivity and, as a result, may exist for very short to
very long times in the atmosphere. Figure 2 shows many of the
important trace gas species relevant to the present discussion of
gas-phase homogeneous atmospheric chemistry. This figure has
divided the complex interactive chemistry into several subsets.
These subdivisions are somewhat artificial, but it is hoped, will
help the presentation.

Those compounds in the four blocks are stable molecular
species and can exist under some conditions for rather long
periods of time. Nevertheless, it must be emphasized that even
with these stable compounds, there is a wide range of atmospheric
lifetimes. For example, NO and NO_2 may only exist for a few
hours to a few days, whereas the residence time of CH_4 is of the
order of 4-7 years and that for CO_2 is probably around a decade
or longer.

The species in the central circle are on the other end of
the residence time spectrum. These are the very reactive free
radicals that are transient and have residence times that can be
measured in seconds, milliseconds, and even shorter times. It is
generally held that these free radicals do not exist to any
appreciable extent at night, since sunlight is required to initi-
ate many of the reactions producing the radicals. There are few
radical generation processes without sunlight, and the radicals
are rapidly consumed by reaction with the stable molecular
species. The subsets contained in the blocks interact through
these free radicals contained in the circle.

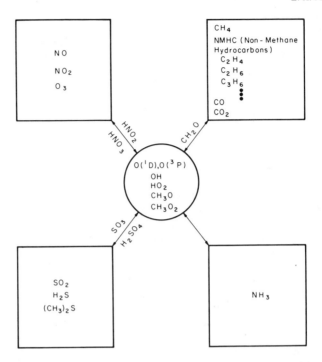

Figure 2.   Separation of important trace gas species
into subsets illustrating the interactions through the
free radicals.

The species listed on the lines, which connect the species
subset blocks with the free radical circle, generally have atmospheric residence times intermediate to the free radicals and the
more stable molecular species. They are generally produced *in
situ* and do not have significant, identifiable natural or anthropogenic sources at ground level. In addition, they have been
formed by interactions between the molecular species and free
radicals, and represent species that have consumed one or more
free radicals. It is also possible that they may generate more
and different free radicals in the chemical processes in which
they are involved.

2.1   Concentrations of Long-Lived Species

In this section, we will briefly summarize the concentrations of some of the more important long-lived species. Obviously, trace species in the atmosphere exhibit a continuum of
residence times, and any such breakdown is quite arbitrary.
Nevertheless, in the present context any species that has a residence time longer than a few hours to a day or so is considered

to be long-lived.

Table 1 summarizes the ranges of concentration of the principal trace gases. It is not an exhaustive and thorough list. Furthermore, the references cited are by no means comprehensive for most of the species and are only examples. The concentrations range from the part per trillion (ppt) level to a few hundred ppm, as is the case with CO_2.

Table 1. Summary of trace gas concentrations in the troposphere for long-lived species.

Gas	Concentration	Reference
CO_2	320 - 340 ppm	--
CH_4	1.4 - 1.7 ppm	Peters and Chameides (1980)
Non-methane hydrocarbons		
C_2H_2	0.1 - 0.5 ppb	Rudolph et al. (1982)
C_2H_4	0.2 - 0.3 ppb	Rudolph et al. (1982)
C_2H_6	0.5 - 5 ppb	Rudolph et al. (1982)
C_3H_6	0.1 - 0.2 ppb	Rudolph et al. (1982)
C_3H_8	0.1 - 1 ppb	Rudolph et al. (1982)
C_4H_{10}	0.01 - 0.2 ppb	Rudolph et al. (1982)
CO	0.05 - 0.2 ppm	Seiler and Fishman (1981)
O_3	0.02 - 0.10 ppm	Fishman and Crutzen (1978)
NO_x	0.01 - 1 ppb	Kelly and Stedman (1979); Kley et al. (1981)
NH_3	0.01 - 20 ppb	Tanner (1982); Ayers and Gras (1982)
SO_2	0.02 - 0.3 ppb	Maroulis et al. (1980)
OCS	0.5 ppb	Rasmussen et al. (1982)
$(CH_3)_2S$	< 0.03 - 0.085 ppb	Maroulis and Bandy (1977)
H_2S	0.03 - 0.40 ppb	Jaeschke et al. (1978)
CH_3CCl_3	0.01 - 0.16 ppb	Logan et al. (1981)

2.2 Concentrations of Short-Lived Species

The determination of the concentrations of short-lived species has frequently been based on objectives for specific

field studies. As a result, there have not been extensive field
monitoring programs to develop a climatology of their abundances.
However, three species that are quite soluble in water have been
studied in somewhat more detail because of their importance in
atmospheric chemistry. These are CH_2O, H_2O_2, and HNO_3. Table 2
summarizes some of these observations.

Table 2. Summary of trace gas concentrations in the
troposphere for CH_2O, H_2O_2, and HNO_3.

Gas	Concentration	Reference
CH_2O	1×10^{10} molecules/cm^3	Zafiriou et al. (1980)
H_2O_2	8×10^9 - 8×10^{10} molecules/cm^3	Kelly and Stedman (1979)
HNO_3	$< 5.4 \times 10^8$ - 1.3×10^{11} molecules/cm^3	Huebert (1980); Huebert and Lazrus (1978); and Kelly and Stedman (1979)

2.3   Concentrations of Free Radicals

There have been relatively few attempts to measure the concentrations of free radicals in the atmosphere. Their measurement is quite difficult due to their high reactivity and extremely low concentration.

The only substantial attempts have been to measure the OH
concentration. Perner et al. (1976) reported maximum values in
the range of 10^6 - 10^7 cm^{-3}, while Wang et al. (1975) reported
OH concentrations greater than 10^7 cm^{-3}. Both studies reported
ground level measurements that could be affected by anthropogenic
activity. Davis et al. (1976) made airborne measurements to find
the OH radical concentration in an unpolluted atmosphere and
consistently observed daytime concentrations between 10^6 and
10^7 cm^{-3}. More recent reports (Campbell et al., 1982; Hubler
et al., 1982; and Davis et al., 1982) continue to suggest daytime
OH concentrations in the 10^5 - 10^7 cm^{-3} range. It might be mentioned that background tropospheric chemistry models predict peak
OH values in the 10^6 - 10^7 cm^{-3} range (cf., Peters and Chameides,
1980).

Preliminary measurements of RO_2 concentrations using electron
spin resonance recently have been reported by Mihelcic et al.
(1982). They reported values from about 10^8 cm^{-3} to 10^{10} cm^{-3},
during airplane flights over Germany. Their measurements could

not distinguish among the various RO_2 radicals, and the concentrations represent the sum of all such radicals. Concentrations of RO_2 in the marine atmosphere apparently have not been determined, and may be different from those measured in the above study.

3. HOMOGENEOUS MARINE AIR CHEMISTRY

Extensive analysis and experimental work have been done on the important gas-phase reactions occurring in the atmosphere. However, considerable work remains to complete our understanding of many of the important chemical processes occurring at background atmospheric concentrations. In this section, the subsets of three of the chemical systems illustrated in Figure 2 will be discussed. For the fourth subset, it appears that the chemistry of NH_3 in the atmosphere is dominated by heterogeneous and aqueous phase processes. Consequently, there will be no discussion of the homogeneous gas-phase chemistry of NH_3 in this section.

3.1 Photochemical Considerations

In the troposphere, the temperature varies from about 215 K to about 315 K; a relatively narrow range of temperatures compared, for example, to the temperature ranges for chemical reactions in combustion processes. Over this rather limited range, chemical reaction rates will generally exhibit relatively small change. These changes may be only over one or two orders of magnitude, unlike many other applications of scientific interest. In addition, many important gas-phase chemical processes occurring in the troposphere do not involve catalysts.

As a result of both of the above restrictions, the energetics for breaking chemical bonds and initiating chemical change in the atmosphere are frequently derived from the sun through photochemical reactions. In order for a species to be photochemically active, the species must absorb electromagnetic radiation from the incident solar radiation reaching the Earth's atmosphere. In the troposphere, this radiation is contained in a reasonably narrow range (wavelengths 3,000 – 7,000 Å), and relatively few species absorb in this range. Notable examples are NO_2 and O_3. Since the sun is a significant energy source for initiating many atmospheric chemical processes, numerous reaction sequences only occur during the day and are non-existent at nighttime.

As an example, we will examine in somewhat more detail the photolysis of NO_2 since this reaction is of central importance to tropospheric chemistry. NO_2 absorbs strongly in the ultraviolet and visible region, and a fraction of the molecules that absorb also dissociate for wavelengths shorter than about 4400 Å.

$$NO_2 \xrightarrow{\quad h\nu \ (< 4400 \text{ Å}) \quad} NO + O(^3P) \qquad (1)$$

The rate of dissociation in the atmosphere depends on the intensity of the solar radiation and is the product of the rate of
absorption and the quantum yield (i.e., the fraction of absorbing
molecules which actually dissociate). Since the rate of this
process depends on the solar flux, it varies with latitude, time
of year, time of day, and cloud cover. In Figure 3, the rate
coefficient for NO_2 photolysis is shown as a function of solar
zenith angle. Obviously, the rate is zero (or close to zero) at
nighttime. The maximum rate constant corresponds to local noon
at the equator under clear skies and equinoctial conditions.
Other photolysis reactions (e.g., photolysis of O_3, HNO_2, CH_2O,
H_2O_2) show similar behavior, although the magnitude of the specific rate constant would be different.

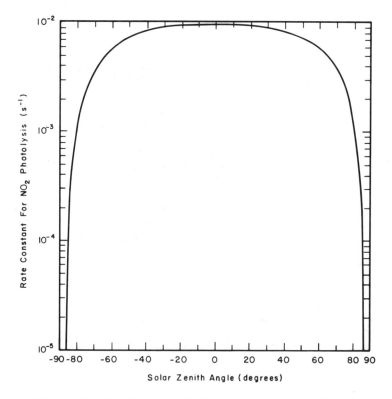

Figure 3. Variation of the rate constant for the
photolysis of NO_2 ($NO_2 \xrightarrow{h\nu} NO + O$) with solar zenith
angle.

3.2 Pseudo-Steady State

Many chemical reactions of atmospheric importance involve
free radical species. These free radicals are generally very
reactive and exist at quite low concentrations. Under such conditions, their rates of formation and destruction by chemical
reactions are nearly in balance. Let R represent any general
free radical species. Then the change of concentration of R with
time by chemical processes can be expressed as the difference of
these formation and destruction processes; i.e.,

$$\frac{d[R]}{dt} = \text{(Formation Rate)} - \text{(Destruction Rate)}. \qquad (2)$$

Note that there may be several formation and/or several destruction pathways, which dominate under differing atmospheric conditions. The destruction rate depends directly on the free radical
concentration and invariably can be expressed as a destruction
rate coefficient multiplied by the free radical concentration;
i.e.,

$$\text{(Destruction Rate)} = \text{(Destruction Rate Coefficient)}[R]. \quad (3)$$

The formation and destruction pathways for the radicals
frequently depend on concentrations of more stable species, and
adjust very rapidly to changes in these. If they rapidly adjust,
the formation and destruction rates are in close balance, so that
$d[R]/dt \cong 0$, and the pseudo-steady state approximation is said to
apply. For the case where the formation rate does not depend on
R (which is frequently but not always true), then the free radical concentration is

$$[R] = \frac{\text{(Formation Rate)}}{\text{(Destruction Rate Coefficient)}}. \qquad (4)$$

Under the pseudo-steady state approximation, (2) is an algebraic
equation (although it may be non-linear) even if the formation
rate depends on [R].

The pseudo-steady state approximation can only be applied
to species that respond to system changes on a time scale much
shorter than the time scales of interest. Furthermore, the
pseudo-steady state approximation does not provide details of the
dynamic response of the free radical concentration on the very
short time scale. These limitations are not very restrictive for
most problems of interest in atmospheric chemistry; and the
pseudo-steady state approximation, if carefully applied, can
alleviate mathematical stiffness problems (a term used to characterize difficulties in numerically solving a set of differential
equations that contain many different time scales) and can reduce

the number of species that must be described by differential
equations. We will see applications of the pseudo-steady state
in our subsequent discussions.

3.3   The NO-NO$_2$-O$_3$ Subsystem

As previously mentioned, there are a number of chemical
species that can usefully be considered together as a subsystem
of the atmosphere. As a first approximation, these can be analyzed without considering other species. The NO-NO$_2$-O$_3$ subsystem
is one of these. It also provides a meaningful and simple
application of the pseudo-steady state approximation.

Consider a volume of the atmosphere that is quiescent and
homogeneous in composition, temperature, and solar flux. During
the daytime, the NO$_2$ in this volume is photolyzed by

$$NO_2 \xrightarrow{k_1} NO + O(^3P), \tag{5}$$

where k_1 depends on the time of day, latitude, etc. (The $O(^3P)$
is frequently written as O.) The oxygen atoms are a very reactive species and can immediately react with a number of compounds.* Molecular oxygen is in great abundance, and the largest
removal rate of O in the background troposphere comes from combination with O$_2$ in the presence of a third body M to form O$_3$.

$$O + O_2 + M \xrightarrow{k_2} O_3 + M \tag{6}$$

Reaction (6), which is the principal ozone formation pathway in
the background troposphere, depends on the temperature and on the
concentration of M. The O$_3$ that is produced principally reacts
with NO by the following temperature dependent reaction.

$$NO + O_3 \xrightarrow{k_3} NO_2 + O_2 \tag{7}$$

Reaction (5) consumed one molecule of NO$_2$, and (7) has regenerat-
ed a molecule of NO$_2$. Similarly, (6) has consumed a molecule of
O$_2$, and (7) has generated one.

There are six species represented in these reactions, and

---

* In the stratosphere where there is short wavelength radiation,
   photo-dissociation of O$_2$ is a significant source of O. In
   fact, the absorption of short wavelength radiation by O$_2$ and
   O$_3$ in the stratosphere acts as a filter for radiation reaching
   the Earth's surface.

two of these (O_2 and M) are in great excess, so that generation and/or loss by (5) through (7) negligibly change their concentrations. However, rate expressions can be written for the remaining four species as follows.

$$\frac{d[NO_2]}{dt} = -k_1[NO_2] + k_3[NO][O_3] \qquad (8)$$

$$\frac{d[NO]}{dt} = k_1[NO_2] - k_3[NO][O_3] \qquad (9)$$

$$\frac{d[O_3]}{dt} = k_2[O][O_2][M] - k_3[NO][O_3] \qquad (10)$$

$$\frac{d[O]}{dt} = k_1[NO_2] - k_2[O][O_2][M] \qquad (11)$$

Note for this limited cycle that $d[NO_2]/dt = -d[NO]/dt$, which simply implies that $[NO_x] = [NO] + [NO_2]$ is conserved. This is true as a first approximation, but expanded chemistry would show that other reactions limit that result to simple situations. Since O is very reactive and in low concentration, it is safe to assume that it is in pseudo-steady state. The result from (11) is

$$[O] = \frac{k_1[NO_2]}{k_2[O_2][M]}. \qquad (12)$$

In many cases, it is also safe to assume that O_3 is at steady state. Then $d[O_3]/dt \approx 0$, and

$$[O_3] = \frac{k_2[O][O_2][M]}{k_3[NO]}, \qquad (13)$$

which, after substituting for [O], becomes

$$[O_3] = \frac{k_1[NO_2]}{k_3[NO]}. \qquad (14)$$

This relation expresses the steady state relationship that is sometimes assumed to exist among these three species. Since k_1 varies diurnally, the concentrations of these species can also change throughout the day. Equation (14) does not specify the concentrations of the three species, but it does provide a constraint and a procedure for inferring, as a reasonable estimate, the concentration of one species through measurement of the other two. Equation (14) is sometimes rewritten by defining $\lambda \equiv k_3[NO][O_3]/k_1[NO_2]$, which should be unity if the limited steady state for O_3 applies. Field measurements generally have shown λ to be different from unity. Possible explanations include: (a) Other reactions are important; (b) The rates of the chemical changes are limited by mixing of NO and O_3; or (c) The deposition rates at ground level are different for O_3, NO, and NO_2.

3.4 The CH_4-CO Oxidation Sequence

Methane and carbon monoxide are two important trace gases which have substantial natural and anthropogenic sources. These species are connected via gas-phase chemistry which was originally outlined in the papers of Weinstock (1969), Levy (1971), and McConnell et al. (1971). Since that time, there has been considerable re-examination of the mechanism and revisions to rate constants for some of the crucial reactions. Here we will only outline the chemistry of the CH_4-CO oxidation, since considerably greater detail can be found elsewhere (cf., Wofsy, 1976; Peters and Chameides, 1980; and Logan et al., 1981).

Figure 4 illustrates the chemical pathways that are currently thought to be important in the oxidation of CH_4 to CO and CO_2. The reactions begin with hydroxyl radical attack of CH_4.

$$CH_4 + OH \rightarrow CH_3 + H_2O \tag{15}$$

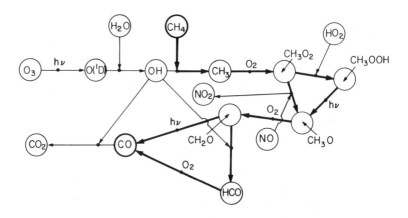

Figure 4. Schematic description of the CH_4-CO oxidation sequence. Bold Lines indicate the reaction flow from CH_4 to CO (adapted from Luther and Peters, 1982).

We will temporarily defer the processes responsible for generation and destruction of OH. The CH_3 radical, which is very reactive, rapidly combines with O_2 in a chain propagation sequence.

$$CH_3 + O_2 + M \rightarrow CH_3O_2 + M \qquad (16)$$

$$CH_3O_2 + NO \rightarrow CH_3O + NO_2 \qquad (17)$$

$$2\ CH_3O_2 \rightarrow 2\ CH_3O + O_2 \qquad (18)$$

$$CH_3O + O_2 \rightarrow CH_2O + HO_2 \qquad (19)$$

$$CH_3O_2 + HO_2 \rightarrow CH_3OOH + O_2 \qquad (20)$$

In this sequence, methyl peroxide (CH_3OOH) and formaldehyde
(CH_2O) are stable but rather short-lived molecular species in the
atmosphere. Also, it is interesting and important to note that
both species are very soluble in water, which would indicate
that, in the presence of clouds and precipitation, there could be
major interactions between these species and the liquid-phase.
These species are also removed via the following gas-phase chemical reactions.

$$CH_2O \xrightarrow{h\nu} HCO + H \qquad (21)$$

$$CH_2O \xrightarrow{h\nu} H_2 + CO \qquad (22)$$

$$CH_2O + OH \rightarrow HCO + H_2O \qquad (23)$$

$$CH_3OOH \xrightarrow{h\nu} CH_3O + OH \qquad (24)$$

$$CH_3OOH + OH \rightarrow CH_3O_2 + H_2O \qquad (25)$$

Reaction (22) finally shows one pathway by which CO is produced
beginning with a molecule of CH_4. But CO can also result from
reaction of O_2 with HCO, which is produced in (21) and (23).

$$HCO + O_2 \rightarrow CO + HO_2 \qquad (26)$$

The reaction chain is terminated by oxidation of CO by OH,

$$CO + OH \rightarrow CO_2 + H, \qquad (27)$$

and ideally one molecule of CH_4 has been converted to one molecule of CO_2. However, there are numerous side paths that can
short-circuit the oxidation of CH_4 to CO to CO_2, but these cannot be discussed in detail here.

3.5 The HO_x and RO_x Radical Chemistry

The hydroxyl and hydroperoxy free radicals, OH and HO_2,
play an important role in the CH_4-CO oxidation sequence and
several other atmospheric chemistry subsets. In this section,
we will discuss the principal formation and destruction pathways
for OH and HO_2, since they are central to understanding the
chemistry of molecular species as shown in Figure 2.

As one might anticipate, photochemistry plays an important
role in the formation of OH radicals. The photolysis of O_3
initiates the production of OH.

$$O_3 \overset{h\nu}{\to} O_2 + O(^1D) \tag{28}$$

$$O(^1D) + H_2O \to 2 \text{ OH} \tag{29}$$

In addition, there are other less evident formation paths. For
example, the H radical formed in (21) and (27) immediately combines with O_2 to form HO_2.

$$H + O_2 \to HO_2 \tag{30}$$

Reaction (19) also forms HO_2. The hydroperoxy radical can lead
to formation of OH by reaction with molecular species.

$$NO + HO_2 \to NO_2 + OH \tag{31}$$

$$O_3 + HO_2 \to 2 \ O_2 + OH \tag{32}$$

Analogous reactions are believed to hold for the alkylperoxy
radicals, RO_2.

$$NO + CH_3O_2 \to NO_2 + CH_3O \tag{33}$$

$$O_3 + CH_3O_2 \to 2 \ O_2 + CH_3O \tag{34}$$

Finally, radical-radical combinations can either terminate chains
or can lead to additional free radicals. More specifically, the
following seem to be important.

$$HO_2 + OH \rightarrow H_2O + O_2 \tag{35}$$

$$HO_2 + HO_2 \rightarrow H_2O_2 + O_2 \tag{36}$$

$$H_2O_2 \xrightarrow{h\nu} 2\ OH \tag{37}$$

$$CH_3O_2 + HO_2 \rightarrow CH_3OOH + O_2 \tag{38}$$

$$CH_3OOH \xrightarrow{h\nu} CH_3O + OH \tag{39}$$

Reaction (35) is an example of the radical-radical chain terminating variety, whereas Reactions (36) and (37) and Reactions
(38) and (39) are schemes whereby radical-radical combinations
lead to an intermediate molecular species which can then be
photolyzed to form new but different radicals. It is apparently
not clear, but (18) may also be described by a two-step mechanism
analogous to (36) and (37).

It is valuable to point out that H_2O_2 is also very soluble
in water so that precipitation and cloud activities could be
important removal mechanisms under such conditions. Additionally,
there are other molecular species which are important in the
$OH-HO_2$ balance and are also very soluble in water; these are the
NO_x compounds. Two important processes are

$$NO + OH \rightarrow HNO_2, \tag{40}$$

and

$$NO_2 + OH \rightarrow HNO_3. \tag{41}$$

The products of these reactions, nitrous and nitric acids, are
extremely soluble in water. However, both can also be photolyzed
as follows.

$$HNO_2 \xrightarrow{h\nu} NO + OH \tag{42}$$

$$HNO_3 \xrightarrow{h\nu} NO_2 + OH \tag{43}$$

The relative rates of the photolysis and precipitation scavenging
processes are important in establishing which removal pathway
dominates.

This has been a brief description of the reaction processes affecting the HO_x-RO_x free radical chemistry. It has been intended only to highlight some of the more important reactions.

3.6 The Chemistry of SO_2

Current research seems conclusive in establishing that sulfate in the atmosphere primarily comes from the oxidation of SO_2. Both homogeneous and heterogeneous chemical reactions contribute to the conversion of SO_2 to sulfate, although our present knowledge of the heterogeneous processes (e.g., on particles and in condensed cloud water) is quite limited.

The conversion of SO_2 to sulfate in the atmosphere is quite variable and shows substantial differences between day and night. Through the day, photochemical processes ultimately leading to reactive free radicals are judged to dominate. The hydroxyl and hydroperoxy radicals are the most important in the oxidation of SO_2.

$$SO_2 + OH \rightarrow HSO_3 \tag{44}$$

$$SO_2 + HO_2 \rightarrow SO_3 + OH \tag{45}$$

Reaction (44) apparently dominates, and the fate of the HSO_3 radical is not fully understood, although it is generally assumed that sulfate is the ultimate product. Reaction with O_2 to form HSO_5 has been suggested as the next step in the oxidation process.

Reaction (45) is quite efficient since it has not only oxidized an SO_2 molecule but has also generated another free radical, which can oxidize another molecule of SO_2 or other species. The SO_3 that is produced is short-lived since it can rapidly combine with water vapor,

$$SO_3 + H_2O \rightarrow H_2SO_4, \tag{46}$$

and nucleate to form sulfate aerosol.

The alkoxyl and alkylperoxy radicals are thought to play an analogous role to OH and HO_2 via the following reactions.

$$SO_2 + RO \rightarrow RSO_3 \tag{47}$$

$$SO_2 + RO_2 \rightarrow SO_3 + RO \tag{48}$$

Their roles are as yet uncertain, but there is strong evidence

that these reactions can be very important in atmospheres with
high hydrocarbon concentrations.

During the daytime, SO_2 conversion rates in clear air conditions are probably around $0.5 - 3\%$ h^{-1}. In-cloud processes can
substantially increase the conversion rate. Hegg and Hobbs
(1981) have reported rates as high as $4 - 300\%$ h^{-1} (i.e., $0.067 -$
5% min^{-1}) under such conditions. We will look at these phenomena
in greater detail in Section 7. During the night, heterogeneous
processes are believed to dominate. These may be non-catalytic
or catalytic, and may occur on solid particles or liquid (or
liquid covered) particles. We will not review this complex area
of SO_2 oxidation.

While heterogeneous processes have been considered most
frequently for nighttime oxidation, homogeneous gas-phase processes can possibly dominate under certain conditions. High O_3
and olefinic hydrocarbon concentrations near sunset represent one
such case. The high O_3 can rapidly consume NO, with any excess
O_3 attacking the olefins producing HO_x and RO_x radicals. These
can then oxidize SO_2 by Reactions (44), (45), (47), or (48),
which can continue until the O_3 or olefins (most likely) are
totally consumed. It should be pointed out that this has not
been determined from field experiments, but model calculations
seem to confirm such a phenomena (Balko and Peters, 1982).

4.   EXCHANGE OF SIMPLE SOLUBLE GASES WITH HYDROMETEORS

In the previous section, we observed that there exist a number of gases in the troposphere that are quite soluble in aqueous
solutions. The rate of exchange of these gases with cloud drops
and rain drops is affected by several factors. Some of these are:
diameter of the drop, solubility of the gas in the aqueous phase,
velocity around the drop, circulation within the drop, occurrence
of chemical reactions in the liquid-phase, and species concentrations within the drop such as pH. In this section, we will introduce this subject by studying gases that simply dissolve in the
aqueous phase with no subsequent chemical processes. More complex
chemical systems will be discussed in subsequent sections.

4.1 Basic Principles

The overall resistance to gas absorption by hydrometeors (a
term used to describe any form of precipitation) can be conceptually divided into three resistances - one in the air-phase, one
in the liquid-phase, and a surface resistance. In other chapters,
this concept has already been discussed in some detail relative
to the direct exchange of gases between the air and ocean, but
the equations used to estimate these resistances in the case of

hydrometeors are somewhat different.

The surface, or interfacial, resistance generally has been assumed to be negligible, although there appear to be no field experiments that confirm this opinion. At least two factors can contribute to a non-zero surface resistance. First, the absorption rate may be limited by the rate at which the gas molecules impinge on the drop surface for extremely short contact times. In the atmosphere, relatively long contact times are expected so that this factor should not introduce a surface resistance. Second, the presence of surface films on the drops could retard the absorption. However, the circulation within the drop (see Section 6) tends to disrupt such films, minimizing this effect. Therefore, in the absence of evidence to the contrary, the surface resistance is neglected, so that equilibrium is assumed to exist at the interface and the total resistance just resides in the air and liquid phases.

The air-phase resistance to the exchange depends on the relative velocity between the hydrometeor and air-phase. However, as the simplest case, consider a hydrometeor of radius R that is in a quiescent air-phase of infinite extent. The hydrometeor contains some species at a concentration of C_{lw}, and the concentration in the air-phase, an infinite distance away, is $C_{a,o}$. Let us further assume that the extent of absorption does not substantially alter its concentration in the hydrometeor; and since the air-phase is infinite in extent, the absorption does not change its abundance in the air. This situation is illustrated in Figure 5.

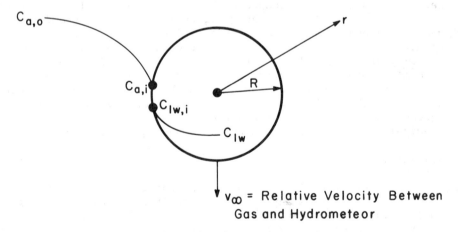

Figure 5. Schematic diagram showing the air-phase and hydrometeor-phase concentrations and the concentration gradients.

For a steady state situation, the diffusion equation in the air-phase becomes (recall that $v_\infty = 0$ for this simple case)

$$\frac{d}{dr} \left(r^2 \frac{dC_a}{dr} \right) = 0. \tag{49}$$

Assuming equilibrium at the interface, the concentration on the air-phase side is related to C_{lw} by Henry's law as

$$C_{a,i} = H C_{lw,i}. \tag{50}$$

For the current simplified case $C_{lw,i} = C_{lw}$; i.e., there are no concentration gradients in the liquid-phase. The boundary conditions for (49) are as follows.

i) at $r = R$, $C_a = C_{a,i}$

ii) at $r \to \infty$, $C_a = C_{a,o}$

The solution for C_a can be easily obtained as

$$C_a = C_{a,o} - (C_{a,o} - C_{a,i}) \frac{R}{r}. \tag{51}$$

Frequently, one is interested in the flux of the species into the hydrometeor. Using Fick's law of diffusion (and recognizing that the flux into the hydrometeor is in the negative r-direction), the magnitude of this flux is

$$F = -\vec{F}\big|_{r=R} = \mathcal{D}_a \frac{dC_a}{dr}\Big|_{r=R}. \tag{52}$$

This becomes

$$F = \frac{\mathcal{D}_a (C_{a,o} - C_{a,i})}{R}. \tag{53}$$

Defining the mass transfer coefficient for the air-phase, k_a, to describe the flux, one can write

$$F = k_a (C_{a,o} - C_{a,i}) = \frac{\mathcal{D}_a (C_{a,o} - C_{a,i})}{R}. \tag{54}$$

This shows very simply that for mass transfer to a sphere in a quiescent infinite air-phase, the dimensionless mass transfer coefficient known as the Sherwood number is

$$Sh_a = \frac{2k_a R}{\mathcal{D}_a} = 2. \tag{55}$$

This result has been confirmed by experiment on numerous occasions (cf., Skelland, 1974; Sherwood et al., 1975; and Clift et al., 1978).

It is true, of course, that the relative velocity between the air and hydrometeor is seldom zero in the atmosphere. Extension of the relationship to higher velocities is most readily obtained by experimentation in conjunction with dimensional analysis. One can easily show that the Sherwood number depends on the Reynolds and Schmidt numbers (cf., the chapters in this volume by Liss and Slinn); i.e.,

$$Sh_a = f(Re = \frac{2\rho R v_\infty}{\mu}, Sc = \frac{\mu}{\rho \mathcal{D}_a}), \tag{56}$$

where $\mu/\rho = \nu$ is the kinematic viscosity, which for air at STP is approximately 0.15 cm^2 s^{-1}. A frequently used correlation is that due to Frössling (1938), which also shows the limiting behavior obtained on theoretical grounds as expressed by (55).

$$Sh_a = 2 + 0.552 \ Re^{1/2} Sc^{1/3} \tag{57}$$

Other correlations have been suggested and several of these are summarized in Table 3 and Figure 6 (Skelland, 1974). One should

Table 3. Some experimental correlations of forced-convection mass transfer for single spheres (adapted from Skelland, 1974).

Equation	Range of Variables	Reference
$Sh_a = 2 + 0.552 \ Re^{1/2} Sc^{1/3}$	$2 \leq Re \leq 800$ $0.6 \leq Sc \leq 2.7$	Frössling (1938) Maxwell and Storrow (1957)
$Sh_a = 2 + 0.60 \ Re^{1/2} Sc^{1/3}$	$2 \leq Re \leq 200$ $0.6 \leq Sc \leq 2.5$	Ranz and Marshall (1952)
$Sh_a = 2 + 0.544 \ Re^{1/2} Sc^{1/3}$	$50 \leq Re \leq 350$ $Sc = 1$	Hsu, Sato, and Sage (1954)
$Sh_a = 2 + 0.95 \ Re^{1/2} Sc^{1/3}$	$100 \leq Re \leq 700$ $1200 \leq Sc \leq 1525$	Garner and Suckling (1958)
$Sh_a = 2 + 0.575 \ Re^{1/2} Sc^{0.35}$	$1 < Re$ $1 \leq Sc$	Griffith (1960)
$Sh_a = 2 + 0.79 \ Re^{1/2} Sc^{1/3}$	$20 \leq Re \leq 2000$	Rowe, Claxton, and Lewis (1965)

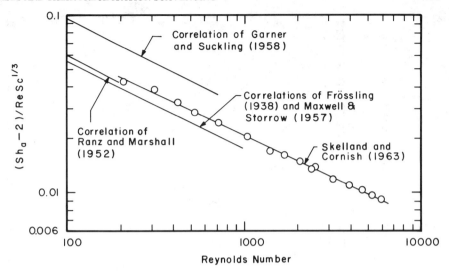

Figure 6. Comparison of mass transfer correlations for
forced-convection mass transfer for single spheres
(adapted from Skelland, 1974).

note that for low molecular weight gases in air, $\mathcal{D}_a \approx \nu$ and $Sc \approx$
1. Therefore, correlations applicable to low Schmidt numbers are
preferred for estimating the air-phase mass transfer coefficient
for low molecular weight gases to hydrometeors. Finally, Figure
7 shows the air-phase mass transfer coefficient for a hydrometeor
falling at terminal velocity and for Schmidt number of 1.

Let us now turn our attention to determining the resistance
in the liquid. The hydrometeor represents a volume of liquid
which has a limited capacity for the gaseous species being absorbed. This implies, of course, that the bulk-average concentration in the drop will change with time. This solute accumulation
makes the use of a mass transfer coefficient on the liquid-phase
side somewhat artificial. Only approximate mathematical solutions
have been obtained for the general case where the external and internal resistances are both important (Clift et al., 1978).

The rates of transfer within the drop depend on molecular
diffusion and the circulation which results from the drag exerted
on the surface due to the relative velocity between the hydrometeor and air. The circulation in very small drops is limited,
and the assumption of a stagnant liquid is reasonable. This
limiting case can provide a good first order approximation to the
mass transfer rate. The role of circulation is discussed in
Section 6.

Consider a stagnant liquid sphere exposed to a gas-phase

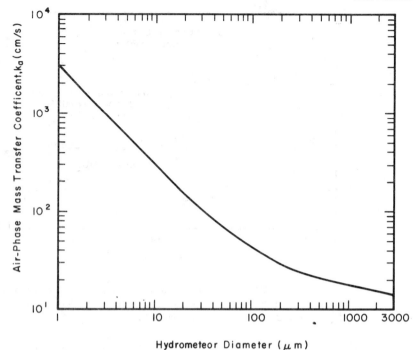

Hydrometeor Diameter (μm)

Figure 7. Air-phase mass transfer coefficient for
hydrometeors with the relative velocity corresponding
to the terminal settling velocity and for Sc = 1.

maintained at a concentration, C_a, such that the interfacial concentration on the liquid side is constant at $C_{lw,i}$ ($= C_{a,i}/H$).
This implies negligible external resistance. The concentration
initially within the drop is assumed to be uniform at the value
$C_{lw,o}$. The solution for the bulk-average concentration $<C_{lw}>$ can
be shown to be (cf., Sherwood et al., 1975 and Clift et al.,
1978)

$$\frac{C_{lw,i} - <C_{lw}>}{C_{lw,i} - C_{lw,o}} = \frac{6}{\pi^2} \sum_{j=1}^{\infty} \frac{1}{j^2} \exp\left(- \frac{j^2 \pi^2 \mathcal{D}_{lw} t}{R^2}\right), \qquad (58)$$

where $\mathcal{D}_{lw}$ is the diffusivity of the solute in the liquid-phase.
An instantaneous value of the Sherwood number for the internal
resistance can be found to be

$$Sh_{1w} = \frac{2\pi^2}{3} \frac{\displaystyle\sum_{j=1}^{\infty} \exp\left(-\frac{j^2\pi^2 \mathcal{D}_{1w}t}{R^2}\right)}{\displaystyle\sum_{j=1}^{\infty} \frac{1}{j^2} \exp\left(-\frac{j^2\pi^2 \mathcal{D}_{1w}t}{R^2}\right)}. \tag{59}$$

Equation (59) is plotted in Figure 8 as a function $\mathcal{D}_{1w}t/R^2$. The

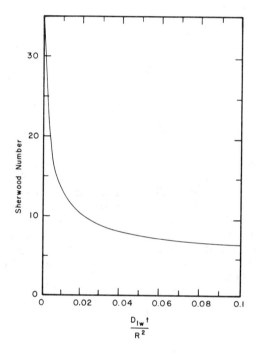

Figure 8. Liquid-phase Sherwood number as a function
of dimensionless time for the stagnant drop case.

result in (59) can be put in a more convenient form by using a
time-averaged Sherwood number; i.e.,

$$\overline{Sh}_{1w} = \frac{1}{\tau} \int_0^\tau Sh\ dt = \frac{2R^2}{3\mathcal{D}_{1w}\tau} \ln\left(\frac{C_{1w,i} - C_{1w,o}}{C_{1w,i} - <C_{1w}>}\right). \tag{60}$$

Hydrometeors are frequently exposed to the air-phase for extended periods of time, and it is interesting to investigate the
limiting value of Sh_{1w} for very large times. As suggested in
Figure 8, this is

$$\lim_{t \to \infty} Sh_{1w} = 6.58. \tag{61}$$

This limiting value of the Sherwood number is approximately reached in times corresponding to $\mathcal{D}_{1w}t/R^2 \doteq 0.1$ as seen in Figure 8. This time would correspond to about 25 seconds for a 1 mm diameter hydrometeor, a rather short time. During such a time interval, this 1 mm diameter drop falling at its terminal velocity would travel about 100 m. The result in (61) is convenient and has provided the basis for various researchers to use 10 – 20% or so of the drop diameter as the effective film thickness, δ_D, to calculate the internal resistance from film theory.

$$k_{1w} = \frac{Sh_{1w} \mathcal{D}_{1w}}{2R} = 3.29 \frac{\mathcal{D}_{1w}}{R} = \frac{\mathcal{D}_{1w}}{\delta_D} \tag{62}$$

Equation (62) shows that the effective thickness is 15.2% of the diameter of the hydrometeor. Thus, for many atmospheric applications the assumption is not too bad.

4.2   The Overall Mass Transfer Coefficient

In the previous sub-section, we discussed procedures by which the flux to hydrometeors can be estimated using either the air- or liquid-phase mass transfer coefficient. It is frequently advantageous to calculate the mass exchange using an overall coefficient based on either equivalent air-phase concentrations or equivalent liquid-phase concentrations. These equivalences can be written as

$$F = k_a (C_a - C_{a,i}) = k_{1w}(C_{1w,i} - C_{1w})$$

$$= K_a (C_a - C_a^*) = K_{1w}(C_{1w}^* - C_{1w}), \tag{63}$$

where upper case K's have been used to designate mass transfer coefficients based on an overall concentration difference. The concentrations C_a^* and C_{1w}^* are fictitious values, which represent respective air- and liquid-phase concentrations that would be in equilibrium with the bulk liquid and gas phases, respectively. Thus,

$$C_a^* = H\, C_{1w}, \tag{64}$$

and

$$C_{1w}^* = \frac{C_a}{H}. \tag{65}$$

These definitions are particularly convenient since C_a^* and C_{1w}^* are either known or are the concentrations of interest, whereas $C_{a,i}$ and $C_{1w,i}$ are rather difficult to determine and frequently of considerably less interest.

Estimations of K_a and/or K_{1w} do not introduce any additional problems, since they can be readily calculated from k_a, k_{1w}, and H for equilibrium relationships that are linear. These are

$$\frac{1}{K_a} = \frac{1}{k_a} + \frac{H}{k_{1w}}, \tag{66}$$

and

$$\frac{1}{K_{1w}} = \frac{1}{k_{1w}} + \frac{1}{Hk_a}. \tag{67}$$

Note that these relationships provide estimates of the resistances in each phase and can accommodate calculations based on either air- or liquid-phase concentrations.* Figure 9 shows graphically

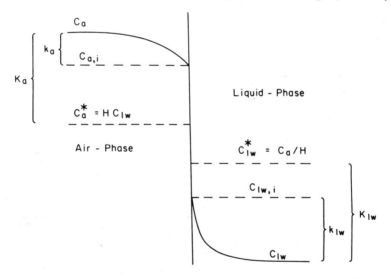

Figure 9. The relationship among the air- and liquid-phase concentrations and the mass transfer coefficients. Equation (63) shows the corresponding equations to calculate the flux.

---

* Equations (66) and (67) can be easily obtained from (63) by noting that $F/K_a = C_a - C_a^* = C_a - C_{a,i} + C_{a,i} - C_a^* = F/k_a + F/k_{1w}\ (C_{a,i} - C_a^*)/(C_{1w,i} - C_{1w}) = F(1/k_a + H/k_{1w})$. Equation (66) results, and (67) can be obtained in a like manner.

the relations among the various air- and liquid-phase concentrations and the mass transfer coefficients.

4.3   Individual Phase-Controlled Processes

Frequently, either the air-phase or liquid-phase provides
the major resistance to mass transfer. These are referred to as
air-phase controlled or liquid-phase controlled, respectively.
For very soluble gases (i.e., $H \to 0$), (66) shows that $K_a \simeq k_a$;
and for very insoluble gases (i.e., $H \to \infty$), (67) reveals that
$K_{lw} \simeq k_{lw}$. Thus, the exchange of extremely soluble gases is
likely to be air-phase controlled, and that for insoluble gases
will be liquid-phase controlled.

In Figure 10, the regions of individual phase-controlled gas

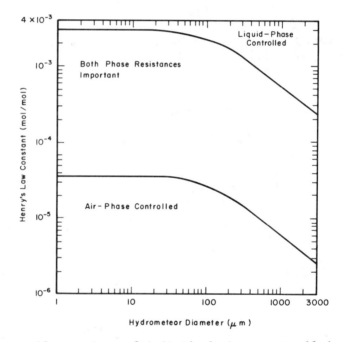

Figure 10.   Regions of individual phase-controlled gas
exchange for hydrometeors falling at their terminal
velocity in air at 25°C and 1 atmosphere pressure (Sc =
1).

exchange are quantified for hydrometeors falling at their terminal settling velocity, but assuming no external circulation. The
calculations are based on the Frössling equation for k_a (Equation
(57)) and $k_{lw} = 3.29 \, \mathcal{D}_{lw}/R$. Specification of the phase-controlled
region is based on the respective phase contributing more than 90%

of the total resistance. Table 4 lists Henry's law constants for
a number of gases. The list shows that many of these have Henry's
law constants which suggest liquid-phase controlled absorption
over the indicated range of hydrometeor diameter, assuming that
$\mathcal{D}_{lw}/\mathcal{D}_a$ is not too different from 10^{-4}. We will see in Sections
5 and 6 that chemical reactions and circulation in the liquid-
phase alter the rates of mass transfer, and thus, change these
phase-controlled regions.

Table 4.   Henry's law constants for trace gaseous species.[a]

Species	$H \ (= C_a/C_{lw})$
CO	37.3
CO_2	0.930
SO_2	7.85×10^{-3} [b]
H_2S	0.322
NO	18.4
N_2O	1.26
NH_3	4.71×10^{-4} [b]
O_3	2.19
CH_4	25.7
C_2H_6	17.2
C_2H_4	6.82
C_2H_2	0.823
C_3H_6	3.86 [c]
HCl	2.82×10^{-9} [b]

NOTES: (a) Abstracted from Perry and Chilton (1973)
             for 15°C temperature.
         (b) Calculated from tabulated data.
         (c) Data at 14°C.

5.   EXCHANGE OF REACTIVE AND IONIZABLE GASES

    When species are absorbed into cloud and rain drops, the
potential exists for liquid-phase chemical reactions to occur.
The chemical reactions might be dissociation or ionization processes, which are typically quite fast, or they might be oxidation processes, which could be considerably slower. Each hydrometeor represents a small liquid-phase reactor, which is being
fed reactants by exchange with its surroundings. For extremely
fast liquid-phase reactions, the conversion processes occurring

in the bulk liquid-phase are limited by the physical transport of
the reactant(s) across the air-phase resistance, surface resistance (usually small), and liquid-phase resistance. On the other
hand, the process of mass exchange is not limiting if the liquid-
phase chemical reactions are extremely slow. In such cases, the
exchange can be approximated using physical absorption theory as
described in Section 4.

    In this section, we will analyze the transport of species
from the air-phase to the liquid-phase when the absorption is
accompanied by chemical reactions in the liquid-phase. The
effects of the simultaneous processes will be quantified with
application to several important atmospheric species. The
limited space precludes a detailed discussion, and the reader is
referred to texts such as Danckwerts (1970) for a more complete
treatment.

5.1 Equilibrium Considerations

    For a simple soluble gas, the phase equilibrium can be
described by Henry's law or some similar set of distribution
data. If the dissolved gas can dissociate or ionize, the equilibrium considerations become only slightly more complex but must
account for dissociation or ionization.

    As an example, let us consider the absorption of CO_2 by pure
water and calculate the resulting concentrations in the liquid-
phase at equilibrium with a given air-phase concentration. Note,
at this point, that we are not considering the transient period
for the liquid to reach equilibrium. The sequence of processes
can be described as follows.

$$CO_2(a) \; \overset{H}{\underset{\leftarrow}{\rightarrow}} \; CO_2(lw) \tag{68}$$

$$CO_2(lw) + H_2O \; \overset{K_o}{\underset{\leftarrow}{\rightarrow}} \; H_2CO_3 \tag{69}$$

$$H_2CO_3 \; \overset{K_1}{\underset{\leftarrow}{\rightarrow}} \; H^+ + HCO_3^- \tag{70}$$

$$HCO_3^- \; \overset{K_2}{\underset{\leftarrow}{\rightarrow}} \; H^+ + CO_3^= \tag{71}$$

For strict physical absorption, only (68) and the associated
Henry's law constant are required. In the more general case,
the following relations can be written by considering these processes to be at equilibrium.

$$\frac{[H^+][HCO_3^-]}{[CO_2(a)]} = \frac{K_o K_1}{H} \tag{72}$$

$$\frac{[H^+][CO_3^=]}{[HCO_3^-]} = K_2 \tag{73}$$

$$[H^+] = [HCO_3^-] + 2\ [CO_3^=] \tag{74}$$

Equations (72) and (73) have been written using the concentrations rather than activities, by assuming that the activity coefficients are unity. Equation (74) follows from the necessity for electroneutrality of the solution to be maintained.

Available data at 20°C for the CO_2 system shows that $H = 25.6$ atm/mol ℓ^{-1}, $K_o K_1 = 4.16 \times 10^{-7}$ g-equiv ℓ^{-1}, and $K_2 = 4.2 \times 10^{-11}$ g-equiv ℓ^{-1}. Equations (72) through (74) can be readily solved. For an atmospheric CO_2 concentration of 330 ppm, $[H^+] = 2.32 \times 10^{-6}$ g-equiv ℓ^{-1} (pH = 5.64), $[HCO_3^-] = 2.31 \times 10^{-6}$ g-equiv ℓ^{-1}, and $[CO_3^=] = 4.2 \times 10^{-11}$ g-equiv ℓ^{-1}. In addition, by knowing $K_o = 2.6 \times 10^{-3}$, one finds that the dissolved carbon in the form of H_2CO_3 is 3.35×10^{-8} mol ℓ^{-1}.

It is important to note that the total dissolved CO_2 in all of the species is 1.29×10^{-5} mol $CO_2(lw)$ $\ell^{-1} + 3.35 \times 10^{-8}$ mol $H_2CO_3 \ell^{-1} + 2.31 \times 10^{-6}$ g-equiv HCO_3^- $\ell^{-1} + 4.2 \times 10^{-11}$ g-equiv $CO_3^= \ell^{-1} = 1.52 \times 10^{-5}$ mol ℓ^{-1}. This is 18% greater than that from just physical absorption. Finally, it is worth noting that frequent references are made to a "normal" rain pH of 5.6. This, of course, assumes the rain to contain no dissolved species other than those from CO_2, which is not realistic.

5.2   Enhancement of Mass Transfer Rate

When chemical reactions occur in the liquid, not only does the capacity for the species dissolving increase, but the rate at which the exchange proceeds can also be affected. This enhancement of the transfer rate has received considerable attention. To more fully appreciate the underlying phenomena, we will analyze a simple case.

Refer to Figure 11 which shows a planar liquid surface in contact with an air-phase that contains a species A that is soluble in the liquid. Within the liquid phase, species B is also dissolved and A reacts with B. It is assumed that the resistance in the air-phase to mass transfer is negligible, so that

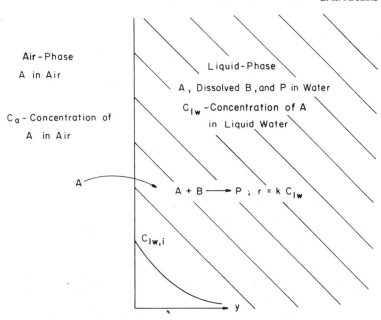

Figure 11. Schematic of simultaneous gas absorption
and first order chemical reaction. Note that B is
assumed to be in great excess, so that the reaction
does not significantly change the concentration of B.

the concentration at the interface can be expressed using Henry's
law; i.e., $C_{1w,i} = H^{-1} C_a$. It is also assumed that the dissolved
B is in sufficient excess that the rate of the reaction is a
pseudo-first order process, and the presence of the product P
does not affect the transfer of A. Finally, assume that the
liquid-phase initially contains no species A.

For a sufficiently extensive interface, the diffusion of A
is one-dimensional and is described by the following.

$$\frac{\partial C_{1w}}{\partial t} = D_{1w} \frac{\partial^2 C_{1w}}{\partial y^2} - k\, C_{1w} \tag{75}$$

i)   at $t \leq 0$, $C_{1w} = 0$ for $y > 0$

ii) at $y = 0$, $C_{1w} = C_{1w,i}$ for $t > 0$

iii) at $y \to \infty$, $C_{1w} = 0$ for $t > 0$

Without considering the details of the solution, which is

straightforward, it is (Danckwerts, 1950)

$$C_{1w} = \frac{C_{1w,i}}{2} \left[\exp\left(-\sqrt{\frac{k}{\mathcal{D}_{1w}}}\, y\right) \text{erfc}\left(\frac{y}{2\sqrt{\mathcal{D}_{1w}t}} - kt\right) \right.$$

$$\left. + \exp\left(\sqrt{\frac{k}{\mathcal{D}_{1w}}}\, y\right) \text{erfc}\left(\frac{y}{2\sqrt{\mathcal{D}_{1w}t}} + kt\right) \right], \qquad (76)$$

where erfc is the complementary error function. For no reaction (k = 0), (76) reduces to the obvious limit.

$$C_{1w} = C_{1w,i}\, \text{erfc}\left(\frac{y}{2\sqrt{\mathcal{D}_{1w}t}}\right) \qquad (77)$$

The instantaneous flux can be calculated for the cases with and without reaction, and the cumulative amount of A transferred can be determined by integrating this flux over time t. An enhancement factor, α, can then be defined as the ratio of the cumulative mass transferred with the chemical reaction to that when a chemical reaction does not occur. Thus, for the case illustrated in Figure 11

$$\alpha = \frac{1}{2}\sqrt{\frac{\pi}{kt}} \left[\left(kt + \frac{1}{2}\right) \text{erf}\left(\sqrt{kt}\right) + \sqrt{\frac{kt}{\pi}} \exp(-kt) \right], \qquad (78)$$

where erf is the error function. Equation (78) is plotted in Figure 12 as a function of dimensionless time. Note that for large times (kt > 10),

$$\alpha = \frac{1}{2}\sqrt{\pi kt}. \qquad (79)$$

Figure 12 and Equation (79) show the extent to which the mass transfer rate is increased by the irreversible first order reaction in the liquid-phase. Although hydrometeors are not one-dimensional and internal circulation may enhance the liquid-phase transfer, estimates of these effects can still be made. If one expects the life of an hydrometeor to be of the order of 10^3 to 10^4 seconds, the first order rate constant must be of the order of 5×10^{-4} to 5×10^{-3} s^{-1} (180 - 1800% h^{-1}) for a doubling of the total absorption caused by chemical reactions. This is an exceedingly high reaction rate for many oxidation processes, but not for dissociation phenomena.

There have been a number of other reaction systems studied. These have included irreversible reactions of general order

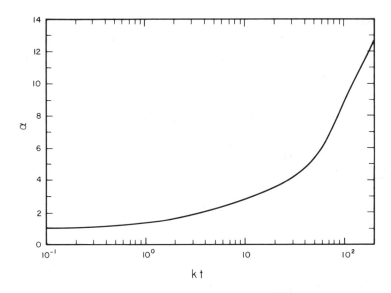

Figure 12. Enhancement coefficient as a function of
dimensionless time for a first order irreversible chemical reaction in the liquid-phase.

(Brian, 1964 and Hikita and Asai, 1964) as well as reversible
reactions. Earlier, Sherwood and Pigford (1952) studied reversible first order reactions,

$$A \underset{k_r}{\overset{k_f}{\rightleftarrows}} B, \tag{80}$$

where $K = k_f/k_r$. The liquid-phase was assumed to initially contain uniform, equilibrium concentrations of A and B. Species A
was in the air-phase, and its absorption disturbed the equilibrium distribution in the liquid-phase. The species B was not
desorbed. They found the enhancement coefficient to be described
as follows:

$$\alpha = 1 + \frac{K^2}{K^2 - 1} \frac{\sqrt{\pi}}{2\gamma} \exp{(\gamma^2)} \left[\text{erf} (K\gamma) - \text{erf} (\gamma) \right]$$

$$- \frac{K}{2\gamma} \sqrt{\frac{\pi}{K^2 - 1}} \, \text{erf} \left(\gamma \sqrt{K^2 - 1} \right), \text{ for } K > 1; \tag{81}$$

and

$$\alpha = 1 + \frac{K^2}{K^2 - 1} \frac{\sqrt{\pi}}{2} \exp(-\gamma^2) [\mathrm{erf}(K\gamma) - \mathrm{erf}(\gamma)]$$

$$+ \frac{K}{2\gamma} \sqrt{\frac{\pi}{1 - K^2}} \mathrm{erf}(\gamma\sqrt{1 - K^2}), \text{ for } K < 1; \qquad (82)$$

where

$$\gamma = \sqrt{\frac{k_f t}{K|1-K|}} \ .$$

Finally, Secor and Beutler (1967) have determined enhancement coefficients for reversible reactions of general order (aA + bB $\rightleftarrows$ cC + dD).

It is hoped that this brief discussion of enhancement by liquid-phase reaction has illustrated the diversity of the results. There is an extensive literature that can be consulted. Broad generalizations are not possible, and the reaction sequence for a specific process must be considered. The SO_2 system will be described in some detail in the next section.

5.3 SO_2 Absorption by Water

The absorption of SO_2 by water has been studied rather extensively because of its suspected importance in the atmospheric sulfur balance and the acid rain issue. This absorption is complex, involving the transport of the gas to the air-water interface, the absorption of gaseous SO_2 into the liquid, and its simultaneous partial hydrolysis in the water. In natural processes, the gaseous transport will involve both convective and diffusive processes.

There have been a number of studies describing SO_2 uptake by aqueous solutions. In most of these studies, either a zero interfacial concentration of SO_2, a linearized solubility relationship, or the predominance of a specific sulfur-bearing species has been assumed. The first case considers SO_2 uptake to be irreversible with the solutions acting as perfect absorbers of SO_2. The second neglects the dissociation of the absorbed SO_2, while the last neglects the presence of other species and disguises the dependence of solubility on pH. These assumptions are simplifications used in order to estimate the liquid-phase resistances to mass transfer.

Carmichael and Peters (1979) presented a conceptual and generalized treatment of SO_2 absorption by aqueous solutions that

eliminates the need for these assumptions. When the absorption
is accompanied by instantaneous reversible reactions, it is possible to express the absorption rate in terms of solution pH,
appropriate equilibrium coefficients, concentrations of the un-
dissolved species, and diffusivities of the individual species.
With equal diffusivities of all species, the rate of absorption
of SO_2 can be the physical mass transfer coefficient multiplied
by a driving force based on the total liquid-phase concentration
of sulfur species.

Sulfur dioxide exists in solution as physically dissolved
SO_2 and in dissociated form as bisulfite, HSO_3^-, and sulfite,
$SO_3^=$ (Scott and Hobbs, 1967). The following processes describe
this equilibrium.

$$SO_2(a) \underset{\leftarrow}{\overset{H}{\rightarrow}} SO_2(lw) \tag{83}$$

$$SO_2(lw) + H_2O \underset{\leftarrow}{\overset{K_o}{\rightarrow}} H_2SO_3 \tag{84}$$

$$H_2SO_3 \underset{\leftarrow}{\overset{K_1}{\rightarrow}} H^+ + HSO_3^- \tag{85}$$

$$HSO_3^- \underset{\leftarrow}{\overset{K_2}{\rightarrow}} H^+ + SO_3^= \tag{86}$$

The first dissociation reaction, (85), proceeds very rapidly with
forward and reverse rate constants at 20°C on the order of 10^6
s^{-1} and 10^8 ℓ mol^{-1} s^{-1}, respectively (Eigen et al., 1961 and
Beilke and Lamb, 1975). The second dissociation reaction, (86),
is a pure ionic reaction and also proceeds very fast. Erickson
and Yates (1976) report forward and reverse rate constants for
the second dissociation reaction of 10^4 s^{-1} and 10^{11} ℓ mol^{-1} s^{-1},
respectively. Therefore, on the time scale of the diffusion pro-
cesses, the equilibrium can be treated as being instantaneously
established (Lynn et al., 1955).

Since (85) is very fast, there is very little dissolved sulfur in the form of undissociated H_2SO_3, and the total sulfur is
essentially comprised of

$$[S] = [SO_2(lw)] + [HSO_3^-] + [SO_3^=]. \tag{87}$$

Figure 13 shows the mole fractions of the three S(IV) species as
functions of pH at 25°C. A typical pH range during precipitation
events is from 3 to 6, and the bisulfite ion is the predominant

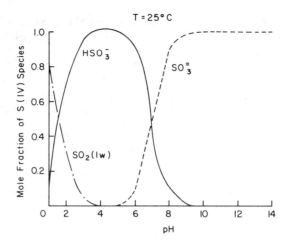

Figure 13. Distribution of S(IV) species mole fractions as a function of solution pH (adapted from
Carmichael and Peters, 1979).

species within this pH range. This result has been the basis for
neglecting the HSO_3^- dissociation, Reaction (86), in the development of some precipitation washout calculations.

Carmichael and Peters used both the film theory and surface-
renewal theory to study SO_2 absorption in the absence of oxidation
of S(IV) to S(VI). Only the results for the surface-renewal
theory will be presented here. The flux can be written in terms
of a concentration based on total dissolved sulfur species, $SO_2(lw)$,
or some other base. The instantaneous flux can be written as

$$F(t) = \sqrt{\frac{\mathcal{D}_{SO_2}}{\pi t} \frac{(1 + \dfrac{\mathcal{D}_{HSO_3^-}}{\mathcal{D}_{SO_2}} \dfrac{K_1}{[H^+]} + \dfrac{\mathcal{D}_{SO_3^=}}{\mathcal{D}_{SO_2}} \dfrac{K_1 K_2}{[H^+]^2})}{1 + \dfrac{K_1}{[H^+]} + \dfrac{K_1 K_2}{[H^+]^2}}} \, ([S]_i - [S]), \quad (88)$$

or

$$F(t) = \sqrt{\frac{\mathcal{D}_{SO_2}}{\pi t} (1 + \frac{\mathcal{D}_{HSO_3^-}}{\mathcal{D}_{SO_2}} \frac{K_1}{[H^+]} + \frac{\mathcal{D}_{SO_3^=}}{\mathcal{D}_{SO_2}} \frac{K_1 K_2}{[H^+]^2})(1 + \frac{K_1}{[H^+]} + \frac{K_1 K_2}{[H^+]^2})}$$

$$x \; ([SO_2(lw)]_i - [SO_2(lw)]). \quad (89)$$

Comparison of (89) with the purely physical absorption case reveals that the enhancement coefficient for SO_2 absorption is

$$\alpha = \sqrt{(1 + \frac{\mathcal{D}_{HSO_3^-}}{\mathcal{D}_{SO_2}} \frac{K_1}{[H^+]} + \frac{\mathcal{D}_{SO_3^=}}{\mathcal{D}_{SO_2}} \frac{K_1 K_2}{[H^+]^2})(1 + \frac{K_1}{[H^+]} + \frac{K_1 K_2}{[H^+]^2})},$$

when the mass transfer rate is based on the concentration difference of dissolved SO_2 in the liquid-phase. At low pH values, α is not very large but increases dramatically as the pH increases. For example, at 25°C α = 2.42 at a pH of 2, but increases to 1420 at a pH of 5. Note that the magnitudes of the diffusion coefficients also affect the enhancement of the mass transfer rate. In (89), this effect results from the electro-neutrality condition, since the diffusion rates of the individual species must adjust to maintain solution electroneutrality at each point.

The diffusivities in (88) and (89) all refer to those for the respective species in the aqueous phase. In order to use these results for SO_2 absorption, these diffusivities are required. Kolthoff and Miller (1941), using a polarographic method in strong acid solutions, found a value of $\mathcal{D}_{SO_2}$ = 2.04 x 10^{-5} cm^2 s^{-1} at 25°C. This value agrees reasonably well with that predicted by the Wilke-Chang correlation (1.71 x 10^{-5} cm^2 s^{-1}; cf., Reid et al., 1977). The diffusivities for HSO$_3^-$ and SO$_3^=$ can be estimated using the Nernst equation (Reid et al., 1977), in conjunction with limiting ionic conductances in water (Dobos, 1975). With this procedure, $\mathcal{D}_{HSO_3^-}$ = 1.33 x 10^{-5} cm^2 s^{-1} and $\mathcal{D}_{SO_3^=}$ = 1.92 x 10^{-5} cm^2 s^{-1} can be calculated. These results provide $\mathcal{D}_{HSO_3^-}/\mathcal{D}_{SO_2}$ = 0.652 and $\mathcal{D}_{SO_3^=}/\mathcal{D}_{SO_2}$ = 0.941, which should be substantially independent of temperature.

With these diffusivity ratios and a procedure to estimate the liquid-phase mass transfer coefficient, absorption rates into liquid bodies can be predicted. Hikita et al. (1978) conducted SO_2 absorption studies at atmospheric pressure and temperatures of 15°, 25°, 35°, and 45°C. The experiments were carried out using a wetted wall column, with precautions to prevent rippling of the falling film and to eliminate end effects. The exposure time of the liquid to the gas was varied from 0.049 to 0.97 s by changing the liquid flow rate and the film height. Their results were reported in terms of the average SO_2 absorption flux. Using (89), this average flux is

$$F = \frac{1}{\tau} \int_0^\tau F(t)dt$$

$$= 2 \sqrt{\frac{\mathcal{D}_{SO_2}}{\pi\tau}} \sqrt{\frac{1 + \dfrac{\mathcal{D}_{HSO_3^-}}{\mathcal{D}_{SO_2}} \dfrac{K_1}{[H^+]} + \dfrac{\mathcal{D}_{SO_3^=}}{\mathcal{D}_{SO_2}} \dfrac{K_1 K_2}{[H^+]^2}}{1 + \dfrac{K_1}{[H^+]} + \dfrac{K_1 K_2}{[H^+]^2}}} \; ([S]_i - [S]). \qquad (90)$$

Equation (90) can be used to calculate the average SO_2 absorption flux into water according to the theory. Figure 14 compares the

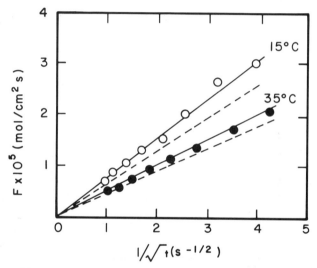

Figure 14. Comparison of theoretical predictions with the experiments of Hikita et al. (1978). The solid lines represent the absorption rates predicted by the theory, and the dotted lines show the values predicted using the Higbie equation for physical absorption. Note the underprediction of the flux when only physical absorption is considered (adapted from Carmichael and Peters, 1979).

fluxes predicted by this model to the measured values at 15°C and 35°C. The solid lines represent the absorption rates predicted by (90), and the dotted lines show the values predicted by the Higbie equation for simple physical absorption; i.e.

$$F = 2 \sqrt{\frac{\mathcal{D}_{SO_2}}{\pi \tau}} \, ([SO_2(1w)]_i - [SO_2(1w)]). \qquad (91)$$

Two observations should be made. First, the model accurately
predicts the SO_2 absorption flux; second, neglecting the liquid-
phase dissociation reactions leads to a low estimate of the rate
of absorption.

Carmichael and Peters also considered the case of liquid
droplets, initially with pH = pH_o, falling through a plume of
constant SO_2 concentration. As the droplet passes through the
plume, it absorbs SO_2 and the pH of the droplet changes until
equilibrium is reached. The amount of S(IV) absorbed versus time
for various drop sizes, air-phase SO_2 concentrations, liquid-film
thicknesses, initial droplet pH, and ratios of ionic to non-ionic
diffusivities were presented. Figure 15 shows the pH for specific

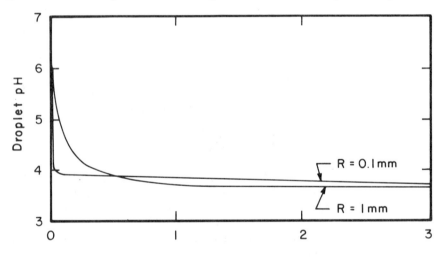

Exposure Time (Seconds)

Figure 15. Total S(IV) concentration and pH as functions of time for different values of drop size. The
conditions are: ([SO_2(a)] = 1 ppm, T = 10°C, pH_o = 7,
film-model with δ_D/R = 0.05) (adapted from Carmichael
and Peters, 1979).

conditions. In general for cases where pH_o > 3.5, the solution
pH reaches an equilibrium value between 3.5 and 4, and the dissociation of absorbed SO_2 supplies the majority of the hydrogen
ions. Terraglio and Manganelli (1967) presented experimental
evidence that follows curves similar to Figure 15, and report an
equilibrium pH for SO_2 absorption into distilled water in the

range of 3.5 - 4. Further experimental support is presented by Hales and Sutter (1973).

For solutions with $4 < pH < 5$, nearly all of the hydrogen ions arise from the first dissociation of H_2SO_3, so that $[H^+] \simeq [HSO_3^-]$. This fact has been used by Barrie (1978) to formulate a simplified SO_2 washout model, assuming $[H^+] = [HSO_3^-]$, and $\mathcal{D}_{HSO_3^-}/\mathcal{D}_{SO_2} = 1$. Absorption curves predicted by Barrie's model are within a few percent of the curves predicted by the analysis presented here, using $\mathcal{D}_{HSO_3^-}/\mathcal{D}_{SO_2} = 1$ and $pH_o > 4$. But for $\mathcal{D}_{HSO_3^-}/\mathcal{D}_{SO_2} > 1$ and $pH_o > 4$, or for $pH_o < 4$, the predictions differ. It can be seen that the simplifying assumptions over-estimate the SO_2 absorption.

6. THE ROLE OF CIRCULATION WITHIN DROPS

The discussion in Section 4 assumed that the hydrometeors were stagnant; i.e., there was no internal circulation. While the effect of internal circulation has minimal effect on the terminal settling velocity of drops up to about 3 mm diameter (Clift et al., 1978 and Pruppacher and Beard, 1970), it can have a pronounced effect on the rate of mass transfer within the drop. Circulation can reduce the liquid-phase resistance to mass trans-fer, because of the convective motion of the fluid within the drop. A cloud or rain drop is a microchemical reactor, which in general is not perfectly mixed, and is being fed by species transfer from the atmosphere. There are substantial concentra-tion gradients, which lead to varying rates of reaction throughout the drop. Internal circulation enhances the mixing, thereby reducing the concentration gradients and the variations in the reaction rates. Only for relatively slow chemical reactions are these gradients sufficiently small to justify the assumption of a uniform concentration.

6.1 The Work of Kronig and Brink

The classic works of Hadamard and Rybczynski (cf., Happel and Brenner, 1965) very early recognized that the viscous shear forces acting on a sphere (e.g., from a sphere settling under the influence of gravity) could set up internal circulation cells. The Hadamard-Rybczynski circulation patterns within a fluid sphere were derived for creeping flow (i.e., Re << 1). This does not apply at higher Reynolds numbers where the fore- and aft-symmetry consistent with creeping flow is not observed. They also ignored contamination from surface active agents. However, even with the Hadamard-Rybczynski solution available, there have been no general analytical descriptions of mass transfer within the drops.

Kronig and Brink (1950) considered a circulating sphere (see Figure 16) with a very large Peclet number (defined as Pe = ReSc = $2\rho v_\infty R/\mu \ \mu/\rho\mathcal{D} = 2v_\infty R/\mathcal{D}$). This means that the time required

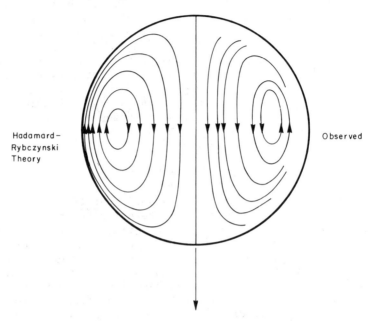

Figure 16. Schematic representation of the circulation cells set up in a sphere (adapted from Spells, 1952).

for diffusion is much greater than the time required for circulation. As a result, the iso-concentration lines coincide with the streamlines, which implies that the diffusion is normal to the streamlines. Kronig and Brink assumed these to correspond to the Hadamard-Rybczynski streamlines.

Kronig and Brink quantified the conditions under which the circulation time, τ_c, is much smaller than the diffusion time, τ_d. The circulation time was established by integrating the time along a Hadamard-Rybczynski circulation streamline. The following condition was obtained

$$\frac{\tau_c}{\tau_d} = \frac{6(3\mu_{1w} + 2\mu_a)\mathcal{D}_{1w}}{0.056 \ g|\rho_{1w}-\rho_a|R^3} \ q(\xi) \ll 1, \qquad (92)$$

where $q(\xi)$ is a function of the specific streamline and decreases very rapidly for positions interior to the droplet surface. For positions not on the hydrometeor surface, an upper limit of $q(\xi)$ of about 10 seems realistic. For typical conditions, $\tau_c/\tau_d \cong$

10^{-4} for a 1 mm drop and $\simeq 10^{-1}$ for a 0.1 mm drop. Consequently, for hydrometeors in air, the conditions for the Kronig-Brink solution are usually satisfied.

Kronig and Brink obtained a solution which Clift et al. (1978) have expressed in terms of the Sherwood number; i.e.,

$$Sh_{1w} = \frac{\frac{32}{3} \sum_{j=1}^{\infty} A_j^2 \lambda_j^2 \exp\left(-\frac{16\lambda_j \mathcal{D}_{1w} t}{R^2}\right)}{\sum_{j=1}^{\infty} A_j^2 \exp\left(-\frac{16\lambda_j \mathcal{D}_{1w} t}{R^2}\right)}. \qquad (93)$$

An approximate solution can be found by using $\lambda_1 = 1.678$, $\lambda_2 = 9.83$, $A_1 = 1.32$, and $A_2 = 0.73$. It is important to note that as $t \to \infty$, $Sh_{1w} = 17.66$, which is 2.68 times the value for the stagnant drop (see Figure 17). For typical drop sizes and aqueous phase diffusivities, this asymptotic value is reached very quickly.

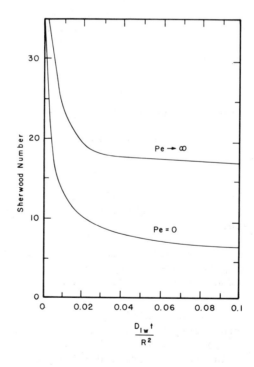

Figure 17. Comparison of the time-dependent Sherwood Number for the stagnant drop and the Kronig-Brink (1950) analysis using the Hadamard-Rybczynski streamlines.

A number of other workers have also used the Hadamard–
Rybczynski streamlines to study mass transfer at intermediate
values of the Peclet number (cf., Chao, 1969; Clift et al.,
1978; and Abramzon and Borde, 1980). Some studies have replaced
the radial coordinate by one that has its origin at the drop surface. Other workers have solved the equations numerically. In
all solutions, the unsteady state value of Sh_{1w} decreases from
very high values at short times to some asymptotic value at long
times. The asymptotic value depends on the Peclet number. At
intermediate times, the Sherwood number apparently oscillates
with a damped amplitude, and the frequency of the oscillations
depends on the circulation time of the fluid within the drop.
(It should be noted that analytical methods for the cases Pe = 0,
Pe → ∞, and those using the transformed y-coordinate do not show
these oscillations.) The asymptotic values for all values of Pe
generally have been reported to fall between 6.58 and 17.66, the
limiting values for the stagnant drop and the Kronig-Brink
analyses.

It appears that there have been few analytical studies
using circulation profiles other than the Hadamard-Rybczynski
streamlines. Previously, we pointed out that the limiting value
of the Sherwood Number for the stagnant drop corresponds to an
effective film thickness of 0.304 R or 15.2% of the drop diameter
(see Section 4.1). For the Kronig-Brink analysis, the effective
film thickness would be 0.113 R, which justifies a value of about
5% of the drop diameter.

6.2 Circulation with Chemical Reaction

Within hydrometeors, circulation and chemical reactions
occur simultaneously. However, there appear to be very limited
studies of such situations. Pruppacher and co-workers (Baboolal
et al., 1981 and Walcek et al., 1981) have recently investigated,
theoretically and experimentally, the absorption of SO_2 by water
droplets. This appears to be, by far, the most comprehensive
study to date.

These investigators solved the time dependent diffusion
equation using the internal and external flow fields established
by LeClair et al. (1972). The following three models were
assumed.

   (a) Air-Phase Controlled - The spherical drop acts as a
       perfect absorber with the only resistance to mass
       transfer residing in the air-phase.
   (b) Liquid-Phase Controlled - The sole resistance to mass
       transfer lies within the drop, and the gas is uniform
       in concentration with no gradients in the air-phase.
   (c) Coupled Transfer - This is the most realistic model.

The internal and external concentration fields are
calculated and coupled via the continuity of mass flux
across the air-drop interface. Furthermore, local phase
equilibrium is assumed to exist at the interface.
Within the droplets, the aqueous phase S(IV) chemistry was
treated in a manner analogous to Equations (83) - (86).

The role of circulation within the drop can be readily
appreciated by evaluating Model (b) and comparing those results
with that for a stagnant drop. Figure 18 shows such a comparison

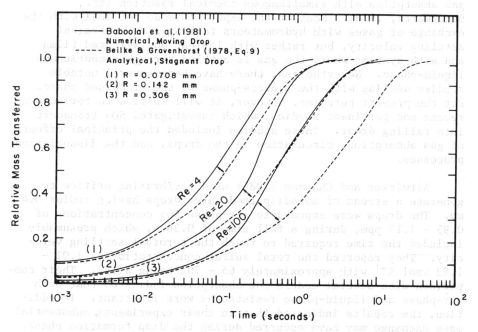

Figure 18. Comparison of the mass transfer rates for
Model (b) with the stagnant drop case (adapted from
Baboolal et al., 1981).

for drop radii of 0.0708 mm, 0.142 mm, and 0.306 mm. In this
figure, the ratio of the mass transferred up to a given time, to
the total amount of species scavenged by the drop at saturation
is plotted as a function of time. The results clearly show that
circulation substantially decreases the time for the drops to
become essentially saturated.

From Figure 18, it is clear that smaller drops more closely
approximate a uniform concentration condition. Beilke and
Gravenhorst (1978) and Freiberg (1979) used stagnant drop conditions to establish that the mass transfer of SO_2 is the rate

limiting step in the oxidation of $SO_3^=$ to $SO_4^=$ for drops with
R > 0.05 mm. As the discussion in this section has clearly demonstrated, circulation enhances the mixing. The results of
Baboolal et al. (1981) indicate that, for freely falling drops,
oxidation is limited by the diffusion processes for drops with
R > 0.5 mm. For hydrometeors smaller than this, the oxidation
step itself is rate limiting.

6.3 Comparison of Theory with Experimental Results

    An extensive literature has been developed which discusses
gas absorption with simultaneous chemical reaction (cf.,
Danckwerts, 1970). Most of the experiments do not deal with the
exchange of gases with hydrometeors falling at the terminal
settling velocity, but rather with laminar and agitated films
and with systems where the gas is bubbled through a continuous
liquid-phase. Nevertheless, there have been several notable
studies dealing with the aqueous-phase as the dispersed phase.
For the present purposes, however, it will suffice to review two
recent and pertinent studies, which investigated SO_2 transport
into falling drops. These studies included the principal effects
of gas absorption, circulation in the drops, and the dissociation
processes.

    Altwicker and Chapman (1981) used a vibrating orifice to
generate a stream of monodispersed water drops having radius 0.42
mm. The drops were exposed to air-phase SO_2 concentrations of
0.95 - 1.17 ppm, during a fall time of 0.36 s, which presumably
included the time required to reach the terminal settling velocity. They reported the total sulfur concentration as 1.01 -
1.93 μmol ℓ^{-1} with approximately 65 - 70% being S(IV). Their comparison with several models for absorption indicated that both
air-phase and liquid-phase resistances were important. In addition, the results indicated that in their experiments substantial
mass exchange may have occurred during the drop formation phase.

    The experiments of Walcek et al. (1981) certainly appear to
be the most comprehensive to date. Water droplets of radii near
0.30 mm were passed through a chamber containing SO_2 at rather
high partial pressures (5 mole % and higher). The length of the
exposure chamber was varied so that exposure times of 0.15 s and
0.72 s were obtained. The results are shown in Figure 19 and
demonstrate that good agreement with the theory can be obtained
if the internal drop circulation and dissociation chemistry are
both included.

    In summary, the experimental verification of the exchange
of gases with freely falling hydrometeors has been somewhat
limited for conditions closely corresponding to those typically
encountered in the atmosphere. In general, the drops have been

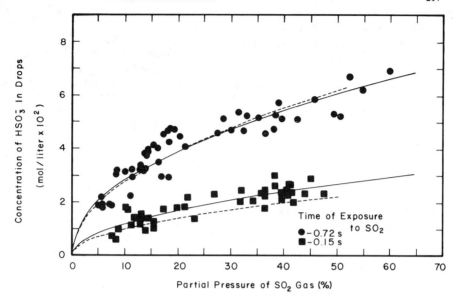

Figure 19. Concentration of HSO_3^- in a freely falling
drop of 0.30 ± 0.02 mm radius as a function of SO_2 concentration in the air-phase: (■) Experiments for 0.15
s exposure time; (●) Experiments for 0.72 s exposure
time. The solid lines are the least-squares fit to experimental data, the dashed lines the theoretical predictions from the theory of Baboolal et al. (1981)
(adapted from Walcek et al., 1981).

stationary (so that the liquid-phase diffusion is through a stagnant drop), the gas-phase concentrations have been quite high,
and/or the drops have not been at terminal velocity prior to
exposure. There is a substantial need for more well-defined and
controlled experiments.

7. HETEROGENEOUS PROCESSES

    Thus far, we have primarily considered gas exchange with
single drops. In the atmosphere, many drops (either in clouds or
during precipitation) are absorbing gases. Thus, the previous
results must be extended to the multiple drop situation. This
section and Section 8 will deal with these cases.

    The interaction of trace gaseous species with hydrometeors
and with aerosol particles obviously can affect the atmospheric
chemistry of the species. These phenomena can introduce additional pathways for intermediate species to follow or short circuit

other pathways. A specific example is in the CH_4 to CO oxidation
sequence illustrated in Figure 4 in Section 3.4. For example,
CH_2O is quite soluble in water, which means that it can be transferred to the liquid-phase in clouds or during precipitation
events. The net result is that gas-phase oxidation to CO and then
to CO_2 will not occur for a CH_2O molecule transferred to the
liquid-phase. Other intermediate species can behave similarly,
and the yield from the oxidation of one molecule of CH_4 is less
than one molecule of CO.

In this section, an analysis to develop rate constants for
removal of the gas species by aerosol particles and hydrometeors
will be presented. The analysis will provide some insight into
the parameters that are required for evaluation of such removal
processes. Upper limit values of heterogeneous loss constants
for particle-gas interactions will be calculated for several trace
species. A similar analysis will be presented for hydrometeor-gas
interactions.

7.1 General Transport Equations

The equations describing the movement, transformation, and
removal processes of trace species in the atmosphere must account
for the heterogeneous processes. Equations can be written for the
gas, cloud-water, and precipitation-water phases.* For a coordinate system fixed in space, the three equations are as follows.

$$\frac{\partial C_a}{\partial t} + \nabla \cdot C_a \vec{v} = \nabla \cdot (\mathbb{K} \cdot \nabla C_a) + r_a \tag{94}$$

$$\frac{\partial C_{cw}(R)}{\partial t} + \nabla \cdot C_{cw}(R)\vec{v} - \frac{\partial [v_g(R)C_{cw}(R)]}{\partial z} = r_{cw}(R) \tag{95}$$

$$\frac{\partial C_r(R)}{\partial t} + \nabla \cdot C_r(R)\vec{v} - \frac{\partial [v_g(R)C_r(R)]}{\partial z} = r_r(R) \tag{96}$$

For simplicity, Equations (94) through (96) have been written
without explicitly using the time-averaged notation. The equation for the air-phase has assumed an eddy diffusion model for
closure of the fluctuating quantities (e.g., $\overline{u'C_a'}$). In (95) and
(96), it has been assumed that the horizontal motions of the
cloud-water and precipitation-water phases are approximated by
the mean winds. Furthermore, these equations emphasize the

---

* Equations can also be written for the aerosol phase, but that
  is beyond the scope of this chapter. See Gelbard and Seinfeld
  (1979) and Gelbard et al. (1980) for such discussions.

dependence on the hydrometeor size, R, and equations for each size interval are required.

The terms r_a, $r_{cw}(R)$, and $r_r(R)$ in these equations represent the generation and/or loss of the species from the respective phase or the drop size interval in the water phases. These terms may result from chemical reaction within the respective phase or may account for the species being transferred between phases. In some literature, these terms have been broken down into components describing each process, but that is not necessary for the general analysis. The chemical and physical phenomena affecting these transformations determine the particular forms for r_a, $r_{cw}(R)$, and $r_r(R)$, which generally depend on the concentrations of the species in one or more of the phases. In the remainder of this section, we will relate these generation and/or loss terms to the scavenging processes of interest.

7.2 Particle-Gas Interaction

Condensation nuclei, which are necessary for cloud formation, can be modified by gas-particle interactions. These interactions change the chemical nature of the nuclei, which obviously affects their nucleating character and the resulting composition of the cloud droplets. In addition, these interactions provide an indirect pathway for gaseous species to be transferred from the air to the ocean.

Calculations by Cadle et al. (1975) indicate that heterogeneous chemical reactions between trace gases and aerosol particles can be important. They argued that the particle concentration in the lower stratosphere is such that collisions of molecules with a particle surface are relatively fast and quite probable. At about the same time, Graedel et al. (1975) made calculations that included the interaction of aerosols and free radicals. They concluded that aerosol effects are a significant part of atmospheric chemical kinetics. However, they made no effort to compare their predicted results with known concentrations, or with results based on other studies. More recently, Baldwin and Golden (1980) suggested that heterogeneous radical reactions would need approximately unit collision-reaction probability (i.e., reaction for each collision) to have any significance in tropospheric chemistry. They measured the collision-reaction probability with sulfuric acid for various species present in the troposphere and found relatively low values (e.g., $\alpha \simeq 5 \times 10^{-4}$ for the OH radical). Their results were probably affected strongly by the chemical nature of the radical species and the simulated aerosol surface.

Luther and Peters (1982) assumed that the aerosol particles are spherical and can be described by the physical and chemical properties of water. This seems reasonable since particles in the troposphere can quite regularly be covered with a layer of water. However, there are several limitations relative to this representation of the heterogeneous processes. First, in-cloud and below-cloud scavenging by hydrometeors were not considered. These phenomena will be analyzed later in this section. Those processes are not always active, and their intermittent nature should be considered. Secondly, the model aerosol particle did not consider absorption by non-aqueous surfaces. While that phenomenon could be included, knowledge of the fraction of particles with surfaces appropriate would be necessary. Finally, adsorption-desorption phenomena were not included. Thus, their analysis did not account for processes such as reactant adsorption and product desorption or product incorporation into the particle.

The heterogeneous removal rate was described using an effective first order rate constant. Two mass conservation balances were used to develop the rate constant expression. With $n(a)$ particles of radius a present, the rate of removal of a species from the gas phase by these particles is

$$\frac{dC_a}{dt} = -4\pi a^2 K_a n(a) (C_a - C_a^*). \tag{97}$$

K_a is the overall mass transfer coefficient based on the air-phase concentration, and C_a^* represents the hypothetical air-phase concentration if the air- and particle-phases are in equilibrium. If one assumes a linear relationship for the species distribution between the air- and particle-phases, then Henry's law can be used

$$C_a^* = H C_p(a), \tag{98}$$

where $C_p(a)$ is the concentration of the species in the aerosol particles of radius a. A balance for the concentration in the particle phase is

$$\frac{d[n(a)V(a)C_p(a)]}{dt} = n(a) \frac{4\pi a^3}{3} \frac{dC_p(a)}{dt}$$

$$= 4\pi a^2 K_a n(a) (C_a - C_a^*). \tag{99}$$

The linear equilibrium relationship can also be used to define $C_p^*(a)$, which represents the equilibrium (or maximum) concentration that can exist in the particle-phase when in contact with the air-phase having a concentration of the gas species of C_a;

i.e., $C_a = H C_p^*(a)$. Using this relationship and rearranging
(99) yields the following.

$$\frac{dC_p(a)}{C_p^*(a) - C_p(a)} = \frac{3K_a H}{a} dt \tag{100}$$

Before integrating (100), the limits of integration must be considered. Initially (at time zero), the concentration of species
in the particle is assumed to be zero. This assumes that the
particle, when it is formed, has none of the species present.
When the residence time of the particle in the atmosphere (designated τ_p) is reached, the concentration in the particle will
be represented as $C_{p,f}(a)$. Therefore, (100) becomes

$$C_{p,f}(a) = C_p^*(a) [1 - \exp(-\frac{3HK_a}{a} \cdot \tau_p)]. \tag{101}$$

The particle population in the troposphere varies in size,
age, and chemical composition. Let n_t be the total number of
particles present, and assume that the distribution can be represented as

$$n(a,\tau) = n_t f(a)g(\tau), \tag{102}$$

where τ is the age of a particle. Equation (102) assumes that
there are no composition variations, an obvious simplification.
By taking an average of $C_p(a)$ over the particle age, the following
is defined

$$\overline{C_p(a)} = \frac{1}{n(a)} \int_o^\infty C_p(a,\tau)n_t f(a)g(\tau) \, d\tau, \tag{103}$$

where the bar superscript represents a particle age-averaged property. Luther and Peters (1982) assumed that

$$g(\tau) = \begin{cases} 1/\tau_p & \tau \leq \tau_p \\ \\ 0 & \tau > \tau_p \end{cases}, \tag{104}$$

which implies that for a given size, particles of all ages
$0 < \tau \leq \tau_p$ are equally probable, and that particles of radius a
cannot have an age greater than τ_p. The expression for $\overline{C_p(a)}$
then becomes

$$\overline{C_p}(a) = \frac{1}{\tau_p} \int_0^{\tau_p} C_p(a,\tau) \, d\tau, \tag{105}$$

where $n(a) = n_t f(a)$ has been used. Substituting for $C_p(a,\tau)$ from (101) for any time gives the average particle-phase concentration for particles of radius a,

$$\overline{C_p}(a) = C_p^*(a) \, \{1 - \frac{a}{3HK_a\tau_p} \, [\exp(-\frac{3HK_a}{a}\tau_p) - 1]\}. \tag{106}$$

Particles of all sizes can of course remove the species from the gas-phase. Thus, (97) must be integrated over the entire size range; i.e.

$$\frac{dC_a}{dt} = -\int_0^\infty 4\pi a^2 K_a n(a) [C_a - H \, \overline{C_p}(a)] \, da, \tag{107}$$

where the age-averaged notation has been included. Substituting for $\overline{C_p}(a)$ from (106) gives

$$\frac{dC_a}{dt} = -\{\int_0^\infty 4\pi a^2 K_a n(a) \, \frac{a}{3HK_a\tau_p} \, [1 - \exp(-\frac{3HK_a\tau_p}{a})] \, da\} C_a. \tag{108}$$

The quantity within the braces can be identified as the heterogeneous loss constant; i.e.,

$$k = \int_0^\infty \frac{4\pi a^3 n(a)}{3H\tau_p} \, [1 - \exp(-\frac{3HK_a\tau_p}{a})] \, da. \tag{109}$$

The limiting cases of very insoluble gases ($H \to \infty$), very soluble gases ($H \to 0$), and very short and long particle residence times, provide some insight to these interactions. These limits can be summarized as follows.

$$H \to 0 \text{ or } \tau_p \to 0, \qquad k = \int_0^\infty 4\pi a^2 n(a) k_a \, da$$

$$H \to \infty \text{ or } \tau_p \to \infty, \qquad k = 0$$

These results show that the heterogeneous removal rate constant is controlled by the air-phase exchange resistance for infinitely soluble gases or for particles with a very short residence time. Neither limit is surprising. The case of $\tau_p \to 0$ corresponds to an instantaneous renewal of particle surface. On the other hand,

the heterogeneous loss constant approaches zero for insoluble
species or for very long residence time particles.

Equation (109) shows that information must be known about
the equilibrium distribution of the species between the air- and
particle-phases (H) and the rate of transfer to the surface (K_a).
Such information is not always available. In addition, the residence times and size distribution of the particles must be known.
Luther and Peters (1982) used the residence time distribution
function for particles in the troposphere given by Jaenicke
(1978) to evaluate the heterogeneous rate constant. The aerosol
size distribution (and the corresponding surface area) is also
important. The clean continental background troposphere typically has a surface area of about 27 $\mu m^2/cm^3$ (Willeke and Whitby,
1975). The oceanic aerosol has a lower surface area and a value
of 5.2 $\mu m^2/cm^3$ was used to evaluate the particle-gas system for
background conditions.

It is standard to divide the aerosol particles into size
regimes according to the Knudsen Number (Kn = $\lambda/2a$) to evaluate
the mass transfer processes. For the free-molecule regime (Kn >
10), the rate of absorption of a species into the particle can be
analyzed using gas-kinetic theory. The air-phase mass transfer
coefficient for this regime can be written as

$$k_{a,n} = \frac{\alpha \bar{c}}{4},$$ (110)

where α is the accommodation coefficient dependent on the fraction of collisions of molecules with the particles that are effective, and $\bar{c}$ is the mean thermal speed of the molecules. The upper
limit on α is unity, and can be used to determine the maximum
effect that the heterogeneous processes might have.

The second regime for relatively large particles is the
continuum regime (Kn < 0.1); then the rate of absorption of a
species into the particle can be calculated by classical diffusional mass transfer theory. Assuming the net velocity between
the air and particles is small, the air-phase mass transfer coefficient for this regime corresponds to $Sh_a = 2$; i.e.,

$$k_{a,c} = \frac{\mathcal{D}_a}{a}.$$ (111)

The third size regime is the transition region (0.1 < Kn <
10) which matches the non-continuum and continuum regimes in a
smooth manner. Fuchs and Sutugin (1971) used the results of
Sahni (1966) to obtain

$$k_{a,t} = \left(\frac{1}{1+\ell Kn}\right) k_{a,c},$$ (112)

where

$$\ell = \frac{4/3 + 0.71 \ Kn^{-1}}{1 + Kn^{-1}}.$$ (113)

Using diffusional theory for liquid-phase mass transfer, the liquid-phase mass transfer coefficient can be approximated by

$$k_{1w} = \frac{\mathcal{D}_{1w}}{a}.$$ (114)

Table 5 shows some selected heterogeneous loss constants calculated by Luther and Peters (1982). The largest values represent species where the liquid-phase resistance would be negligible. These heterogeneous removal constants were incorporated in a CH_4-CO model by comparing the predicted concentrations

Table 5. Representative heterogeneous loss constants for aerosol particle-gas interactions (adapted from Luther and Peters, 1982).

Species	Heterogeneous Loss Constant (s^{-1}) [a]
NO	1.41×10^{-8}
CO	6.25×10^{-9}
O_3	6.70×10^{-8}
CH_4	7.42×10^{-9}
H_2O_2	6.20×10^{-4}
CH_2O	7.23×10^{-4}
CH_3OOH	6.07×10^{-4}
HNO_2	7.31×10^{-4}
HNO_3	6.53×10^{-4}

NOTE: (a) The calculated loss constants assume $\alpha = 1$, H as given in Table 4 (for aqueous solutions), the residence time distribution according to Jaenicke (1978), and an aerosol size distribution according to Junge (1963) with a surface area of $5.2 \ \mu m^2/cm^3$.

of CH_2O, H_2O_2, and HNO_3 with some of the limited field observations available. In all cases, the heterogeneous removal rate seemed to be too large, implying that the upper limit values based on $\alpha = 1$ are too high. However, such a mechanism could explain some of the observed nitrate concentrations on aerosol particles (see Buat-Menard, this volume). Assume that the heterogeneous rate constant for HNO_3 vapor is about 5×10^{-7} to 5×10^{-6} s^{-1} (2 to 3 orders of magnitude lower than in Table 5), and that the HNO_3 vapor concentration is 10^9 to 10^{10} molecules cm^{-3} (see Table 2). If the average particle age is 2 days or so, then 0.01 to 1 $\mu g\,m^{-3}$ of HNO_3 would be deposited on the aerosol particle surfaces, which brackets many of the observations. Unfortunately, present knowledge does not permit more accurate estimates.

More theoretical and experimental work are required to establish the role of gas-particle interactions in atmospheric chemistry. In the present context, the early stage in the growth of condensation nuclei appears to be a particularly fruitful area of research to determine the significance of that pathway as a removal mechanism for gaseous species from the atmosphere.

7.3 Hydrometeor-Gas Interaction

The effect of hydrometeors on the air-phase chemistry can occur with two common atmospheric phenomena: cloud formation and precipitation. Consider cloud systems first, and for simplicity, assume that there are no spatial variations in the size distribution of the drops. This implies that, as a first approximation, only temporal changes in these parameters will occur. This, of course, means that the cloud droplet composition and size distribution may change with time, but the cloud is well-mixed. Polluted air is pumped-up from the boundary layer and through the cloud with the highest air-phase concentrations at the base of the cloud. A lateral influx of air at the cloud sides also may occur, and generally these concentrations of gaseous species will be different. As the air moves through the cloud, some of the gaseous species are removed and their concentrations are reduced, being lowest at the cloud top. The vertical air motion is assumed to be sufficient to offset the settling velocities preventing precipitation.

Within the cloud, the drop distribution depends on size, chemical composition of each species, and time. This is designated as $n[R(t), C_{cw}(R,t)]$. If we consider only one-dimensional transport for simplicity, the entering air-phase composition from the boundary layer is $C_{a,b}$, moving at a vertical velocity w. This entering concentration is considered constant. At the top, the concentration is a function of time, since the degree of absorption depends on the liquid-phase concentration which is constantly changing; this is designated as $C_{a,top}(t)$. The simplified system is shown in Figure 20.

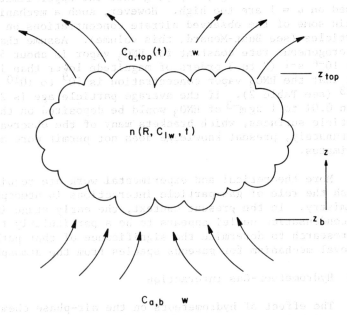

Figure 20.   Schematic representation of the pumping of gas species from the boundary layer through a cloud.

If we ignore lateral effects, the air-phase concentration is described by the following integro-differential equation,

$$w \frac{\partial C_a}{\partial z} = -\int_0^\infty 4\pi R^2 n[R(t), C_{cw}(R,t)] K_a[R(t)][C_a - C_a^*(R,t)]\, dR, \quad (115)$$

with the associated boundary condition

$$\text{at } z = z_b,\ C_a = C_{a,b} \quad \text{for } t > 0. \quad (116)$$

Note that the mass transfer coefficient depends on time, since the drop radius may change with time. In order to close this problem, an equation must be constructed for the cloud-water concentrations. These concentrations are the accumulated solute transferred into each drop

$$C_{cw}(R,t) = C_{cw,o} + \int_0^t \frac{3}{R} K_a(R,t)[C_a - C_a^*(R,t)]\, dt, \quad (117)$$

where $C_{cw,o}$ is the concentration of the species in the drop at

time zero. The concentration $C_{cw,o}$ may not be zero for some
species, depending on the chemical composition of the condensation nucleus at the core of the drop.

Scott (1982) has developed a three-dimensional analysis for
in-cloud scavenging of sulfate that solves the time dependent
equations for the sulfate concentration in the air-phase, cloud-
water, and precipitation-water. All horizontal motions are
described by the horizontal wind field, and only the precipitation-water has a vertical velocity different from the air. Pro-
vision is made for production of sulfate by SO_2 interactions with
the aerosol, the cloud-water, and the precipitation-water, and
sulfate can be exchanged among the phases.

The analysis assumes that the sulfate deposited in precipitation originates from nucleation of pre-existing sulfate particles, or from SO_2 conversion in the liquid-phase occurring
within the cloud. Thus, below-cloud scavenging has been ignored.
The aerosol ingested by a storm is assumed to contain sulfates
and nitrates of known composition.

The model was run for two idealized situations indicative
of average winter and summer storms in the Northeastern United
States. With the restrictions noted, more than 41% and 55% of
the sulfate deposited for the winter and summer storms, respectively, were the result of in-cloud conversion processes. This
conversion contribution never exceeded 70% for either the summer
or winter simulations. Scott pointed out that these model results indicate that the majority of the summer storm acidity
originated during in-cloud transformation of SO_2, but HNO_3
appears to substantially contribute to winter precipitation.

The predictions of conversion rates depended on the boundary
layer concentrations of ingested SO_2 and sulfate. The typical
conversion rates were about 12% h^{-1} for the summer storm and 7%
h^{-1} for the winter storm simulation, although these rates did vary
by as much as a factor of three for different boundary layer concentrations. Most of the ingested sulfate was removed by the
simulated storms ($\approx$ 80% in winter and 90% in summer).

This study by Scott is one of the few that has attempted a
detailed analysis of in-cloud transformation processes. Nevertheless, the results cannot be generalized since the chemical and
kinematic conditions were only typical of average conditions in
one region.

Below-cloud scavenging is the second important component of
gas interaction with hydrometeors. The removal of a gaseous
species from the atmosphere can be written in a manner similar to
(97) with n(a) being replaced by the number density of rain drops

and the result integrated over the entire drop distribution; i.e.,

$$\frac{dC_a}{dt} = - \int_0^\infty 4\pi R^2 N(R) K_a [C_a - C_a^*(R)] \, dR. \tag{118}$$

Note that in addition to $K_a(R)$ and $N(R)$ being functions of radius, $C_a^*(R)$ may also depend on the drop radius since the prior uptake of species by the hydrometeors depends on the size. Thus, even if all of the drops initially have none of the species present, $C_a^*(R)$ [$= H \, C_{lw}(R)$] at any time depends on the prior history of the transfer. Again, as with (115) and (116), an equation similar to (117) for $C_r(R)$ must be developed.

In many cases, the drops have been assumed to be a perfect absorber, and $C_a^*(R) = 0$. As a result,

$$\frac{dC_a}{dt} = - [\int_0^\infty 4\pi R^2 N(R) K_a dR] \, C_a. \tag{119}$$

If the total derivative is taken to be the substantial derivative (i.e., $dC_a/dt = DC_a/Dt = \partial C_a/\partial t + \vec{v} \cdot \nabla C_a$), the quantity within brackets is termed the scavenging coefficient, frequently designated Λ (cf., Hales, 1972, 1978). Note that the concept of the scavenging coefficient only has the conventional meaning for irreversible mass exchange when the drop is a perfect absorber (see Slinn, this volume).

If one uses a mean drop radius $\bar{R}$, the scavenging coefficient can be approximated as $PK_a/(3\bar{R}v_g)$, where P is the precipitation rate, and K_a and v_g are evaluated using the mean radius. As an example, consider a precipitation rate of 1 cm h^{-1} and a mean drop radius of 0.6 mm. Assuming that the absorption is air-phase controlled, $\Lambda \cong 5 \times 10^{-5}$ s^{-1} for this case. If one considers a rather light precipitation of 1 mm h^{-1} and a mean drop radius of 0.35 mm, $\Lambda \cong 2 \times 10^{-5}$ s^{-1} for air-phase controlled absorption.

As a final note in this section, the similarities among the three heterogeneous processes of gas-particle, in-cloud, and below-cloud scavenging should be apparent. The principal differences are in the amount of particle or hydrometeor surface area for mass exchange, and the equations used to estimate K_a. Cloud and precipitation drop sizes are such that continuum mass transfer theory is applicable. On the other hand, gas-aerosol particle interactions span the range from free-molecule to continuum mass transfer theory.

8. SCAVENGING IN COMPLEX SYSTEMS

In this last section, we will consider the scavenging of gases by hydrometeors in some complex chemical and physical systems. In the atmosphere, complexities arise from simultaneous absorption of many species, along with the well-developed circulations in the drops. Both effects have been considered only for the simplest situations. Ignoring one or the other can lead to substantial errors, but some significant insights can be gained by investigating limiting cases.

The scavenging of gases from the atmosphere probably never occurs by an uncontaminated water droplet. The drop begins with a condensation nucleus which, while it may produce a neutral aqueous solution, can affect the mass transfer within the drop by altering the solubility, the dissociation chemistry, or conditions for electroneutrality. In addition, the hydrometeor once formed may evaporate, leaving a condensation nucleus that is chemically changed. Condensation-evaporation cycles may repeatedly occur. Each condensation nucleus has a distinct history, and its history is expected to have a pronounced effect on its subsequent growth and absorption behavior.

Spatial non-homogeneities of air-phase concentrations also introduce complications in the analysis of gas scavenging by hydrometeors. Since cloud and/or precipitation drops will invariably have a velocity different from the air, the drops can be exposed to substantially varying gas-phase concentrations.

8.1 Spatial Effects and Equilibrium and Reversible Scavenging

During cloud formation and precipitation events, an individual hydrometeor generally experiences a time-varying air-phase concentration from depletion of the species from the air-phase or from passage through different air masses. Consider the equation describing the liquid-phase concentration of the hydrometeor for the case of exposure to a simple soluble gas,

$$\frac{4}{3} \pi R^3 \frac{dC_{1w}}{dt} = 4\pi R^2 K_{1w} \left[\frac{C_a(t)}{H} - C_{1w} \right], \qquad (120)$$

where the air-phase concentration has been written as $C_a(t)$, to emphasize its time-varying nature. (Obviously, the time coordinate can be recast using $v_g = -dz/dt$ for a droplet that is not evaporating or growing.) At any instant of time, the right-hand side of (120) may be either positive or negative.

The time scale for mass transfer can be expressed as $R/3K_{1w}$, which can be designated τ_m. The time scale associated with the variations of the air-phase concentration that the drop is exposed to can be established as $C_a(t)/[dC_a(t)/dt] = [dlnC_a(t)/dt]^{-1}$, which can be rewritten as $\tau_g = |v_g \, dlnC_a(z)/dz|^{-1}$. In the case of $\tau_m/\tau_g \ll 1$, the time scale required for uptake of the gas by the drop is very small relative to the changes in the air-phase concentration it experiences. This can be summarized as

$$\frac{Rv_g |dlnC_a(z)/dz|}{3K_{lw}} \ll 1. \tag{121}$$

Equation (121) is satisfied for small drops and weak concentration gradients. For example, consider a drop radius of 0.1 mm and liquid-phase controlled absorption; then, $|dlnC_a(z)/dz|$ must be less than 4×10^3 km^{-1}, which corresponds to substantial changes in the concentration over a 1 km vertical distance. On the other hand, $|dlnC_a(z)/dz|$ must be less than 4 km^{-1} if (121) is to apply for a 1 mm drop radius under liquid-phase controlled absorption. Such modest gradients can be encountered in the atmosphere.

For cases satisfying (121), the air- and liquid-phases are in near-equilibrium, and

$$C_{lw} = \frac{C_a(t)}{H}. \tag{122}$$

This situation is referred to as equilibrium scavenging. For such cases, wet deposition rates can be estimated on the basis of air-phase concentrations, precipitation rate, and solubility data.

Equation (120) also suggests another possible phenomenon referred to as reversible scavenging. Reversible absorption can occur if the dissolved gas does not undergo any chemical reaction and can be desorbed into the air-phase in its original state. Then, there can be a redistribution of the gas in the atmosphere. Imagine that the air-phase concentration increases with height above ground; i.e., $dC_a(a)/dz > 0$. For a high mass transfer rate, gas absorbed at higher levels can be desorbed at lower altitudes. As a result, high concentrations of a trace gas can be distributed downward. In the case of equilibrium-reversible scavenging, the air-phase concentration of importance is that at ground level, since it ultimately establishes the concentration in the hydrometeor.

8.2 Simultaneous Absorption of Several Gases

As already mentioned, hydrometeors normally absorb several gases simultaneously, and dissociation and oxidation reactions

can be expected to occur in the liquid-phase. For example, S(IV)
can be oxidized in the liquid phase, and the reaction rate varies
considerably depending on such factors as the presence of NH_3,
O_3, CO_2, and catalysts. The product of this oxidation, $SO_4^=$, is
undoubtedly a contributor to the acidity of precipitation,
although the role of other acid gases such as HNO_3 cannot be ignored. The oxidation of S(IV) in the aqueous phase has been experimentally studied for many years (cf., Junge and Ryan, 1958;
Van den Heuval and Mason, 1963; Cheng et al., 1971; Miller and
dePena, 1972; Penkett, 1972; Erickson and Yates, 1976; and Martin
and Damschen, 1981 as a few examples). Space does not permit a
comprehensive survey of the voluminous literature and the
numerous discrepancies noted among the studies.

Modeling studies have also been attempted. The work of
Scott and Hobbs (1967) was one of the first attempts to analyze
the influence of other gases on sulfate formation. More recently,
studies have been directed at increasing the complexity and supposed representativeness of the liquid-phase chemistry. The
research of Overton et al. (1979) is one such example. In their
study, a cloud is assumed to form above a polluted air mass, the
raindrops fall through the air mass, and SO_2, O_3, NH_3, and CO_2
are absorbed into drops containing a dissolved metal catalyst
(Fe(III)). Although their analysis considered below-cloud scavenging, the situation would be similar to an air mass passing
through a cloud, although the hydrometeor size would typically
be smaller. This study made the following important assumptions:
(a) Isothermal behavior (25°C); (b) The drops are perfectly mixed
with no concentration gradients; (c) The droplet diameter is constant; (d) The Frössling equation (Equation (57)) is used to calculate the mass transfer rate (i.e., the mass exchange process
is assumed to be air-phase controlled); and (e) There is no depletion of the species from the air mass. The aqueous phase
dissociation and oxidation processes which were considered are
listed in Table 6. Drop radii from 0.02 to 2.8 mm were considered,
and the air-phase concentrations were $[SO_2]$ = 10 ppb, $[O_3]$ = 0 or
100 ppb, $[NH_3]$ = 0, 5, or 20 ppb, and $[CO_2]$ = 320 ppm.

The initial pH of the drop was taken as 5.56, and not surprisingly, the final pH depended on drop size, the fall distance
(equivalent to exposure time), and ambient concentrations. In
all cases, the final pH was between 4.2 and 6.6, as shown in
Figure 21. The presence of NH_3, of course, raised the pH. The
sulfate concentrations also depended on the same parameters
ranging from 2 - 2000 μmol ℓ^{-1}, and the presence of NH_3 increased
the sulfate concentration. This latter phenomenon was apparently
caused by the higher concentration of $SO_3^=$ at the higher pH, and
the higher oxidation rate of this species.

Although the model is based on a number of serious limitations,

Table 6. Aqueous phase mechanism for dissociation and oxidation considered by Overton et al. (1979).

Dissociation Mechanism

$H_2O \rightleftharpoons H^+ + OH^-$

$CO_2 + H_2O \rightleftharpoons H^+ + HCO_3^-$

$CO_2 + OH^- \rightleftharpoons HCO_3^-$

$SO_2 + OH^- \rightleftharpoons H^+ + HSO_3^-$

$HSO_3^- \rightleftharpoons H^+ + SO_3^=$

$NH_3 + H_2O \rightleftharpoons OH^- + NH_4^+$

$NH_3 + H^+ \rightleftharpoons NH_4^+$

Oxidation Mechanism

$SO_2 + O_3 \rightarrow 2\ H^+ + SO_4^=$

$SO_3^= + O_3 \rightarrow SO_4^=$

$HSO_3^- + O_3 \rightarrow H^+ + SO_4^=$

$Fe(III) + HSO_3^- \rightarrow H^+ + SO_4^= + Fe(II)$

$Fe(III) + SO_3^= \rightarrow SO_4^= + Fe(II)$

$Fe(II) + O_2 \rightarrow Fe(III)$

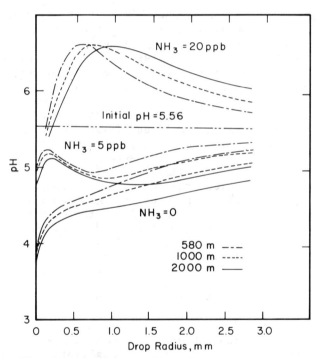

Figure 21. Drop pH vs. radius for various fall distances and ambient NH_3 concentrations. Initial liquid-phase conditions of the drop: pH = 5.56, $[CO_2(1w)]$ = 0.11 x 10^{-4} mol ℓ^{-1}, $[HCO_3^-]$ = 0.29 x 10^{-5} mol ℓ^{-1}, and $[Fe(III)]$ = 10^{-5} mol ℓ^{-1}. Ambient gas-phase concentrations: $[SO_2]$ = 10 ppb, $[O_3]$ = 100 ppb, $[NH_3]$ = 0, 5, and 20 ppb, and $[CO_2]$ = 320 ppm (adapted from Overton et al. (1979).

the values are essentially in the range of values reported in
the literature. Individual rain events have shown pH values from
3 to about 9 (cf., Likens, 1976; Jacobsen et al., 1976; and
Semonin, 1976). The average sulfate values have been as low as
20 μmol ℓ^{-1} and as high as 150 μmol ℓ^{-1} for SO_2 concentrations
in the 1 - 10 ppb range (Semb, 1976).

More recently, Adewuyi and Carmichael (1982) used a similar
approach, but with slightly different parameters. The liquid-
phase oxidation to produce $SO_4^=$ was not included, but absorption
was extended to include HNO_3 and HCl, in addition to SO_2, NH_3,
and CO_2. Furthermore, enhancement of the mass transfer by
aqueous-phase processes was incorporated, using the two-film
theory with δ_D = 0.05 R assumed. Finally, air-phase depletion
of the species was included. Air-phase concentrations similar
to those of Overton et al. (1979) were used.

Adewuyi and Carmichael's results show similar dependencies
on drop size, fall distance, and air-phase concentrations. The
final pH values ranged from about 4.5 to 6.0. SO_2 was found to
be a major contributor to the acidity by virtue of its relatively
high ambient concentration. On the other hand, the acidity due
to HNO_3 was attributed to its high solubility and rapid dissociation. The presence of NH_3 again resulted in increased solution
pH and drop capacity for S(IV). However, the effect was very
dependent on concentration, and the influence of NH_3 can substantially change during precipitation, since it is rapidly removed
from the atmosphere by the hydrometeors. In the simulations that
considered gas species depletion, the concentrations in the
liquid-phase were as much as a factor of six lower than for the
cases with constant air-phase concentrations. However, the pH
appeared to be substantially insensitive to the gas-phase depletion.

The Overton et al. (1979) and Adewuyi and Carmichael (1982)
studies are idealizations of a complex physical-chemical system.
Despite the limitations of these studies, some useful insights
into the mechanism and formation of acid rain were illustrated.
However, the current, significant lack of controlled laboratory
experiments for such complex conditions inhibits definitive comparisons among mechanisms and knowledgeable choices for improved
models. The recent paper by Charlson and Rodhe (1982) demon-
strates that, based simply on equilibrium considerations, the pH
of natural rainwater might range from 4.5 to 5.6 due only to
variability of the sulfur cycle.

Certainly, there have been careful studies and analysis of
specific systems (cf., Lee and Schwartz, 1981), but extension is
needed to more complex systems. Such studies must be forthcoming
if we are to establish the significance of the liquid-phase

234 L. K. PETERS

oxidation mechanisms being proposed. Some of the more speculative models are now addressing the role of free radical chemistry
in cloud droplets (Chameides and Davis, 1982), but without definitive experiments, such phenomena will remain speculative.

ACKNOWLEDGEMENTS

The author gratefully acknowledges support from NASA through
Research Grants NSG 1424 and NAG 1-36, and from the MAP3S Program
through Battelle Pacific Northwest Laboratory, for the preparation
of this chapter. Special thanks are due Ms. Mitzi Gilliam for
her careful typing of the manuscript.

GENERAL REFERENCES

Bird, R. B., W. E. Stewart, and E. N. Lightfoot, 1960: Transport
    Phenomena. John Wiley and Sons, Inc., New York, 780 pp.
Clift, R., J. R. Grace, and M. E. Weber, 1978: Bubbles, Drops,
    and Particles. Academic Press, Inc., New York, 380 pp.
Danckwerts, P. V., 1970: Gas-Liquid Reactions. McGraw-Hill, Inc.,
    New York, 276 pp.

REFERENCES

Abramzon, B. and I. Borde, 1980: Conjugate unsteady heat transfer
    from a droplet in creeping flow. AIChE J., 26, 536-544.
Adewuyi, Y. G. and G. R. Carmichael, 1982: A theoretical investi-
    gation of gaseous absorption by water droplets from SO_2-
    HNO_3-NH_3-CO_2-HCl mixtures. Atmos. Environ., 16, 719-729.
Altwicker, E. R. and E. Chapman, 1981: Mass transfer of sulfur
    dioxide into falling drops: A comparison of experimental
    data with absorption models. Atmos. Environ., 15, 297-300.
Ayers, G. P. and J. L. Gras, 1982: The concentration of ammonia
    gas in the southern ocean air. 2nd Symposium, Composition
    of the Nonurban Troposphere, Am. Meteorol. Soc., 45 Beacon
    Street, Boston, Mass., 36-39.
Baboolal, L. B., H. R. Pruppacher, and J. H. Topalian, 1981: A
    sensitivity study of a theoretical model of SO_2 scavenging
    by water drops in air. J. Atmos. Sci., 38, 856-870.
Baldwin, A. C. and D. M. Golden, 1980: Heterogeneous atmospheric
    reactions: Atom and radical reactions with sulfuric acid.
    J. Geophys. Res., 85, 2888-2889.
Balko, J. A. and L. K. Peters, 1982: A modeling study of SO_x-NO_x-
    hydrocarbon plumes and their transport to the background
    troposphere. Atmos. Environ., submitted for publication.
Barrie, L. A., 1978: An improved model of reversible SO_2 washout
    by rain. Atmos. Environ., 12, 407-412.

Beilke, S. and G. Gravenhorst, 1978: Heterogeneous SO_2 oxidation in the droplet phase. Atmos. Environ., 12, 231–239.

Beilke, S. and D. Lamb, 1975: Remarks on the rate of formation of bisulfite ions in aqueous solution. AIChE J., 21, 402–404.

Brian, P. L. T., 1964: Gas absorption accompanied by an irreversible reaction of general order. AIChE J., 10, 5–10.

Cadle, R. D., P. Crutzen, and D. Ehhalt, 1975: Heterogeneous chemical reactions in the stratosphere. J. Geophys. Res., 80, 3381–3385.

Campbell, M. J., J. C. Sheppard, F. J. Hopper, and R. Hardy, 1982: Measurements of tropospheric hydroxyl radical concentrations by the ^{14}C tracer method. 2nd Symposium, Composition of the Nonurban Troposphere, Am. Meteorol. Soc., 45 Beacon Street, Boston, Mass., 314.

Carmichael, G. R. and L. K. Peters, 1979: Some aspects of SO_2 absorption by water-generalized treatment. Atmos. Environ., 13, 1505–1513.

Chameides, W. L. and D. D. Davis, 1982: The free radical chemistry of cloud droplets and its impact upon the composition of rain. J. Geophys. Res., in press.

Chao, B. T., 1969: Transient heat and mass transfer to a translating drop. J. Heat Trans., 273–281.

Charlson, R. J. and H. Rodhe, 1982: Factors controlling the acidity of natural rainwater. Nature, 295, 683–685.

Cheng, R. T., M. Corn, and J. O. Frohliger, 1971: Contribution to the reaction kinetics of water soluble aerosols and SO_2 in air at ppm concentrations. Atmos. Environ., 5, 987–1008.

Clift, R., J. R. Grace, and M. E. Weber, 1978: Bubbles, Drops, and Particles. Academic Press, Inc., New York, 380 pp.

Danckwerts, P. V., 1950: Absorption by diffusion and chemical reaction. Trans. Faraday Soc., 46, 300–304.

Danckwerts, P. V., 1970: Gas-Liquid Reactions. McGraw-Hill, Inc., New York, 276 pp.

Davis, D. D., W. Heaps, and T. McGee, 1976: Direct measurements of natural tropospheric levels of OH via an aircraft borne tunable dye laser. Geophys. Res. Lett., 3, 331–333.

Davis, Jr., L. I., C. C. Wang, H. Niki, and B. Weinstock, 1982: Fluorescence measurements of OH at Niwot Ridge. 2nd Symposium, Composition of the Nonurban Troposphere, Am. Meteorol. Soc., 45 Beacon Street, Boston, Mass., 319–323.

Dobos, D., 1975: Electrochemical Data. Elsevier Scientific Publishing, Amsterdam, 339 pp.

Eigen, M., K. Kustin, and G. Maas, 1961: Die Geschwindigkeit der Hydration von SO_2 in wässriger Lösung. Z. Phys. Chem., 30, 130–135.

Erickson, R. E. and L. J. Yates, 1976: Reaction kinetics of ozone with sulfur compounds. U.S. Environmental Protection Agency, Report No. EPA-60013-76-089.

Fishman, J. and P. J. Crutzen, 1978: The origin of ozone in the troposphere. Nature, 274, 855–858.

Freiberg, J., 1979: Heterogeneous oxidation in the droplet phase.
    Atmos. Environ., 13, 1061-1062.
Frössling, N., 1938: The evaporation of falling drops. Beitr.
    Geophys., 52, 170-216.
Fuchs, N. A. and A. G. Sutugin, 1971: Highly dispersed aerosols.
    Topics in Current Aerosol Research (G. M. Hidy and J. R.
    Brock, editors), Pergamon Press, Inc., New York, 1-60.
Garner, F. H. and R. D. Suckling, 1958: Mass transfer from a
    soluble solid sphere. AIChE J., 4, 114-124.
Gelbard, F. and J. H. Seinfeld, 1979: The general dynamic equa-
    tion for aerosols. J. Coll. Interface Sci., 68, 363-382.
Gelbard, F., Y. Tambour, and J. H. Seinfeld, 1980: Sectional
    representations for simulating aerosol dynamics. J. Coll.
    Interface Sci., 76, 541-556.
Graedel, T. E., L. A. Farrow, and T. A. Weber, 1975: The influence
    of aerosols on the chemistry of the troposphere. Int. J.
    Chem. Kin. Symposium No. 1 (S. W. Benson, editor), John Wiley
    and Sons, New York, 581-594.
Griffith, R. M., 1960: Mass transfer from drops and bubbles. Chem.
    Eng. Sci., 12, 198-213.
Hales, J. M., 1972: Fundamentals of the theory of gas scavenging
    by rain. Atmos. Environ., 6, 635-659.
Hales, J. M., 1978: Wet removal of sulfur compounds from the
    atmosphere. Atmos. Environ., 12, 389-399.
Hales, J. M. and S. L. Sutter, 1973: Solubility of sulfur dioxide
    in water at low concentrations. Atmos. Environ., 7, 997-1001.
Happel, J. and H. Brenner, 1965: Low Reynolds Number Hydrodynamics.
    Prentice-Hall, Inc., Englewood Cliffs, New Jersey, 553 pp.
Hegg, D. A. and P. V. Hobbs, 1981: Cloud water chemistry and the
    production of sulfates in clouds. Atmos. Environ., 15,
    1597-1604.
Hikita, H. and S. Asai, 1964: Gas absorption with (m,n)-th order
    irreversible chemical reaction. Int. Chem. Eng., 4, 332-
    340.
Hikita, H., S. Asai, and H. Nose, 1978: Absorption of sulfur
    dioxide into water. AIChE J., 24, 147-149.
Hsu, N. T., K. Sato, and B. H. Sage, 1954: Material transfer in
    turbulent gas streams. Ind. Eng. Chem., 46, 870-876.
Hubler, G., D. Perner, U. Platt, A. Toennissen, and D. H. Ehhalt,
    1982: Groundlevel OH radical concentration: New measurements
    by optical absorption. 2nd Symposium, Composition of the
    Nonurban Troposphere, Am. Meteorol. Soc., 45 Beacon Street,
    Boston, Mass., 315-318.
Huebert, B. J. and A. L. Lazrus, 1978: Global tropospheric
    measurements of nitric acid vapor and particulate nitrate.
    Geophys. Res. Lett., 5, 577-580.
Huebert, B. J., 1980: Nitric acid and aerosol nitrate measure-
    ments in the equatorial Pacific region. Geophys. Res. Lett.,
    7, 325-328.
Jacobson, J. S., L. I. Heller, and P. Van Leuken, 1976: Acid

precipitation at a site within the northeastern conurbation. Wat. Air. Soil Pollut., 6, 339–349.

Jaenicke, R., 1978: Über die dynamik atmosphärischer Aitkenteilchen. Ber. Bunsenges. Phys. Chem., 82, 1198–1202.

Jaeschke, W., H. W. Georgii, H. Claude, and H. Malewski, 1978: Contribution of H_2S to the atmospheric sulfur cycle. Geofis. Pura. Appl., 116, 465–475.

Junge, C. E. and T. G. Ryan, 1958: Study of the SO_2 oxidation in solution and its role in atmospheric chemistry. Quart. J. Roy. Meteorol. Soc., 84, 46–56.

Junge, C. E., 1963: Air Chemistry and Radioactivity. Academic Press, New York, 382 pp.

Kelly, T. J. and D. H. Stedman, 1979: Measurements of H_2O_2 and HNO_3 in rural air. Geophys. Res. Lett., 6, 375–378.

Kley, D., J. W. Drummond, M. McFarland, and S. C. Liu, 1981: Tropospheric profiles of NO_x. J. Geophys. Res., 86, 3153–3161.

Kolthoff, I. M. and C. S. Miller, 1941: The reduction of sulfurous acid (sulfur dioxide) at the dropping electrode. J. Am. Chem. Soc., 63, 2818–2821.

Kronig, R. and J. C. Brink, 1950: On the theory of extraction from falling droplets. Appl. Sci. Res., A2, 142–154.

LeClair, B. P., A. E. Hamielec, H. R. Pruppacher, and W. D. Hall, 1972: A theoretical and experimental study of the internal circulation in water drops falling at terminal velocity in air. J. Atmos. Sci., 29, 728–740.

Lee, Y-N. and S. E. Schwartz, 1981: Evaluation of the rate of uptake of nitrogen dioxide by atmospheric and surface liquid water. J. Geophys. Res., 86, 11,971–11,983.

Levy, H., 1971: Normal atmosphere: Large radical and formaldehyde concentrations predicted. Science, 173, 141–143.

Likens, G. E., 1976: Acid precipitation. Chemical and Engineering News, November 22, pp. 29–44.

Logan, J. A., M. J. Prather, S. C. Wofsy, and M. B. McElroy, 1981: Tropospheric chemistry: A global perspective. J. Geophys. Res., 86, 7210–7254.

Luther, C. J. and L. K. Peters, 1982: The possible role of heterogeneous aerosol processes in the chemistry of CH_4 and CO in the troposphere. Workshop/Conference on Heterogeneous Catalysis – Its Importance to Atmospheric Chemistry, Albany, New York, June 29 – July 3, 1981.

Lynn, S., J. R. Straatemeier, and H. Kramers, 1955: Absorption studies in the light of the penetration theory. Chem. Eng. Sci., 4, 49–57.

Maroulis, P. J. and A. R. Bandy, 1977: Estimate of the contribution of biologically produced dimethyl sulfide to the global sulfur cycle. Science, 196, 647–648.

Maroulis, P. J., A. L. Torres, A. B. Goldberg, and A. R. Bandy, 1980: Atmospheric measurements of SO_2 on Project Gametag. J. Geophys. Res., 85, 7345–7349.

Martin, L. R. and D. E. Damschen, 1981: Aqueous oxidation of
    sulfur dioxide by hydrogen peroxide at low pH. Atmos.
    Environ., 15, 1615-1621.
Maxwell, R. W. and J. A. Storrow, 1957: Mercury vapor transfer
    studies - I. Chem. Eng. Sci., 6, 204-214.
McConnell, J. C., M. B. McElroy, and S. C. Wofsy, 1971: Natural
    sources of atmospheric CO. Nature, 223, 187-188.
Mihelcic, D., M. Helten, H. Fark, P. Müsgen, H. W. Pätz, M.
    Trainer, D. Kempa, and D. H. Ehhalt, 1982: Tropospheric
    airborne measurements of NO_2 and RO_2 using the technique
    of matrix isolation and electron spin resonance. 2nd Sympo-
    sium, Composition of the Nonurban Troposphere, Am. Meteorol.
    Soc., 45 Beacon Street, Boston, Mass., 327-329.
Miller, J. M. and R. G. dePena, 1972: Contribution of scavenged
    sulfur dioxide to the sulfate content of rain water. J.
    Geophys. Res., 77, 5905-5916.
Overton, Jr., J. H., V. P. Aneja, and J. L. Durham, 1979: Pro-
    duction of sulfate in rain and raindrops in polluted
    atmospheres. Atmos. Environ., 13, 355-367.
Penkett, S. A., 1972: Oxidation of SO_2 and other atmospheric
    gases by ozone in aqueous solution. Nature, 240, 105-106.
Perner, D., D. H. Ehhalt, H. W. Pate, E. P. Roth, and V. Volz,
    1976: OH-radicals in the lower troposphere. Geophys. Res.
    Lett., 3, 466-468.
Perry, R. H. and C. H. Chilton, 1973: Chemical Engineers' Hand-
    book, 5th Edition, McGraw-Hill, Inc., New York.
Peters, L. K. and W. L. Chameides, 1980: The chemistry and
    transport of methane and carbon monoxide in the troposphere.
    Adv. Env. Sci. Eng., III, 100-149.
Pruppacher, H. R. and K. V. Beard, 1970: A wind tunnel investi-
    gation of the internal circulation and shape of water drops
    falling at terminal velocity in air. Quart. J. Roy. Meteor.
    Soc., 96, 247-256.
Ranz, W. E. and W. R. Marshall, 1952: Evaporation from drops.
    Chem. Eng. Prog., 48, 141-146, 173-180.
Rasmussen, R. A., S. D. Hoyt, and M. A. Khalil, 1982: Atmospheric
    carbonyl sulfide (COS): Techniques for measurement in air
    and water. 2nd Symposium, Composition of the Nonurban
    Troposphere, Am. Meteorol. Soc., 45 Beacon Street, Boston,
    Mass., 265-268.
Reid, R. C., J. M. Prausnitz, and T. K. Sherwood, 1977: The
    Properties of Gases and Liquids. McGraw-Hill, Inc., New
    York, 688 pp.
Rowe, P. N., K. T. Claxton, and J. B. Lewis, 1965: Heat and mass
    transfer from a single sphere in an extensive flowing fluid.
    Trans. Inst. Chem. Eng., 43, T14-T31.
Rudolph, J., D. H. Ehhalt, A. Khadim, and C. Jebsen, 1982: Lati-
    tudinal profiles of some C_2-C_5 hydrocarbons in the clean
    troposphere over the Atlantic. 2nd Symposium, Composition
    of the Nonurban Troposphere, Am. Meteorol. Soc., 45 Beacon

Street, Boston, Mass., 284-286.

Sahni, D. C., 1966: The effect of a black sphere on the flux distribution in an infinite moderator. J. Nucl. Energy A/B, 20, 915-920.

Scott, B. C., 1982: Predictions of in-cloud conversion rates of SO_2 to SO_4 based upon a simple chemical and kinematic storm model. J. Atmos. Sci., 16, 1735-1752.

Scott, W. D. and P. V. Hobbs, 1967: The formation of sulfate in water drops. J. Atmos. Sci., 24, 54-57.

Secor, R. M. and J. A. Beutler, 1967: Penetration theory for diffusion accompanied by a reversible chemical reaction with generalized kinetics. AIChE J., 13, 365-373.

Seiler, W. and J. Fishman, 1981: The distribution of carbon monoxide and ozone in the free troposphere. J. Geophys. Res., 86, 7255-7265.

Semb, A., 1976: Measurement of acid precipitation in Norway. Wat. Air. Soil Pollut., 6, 231-240.

Semonin, R. G., 1976: The variability of pH in convective storms. Wat. Air. Soil Pollut., 6, 395-406.

Sherwood, T. K. and R. L. Pigford, 1952: Absorption and Extraction. McGraw-Hill Book Co., New York, 2nd Edition, 478 pp.

Sherwood, T. K., R. L. Pigford, and C. R. Wilke, 1975: Mass Transfer. McGraw-Hill, Inc., New York, 677 pp.

Skelland, A. H. P., 1974: Diffusional Mass Transfer. John Wiley and Sons, Inc., New York, 510 pp.

Skelland, A. H. P., and A. R. H. Cornish, 1963: Mass transfer from spheroids to an air stream. AIChE J., 9, 73-76.

Spells, K. E., 1952: A study of circulation patterns within liquid drops moving through a liquid. Proc. Phys. Soc., B65, 541-546.

Tanner, R. L., 1982: Temporal and spatial variability of ammonia concentrations in the non-urban troposphere. 2nd Symposium, Composition of the Nonurban Troposphere, Am. Meteorol. Soc., 45 Beacon Street, Boston, Mass., 29-32.

Terraglio, F. P. and R. M. Manganelli, 1967: The absorption of atmospheric sulfur dioxide by water solutions. J. Air Pollut. Control Assoc., 17, 403-406.

Van den Heuval, A. P. and B. J. Mason, 1963: The formation of ammonium sulphate in water droplets exposed to gaseous sulphur dioxide and ammonia. Quart. J. Roy. Meteor. Soc., 89, 271-275.

Walcek, C., P. K. Wang, J. H. Topalian, S. K. Mitra, and H. R. Pruppacher, 1981: An experimental test of a theoretical model to determine the rate at which freely falling water drops scavenge SO_2 in air. J. Atmos. Sci., 38, 871-876.

Wang, C. C., L. I. Davis, Jr., C. H. Wu, S. Japar, H. Niki, and B. Weinstock, 1975: Hydroxyl radical concentrations measured in ambient air. Science, 189, 797-800.

Weinstock, B., 1969: Carbon monoxide residence time in the atmosphere. Science, 166, 224-225.

Willeke, K. and K. T. Whitby, 1975: Atmospheric aerosols: Size
    distribution interpretation. J. Air Poll. Control Assoc.,
    25, 529-534.
Wofsy, S. C., 1976: Interactions of CH_4 and CO in the Earth's
    atmosphere. Annual Review of Earth and Planetary Sciences,
    4 (F. A. Donath, F. G. Stehli, and G. W. Wetherill, editors),
    Annual Reviews, Palo Alto, California, 441-469.
Zafiriou, O. C., J. Alford, M. Herrera, E. T. Peltzer, and R. B.
    Gagosian, 1980: Formaldehyde in remote marine air and rain:
    Flux measurements and estimates. Geophys. Res. Lett., 7,
    341-344.

GAS TRANSFER: EXPERIMENTS AND GEOCHEMICAL IMPLICATIONS

Peter S. Liss

School of Environmental Sciences
University of East Anglia
Norwich, NR4 7TJ
United Kingdom

1.   INTRODUCTION

    The processes governing the transfer of gases across the air-sea interface have been reviewed by several authors (e.g. Bolin, 1960; Kanwisher, 1963; Broecker and Peng, 1974; Liss and Slater, 1976; Slinn et al.,1978). Basically, any net gas flux (F) across the interface must be driven by a concentration difference (ΔC) between air and surface water; the magnitude and direction of the flux being proportional to the numerical value and sign of ΔC, i.e.,

$$F = K \, \Delta C \tag{1}$$

The constant of proportionality (K), linking the flux and the concentration difference, has dimensions of a velocity and is variously known as an exchange constant/coefficient or a piston/transfer velocity; the lattermost term will be used here. It is often more convenient to think in terms of the reciprocal of the transfer velocity, which is a measure of the resistance (R) to interfacial gas exchange. Following Danckwerts (1970) and Liss and Slater (1974), (2) - (5) show how the total resistance to gas transfer (R) can be split into resistances in the air (r_a) and in the water (r_w):

$$1/K_w = 1/\alpha k_w + 1/Hk_a \tag{2}$$

Which, in terms of resistances, may be written as

$$R_w = r_w + r_a \tag{3}$$

241

P. S. Liss and W. G. N. Slinn (eds.), Air-Sea Exchange of Gases and Particles, 241–298.
Copyright © 1983 by D. Reidel Publishing Company.

In (2), k_a and k_w are the transfer velocities for chemically unreactive gases in the air and water, respectively. H is the dimensionless Henry's law constant, expressed as the ratio of the concentration of the gas in air to its concentration in un-ionised form dissolved in the water, at equilibrium. α is a factor which quantifies any enhancement in the value of k_w due to chemical reactivity of the gas in the water, and is defined as the ratio of the transfer velocity for the gas of interest to that for an inert one, all other exchange processes remaining constant. On this definition, α for gases chemically unreactive in water (e.g. N_2, The Inert Gases) is unity, but values up to several thousand can occur for gases with very rapid aqueous phase chemistry (e.g. SO_2 - discussed later).

For (2), the driving force, ΔC, has been taken to be $(C_a H^{-1} - C_w)$, where C_a and C_w are the gas concentrations in air and water, respectively. If ΔC had been taken as $(C_a - HC_w)$ then, instead of (2) we would have

$$1/K_a = 1/k_a + H/\alpha k_w \tag{4}$$

and

$$R_a = r_a + r_w \tag{5}$$

From the form of (2) - (5) there is a clear analogy between gas flux across an air-water interface and the flow of electrons through an electrical circuit containing two resistances in series. This is illustrated in Figure 1 a) and b), where i(current) $\equiv$ F, Δe(potential difference) $\equiv$ ΔC, and r(electrical resistance) $\equiv$ r(aerodynamic or hydrodynamic resistance).

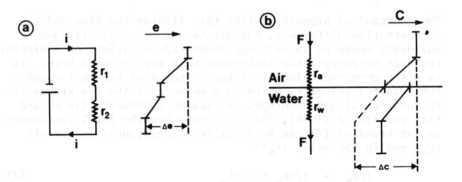

Figure 1.   Analogy between electron flow in an electrical circuit (a), and gas flux across and air-water interface (b). Symbols used are given in the text.

In fact the only respects in which the analogy breaks down are
the presence of H and α in (2) and (4). However, H only appears
because concentrations rather than partial pressures are used to
express ΔC. Further, as will be shown later, α can in fact be
thought of in resistance terms, in this case via parallel
resistance paths in the water boundary layer.

By substituting appropriate values for the various terms in
(2) - (5), it is possible to calculate the magnitude of r_w and
r_a for any particular gas and thus to identify the phase whose
resistance controls its air-sea transfer. When this is done,
many simple gases fall into two distinct groups: i) those for
which $r_a \gg r_w$ (e.g. H_2O, SO_2, SO_3, and NH_3), and ii),those for
which $r_w \gg r_a$ (e.g. O_2, N_2, The Inert Gases, CO_2, CO, CH_4, MeI
and Me_2S). It is apparent that gases whose exchange is controlled
by transfer in the water generally have low solubility and are
chemically unreactive in the aqueous phase: whereas,for gases
of high solubility and/or rapid chemistry in the water it is processes in the air that control their interfacial transfer. There
are a few gases that fall between these two broad classes; for
these the resistances in both the air and the water are important.
The only low molecular weight compounds yet identified for which
this appears to be the case are HCHO and H_2S (for solution pH $\geqslant$ 8),
both of which are referred to later. Air-water exchange of
several intermediate and high molecular weight organic compounds
(e.g. DDT and PCBs) has been predicted to be subject to transfer
resistance from both phases (Hunter and Liss, 1977; Mackay and
Yuen, 1981; Atlas et al., 1982).

In this chapter, I first describe laboratory experiments
(mainly using wind tunnels) aimed at investigating some of the
parameters identified in (2) and (4) as important in air-water
gas transfer. After that, field studies aimed at measuring air-
sea gas fluxes and/or transfer velocities are discussed. Finally,
the findings of the previous two sections are brought together
in an attempt to use our present, albeit imperfect, understanding
of the factors controlling gas fluxes across the sea surface to
calculate the magnitude and importance of such exchange in the
geochemical cycling of natural and man-made substances in the
global environment.

2. LABORATORY STUDIES OF AIR-WATER GAS TRANSFER

The majority of serious laboratory studies of gas exchange
at air-water interfaces have employed wind tunnels in order to
try to simulate, in as meaningful a way as possible, conditions in
the real environment. Here the major wind tunnel studies will
be reviewed and the results compared with model predictions,where
possible; physical and chemical transfer processes will be
considered separately.

2.1   Physical Transfer Processes

Most laboratory effort on transfer by physical processes has been devoted to gases whose exchange is under liquid phase control ($r_w \gg r_a$), and the results can be categorized into the effects of wind, waves, bubbles, and heat transfer.

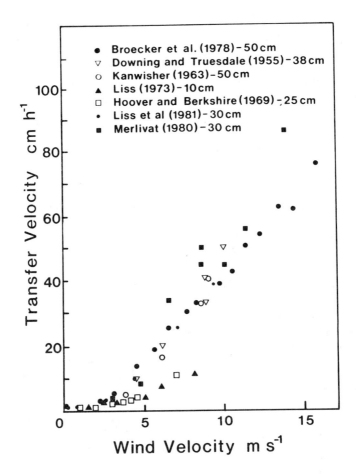

Figure 2.   Relationship between air-water gas transfer velocity and wind velocity found in various wind tunnel studies.   The depth of water in the tunnel is shown in the key.   Heights above the water surface at which the plotted wind velocities were measured are: 5 cm (Downing and Truesdale);   10 cm (Kanwisher, Liss); 17 cm (Liss et al.);   60 cm (Broecker et al.); height of wind measurement not reported by Hoover and Berkshire or Merlivat.   (From Broecker et al., 1978, with later results added.)

2.1.1   Wind.   The effect of wind is illustrated in
Figure 2, which is a composite plot incorporating the findings
of several different wind tunnel studies.   Despite the fact that
the tunnels had very different dimensions and the height at which
the wind velocity was measured varied between studies, the results are in general quantitative agreement, and show a clear
increase in transfer velocity with wind speed.

Closer inspection of the data, however, reveals some differences in detail.   For example, Kanwisher (1963) interprets his
data as showing the increase in transfer velocity as proportional
to the square of the wind speed, whereas the results of Broecker
et al. (1978) closely follow a linear trend, at least above a
wind speed of 2-3 m s^{-1}.   Broecker et al. (1978) remark that
these differences appear to be related more to the depth of
water in the tunnel (given on the figure) than to its length.
This, coupled with the obvious differences between the experimental arrangements, and the fact that the range of wind speeds
differs between studies, means that some disagreement among the
data sets is to be expected.   It probably isn't profitable to
spend a lot of time trying to find reasons for the discrepancies.
Instead, the surprising (in the light of the experimental differences) agreement among the studies should, perhaps, be stressed.

All the data shown in Figure 2 were obtained using linear
wind tunnels operated in either the straight-through or recircu-
lating modes.   However, linear tunnels suffer from the obvious
defects of limited wind fetch and constrained water flow, compared with the open oceans.   Even the largest tunnel for which gas
exchange results have been published (Hamburg - Broecker et al.
1978) has an effective working length of only 18 m, and most of
the tunnels used are substantially smaller.   An elegant alternative approach is to use a circular tunnel, in which the wind
flows continuously over the unconstrained water surface.   A
small-scale facility of this type (outside diameter 75 cm, water
depth and width both 10 cm) has been built in Heidelberg, and
the tunnel and results obtained from it are described in Jahne
et al. (1979).   Some of their data are shown in Figure 3.

The form of the results from the circular tunnel is clearly
bi-linear, with a very gradual increase in transfer velocity up
to a wind speed of about 8 m s^{-1}, and a much more rapid increase
thereafter.   What seems to happen is that, at a wind velocity
near 8 m s^{-1}, there is a jump in the transfer velocity by about
a factor of 5, concomitant with a similar increase in the friction
velocity.   The wave pattern shows a similarly abrupt transition,
from almost calm (wave amplitude < 0.2 cm) below 8 m s^{-1}, to
irregular waves of amplitude up to several centimetres above this
wind speed.   The authors suggest that the discontinuity may be

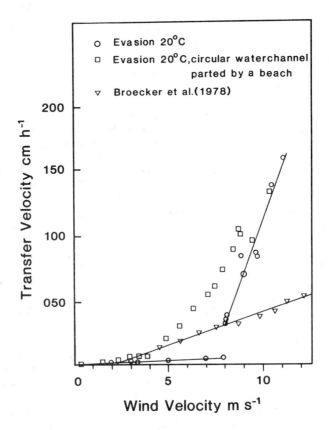

Figure 3.   Relationship between air-water transfer
velocity for CO_2 and wind velocity in the Heidelberg
circular wind tunnel. Results are for evasion of gas,
with and without a beach in the water channel. Also
shown are data from a linear tunnel (Broecker et al.,
1978). (From Jahne et al., 1979.)

caused by a Kelvin-Helmholtz instability, and point out that
similar effects have been observed in studies of wave formation
in annular channels (Francis, 1949).

Also shown in Figure 3 are data from an experiment in which
the water channel was made discontinuous (by insertion of a
beach), and results from the Hamburg linear tunnel (Broecker et
al. 1978). With the beach in place no sharp jump in transfer
velocity is observed, and the quasi-continuous increase is
similar in form (but not necessarily in magnitude - see below)
to that found in several of the studies in linear tunnels. Furthermore, the water surface is covered with rough waves at a wind

velocity of 3 m s^{-1}. It is noteworthy that, at higher wind
speeds, transfer velocities measured in the Hamburg facility
are substantially lower than those in the circular tunnel, with
or without the beach. Although these high transfer velocities
appear to be matched by equivalently large values of the friction
velocity (see Figure 6) they are much higher than values generally
measured at sea (see Section 3). This finding, coupled with the
instability discussed above and the possibility of artefacts
from rotational forces inherent in a circular system, may mean
that the potential benefits of this type of tunnel cannot be
fully realised.

   2.1.2 Waves. In several of the wind-tunnel studies reported in the previous section, the authors have noted the rapid
increase in air-water gas transfer velocity which appears to
occur with the onset of waves on the water surface (Kanwisher,
1963; Broecker et al., 1978; Jahne et al., 1979). However,
there seems to have been very little systematic investigation of
the effects of waves on gas transfer in wind tunnels. Certainly,
the idea of enhanced gas transfer coincident with the appearance
of capillary waves has appeal, because of the increased area for
gas exchange presented by the rippled surface. MacIntyre (1971)
has calculated that for a surface covered with steep Crapper
ripples the enhancement in gas transfer could be 3.5 times the
flat-surface value. However, it is doubtful whether such steep
waves are sufficiently stable to cover any significant area of
the surface (Hasse and Liss, 1980). These authors conclude that,
under natural conditions, the increase in area due to capillaries
is likely to be 50% at the very most, so the effect is probably
not of great quantitative importance. Following on from this, it
could be argued that capillary waves themselves may not be the
cause of the observed enhancement in gas exchange, but may merely
act as indicators of a change in the nature of the gas-transfer
process in the water close to the interface, which may, or may
not, have the same origin as the ripples. Some support for this
idea comes from wind tunnel studies where mechanically generated
waves were created (as well as those produced by wind action),
since the effect of mechanical waves was found to be small or
negative (Downing and Truesdale, 1955; Kanwisher, 1963). In
these studies it was found that in the wind speed range 4-10 m s^{-1},
the imposition of small waves (height 2 cm, Downing and Truesdale -
their Figure 3; height 3 cm, Kanwisher - his Figure 4) actually
led to a significant decrease (of order 25%) in the measured transfer velocity. This is often attributed, probably naively, to the
lee side of the wave being sheltered from the wind.

   It should be borne in mind that, however well wind tunnel
studies are performed, there are certain fundamental problems in
translating the findings to processes at the air-sea interface. F.
Dobson (personal communication) has pointed out that with the

limited water depth in wind tunnels, waves grow faster and break
at lower wind speeds than at sea. Furthermore, reflection of
waves from the walls of the tank produces standing waves which,
again, are more likely to grow and break than those on the
largely unconfined sea surface.

    2.1.3 Bubbles.   Although there is considerable theoretical
and experimental information on the role of bubbles as gas exchangers (see, for example, Thorpe, 1982; Merlivat and Memery,
1983), they have been largely ignored in wind tunnel experiments.
One study in which the importance of bubbles is assessed is that
carried out by Broecker (1980) in the Hamburg facility, the
results of which are shown in Figure 4.   In these experiments,

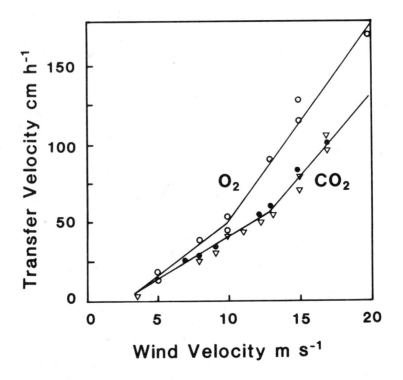

Figure 4.   Dependence of air-water transfer velocities
for O_2 and CO_2 on wind velocity in the Hamburg wind
tunnel. (From Broecker, 1980).

transfer velocities were measured as a function of wind speed,
for both O_2 and CO_2. Beyond a wind speed of 2 m s^{-1}, when
capillary waves begin to be present on the water surface, the
transfer velocity for both gases increases linearly, up to a wind

speed of 8 m s^{-1} for O_2 and 14 m s^{-1} for CO_2, when an abrupt
change in slope occurs. This change in slope is attributed to
the production of bubbles by breaking waves (there were no
artificial bubbles in these experiments). The results clearly
indicate the potential importance of bubbles as gas exchangers,
provided the (wind) conditions are right for their formation.
The effect of bubbles is more pronounced and occurs at a lower
wind speed for O_2 than for CO_2, and this is attributable to the
relative insolubility of oxygen ($H \sim 30$) compared with carbon
dioxide ($H \sim 1$). Wyman et al. (1952) found similar differential
effects for dissolution of gas bubbles containing O_2 and N_2
($H \sim 60$). Recently Merlivat and Memery (1983) have reported on a
detailed theoretical and experimental study in which the Marseille
wind tunnel was used to examine the role of bubbles in the air-
water transfer of Ar and N_2O.

   2.1.4   Heat Transfer. Almost all laboratory studies of air-
water gas exchange have been performed under conditions where
there was net evaporation of water from the liquid surface.
However, in a wind-tunnel study of CO_2 exchange, Hoover and
Berkshire (1969) noted that, if the temperature of the water in
their tunnel was below the dew point of the air, then gas transfer was severely inhibited. This was an unwanted artefact in
their experiments, and Hoover and Berkshire thereafter ensured
that the water temperature was always above the local atmospheric
dew point. This finding led Quinn and Otto (1971) to suggest that
what was being observed was damping of the 'normal' evaporation-
driven convective motions by condensation onto the surface, leading to enhanced thermal stability of the near-surface water.

   A detailed examination of this phenomenon was carried out by
Liss et al. (1981) using the smaller of the two wind tunnels at
Marseille. In this context, a vital feature of these tunnels is
the ability to control not only the temperature of both air and
water, but also the humidity of the air stream. The invasion of
O_2 was monitored at various wind speeds under neutral (i.e. essentially zero heat flux between air and water), as well as net
evaporation and net condensation conditions. The direction of
any flux of sensible heat was in the same sense as the latent heat
flux, but was generally only a small fraction of it. The experimental results are shown in Figure 5.

   No significant difference was found between transfer velocities measured at the same wind speed, but under neutral conditions or with various degrees of evaporation. However, at intermediate wind speeds, condensing conditions produced a significant
(> 30%) decrease in transfer velocity. It was possible to
account for the observed effect in terms of a stratification
parameter, defined as the ratio of the total (latent and sensible) heat flux to the cube of the friction velocity.

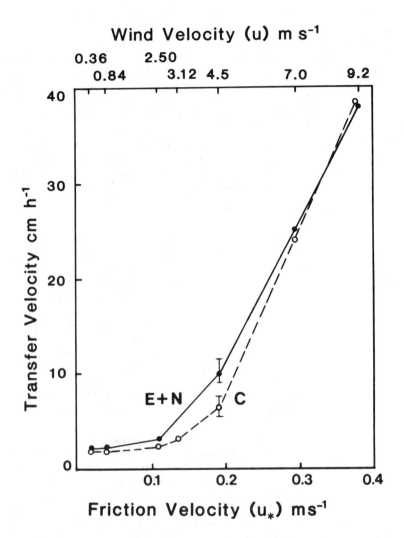

Figure 5.   Oxygen transfer velocity as a function of
friction velocity (u_*) and wind speed (u).   Mean of
results for: evaporative (E) and neutral (N) case experiments (●), condensing (C) experiments (○).   The
maximum and minimum values for each of the two groups
of results are shown for $u_* = 19 \times 10^{-2}$ ms^{-1}. (From
Liss et al., 1981).

    In the environmental context, there are two conclusions to
be drawn from these results.   The first is that convective motion
in the near-surface water, driven by evaporative cooling, is unlikely to be of importance (except possibly under the very calm

conditions), and must generally be overwhelmed by mechanically-
(e.g. wind-) generated turbulence. Even at the lowest wind speed
used in the tunnel no effect of evaporation on O_2 transfer was
observed.

Secondly, the reduction in transfer velocity under condensing conditions may be of some environmental importance. That
is, although evaporation must dominate over condensation for the
oceans as a whole, in some particular areas the reverse may per-
tain. For example, in coastal upwelling regions where cold, deep
water is brought to the surface, condensation may occur and so
lead to reduced transfer of O_2 and other biologically important
gases across the sea surface.

2.2   Modelling Approaches

Several models have been proposed to describe air-water gas
transfer. In the film model (Whitman, 1923) the rate controlling
process is transport by molecular diffusion across a stagnant
layer of fluid adjacent to the interface. The surface renewal
model (Higbie, 1935; Danckwerts, 1951) envisages periodic re-
placement of the stagnant fluid at the interface with material
from the bulk, as the rate determining step. For the film and surface renewal models the transfer velocity is proportional to
molecular diffusivity ($\mathcal{D}$) to the power 1 or $\frac{1}{2}$, respectively.
Neither model is easy to apply to laboratory (or field) studies
since film thicknesses cannot be measured directly and rates of
surface renewal are very hard to quantify.

The above difficulties led Deacon (1977) to apply boundary-
layer theory, as used by micrometeorologists, to describe wind-
tunnel results for air-water gas transfer. He used a treatment
based on Reichardt's (1951) formulation of the velocity profile in
turbulent air flow over a smooth, rigid surface. The major feature
of Deacon's analysis was to use, instead of the friction velocity
in air, $u_{*,a}$, the friction velocity in water, $(\rho_a/\rho_w) u_{*,a}$, which
assumes continuity of stress across the interface, and where ρ_a
and ρ_w are the density of air and water, respectively. By using
the friction velocity in the water, the Reichard formulation then
applies to mass transfer through the water boundary layer, which
constitutes the principal resistance to transfer for gases of
interest here. In Deacon's treatment the water transfer velocity
(k_w) is given by:

$$k_w = 0.082 \ Sc^{-2/3} (\rho_a/\rho_w)^{\frac{1}{2}} \ u_{*,a} \tag{6}$$

where Sc is the Schmidt number (ratio of kinematic viscosity to
molecular diffusivity) in the water.

    The major advantage of Deacon's approach, over the film and
surface renewal models, is that the only unknown quantity in
(6) is $u_{*,a}$, and this can be determined reasonably easily in
both laboratory and field situations. It should be noted that
in (6) k_w is proportional to $Sc^{-2/3}$, i.e. to $\mathcal{D}^{2/3}$, which is in
between the predictions of the film ($\mathcal{D}^1$) and surface renewal ($\mathcal{D}^{\frac{1}{2}}$)
models.

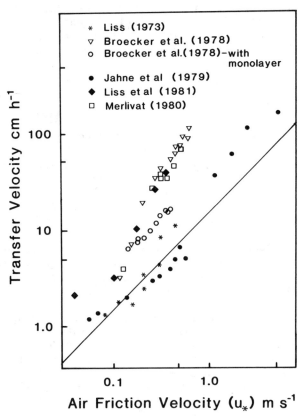

Figure 6. Log-log plot of the relationship between
transfer velocity, as measured in various wind tunnel
studies, and air friction velocity, compared to predictions (shown by the full line) from the smooth-wall
model proposed by Deacon (1977). Experimentally determined transfer velocities have been adjusted to a common Schmidt number of 600 (the value for CO_2 at 20°C), by
assuming that $k \propto Sc^{-2/3}$. The results of Liss (1973)
and Liss et al. (1981) are for O_2; those of Broecker et
al. (1978) and Jahne et al. (1979) are for CO_2; and
those of Merlivat (1980) are for Ar. (After Jahne and
Munnich, 1980).

The most detailed application of (6) to wind-tunnel results for air-water gas transfer has been made by Jahne and Munnich (1980), and their findings are shown in Figure 6. For transfer velocities up to about 4 cm h^{-1} ($u_* \sim 4$ m s^{-1}), the experimental results of Liss (1973), for the Harwell linear tunnel, and Jahne et al. (1979), for the Heidelberg circular tunnel, are in reasonable agreement with the predictions of the Deacon model; but there is considerable underestimation of the transfer velocity at higher wind speeds. In the studies of Broecker et al. (1978), using the Hamburg tunnel, and Merlivat (1980) and Liss et al. (1981), both for the Marseille facility, the experimental transfer velocities are greater than the model predictions at all wind speeds. Since these latter are the biggest of the tunnels so far employed, it is probable that the wave field is better developed, even at low wind speeds, than in the smaller facilities. Thus, although useful, (6) appears to underestimate transfer velocities, as soon as the surface ceases to be smooth.

This deficiency led Kerman (1983) to extend the micrometeorological approach by using Yaglom and Kader's (1974) treatment for transfer to a rough surface. To apply this in the field or laboratory, it is necessary to model the surface roughness caused by breaking gravity and capillary waves, which occur only patchily over the water surface. As with the smooth surface treatment, there is quite reasonable agreement between some wind-tunnel data sets and the model predictions. However, especially for the results from the larger tunnels, significant discrepancies between theory and experiment are apparent.

2.3   Gases for Which $r_a \gg r_w$.

So far only gases whose exchange is under liquid phase control have been considered. Next, gases for which resistance in the water is negligible, and thus, whose transfer is controlled by processes in the gas phase, are briefly discussed. The most obvious gas in this category is H_2O itself and we will confine ourselves to it here. The vast excess of water molecules in an aqueous solution means that, except possibly at a surface covered by a coherent organic film (see Section 4), there is no possibility of a concentration gradient developing, so that all transfer resistance must be in the gas phase.

Results for the transfer velocity for H_2O as a function of wind velocity are shown in Figure 7 for the Harwell wind tunnel (Liss, 1973). There is a clear linear relationship between transfer velocity and wind speed (or u_* - not shown). Deacon (1977) has demonstrated that there is reasonable agreement between the measured transfer velocities for H_2O and those predicted theoretically using the smooth-surface approach. Theoretical approaches for predicting water evaporation rates are reviewed by Merlivat and Coantic (1975).

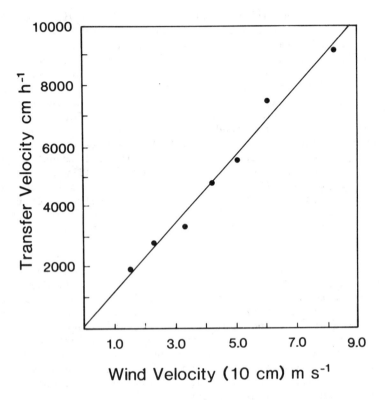

Figure 7. Transfer velocity for H_2O as a function of
wind velocity measured at a height of 10 cm above the
water surface. (From Liss, 1973).

2.4   Chemical Transfer Processes.

     Molecular and turbulent diffusion are processes applicable to
transfer of all gases through the air and water interfacial boundary regions. However, for gases that can react sufficiently rapidly in the aqueous phase, transfer across the water boundary
layer may be enhanced over that due solely to physical mechanisms.
For example, if CO_2 is invading from air to water then there will
be not only a concentration gradient in the near-surface water,
down which $CO_{2(aq)}$ will diffuse, but also the possibility of these
molecules reacting with the water in the boundary region to form
ionic species. The extent to which this takes place depends on:
i) the pH of the water, and ii) the relative time for a molecule
to diffuse through the water boundary layer, as compared to the
time for chemical reaction. These two factors are now briefly
discussed.

     i)   Water pH is important for CO_2 since below a pH of about
5 ionic forms of inorganic carbon will not exist, so that
chemical enhancement of exchange is impossible. At higher pHs

$CO_{2(aq)}$ can be converted to HCO_3^-, either by dissociation of $H_2CO_{3(aq)}$ or by direct reaction with OH^-, i.e.

$$CO_{2(aq)} + H_2O \rightleftarrows H_2CO_{3(aq)} \rightleftarrows H^+ + HCO_3^- \qquad (7)$$

$$\text{or} \qquad CO_{2(aq)} + OH^- \rightleftarrows HCO_3^- \qquad (8)$$

Reaction (8) is dependent on the concentration of OH^- and only occurs to any significant extent at solution pHs > 8.

ii) The time constant for chemical reaction to occur is a property of the interacting species, and will vary with temperature and solution composition, but will not depend on hydrodynamic conditions in the water. In contrast, the time for diffusive transport through the near-surface water will depend on the degree of stirring (turbulence). Essentially, the liquid phase transfer velocity is an inverse measure of the time required for a molecule to traverse the water boundary layer. Thus, at a high wind speed, when the transfer velocity is large, the transit time will be short, and vice versa under calm conditions. These general considerations lead to the conclusions that chemical enhancement is more likely to be apparent under calm, rather than rough, regimes; and for CO_2, only for solution pH > 5.

It is possible to think of the chemical enhancement factor, α, in terms of the electrical resistance analogy developed in Section 1. This approach is illustrated in Figure 8. The transfer resistance in the air is r_a, as previously. However, in this case,

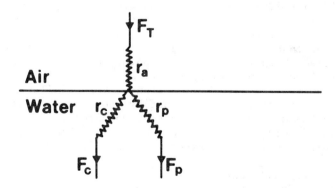

Figure 8. Resistance analogy for enhanced chemical transfer of gas through the aqueous boundary layer near an air-water interface. Symbols used are given in the text.

the resistance to gas transfer through the surface water (r_w) is composed of two parallel resistance paths, one due to the normal diffusive, physical transfer (r_p), and the other by chemical reaction (r_c). The portions of the total gas flux (F_T) carried by the physical and chemical resistance paths are F_p and F_c, respectively. Application of the well known relationships for addition of electrical resistances in series and parallel yields for the total resistance (R),

$$R = (1/r_c + 1/r_p)^{-1} + r_a \tag{9}$$

It is then straightforward to express α in terms of the two parallel resistances in the water, i.e.

$$\alpha = (r_p/r_c) + 1 \tag{10}$$

Equation (10) seems physically realistic since when $r_c \gg r_p$, $\alpha \to 1.0$ (i.e. no chemical enhancement); if $r_c = r_p, \alpha = 2.0$; and for $r_c \ll r_p$ then $\alpha \gg 1.0$ and chemical processes dominate on the liquid side of the interface.

Several laboratory studies have been carried out to try to quantify the chemical-enhancement effect. Liss (1973) used the Harwell wind tunnel to measure simultaneously the transfer velocities of O_2 (which, because of its unreactivity, will not be subject to chemical exchange enhancement) and CO_2, as a function of wind velocity. The pH of the water in these experiments was 6.2 - 6.6. The results are shown in Figure 9, which clearly illustrates the higher transfer velocity for CO_2 over O_2 at low and intermediate wind speeds. The maximum exchange enhancement is 61% ($\alpha = 1.61$), at the lowest wind speed, and decreases with increasing wind velocity until, at $u \sim 5$ m s^{-1}, the transfer velocities for the two gases are indistinguishable. The lack of exchange enhancement above a certain wind speed is as predicted, and arises because, above this point, diffusion dominates over chemical reaction. In contrast, at lower wind speeds chemical transfer is increasingly important. Very similar results to those shown in Figure 9 were obtained by Hoover and Berkshire (1969) in a wind tunnel study in which they compared the transfer velocity for CO_2 evading from water of pH < 4 (no chemical enhancement expected), with that for water with pH 6.3 - 6.8.

Hoover and Berkshire (1969) derived the following equation to predict the value of the enhancement factor (α):

$$\alpha = |\tau(k^*\tau/D)^{\frac{1}{2}}(D/k_w)| / |(\tau-1)(k^*\tau/D)^{\frac{1}{2}}(D/k_w)$$
$$+ \tanh\{(k^*\tau/D)^{\frac{1}{2}}(D/k_w)\}| \tag{11}$$

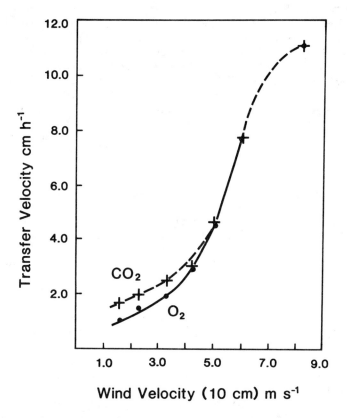

Figure 9. Simultaneously determined values of the transfer velocity of O_2 and CO_2 as a function of wind velocity measured at a height of 10 cm above the water surface. (From Liss, 1973).

Where τ is the ratio of total to ionic forms of inorganic carbon in the water, k^* is the hydration rate constant, $\mathcal{D}$ is the co-efficient of molecular diffusion of CO_2 in the water, and k_w is the transfer velocity for an inert gas. The authors found that (11) would predict the observed enhancements to within a few per cent, and the agreement for the results shown in Figure 9 is al-most as good.

Although apparently successful in explaining quantitatively the laboratory results, (11) has been criticised by Quinn and Otto (1971), since in deriving it the assumption is made that the pH is constant in the near-surface water, despite the fact that CO_2 (and hence, at pH > 5, H^+) is being added or subtracted through chemical reaction. Quinn and Otto presented an alternative method for predicting α by making the assumption that, whatever chemical

changes occur, electrical neutrality will be maintained in the
water. However, except at very low transfer velocities (< 1 cm
h^{-1}), predictions based on constant pH or electrical neutrality
assumptions do not differ significantly, so that the computat-
ionally simpler relationship of Hoover and Berkshire (11) is almost always used. For waters having high pHs (9-10), such as are
found in some eutrophic lakes in summer, Emerson (1975) has developed a prediction scheme for α which pays particular attention
to reaction between CO_2 and OH^- (reaction 8), since this dominates
chemical-exchange enhancement in highly alkaline waters.

Equation (11) can be used to estimate α for CO_2 under average oceanic conditions. What is required are numerical values
for $k^*(\sim 10^{-2} \text{ s}^{-1})$, $\mathcal{D}(\sim 10^{-5} \text{ cm s}^{-1})$, τ (which is calculated from
the dissociation constants for carbonic acid and the pH of sea-
water), and k_w for which the global average can be assumed to be
$\sim 20 \text{ cm h}^{-1}$ (discussed later). If this is done, then the value
of α turns out to be 1.02-1.03; i.e. CO_2 transfer is enhanced by
only 2-3% over that for an unreactive gas. The reason for this
very small enhancement is because conditions at sea are too rough,
so that chemical reaction is largely overwhelmed by diffusion.
It should be noted that in the wind-tunnel studies, even under
calm conditions, values of α for CO_2 are generally less than 2;
unless the pH is high enough for reaction (8) to be of importance,
when somewhat higher values are possible. The basic reason for
these small (relative to other gases such as SO_2, discussed
shortly) exchange enhancements for CO_2 is because of its slow
rate of hydration ($k^* \sim 10^{-2} \text{ s}^{-1}$). The hydration/dehydration reaction is susceptible to catalysis; for example, in human and
other mammalian lungs the enzyme carbonic anhydrase catalyses the
reaction and enables CO_2 to be exhaled quickly, preventing the
blood from becoming too acidic. Berger and Libby (1969) presented some experimental evidence that they interpreted as indicating the existence of carbonic anhydrase in seawater. This
initial observation doesn't appear to have been followed up by
others, and must be regarded as unsubstantiated.

The Hoover and Berkshire equation (11) has also been used
to calculate α for other gases. For example, Liss (1971) used it
to predict the degree of chemical enhancement to be expected for
SO_2, and found that, under typical oceanic conditions, the value
of α was almost 3000. The fundamental reason for this very high
value is the extremely fast hydration rate for SO_2 ($\sim 10^6 \text{ s}^{-1}$,
i.e. 10^8 times that for CO_2). The large value of α for SO_2,
coupled with its high solubility ($H \sim 4 \times 10^{-2}$), lead to its
transfer across air-water interfaces in the natural environment
almost invariably being under gas phase control ($r_a \gg r_w$).
Experimental confirmation of the appropriateness of (11) for predicting α for SO_2 comes from the good agreement between the
predicted value of the total air-water transfer resistance as a

function of solution pH (Liss, 1971), with that measured in an
experimental study (Brimblecombe and Spedding, 1972), as shown
in Figure 10. Equation (11) has recently been used to predict

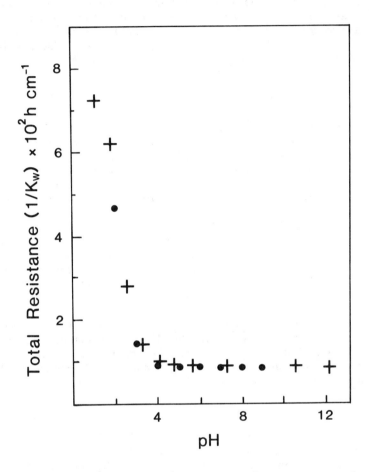

Figure 10. Total resistance to air-water transfer of
SO_2 as a function of solution pH. ● - Values predicted by Liss (1971); ✚ - Experimental values
measured by Brimblecombe and Spedding (1972). (From
Brimblecombe and Spedding, 1972).

α for H_2S in a laboratory study conducted by Balls and Liss
(1983), and again, reasonable concordance is found between the
measured and calculated values. The results indicate that the value
of α for H_2S ranges from about 1.0 at pH < 7 to almost 200 at pH
10. This leads to a decrease in the importance of r_w, which is rate
controlling at pH < 7, but constitutes 90% and 40% of the total resistance of pHs 8 and 10, respectively.

3.   AIR-SEA GAS TRANSFER IN THE FIELD

In this section, field studies of air-sea gas transfer are reviewed. First, gases whose exchange is under liquid phase control are treated in terms of the field methods used to measure their air-water transfer, and this is followed by an assessment of the role of bubbles in the exchange process. Then, a less detailed account is given of transfer of those gases for which processes in the air are dominant.

3.1 Gases for Which $r_w \gg r_a$

3.1.1 Direct Flux Method. This involves covering the sea surface with an enclosure and measuring directly the evasion or invasion of gas. The method suffers from two important defects: the flux chamber substantially alters the turbulence regimes in both air and water, as well as the interactions between them; and there is great difficulty in deploying such structures at sea under anything other than very calm conditions. A few measurements for CO_2 were made by Sugiura et al. (1963) in the Gulf of Mexico, but the only extensive series of observations is for H_2S release from coastal sites in Denmark (Hansen et al. 1978; Ingvorsen and Jorgensen, 1982). The method has found more utility in measuring fluxes of various biogenic sulphur gases from soils (Adams et al., 1979; Hill et al., 1978).

3.1.2 Oxygen Balance Approach. This consists of making a time series of measurements of the dissolved gas of interest (generally O_2, but the method has also been used once for CO_2) in near-surface seawater. Then the change in O_2 concentration between two measurement times (ΔO_2^t) can be expressed as,

$$\Delta O_2^t = \Delta O_2^e + \Delta O_2^b \tag{12}$$

Where ΔO_2^e is the change caused by transfer across the sea surface, and ΔO_2^b is that produced by biological processes. Redfield (1948) was the first to employ this approach, and he also introduced the idea of using parallel measurements of the plant nutrient element phosphorous, as a way of quantifying the term O_2^b. Redfield had discovered that, during the formation and decomposition of plant material in the sea, O_2 and phosphorus (as well as other nutrients) tend to be taken up/released in constant ratios (the Redfield ratios), as shown by the following equation,

$$(CH_2O)_{106}(NH_3)_{16}H_3PO_4 + 1380_2 \underset{\text{photosynthesis}}{\overset{\substack{\text{respiration/}\\\text{decomposition}}}{\rightleftarrows}} 106CO_2$$
$$+ 16HNO_3 + H_3PO_4 + 122H_2O \tag{13}$$

For example, in photosynthesis, for each mole of phosphorus used by plants, 138 moles of O_2 are released, with the opposite change for respiration/decomposition. Thus, from the change in phosphate observed in the water between measurements, the term ΔO_2^b can be evaluated. Then, through (12), the total change in dissolved oxygen (ΔO_2^t) between measurements can be corrected for biological uptake/release of O_2 (ΔO_2^b), to yield the change in gas concentration attributable to transfer across the air-sea interface (ΔO_2^e). Knowing the depth to which the observed change in dissolved O_2 extends, ΔO_2^e can be converted into a flux across the surface. Furthermore, since the driving force for the air-water transfer is known (from the difference between the measured O_2 concentration in the surface-most sample and the 100% saturation value), the calculated fluxes can be converted into corresponding transfer velocities. One important assumption of the method is that there is no significant horizontal or vertical transport, into the study area, of water containing different concentrations of dissolved oxygen from those being measured.

This approach has been applied in 4 locations, as detailed in Table 1. For present purposes, the results are given in terms

Table 1. Transfer velocities measured using the oxygen balance technique at various field sites.

Location Length of record Water Depth	Transfer Velocity[*] cm h^{-1}		Reference
	Summer	Winter	
Gulf of Maine 15 months (1933-34) 200-250 m	14.5	54	Redfield (1948)
Off Oregon coast Summer 1962 >45 m	10.8	-	Pytkowicz (1964)
Stuart Channel, B.C. 15 days (July 1976) ∿ 40 m	5.8 7.9 (CO_2)	-	Johnson et al. (1979)
Funka Bay, Japan 7 years 92 m	14	43	Tsunogai and Tanaka (1980)

[*] For O_2 unless otherwise stated.

of transfer velocities. Johnson et al. (1979) extended the
treatment to include CO_2 exchange, by making measurements of
pH and alkalinity from which the partial pressure of CO_2 was
calculated.

The two longest records (Redfield; Tsunogai and Tanaka)
cover all seasons of the year and a clear increase in transfer
velocity for winter over summer is apparent. For the four data
sets taken together the summer values cover a rather small range
(5.8 - 14.5 cm h^{-1}). The only study for which wind speed records are available is that of Tsunogai and Tanaka (1980) and
then the meteorological observations were made about 30 km from
the water-sampling point. In Figure 11 transfer velocities from

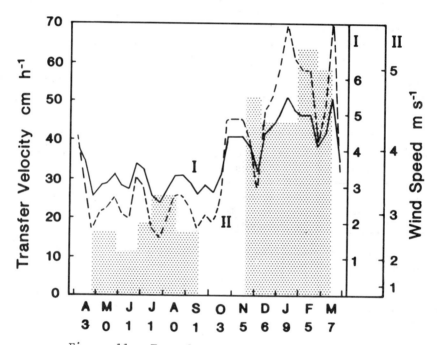

Figure 11. Transfer velocity (shaded bars) and wind
speed (line I) in Funka Bay, Japan for the period
1979-80. Curve II is proportional to the square of the
wind speed averaged for each 10-day period. Numbers
below the letter for each month indicate the number of
days that month when the wind velocity exceeded 10 m
s^{-1}. (After Tsunogai and Tanaka, 1980.)

a limited part of this data set (1979-1980) are plotted against
time of year. Again the transfer velocities in winter are substantially elevated over the summer values. Corresponding wind
speeds and their squares are also plotted. The authors comment

that although the transfer velocities in winter are 3-4 times
larger than in summer, wind speeds increase by only a factor of
1.5, and that better correlation is obtained between transfer
velocity and the square of the wind speed, as suggested by the
results of some of the wind tunnel studies reported in Section
2.1.1.

   3.1.3 Profile and Eddy Correlation Techniques. The profile
method uses measured, vertical gradients of gases in the air above
the surface to obtain fluxes. Although useful in air-sea measurements for H_2O and, possibly, other gases for which $r_a \gg r_w$ (see
Section 3.3), it is not applicable to gases for which the water
controls the transfer rate. For these, virtually all the resistance, and hence by far the largest concentration gradient,
will occur on the water side of the interface. As Deacon (1977)
has shown, for CO_2 transfer a typical concentration gradient between 1 and 10 m in the air would be a mere 0.05 ppm, i.e. less
than 0.02% of the atmospheric concentration.

   The eddy correlation technique, in which fluxes are obtained
from the time-averaged product of the measured gas concentration
and the vertical component of the wind, is potentially more useful. The chief instrumental requirement is for gas sensors with
sufficient sensitivity to measure small concentration changes and
with a time resolution of better than 1 second.

   Such sensors have been made for CO_2, and Jones and Smith
(1977) have reported some preliminary measurements made off Sable
Beach, Nova Scotia. A more detailed study of Wesley et al. (1982),
conducted at the Fowey Rocks lighthouse, Florida, shows a relationship between transfer velocity for CO_2 and friction velocity similar in form to that found in wind tunnel studies (Figure 2), but
the values are much higher than those typically observed in wind
tunnels or at sea by other methods. The range of transfer velocities found by Wesley et al. is from 85 cm h^{-1} (at u_* $\sim$ 8 cm s^{-1})
up to 720 cm h^{-1} (at u_* = 38 cm s^{-1}), which latter value is at
least 40 times higher than typical numbers from the Radon Deficiency method (see Section 3.1.5), and more than an order of magnitude greater than the highest figure obtained using the O_2
Balance Approach (Table 1). It isn't possible to calculate a
transfer velocity from the eddy correlation results of Jones and
Smith (1977),since no air-sea CO_2 concentration difference is
stated, but assuming ΔC to be about 100 ppm then a value of 65 cm
h^{-1} is obtained (with u = 7.5 m s^{-1}); i.e. lower than the results
of Wesley et al. but still high relative to laboratory and other
field studies. The reason for the large discrepancy between results from eddy correlation and other techniques is presently un-
resolved. Two types of problem have been identified. One comes
about from the need to correct eddy correlation measurements for
the effects of fluxes of heat and water vapour (Webb et al.,

1980). The other arises because application of the eddy correlation (and profile) technique assumes the measurements were made
under constant-flux conditions. Slinn (1983 and this volume)
argues that this condition has not been established for many of
the results so far reported.

Recently, Lenschow et al. (1982) have used the eddy correlation technique to measure rates of uptake of ozone by the oceans
and the results they obtain are in good agreement with those
obtained by other techniques (see Section 5.5.1).

3.1.4 Natural and Bomb-Produced 14-C. Broecker and Peng
(1974) have summarised how the 14-CO_2 from both natural and bomb
sources can be used to obtain global-average values for the air-
sea transfer velocity. In the case of 14-C produced naturally in
the air, and assuming steady-state in the ocean-atmosphere system,
the rate of input across the sea surface must balance the rate of
decay of the isotope in the ocean. Substitution of measured
values of 14-C activity in air and sea into the steady-state
equation yields an estimate of the global value of the transfer
velocity. Radiocarbon from bomb detonation is also introduced
into the atmosphere and enters the oceans across the sea surface.
From the many air measurements made since bomb testing started in
1954, the geographical and temporal variability of bomb-produced
14-C entering the oceans is well known. The rate of invasion,
and hence the transfer velocity for CO_2 entering the sea, can be
obtained by integration of the excess 14-C observed in any depth
profile. Transfer velocities obtained using data from either
natural or bomb-produced 14-CO_2 yield values in close agreement,
which average globally to about 20 cm h^{-1}.

3.1.5 The Radon Deficiency Method (Broecker, 1965). In the
oceans, away from either the air-sea or sediment-seawater interfaces, the isotope 226-Ra (half-life 1622 yr) and its daughter
222-Rn (half-life 3.8 d) are found to be at radioactive equilibrium.
However, since Rn occurs as a gas and seawater is greatly supersaturated with respect to the activity of 222-Rn found in the
atmosphere, this isotope will tend to be lost across the air-sea
interface. Observationally, this is found to be the case in the
oceans, as is illustrated in Figure 12. In the figure the dashed
line indicates the 222-Rn activity to be expected from radioactive equilibrium with the 226-Ra in the water. At shallower
depths there is a deficiency of 222-Rn relative to its parent
226-Ra activity, the missing radon having escaped to the atmosphere across the sea surface. Assuming steady-state, the rate
of loss of radon across the interface must balance the depth
integrated deficiency in 222-Rn activity. By solving this
equality for each measured profile, values of the transfer velocity for Rn gas are obtained, i.e.

$$k_{Rn} \ (A_i - A_a) \ = \ \lambda \int_o^\infty (A_e - A_z) \ dz \qquad\qquad (14)$$

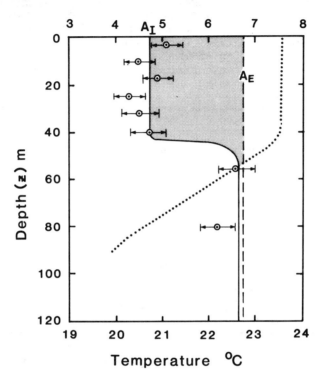

Figure 12. Activity of 222-Rn as a function of depth
at GEOSECS station 57 in the Atlantic (24°S, 35°W).
The dashed line is the 222-Rn activity to be expected
if the isotope is in equilibrium with its parent 226-Ra.
The shaded area represents the amount of 222-Rn lost
to the air. The dotted line is the temperature profile.
(After Broecker, 1974).

Where k_{Rn} is the required transfer velocity, A_i is the measured
activity of 222-Rn in the surface-most water, A_a is the activity
to be expected in this water if at equilibrium with the amount
in air (generally taken as zero in view of the small atmospheric
activity and the low solubility of the gas in water), A_e is the
activity of 222-Rn expected in the water column if equilibrium is
achieved with dissolved 226-Ra, A_z is the measured activity of
222-Rn at some depth z, and λ is the decay constant of 222-Rn

(0.18 d^{-1}). The method further assumes that the profile is not affected by horizontal or vertical water movements.

The radon deficiency method was first used in the field on Project BOMEX (Broecker and Peng, 1971), and then at Station Papa in the N. Pacific (Peng, et al., 1974); but by far the largest data set was gathered during the GEOSECS cruises in the Atlantic and Pacific Oceans, when about 100 profiles were measured (Peng et al., 1979). These last results are plotted in Figure 13 as a function of wind speed measured 0-24 hr prior to the water samples being collected. Several interesting points emerge.

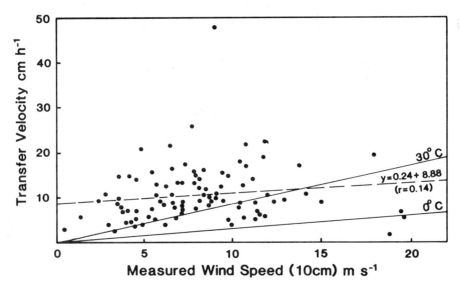

Figure 13. Transfer velocity determined on GEOSECS Atlantic and Pacific cruises by Radon Deficiency Method (Peng et al., 1979), plotted against wind speed measured 0-24 h prior to sample collection. Wind speeds at nominal 10 m height reduced to 10 cm height assuming a logarithmic wind profile. Full lines - predictions based on approaches of Deacon (1977) and Hasse (1971); broken line - least squares fit through data points. (After Hasse and Liss, 1980.)

i) The mean transfer velocity for all the GEOSECS measurements is 11.9 cm h^{-1} (from Peng et al., 1979). Although this is low compared with the global values from the 14-CO$_2$ data, discussed previously, a correction should be applied to account for the different diffusivities of the two gases. According to the film model this correction would be linearly related to the gas diffusivities, but boundary-layer theory predicts a diffusivity

to the $\frac{2}{3}$ dependence. Applying these two possibilities, the
mean transfer velocity of the GEOSECS data increases from 11.9
to either 16.3 (film) or 14.6 ($D^{\frac{2}{3}}$) cm h^{-1}. Although both values
are somewhat lower than the transfer velocity derived from 14-C
(20 cm h^{-1}), in view of the assumptions involved and the spread
of the observations, for both methods, the agreement between
them is probably acceptable. Results from the oxygen balance
approach from Tsunogai and Tanaka (1980) yield a mean value for
k_{O_2} of about 28 cm h^{-1}, which with a linear diffusivity correction is equivalent to a value of 22 cm h^{-1}. Putting the results
from these three approaches together suggests a _global_ value of
the transfer velocity for CO_2 of about 20 cm h^{-1}.

    ii) The data plotted in Figure 13 show little or no relation-
ship between transfer velocity and wind speed, as confirmed by
the low slope and correlation coefficient of the least-squares
line.   Similarly, Kromer and Roether (1983) have recently reported only poor correspondence between k_w, obtained using the
radon deficiency method, and wind speed in measurements made as
part of the JASIN 1978 (N.E. Atlantic) and FGGE 1979 (Equatorial
Atlantic) experiments. These results are very surprising in
view of the findings of the laboratory and field studies described earlier, and go against expectations that increasing wind
favours all the processes (mixing, waves, bubbles, etc.,) that
are likely to enhance transfer across the surface. One possible
explanation for the discrepancy is that, since the radon method
has a half response time of 2-3 days, it may be incorrect to seek
a wind-speed dependence using the wind speed 0-24 prior to sample
collection. However, no greater relationship between the variables is obtained if the wind speed used is from the period 24-48
h earlier. Since all wind observations were made on the ship then,
in this latter case, wind speeds would have been measured some
distance (usually about 200 km) from the actual station.

    A more thorough examination of the GEOSECS data along the
above lines has been done by Deacon (1981). He argues that since
it requires some days for the 222-Rn profile to adjust to a change
in conditions, only results from stations at which the wind had
been blowing steadily prior to occupation should be included in
the analysis. The criterion he employs is that the 0-24 and
24-48 h winds should not differ by more than 30%. Again, difficulties arise since the ship was generally moving between stations during this period. Deacon also omits the results from 10
profiles far removed from the ideal type shown in Figure 12, and
two because there is reason to think that they may have been affected by horizontal water movements. After this selection process,
of the original 100 data points, only 37 remain. The outcome is to
reduce the scatter on a plot of transfer velocity against wind
speed, but the correlation coefficient, although significant at
the 1% level, is rather small (r = 0.46). Furthermore, the

slope of the plot, although greater than in Figure 13, is still
about an order of magnitude less than that found in wind tunnel
studies (Figure 2). Other possible explanations are clearly in
need of investigation.

One assumption of the radon deficiency method is that,
during the several days required for the near-surface radon to
reach steady-state following some perturbation, there is no vertical exchange of water across the base of the 222-Rn profile.
However, Slinn (1983) has pointed out that this assumption may be
far from justified, and changes in mixed-layer depth will tend to
modify calculated values of k_{Rn}. This may be illustrated by reference to (14) for a situation in which the wind speed is increasing. As the wind freshens, so radon transfer across the sea surface will increase (as expected from the wind-tunnel experiments),
and the integral on the right-hand side of (14) will become larger.
An increased value of the integral clearly leads to a larger calculated value of k_{Rn}, *all other things being equal*. However, if
the stronger wind increases the depth of the mixed layer, then the
shape of the profile of radon deficit will alter, to become deeper
and less wide (i.e. with a higher value of A_i), compared to the
situation with no deepening of the profile. Any increase in A_i
will produce a corresponding change in the left-hand side of (14),
and this will tend to cancel the increase in the integral on the
opposite side of the equation, produced by the larger transfer
velocity. Thus, values of k_{Rn}, calculated from profiles affected
in this way, will tend to show a smaller-than-expected change
with wind speed.

3.2 The Role of Bubbles

Although it seems clear that bubbles must be important as gas
exchangers in the oceans for conditions under which a substantial
bubble population exists, quantitative assessment of the effect
is extremely difficult. Not only is there the general problem of
specifying the number and size distribution of bubbles as a function of depth, under any particular set of conditions, but there
are also difficulties related specifically to bubbles as gas exchangers. The ability of bubbles to adsorb/desorb gases differ-
entially, due to solubility effects, has already been discussed
in Section 2.1.3. Other factors include asymetrical behaviour
of bubbles between invasion and evasion (Atkinson, 1973), changes
in gas pressure within bubbles with depth (due to hydrostatic
pressure) and bubble radius (r) arising from the surface tension
(γ) parameter ($2\gamma/r$). This latter term clearly varies in importance inversely with bubble size and is only of importance for
very small bubbles. In addition, in the marine environment
bubbles will almost certainly be covered with an organic film.
Although this will probably not act as a direct resistance barrier
to gas exchange across the water/bubble interface (Section 4.1),

it may well alter the hydrodynamic, and therefore gas transfer,
properties of the aqueous boundary layer around the bubble.

In view of these many complications, it is not surprising
that the attempts which have been made to address the problem for
the field situation involve many assumptions and have reached
only very general conclusions. Atkinson (1973) tried to assess
the importance of bubbles for the invasion of oxygen into seawater. He found that, for wind speeds above 10 m s^{-1}, the flux
appears to be substantial; and that with the atmospheric pressure
changes occurring with the passage of a storm, bubble solution
could account for about half the oxygen flux required to maintain
equilibrium between air and water. Recently, Thorpe (1982) has
presented a detailed analysis of bubble spectra in fresh (Loch
Ness) and sea waters, and has applied this to calculation of the
role of bubbles as gas exchangers. He finds that most of the flux
will occur in the upper 2 m of the water column, where the bubble
density is greatest,which implies that the hydrostatic-pressure
term in the bubble-dissolution equation increases the gas pressure inside the bubble by about 20%. Furthermore, Thorpe cal-
culates that the contribution of bubbles to the air-water gas
flux is probably negligible in Loch Ness, but at sea will become
significant at wind speeds in excess of 12 m s^{-1}; which is in
reasonable agreement with Atkinson's earlier predictions.

3.3   Gases for Which $r_a \gg r_w$

The gradient and eddy correlation techniques, described in
Section 3.1.3, are both in principle applicable to the measurement of air-sea fluxes for gases for which r_a is the dominant
transfer resistance. In practice the only gas for which these
two techniques have been used in the marine environment is H_2O.
For example, Pond et al. (1971) and Hasse et al. (1978), have
used the eddy correlation and gradient techniques, respectively,
to measure air-sea fluxes of water vapour. Both groups found a
strong linear relationship between transfer velocity and wind
speed, as expected from laboratory results and theoretical
approaches, discussed earlier.

Sulphur dioxide is a gas for which the gradient technique
has been widely used to obtain fluxes over land surfaces. However, although there are a few measurements for deposition to
fresh waters (e.g. Garland, 1977), it does not appear to have
been used over the oceans.

Theoretical approaches have been employed to calculate transfer velocities for gases for which r_a is the dominant resistance.
Since k_a (often called a deposition velocity in the present context) is known to vary linearly with wind speed (u), a relationship
of the following form can be used,

$$k_a = C_X \cdot u \tag{15}$$

where the dimensionless constant of proportionality, C_X, is called the drag coefficient and the subscript, X, identifies the substance (or property) being transferred. The drag coefficient over the oceans has a value of approximately 1.3×10^{-3} (Hasse, this volume). Effects of atmospheric stability can be incorporated by use of either the Richardson number or the ratio z/L, where L is the Obukhov scale length of turbulence. An example of the application of these ideas to calculation of the transfer velocity for SO_2, and other gases for which $r_a \gg r_w$, across the air-sea interface is given in Hicks and Liss (1976).

4. POSSIBLE ROLE OF NATURAL AND POLLUTANT FILMS IN AIR-SEA GAS TRANSFER

The effects of films at the sea surface on gas transfer can be divided into two types: i) Direct or Static, where the film directly affects exchange by presenting a significant additional resistance to transfer, as compared to the 'clean' surface, and ii) Indirect or Dynamic, where the film material although possessing little or no intrinsic resistance, alters some property of the surface water which is important in bringing about gas transfer. Here, laboratory studies of these two types of effects are briefly reviewed prior to a discussion of the existence and possible role of natural and pollutant films on gas transfer at sea. More comprehensive treatments of the chemistry of sea surface films, their physical properties, and potential effects on air-sea gas exchange can be found in Garrett (1972), MacIntyre (1974a), Liss (1975 and 1977), Wangersky (1976), Hunter and Liss (1981) and Lion and Leckie (1981).

4.1 Laboratory Studies

4.1.1 Direct or Static Effects. The effect of oil films of various thicknesses on air-water gas transfer has been examined in the laboratory by Liss and Martinelli (1978). Their results for simultaneous transfer of O_2 and H_2O are shown in Figure 14. These two gases were chosen for study since, under normal conditions, H_2O has $r_a \gg r_w$, whereas O_2 has $r_w \gg r_a$. It is apparent that the effect of the film is more pronounced for transfer of H_2O than it is for O_2. This must arise because, in the first case, a liquid-phase resistance is being added where none existed before; whereas for O_2 one is merely adding a relatively small amount to the already dominant resistance presented by the aqueous phase. It is noteworthy that even under these ideal conditions, where there was no mechanical disruption of the oil, no effect on transfer was detectable for H_2O until the film was

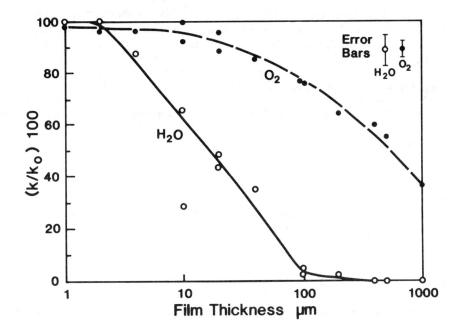

Figure 14. Ratio (%) of transfer velocities in the presence (k) and absence (k_o) of an interfacial oil film as a function of film thickness, for H_2O and O_2. (From Liss and Martinelli, 1978.)

about 2-3 μm thick; for O_2, a thickness of 10 μm was required. A more extensive set of results for oil films is to be found in Martinelli (1979), where simultaneously determined data for four gases (H_2O and SO_2 (r_a dominant) and O_2 and CO_2 (r_w dominant)) are given, together with a 3-film model that provides a reasonable description of the observations.

The other class of film-forming substances, whose ability to directly retard gas transfer has been tested in the laboratory, are highly surface-active compounds, which form monolayers at the interface. Unlike the oil films just discussed, which remain at the surface because they are lighter than water, these substances form interfacial films because one part of the molecule is essentially insoluble in water (hydrophobic), whereas the rest is soluble (hydrophilic). The molecules are oriented at the interface with the hydrophilic moiety in the water and the hydrophobic part protruding into the air. Typical compounds of this type (which MacIntyre, 1974b terms 'dry' surfactants) are straight-chain fatty acids and alcohols, having from 12-22 carbon atoms in the hydrophobic chain.

Results from laboratory experiments show that, for gases
whose exchange is under liquid-phase control, dry surfactant films
present no significant, direct resistance to transfer (Petermann,
1976; Martinelli, 1979). Furthermore, for gases with $r_w \gg r_a$,
neither soluble ('wet') surfactants (which are long-chain organic
compounds having considerable water solubility and whose surface
activity arises from the presence of occasional hydrophobic groups
on the molecule) nor material from the sea-surface microlayer (top
few hundred microns collected using a Garrett (1965) screen) provide any measurable barrier to gas transfer (Martinelli, 1979).

In contrast, for gases for which r_a is the dominant resistance, clear effects on transfer are observed in the laboratory.
Most of the studies have been for evaporation of water, which is
found to be significantly reduced with a monolayer of dry surfactant at the surface (Jarvis et al., 1962; Frenkiel, 1965; La Mer
and Healy, 1965; Garrett, 1971). The generality of the results
has been shown by Martinelli (1979), who found similar behaviour
for both SO_2 and H_2O. As pointed out by Jarvis et al. (1962), for
a monolayer to be able to retard evaporation, the molecules must
be linear and capable of close packing in a compressed layer.
The presence of double bonds, ionised polar groups, or branching
of the hydrocarbon chain greatly reduces the ability of the film
to reduce evaporation. For example, the presence of only 1% of
non-linear impurities can lower the effectiveness of the film in
reducing evaporation by 90-99% (La Mer, 1962).

In a practical context, the ability of a continuous and coherent monolayer to inhibit transfer of water molecules has been
tested on water-storage reservoirs, in attempts to reduce losses
by evaporation. Although by proper choice of monolayer a substantial effect can be achieved, practical problems arise, under
even only moderately windy conditions, in preventing the film
from breaking up and being blown down-wind.

4.1.2 Indirect or Dynamic Effects. The major indirect effect
of films is through their ability to damp capillary waves since,
as suggested in Section 2.1.2, there is some evidence from wind-
tunnel studies for the importance of capillary waves in gas transfer. Laboratory studies by Garrett and Bultman (1963) and Garrett
(1967) show that dry surfactants can damp capillary waves, provided the film pressure (= surface-tension reduction caused by the
presence of the film) is about 1×10^{-3} N m^{-1} (1 dyne cm^{-1}) or
more; see Figure 15. Furthermore, experiments in the Hamburg
wind tunnel, the results of which are summarised in Figure 16,
clearly demonstrate that a monolayer of the dry surfactant oleyl
alcohol ($C_{17}H_{33}CH_2OH$) can cause a large reduction in the transfer
velocity for CO_2, presumably because of capillary wave suppression,
which was visually obvious. The film itself provides no significant resistance, since the transfer velocities for normal and

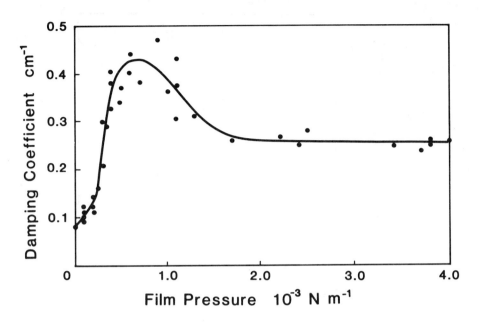

Figure 15. Variation of capillary wave damping co-efficient (k) with film pressure. k is defined by $a = a_o e^{-kx}$, where a_o is the wave amplitude at its source, and a is the wave amplitude of distance x from the source. (After Garrett, 1967).

monolayer-covered surface are identical at low wind speeds, when no capillary waves are present. Recently, Brockmann et al. (1982) have reported on a field experiment, conducted in the North Sea, in which significant decrease in the CO_2 evasion rate was measured after an artificial film of oleyl alcohol had been spread on the sea surface.

Another possible indirect effect, albeit a somewhat negative one, is through the laboratory finding (discussed in Section 2.1.4) that condensation of water vapour can lead to a decrease in air-water transfer velocity for O_2. If a film capable of significantly affecting H_2O transfer exists at the interface, then presumably it would tend to decrease any reduction in transfer occurring at the clean surface due to condensation.

There is some evidence from laboratory tank experiments that soluble surfactants, although possessing no intrinsic transfer resistance, can lead to significant reduction in the transfer velocity for O_2 under conditions of high stirring in the aqueous phase (Mancy and Okun, 1965; Martinelli, 1979).

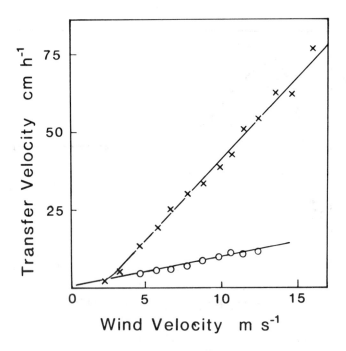

Figure 16. Transfer velocity for CO_2 as a function of
wind velocity with (O) and without (x) a monolayer of
oleyl alcohol on the water surface. (After Broecker
et al., 1978)

4.2 The Role of Films in Gas Transfer at the Sea Surface

     In summary, to have a direct effect on air-water gas transfer,
the film must apparently form a continuous, chemically-coherent
layer at the interface. To be coherent, the molecules of the
film must be capable of close packing, a requirement best met when
the molecules are identical (as in a monolayer or oil film of pure
material). For the film to be continuous, there must be a sufficient supply of film-forming material, and conditions must be
calm enough for the film not to be disrupted.

     Several lines of argument indicate that the conditions of
coherence and continuity are unlikely to be met, at least over
most of the oceans. i) Analysis of the organic composition of
unpolluted, sea-surface material indicates that it consists of a
large number of different compounds. A small amount (< 10%) is
dry surfactants, which have been characterised reasonably well,
but the majority is wet surfactants, whose composition is largely
unknown and which are certainly very inhomogeneous (Hunter and
Liss, 1981; Sieburth, this volume). ii) Film pressure measurements

(summarised in Table 2), made at sea using the spreading drop technique, show that only in coastal waters or areas with obvious slicks are pressures greater than 1×10^{-3} N m^{-1} common. In the

Table 2. In situ measurements of film pressures at the sea surface. (After Hunter and Liss, 1977, with later additions.)

Investigators	Area	Condition of water surface	Number of measurements	Film pressure 10^{-3} N m^{-1}
Lumby and Folkard (1956)	Monaco Bay	ruffled smooth	13 16	2-2.5 2->22.5
Sieburth and Conover (1965)	Sargasso Sea	rippled slicked	9 9	<1-2 <1-4.5
Garrett (1965)	Chesapeake Bay	slicked	-	3-12
Sturdy and Fischer (1966)	Kelp Beds S.California	-	-	2->24
Barger et al. (1964)	Mission Beach S.California (0.7 mi offshore)	rippled slicked	>150 18	1 or less 1-23
Huhnerfuss et al. (1977)	North Sea	*	17	6.5-21
	Tropical Atlantic	*	54	2-12

* Sea surface condition not given for each measurement, but high film pressures appear to correspond to slick-covered water.

large number of observations made by Barger et al. (1974) off S. California the film pressure was always at or below the limit of sensitivity of the method (1×10^{-3} N m^{-1}), when capillary waves were present on the water surface. iii) Laboratory measurements of the relationship between film pressure and film area indicate that sea surface material does not become close-packed until the film pressure reaches a value in excess of 1×10^{-3} N m^{-1}, which generally only happens in obviously slicked water (Table 2).

From the evidence, it seems reasonable to conclude that, over most of the oceans, there will not be a significant, direct effect of organic films on gas transfer. Essentially, this arises from

the absence of sufficient material with the right chemical properties, and the difficulty of keeping the film intact in the face
of dispersive forces of wind and waves. In areas of oil spills
or sheltered coastal waters, where high biological production
creates an abundance of organic matter, the possibility of direct
reduction in gas transfer cannot be ruled out. As stated before,
any effect will be much more pronounced for gases whose exchange
is controlled by atmospheric processes.

Indirect effects of films on air-sea gas exchange (essentially capillary wave damping) appear to have less stringent requirements in terms of film composition and properties. Once
the film achieves a pressure of 1 x 10^{-3} N m^{-1}, capillaries are
substantially damped and exchange may be inhibited. Since suppression of capillary waves is obvious from the 'slicked' appearance of the water surface, areas where gas-transfer reduction may
be occurring are easy to identify by eye. Slicks are found sometimes over the open ocean but more frequently in coastal regions,
especially where biological production or anthropogenic input of
organic matter is high and winds are light. In the case of this
indirect effect of capillary wave damping, the gases whose exchange is likely to be reduced are those for which liquid phase
resistance is dominant. This follows because capillary waves do
not appear to influence the evaporation and condensation of water,
and, therefore, they probably do not influence the transfer of other
gases whose exchange is under gas phase control. For example, in
Figure 7 there is no change in the slope of the graph with the
onset of capillary waves.

5.   THE ROLE OF AIR-SEA GAS EXCHANGE IN GEOCHEMICAL CYCLING

5.1 Introduction

The aim in this section is to present the results of attempts
to estimate the fluxes of various gases across the sea surface and
to indicate the quantitative importance of such cycling for global
geochemistry. Many of the estimates are based on ideas outlined
earlier in the chapter. Basically, (1) shows that any net flux
(F) across the sea surface must be driven by a concentration difference (ΔC), with the rate of exchange being controlled by the
value of the transfer velocity (K). Expansion of the overall tran-
fer velocity (K) into its individual air- and water-phase components is achieved by the use of (2) or (4). Several of the terms in
these equations have already been discussed, e.g. the chemical enhancement factor (α), and k_w for which a value of near 20 cm h^{-1}
(for CO_2) seems appropriate for conditions at the sea surface, averaged spatially and temporally. Factors influencing k_a have also been
mentioned, and a global average value of 3000 cm h^{-1} (for H_2O) is
often used (Schooley, 1969). A somewhat higher value (3650 cm h^{-1})

is obtained if a reasonable estimate of the mean scalar wind over
the oceans (15 knots or 7.8 m s^{-1} - Broecker and Peng, 1974) is
substituted into (15), with the drag coefficient taken to be
1.3 x 10^{-3}. Approaches to correcting these transfer velocities
for gases other than CO_2 and H_2O have already been mentioned, i.e.
through a linear (film model), $\frac{2}{3}$ (boundary-layer theory) or $\frac{1}{2}$
(surface-renewal model) power dependence on gas diffusivities.
As well as problems associated with which model is appropriate,
difficulties also arise because there are considerable uncertainties over the diffusivity values and their variation with temperature (and salinity). Similar difficulties arise over the
Henry's law constant (H) in (2) and (4), since again our knowledge is unsatisfactory for several gases (e.g. NH_3, Halocarbons,
see later).

    In order to use (1) to calculate F with a globally-averaged
value for the transfer velocity means that the term ΔC must also
be available in a similarly averaged form. Since for many of the
gases of interest the magnitude (and sometimes even the sign) of
the air-sea concentration difference is changing both geographically and on various time scales (diel, seasonal), a large number
of observations are required in order to obtain an accurate mean
value. Furthermore, the technical difficulties of many of the
measurements make the acquisition of sufficient data extremely
hard. Thus, for many gases specification of ΔC is a very important constraint on the reliability of the computed global flux.

    In Table 3 are assembled some of the best estimates of gas
fluxes across the sea surface. Compounds are tabulated under one
of their constituent elements. The choice of which element is
somewhat arbitrary, but generally the central atom and/or the most
geochemically active one is used. Comments on how the flux numbers
have been obtained and their geochemical importance are given in
the following sections.

5.2   Hydrogen

    5.2.1 Molecular Hydrogen (H_2). Surface seawater is gener-
ally 2-3 times supersaturated with respect to the concentration of
hydrogen in the overlying atmosphere, according to Schmidt and
Kulessa (1981) who calculate a total global sea-to-air flux of
about 10^{12} g H_2 yr^{-1}. This is in rather good agreement with the
earlier estimate for the Northern Hemisphere oceanic source
strength of 0.33 x 10^{12} g yr^{-1} (Herr and Barger, 1978). However,
since the concentration data is rather limited geographically, it
seems quite likely that the tabulated flux estimate will have to
be revised in the future. A total ocean source strength of $\sim 10^{12}$
g yr^{-1} accounts for only a few percent of estimates of the total
global H_2 production rate, most of which is thought to be anthro-

Table 3. Net global gas fluxes across the air-sea interface

Element	Compound	Controlling Resistance	Global Air-Sea Flux Direction*	Magnitude+	Method+	Reference
H	H_2	r_w	+	10^{12}	✓	Schmidt and Kulessa (1981)
C	HCHO	$r_a \sim r_w$	−	10^{13}	✓	Zafiriou et al. (1980a)
	CH_4	r_w	+	$10^{12}-10^{13}$	✓	Ehhalt & Schmidt (1978)
	C_2H_6	r_w	+	10^{12}	✓	Rudolph and Ehhalt (1981)
	C_3H_8	r_w	+	10^{12}	✓	Rudolph and Ehhalt (1981)
	C_2H_4	r_w	+	10^{12}	✓	Rudolph and Ehhalt (1981)
	C_3H_6	r_w	+	10^{12}	✓	Rudolph and Ehhalt (1981)
	CO	r_w	+	10^{14}	✓	Seiler (1974)
	CO_2 man made	r_w	−	6×10^{15}	X	Broecker et al. (1979)
N	N_2O	r_w	+	6×10^{12}	✓	Cohen & Gordon (1979) Seiler (1981)
	NO	r_w	+(?)		✓	Zafiriou et al. (1980b)
	NO_2	?	+(?)			Helas et al. (1981)
	NH_3	r_a	+(?)	?	✓	Georgii & Gravenhorst (1977)
O	O_3	r_w	−	6×10^{14}	✓	This work
S	Total volatile S		+	$34-170 \times 10^{12}$ (as S)	X	Granat et al. (1976) Eriksson (1963)

Group	Compound		Flux	Value	Technique	Reference
	H_2S	r_w	+	15×10^{12}	X	Graedel (1979)
	$(CH_3)_2S$	r_w	+	$30-50 \times 10^{12}$	√	Barnard et al. (1982) Nguyen et al. (1978)
	CS_2	r_w	+	0.3×10^{12}	√	This work
	COS	r_w	+	0.8×10^{12}	√	Rasmussen et al. (1982a)
	SO_2	r_a	−	5×10^{12}	√	This work
Cl	CH_3Cl	r_w	+	$3-8 \times 10^{-2}$	√	Singh et al. (1979) Watson et al. (1980)
	CCl_4	r_w	−	10^{10}	√	Liss & Slater (1974)
	CCl_4	r_w	=	~ 0	√	Hunter-Smith et al. (1983)
	CCl_3F	r_w	−	5×10^{9}	√	Liss & Slater (1974)
	CCl_3F	r_w	=	~ 0	√	Gammon et al. (1982)
	$CHCl_3$	r_w	+	7×10^{11}	√	Rasmussen (pers. comm., 1982)
I	CH_3I	r_w	+	3×10^{11}	√	Liss & Slater (1974)
	CH_3I	r_w	+	13×10^{11}	√	Rasmussen et al. (1982b)

*   +, Sea → Air; −, Air → Sea; =, no net flux

\+   Units are g (of the compound) yr^{-1}, except as indicated

†   √ calculated by KΔC technique; X calculated by other technique − details in text.

pogenic (Schmidt, 1974).

5.3 Carbon

5.3.1 Formaldehyde (HCHO).   The physical chemistry of the
air-sea transfer of HCHO is interesting because this appears to
be one of the few simple compounds for which r_a and r_w are of
similar magnitude, a situation brought out clearly by the two
attempts to estimate r_a and r_w for HCHO. Zafiriou et al. (1980a)
calculate the total resistance to be partitioned 2/3 in the liquid phase and 1/3 in the gas phase, using $H = 0.07$ and $\alpha = 7$.
In contrast, Thompson (1980) suggests that the gas phase is just
dominant ($r_a : r_w = 2/3 : 1/3$), taking $H = 0.008$ and $\alpha = 2.2$. Both
calculations are based on theory, and the situation clearly requires experimental verification. However, it is obvious that the
resistance of both phases must be included in any calculation of
the air-sea flux of this gas.

The flux given in Table 3 for HCHO is calculated from the
results given by Zafiriou et al. (1980a), since their concentration measurements are from a remote oceanic area (Enewetak Atoll
in the central equatorial Pacific). Thompson (1980), using similar data from a coastal site (Woods Hole, Mass.), predicts an
air-sea flux about 50% greater. In both of these papers fluxes
are given per unit area of sea surface. For present purposes the
concentration data of Zafiriou et al. are assumed to be representative of the whole of the oceans. This may well not be too gross
an assumption since recent measurements in the N. and S. Atlantic
give an HCHO air concentration of 0.22 ppbv (Lowe and Schmidt,
1981), which is reasonably close to the value of 0.4 ppbv obtained
by Zafiriou et al. at Enewetak. The work of Lowe and Schmidt
also lends credance to the assumption made by Zafiriou et al.
that surface seawater contains zero HCHO, since although the
former authors were just able to detect HCHO in surface waters,
they found them vastly undersaturated with respect to the overlying air.

Both Zafiriou et al. and Thompson found some transfer of
HCHO to the sea surface by wet deposition. In the case of the
open Pacific measurements, precipitation accounted for only about
20% of the total flux; whereas, at the Woods Hole site, gaseous
and wet deposition were of roughly equal magnitude.

5.3.2 Methane (CH_4) and other Light Hydrocarbons (C_2H_6,
C_3H_8, C_2H_4 and C_3H_6.   Ehhalt and Schmidt (1978) calculate the
sea-to-air flux of CH_4 to be in the range $1.3 - 16.6 \times 10^{12}$ g yr^{-1}.
Most open-ocean estimates are near the lower end of this range:
Weiss (1981); Scranton and Brewer (1977) whose calculated flux
for the western subtropical N. Atlantic extrapolates to a global
value of about 5×10^{12} g yr^{-1}; and Liss and Slater (1974), 3×10^{12}

g yr^{-1}. Supersaturations are often considerably greater in
shallow coastal areas, leading to substantially higher fluxes
from such waters. In spite of this, production by the oceans is
probably at most only a few percent of the total amount of CH_4
formed by natural terrestrial sources ($\sim 10^{15}$ g yr^{-1}, Ehhalt and
Schmidt, 1978; Sheppard et al., 1982).

Recently, Rudolph & Ehhalt (1981) used their measurements of
C_2 and C_3 hydrocarbons in the air over the Atlantic, coupled with
data for surface-seawater concentrations from Swinnerton &
Lamontagne (1974), to calculate sea-to-air fluxes for ethane
(C_2H_6), propane (C_3H_8), ethene (C_2H_4) and propene (C_3H_6). On
the assumption that these measurements are representative for all
oceans, the total global fluxes have been calculated, and are included in Table 3. The flux of all four gases is similar and
close to 10^{12} g yr^{-1}, which is only marginally less than the values for CH_4.

5.3.3 Carbon Monoxide (CO). Carbon monoxide shows supersaturations in surface seawaters of several tens of times, and
this leads to the large flux shown in Table 3. Although the figure given was published some years ago, more recent measurements
of the gas in both surface seawaters (Conrad et al., 1982) and the
marine atmosphere (Heidt et al., 1980) are in rather good agreement with the data used by Seiler in his 1974 paper, so that his
calculated flux still stands. There is evidence that the CO
found in surface seawater is formed both by photooxidative as
well as microbiological processes (Wilson et al., 1970; Conrad
et al., 1982; Setser et al., 1982).

The sea-to-air flux of CO is of about the same size as the
flux from man-made sources caused by the incomplete combustion of
fossil fuels. However, the efflux of CO from the oceans appears
to be between one to two orders of magnitude less than the amount
formed in the atmosphere, by the photochemical oxidation of CH_4
by OH radicals. An intermediate stage of this latter reaction
produces much of the formaldehyde which was calculated earlier to
transfer from the atmosphere into the oceans.

5.3.4 Carbon Dioxide (CO_2). The net flux of anthropogenic
CO_2 into the oceans given in Table 3 is equivalent to only about
2% of the natural rate of exchange of this gas across the sea
surface. Averaged globally and annually, the natural exchange is
in balance with about 330 x 10^{15} g yr^{-1} of atmospheric CO_2
entering the oceans and an equal amount returning to the atmosphere. In view of this very large natural cycling, it is presently not possible to substitute into (1) - hereafter called the
KΔC technique - to estimate the uptake of man-made CO_2 by the
oceans, and other techniques have to be used instead (Siegenthaler
and Oeschger, 1978; Broecker et al., 1979).

Observations of the partial pressure of CO_2 in both air and surface seawater show that the two phases are rarely at equilibrium. However, as expected, the size and sign of the air-sea CO_2 concentration difference is found to be highly variable, both geographically and on various time scales (diel, seasonal). Thus, to use the KΔC approach to extract the net anthropogenic input signal from the much larger natural two-way flux, a very large number of observations would be required. As well as this, knowledge of the appropriate transfer velocity for each set of measurements would be needed. As is apparent from Section 3.1.5, although the global-mean value for k_w may be well known, its value at any particular time and place in the oceans cannot now be reliably specified.

It is hardly necessary to stress the importance of being able to accurately quantify the amount of man-mobilised CO_2 taken up by the oceans. The global budget for such CO_2 is at present either difficult or near impossible (depending on your view of the importance of land-use change in affecting atmospheric CO_2) to balance. However, the oceans are well accepted to be the major sink for anthropogenic CO_2 released into the atmosphere, so that predictions of the time scale of any climatic warming caused by elevated atmospheric levels of the gas are clearly dependent on accurate assessments of the role of the oceanic sink.

The air-sea fluxes of various carbon compounds discussed in Section 5.3 are summarised in Figure 17.

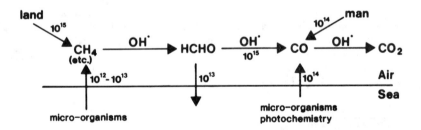

Figure 17. Air-sea and other fluxes (in g yr^{-1}) of various carbon compounds, and their atmospheric transformations.

5.4 Nitrogen

5.4.1 Nitrous Oxide (N_2O). Earlier estimates of the flux
of N_2O from the sea to the atmosphere are about an order of magnitude greater than the present best value given in Table 3. This
reflects the poor geographical coverage of the early surface-water
data, which were largely confined to the N. Atlantic. The extent
of supersaturation of surface waters is only 1-2%, for large parts
of the oceans, but with substantially higher concentrations in
coastal water and areas of upwelling and vertical mixing (Weiss,
1981). The flux given in the table corresponds to an average
supersaturation for all surface seawaters of approximately 4%.
The size of the sea-to-air flux is about an order of magnitude
less than the probable input of N_2O to the atmosphere from terrestrial sources (Hahn, 1979).

5.4.2 Nitric Oxide (NO). No one has yet used the $K\Delta C$ approach to calculate the flux of NO across the sea surface. However,
it is probable that NO can be formed in near-surface seawaters by
photolysis of nitrite ions, and Zafiriou et al. (1980b) present
results of direct measurements, made in the equatorial Pacific,
which appear to show that this is the case. These authors calculate that, in the area where their measurements were performed,
the sea is clearly a net source of NO for the atmosphere. Extra-
polation of this result to other oceanic areas is difficult, because formation of NO requires both light in the correct wavelength range (295-410 nm) and the presence of precursor NO_2^- ions,
whose concentration tends to be very variable. Further measurements of NO in surface waters are needed to resolve the matter.
Since NO is rather insoluble in water ($H = 33$, Zafiriou and
McFarland, 1980) its air-water transfer is very probably under
liquid-phase control.

In the photolytic production of NO from NO_2^-, hydroxyl radicals are also formed, and these are likely to react with many
different species in surface seawater. Although this possibility
has apparently received little attention, it is potentially of
considerable importance in the context of interaction between
geochemical cycles.

5.4.3 Nitrogen Dioxide (NO_2). As with nitric oxide, no
attempt has yet been made to calculate an air-sea flux for NO_2,
although Helas et al. (1981), from their measurements of diurnal
variations in NO_2 concentrations in marine air, suggest it is
possible that the ocean acts as a source.

The air-water exchange of NO_2 is often assumed to be under
gas-phase control, due to its presumed high solubility in water,
and there is some theoretical (Kabel, 1976) and experimental
(Slater, 1974) evidence to support this. However, recently, Lee

and Schwartz (1981a) have redetermined the Henry's law constant
for NO_2 and find it to be much less soluble than previously
thought. This, coupled with the slow kinetics of the reactions
of NO_2 in solution, leads them to conclude that air-water transfer of the gas will be under liquid phase control (Lee and Schwartz,
1981b).

5.4.4 <u>Ammonia</u> (NH_3). Georgii and Gravenhorst (1977) tried
to use the $K\Delta C$ approach to calculate the air-sea transfer of ammonia; they obtain an upward flux of 5×10^{-2} μg m^{-2} h^{-1}. However,
the result must be treated with caution, because it is based on
very few observations and because the authors appear to have assumed NH_3 exchange is under liquid-phase control. The validity of
this assumption is questionable in view of the high solubility of
the gas in water ($H = 3 \times 10^{-4}$ in pure water, although according
to Hales and Drewes (1979) there is uncertainty over the value in
water containing CO_2). Further difficulties arise because water
concentrations are obtained by calculation from measurements of
pH and total dissolved ammonia ($NH_3 + NH_4^+$), which is quite variable. All that can really be said is that marine air and surface
seawater do not appear to be greatly out of equilibrium with respect to NH_3, a conclusion supported by recent measurements of
ammonia gas over the Southern Ocean by Ayers and Gras (1980).

Since NH_3 is the only well established alkaline gas in the
atmosphere, it is clearly important to understand the quantitative
role of the oceans as a source/sink, particularly with respect to
understanding what controls the pH of rain in remote marine areas.

The air-sea fluxes of various nitrogen compounds discussed
in Section 5.4 are summarised in Figure 18.

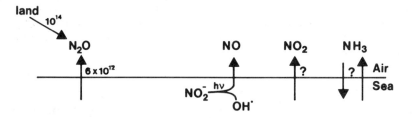

Figure 18. Air-sea and other fluxes (in g yr^{-1}) of
nitrogen compounds.

5.5   Oxygen

5.5.1   Ozone (O_3).   Previous estimates of ozone uptake by the oceans have ranged from 3 to 7 x 10^{14} g yr^{-1} (Fabian and Junge, 1970; Tiefenau and Fabian, 1972).   Recent measurements of the total surface resistance (R_a) of the oceans to uptake of O_3 are in good agreement at 1.8 x 10^3 s m^{-1} (Galbally and Roy, 1980; Lenschow et al., 1982).   Using this value for R_a, a mean concentration of O_3 in air over the oceans of 100µg m^{-3}, and assuming the sea to be a perfect sink (because of the high reactivity of sea-water components with ozone), yields a total flux into the oceans of 6 x 10^{14} g yr^{-1}, in excellent agreement with the earlier estimates.

5.6   Sulphur

5.6.1   Volatile Sulphur.   Over the last 20 years several global sulphur budgets have been published.. Each of them has a flux of volatile sulphur from the oceans (and land surfaces) to the atmosphere whose magnitude is calculated by difference, i.e. to make the budget balance.   Since such an approach is inherently unreliable, it is not surprising that the size of the flux varies substantially between budgets (see Table 3).   Furthermore, we do not know at all well how the total flux is partitioned among the various possible volatile sulphur compounds (H_2S, $(CH_3)_2S$, COS, CS_2, etc.).   Present knowledge (or lack of it) concerning the transfer of these individual compounds is discussed in subsequent sections.

However, balancing of the global cycle is not the only reason why the sea-to-air flux of volatile sulphur is important. For example, once in the atmosphere these compounds will be subject to photochemical oxidation leading to the production of SO_2 (Logan et al., 1979).   A fraction of the SO_2 may be reabsorbed in the ocean (see later), but much will probably remain in the atmosphere.   There, it can have at least two environmentally important fates: i) incorporation into rain drops whose acidity will thereby increase, thus playing a potentially important role in controlling the pH and excess sulphate of marine precipitation; ii) formation of stratospheric sulphate aerosol, with implications for the radiation balance and hence the climate of the Earth (Sze and Ko, 1979;   Turco et al., 1980).

5.6.2   Hydrogen Sulphide (H_2S).   Until quite recently, it was assumed that H_2S was the compound which carried the whole of the sea-to-air flux of volatile sulphur.   However, this is unlikely, since H_2S is thermodynamically unstable in oxygenated waters, and even if formed, is very rapidly oxidised in such environments.   Thus, the only areas where H_2S is at all likely to be a transferring species are shallow coastal waters overlying anoxic sediments. Release of H_2S has indeed been shown in such environments (e.g. Hansen et al., 1978; Ingvorsen and Jorgensen, 1982), but, as expected, it does not seem to have been found in open-ocean surface waters.

However, H_2S does occur in remote marine air, and this led Graedel (1979) to compute, from a knowledge of its tropospheric photochemistry, the size of the sea-to-air flux required to maintain the observed air concentration. This value is given in Table 3, but, since it is obtained only indirectly, it should be regarded as unconfirmed. Indeed, it could be that the open oceans act as a net sink for atmospheric H_2S.

5.6.3 Dimethyl Sulphide (DMS-$(CH_3)_2S$). Using the concentrations of DMS measured by Lovelock et al. (1972) in the Atlantic, Liss and Slater (1974) calculated the global sea-to-air flux of this gas to be 7×10^{12} g yr^{-1}. More recent concentration measurements have led to a substantial upward revision of the flux, and these new values are the ones given in Table 3.

The value of 50×10^{12} g yr^{-1} calculated by Nguyen et al., (1978) is based on sea surface microlayer (top 400 μm) samples, which they found to be up to five times enriched in DMS compared to conventional 'surface' waters.

5.6.4 Carbon Disulphide (CS_2) and Carbonyl Sulphide (COS). Using the data given by Lovelock (1974) for concentration of CS_2 in the open Atlantic (50°N to 65°S), it is possible to calculate the global sea-to-air flux of the gas as approximately 0.3×10^{12} g yr^{-1}. This implies that CS_2 is about 100 times less effective than DMS in transferring vapour-phase sulphur from the oceans to the atmosphere.

Recently Rasmussen et al. (1982a) have measured COS in surface ocean samples and calculate a global flux of 0.8×10^{12} g yr^{-1} from sea-to-air, i.e. about twice the size of the flux calculated here for CS_2, but still substantially less than the flux of DMS.

5.6.5 Sulphur Dioxide (SO_2). Unlike the sulphur compounds discussed so far, the direction of the net flux for SO_2 is into the oceans. Any possibility of a reverse flux can be ruled out, in view of the high pH of seawater and its ability to oxidise sulphite to sulphate.

The magnitude of the flux from air to sea was calculated by Liss and Slater (1974) to be 1.5×10^{14} g yr^{-1}, based on a concentration of SO_2 in marine air of 3 μg m^{-3}. More recent measurements give substantially lower SO_2 concentrations over the oceans, generally close to 0.1 μg m^{-3} (Meszaros, 1978; Maroulis et al.,

1980; Ockelmann et al., 1981). This leads to a much reduced flux, as given in Table 3.

Much of the SO_2 taken up by the oceans is probably derived from sulphur compounds evolved from seawater and oxidised in the atmosphere, although some may be from combustion sources, especially near heavily-industrialised land areas.

The air-sea fluxes of various sulphur compounds discussed in Section 5.6 are summarised in Figure 19.

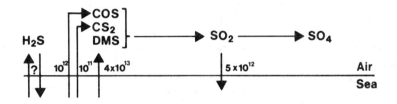

Figure 19. Air-sea fluxes (in g yr^{-1}) of sulphur compounds, and their atmospheric transformations.

5.7 Chlorine

5.7.1 Methyl Chloride (CH_3Cl). Since few measurements have been made of the concentration of CH_3Cl in seawater, and these show rather variable results, the sea-to-air flux in Table 3 must be regarded as tentative. Recent atmospheric measurements made during project GAMETAG (Rasmussen et al., 1980), which give global tropospheric CH_3Cl levels in the range 755-815 pptv (compared with the value of 613 pptv used by Singh et al., 1979), do not significantly alter the calculated flux.

The mechanism for the production of CH_3Cl in seawater is not known, although it is assumed to be natural. It has been suggested that CH_3Cl is formed by reaction between the abundant Cl^- ions in seawater and methyl iodide produced by algae (Zafiriou, 1975). As will be discussed, surface seawater also appears to be a source of methyl iodide to the atmosphere ($\sim 10^{12}$ g yr^{-1}) so that if the mechanism proposed by Zafiriou is correct, then the algae (or some inorganic process - see later) must be producing substantially more CH_3I than is required to account for the sea-to-air flux of this compound alone.

The sea appears to be the largest source of CH_3Cl to the atmosphere, with biomass burning as the next most prolific con-

tributor (Watson et al., 1980). The reason for estimating the
flux of this compound to the atmosphere is because it is probably
comparable to the man-made Freons as a natural source of chlorine
atoms to the stratosphere, and may also be of importance for the
behaviour of gases in the troposphere (Rasmussen et al., 1980).

5.7.2 Carbon Tetrachloride (CCl_4) and Trichlorofluoromethane
(CCl_3F - Freon-11). Using the results of Lovelock et al. (1973)
for CCl_4 and Freon-11 concentrations in marine air and surface
seawater in the N. and S. Atlantic, Liss and Slater (1974) applied
the $K\Delta C$ approach to calculate global air-to-sea fluxes of these
two gases of 1.4×10^{10} and 5.4×10^9 g yr^{-1}, respectively. The
mean undersaturation of the seawater with respect to CCl_4 in the
air was about 10%. A decade later Hunter-Smith et al. (1983)
sampled on an almost identical cruise track and found the air and
seawater to be essentially at equilibrium, implying no net flux.
Similarly, recent measurements of Freon-11 in the N. Pacific
(Gammon et al., 1982) indicate almost precise saturation equilibrium between atmosphere and surface ocean. In view of the level-
ling off and recent decrease in release of Freon-11 to the atmosphere, it may well be that the change found in the air-to-sea flux
over the last decade is real. At the beginning of the period
atmospheric releases (mainly in the N. hemisphere) were still increasing and this led to a significant decrease in air concentrations between the northern and the southern hemispheres. In the
circumstances, a flux into the oceans was to be expected - the
figure given in Table 3 corresponds to approximately 2% of the
anthropogenic production rate at that time. The recent changes
in Freon-11 release rates have led to a decrease in the N-S difference in air levels (Hunter-Smith et al., 1983), and may have
allowed the surface oceans and the atmosphere to come close to
equilibrium. Both sets of flux estimates are given in Table 3.

Several reports of surface seawater apparently being substantially supersaturated with respect to Freon-11 in the atmosphere (Rasmussen et al., 1976; Hahne et al., 1978; Hammer et
al., 1978; Singh et al., 1979) can be explained on the basis
that incorrect values (Zeininger, 1975) were used for the Henry's
law constant for Freon-11 in seawater. When recent measurements
of H for Freon-11 (Hunter-Smith et al., 1983) are used, the
surface waters are found to be either close to equilibrium, or
somewhat undersaturated.

5.7.3 Chloroform ($CHCl_3$). Very recently Rasmussen and his
co-workers (R.A. Rasmussen, personal communication) have used
measurements they have made of chloroform in marine air and surface seawater to calculate the flux of this compound across the
sea surface. They obtain a flux of 0.7 (S.D. 0.3) $\times 10^{12}$ g yr^{-1}
from the sea to the air. This result may imply that much of the
chloroform in the atmosphere is natural and not anthropogenic, as
thought previously.

5.8 Iodine

5.8.1 Methyl Iodide (CH₃I). For balance, the global geo-chemical budget for iodine, as for sulphur, requires a flux of some volatile species from the ocean to the atmosphere, and both I_2 and CH_3I have been suggested as possibilities. Laboratory studies seem to show that I_2 can be generated in seawater by either u.v. irradiation (Miyake and Tsunogai, 1963) or the action of O_3 (Garland and Curtis, 1981), although the applicability of such laboratory results to the real oceans is unknown. If methyl iodide is the transferring species, then field data are available for marine air and surface seawater concentrations in the Atlantic (Lovelock et al., 1973), and Liss and Slater (1974) have used these to calculate the oceanwide flux of CH_3I out of the oceans as 3×10^{11} g yr^{-1}. This is about half the amount needed to balance the global iodine cycle. Recently, Rasmussen et al. (1982b) have estimated the sea-to-air flux of CH_3I and obtain a value for the open oceans in close agreement with that of Liss and Slater. However, Rasmussen et al. suggest that if areas of high biological production are included, then the global flux might be four times higher (i.e. 1.3×10^{12} g yr^{-1}). Both values are given in Table 3.

5.9 Conclusions

It should be stressed that the global air-sea fluxes listed in Table 3 have considerable uncertainties and limitations assoc-iated with them. As indicated earlier, the KΔC approach can give results only as reliable as the transfer velocity and concentration data used. Certainly, better knowledge of global, average values for ΔC has led to significant alteration in calculated fluxes for a number of gases. For example, comparing air-sea fluxes in Liss & Slater (1974) with those in Table 3 indicates that over this per-iod the calculated N_2O flux has decreased 20-fold, whereas that for DMS has increased by a factor of about six. Uncertainty also arises over appropriate values for the transfer velocity. Although globally averaged values may appear to be reasonably well known (see Section 5.1), it may be that much of the exchange occurs under high wind regimes, as the wind tunnel results imply. These are just the conditions which are, almost inevitably, most poorly studied with regard to transfer velocity determination (and con-centration measurements).

Estimation of air-sea fluxes on smaller space and time scales than the global, annual averages given in Table 3, may be some-what easier in terms of specification of ΔC, but may well be more difficult in view of our present ignorance of how k_w varies with measurable parameters, such as wind speed.

What is clear is that in order to improve future estimates of
air-sea gas fluxes many more concentration measurements will have
to be made, and knowledge of the physico-chemical, and especially
the dynamic, factors controlling transfer velocities will need to
be better understood. The transfer mechanism and concentration
measurement studies will need to proceed in tandem, and will prove
a stimulating area for interdisciplinary marine research in the
years to come.

ACKNOWLEDGEMENTS

I thank the many participants in the Advanced Study
Institute who contributed to the discussions of the lectures I
gave in Durham. Comments from Blayne Hartman, Lutz Hasse, Peter
Hyson, and especially George Slinn, were particularly helpful in
preparing this chapter.

REFERENCES

Adams, D.F., S.O. Farwell, M.R. Pack and W.L. Bamesberger, 1979:
    Preliminary measurements of biogenic sulfur-containing gas
    emissions from soils. J. Air Pollut. Control Ass., 29,
    380-383
Atkinson, L.P., 1973: Effect of air bubble solution on air-sea
    gas exchange. J. Geophys. Res., 78, 962-968.
Atlas, E., R. Foster and C.S. Giam, 1982: Air-sea exchange of
    high molecular weight organic pollutants: laboratory
    studies. Environ. Sci. Technol., 16, 283-286.
Ayers, G.P. and J.L. Gras, 1980: Ammonia gas concentrations over
    the Southern Ocean. Nature, 284, 539-540.
Balls, P.W. and P.S. Liss, 1982: Exchange of H_2S between water
    and air. Atmospheric Environment, in press.
Barger, W.R., W.H. Daniel and W.D. Garrett, 1974: Surface chem-
    ical properties of banded sea slicks. Deep-Sea Res., 21,
    83-89.
Barnard, W.R., M.O. Andreae, W.E. Watkins, H. Bingemer and
    H.W. Georgii, 1982: The flux of dimethylsulfide from the
    oceans to the atmosphere. J. Geophys. Res., 87, 8787-8793.
Berger, R. and W.F. Libby, 1969: Equilibration of atmospheric
    carbon dioxide with sea water: possible enzymatic control of
    the rate. Science, 164, 1395-1397.
Bolin, B., 1960: On the exchange of carbon dioxide between the
    atmosphere and the sea. Tellus, 12, 274-281.
Brimblecombe, P. and D.J. Spedding, 1972: Rate of solution of
    gaseous sulphur dioxide at atmospheric concentrations.
    Nature, 236, 225.
Brockmann,U.H., H. Huhnerfuss, G. Kattner, H.-C. Broecker and G.
    Hentzschel, 1982: Artificial surface films in the sea area
    near Sylt. Limnol. Oceanogr., 27, 1050-1058.

Broecker, H.C., 1980: Effects of bubbles upon the gas exchange between atmosphere and oceans. In: Berichte SFB 94, Universität Hamburg, 127-139.

Broecker, H.C., J. Petermann and W. Siems, 1978: The influence of wind on CO_2-exchange in a wind-wave tunnel, including the effects of monolayers. J. Mar. Res., 36, 595-610.

Broecker, W.S., 1965: An application of natural radon to problems in ocean circulation. In: Symposium on Diffusion in Oceans and Fresh Water (T. Ichiye, ed.), Columbia University, 116-145.

Broecker, W.S., 1974: Chemical Oceanography. Harcourt Bruce Jovanovitch, 214 pp.

Broecker, W.S. and T.H. Peng, 1971: The vertical distribution of radon in the Bomex Area. Earth Planet. Sci. Lett., 11, 99-108.

Broecker, W.S. and T.H. Peng, 1974: Gas exchange rates between air and sea. Tellus, 26, 21-35.

Broecker, W.S., T. Takahashi, H.J. Simpson and T.H. Peng, 1979: Fate of fossil fuel carbon dioxide and the global carbon budget. Science, 206, 409-418.

Cohen, Y. and L.I. Gordon, 1979: Nitrous oxide production in the ocean. J. Geophys. Res., 84, 347-353.

Conrad, R. and W. Seiler, 1982: Carbon monoxide in seawater (Atlantic Ocean). J. Geophys. Res., 87, 8839-8852.

Danckwerts, P.V., 1951: Significance of liquid-film coefficients in gas absorption. Ind. Engng. Chem., 43, 1460-1467.

Danckwerts, P.V., 1970: Gas-Liquid Reactions. McGraw-Hill, 276 pp.

Deacon, E.L., 1977: Gas transfer to and across an air-water interface. Tellus, 29, 363-374.

Deacon, E.L., 1981: Sea-air gas transfer: the wind speed dependence. Boundary-Layer Met., 21, 31-37.

Downing, A.L. and G.A. Truesdale, 1955: Some factors affecting the rate of solution of oxygen in water. J. Appl. Chem., 5, 570-581.

Ehhalt, D.H. and U. Schmidt, 1978: Sources and sinks of atmospheric methane. Pure App. Geophys., 116, 452-464.

Emerson, S., 1975: Chemically enhanced CO_2 gas exchange in a eutrophic lake: a general model. Limnol. Oceanogr., 20, 743-753.

Eriksson, E., 1963: The yearly circulation of sulfur in nature. J. Geophys. Res., 68, 4001-4008.

Fabian, P. and C.E. Junge, 1970: Global rate of ozone destruction at the earth's surface. Arch. Met. Geoph. Biokl. Ser. A., 19, 161-172.

Francis, J.R.D., 1949: Laboratory experiments on wind-generated waves. J. Mar. Res., 8, 120-131.

Frenkel, J., 1965: Evaporation Reduction. UNESCO, 79 pp.

Galbally, I.E. and C.R. Roy, 1980: Destruction of ozone at the earth's surface. Quart. J. Roy. Met. Soc., 106, 599-620.

Gammon, R.H., J. Cline and D, Wisegarver, 1982: Chlorofluoro-
    methanes in the northeast Pacific Ocean: measured vertical
    distributions and application as transient tracers of upper
    ocean mixing. J. Geophys. Res., 87, 9441-9454.
Garland, J.A., 1977: The dry deposition of sulphur dioxide to
    land and water surfaces. Proc. R. Soc. Lond. A., 354,
    245-268.
Garland, J.A. and H. Curtis, 1981: Emission of iodine from the
    sea surface in the presence of ozone. J. Geophys. Res., 86,
    3183-3186.
Garrett, W.D., 1965: Collection of slick-forming materials from
    the sea surface. Limnol. Oceanogr., 10, 602-605.
Garrett, W.D., 1976: Damping of capillary waves at the air-sea
    interface by oceanic surface-active material. J. Mar. Res.,
    25, 279-291.
Garrett, W.D., 1971: Retardation of water drop évaporation with
    monomolecular surface films. J. Atmos. Sci., 28, 816-819.
Garrett, W.D., 1972: Impact of natural and man-made surface films
    on the properties of the air-sea interface. In: Nobel
    Symposium 20 - The Changing Chemistry of the Oceans, Wiley
    Interscience, 75-91.
Garrett, W.D. and Bultman, J.D., 1963: Capillary-wave damping by
    insoluble organic monolayers. J. Colloid. Sci., 18, 798-801.
Georgii, H.W. and G. Gravenhorst, 1977: The oceans as source or
    sink of reactive trace-gases. Pure Appl. Geophys., 155,
    503-511.
Graedel, T.E., 1979: Reduced sulfur emission from the open
    oceans. Geophys. Res. Letts., 6, 329-331.
Granat, L., H. Rodhe, and R.O. Hallberg, 1976: The global
    sulphur cycle. Ecol. Bull. (Stockholm), 22, 89-134.
Hahn, J. 1979: The cycle of atmospheric nitrous oxide. Phil.
    Trans. R. Soc. Lond. A, 209, 495-504.
Hahne, A., A. Volz, D.H. Ehhalt, H. Cosatto, W. Roether, W. Weiss,
    and B. Kromer, 1978: Depth profiles of chlorofluoromethanes
    in the Norwegian Sea. Pure App. Geophys., 116, 575-582.
Hales, J.M. and D.R. Drewes, 1979: Solubility of ammonia in
    water of low concentrations. Atmos. Environ., 13, 1133-1147.
Hammer, P.M., J.M. Hayes, W.J. Jenkins and R.B. Gagosian, 1978:
    Exploratory analyses of trichlorofluoromethane (F-11) in
    North Atlantic water columns. Geophys. Res. Lett., 5,
    645-648.
Hansen, M.H., K. Ingvorsen and B.B. Jorgensen, 1978: Mechanisms
    of hydrogen sulfide release from coastal marine sediments to
    the atmosphere. Limnol. Oceanogr., 23, 68-76.
Hasse, L., 1971: The sea surface temperature deviation and the
    heat flow at the sea-air interface. Boundary-Layer Met., 1,
    368-379.
Hasse, L. and P.S. Liss, 1980: Gas Exchange across the air-sea
    interface. Tellus, 32, 470-481.

Hasse, L., M. Grunewald, J. Wucknitz, M. Dunckel and D. Schriever, 1978: Profile derived turbulent fluxes in the surface layer under disturbed and undisturbed conditions during GATE. "Meteor" Forsch.-Ergebnisse B 13, 24-40.

Heidt, L.E., J.P. Krasnec, R.A. Lueb, W.H. Pollock, B.E. Henry and P.J. Crutzen, 1980: Latitudinal distributions of CO and CH_4 over the Pacific. J. Geophys. Res., 85, 7329-7336.

Helas, G., A. Broll and P. Warneck, 1981: NO_2 measurement in marine air. Abstracts ROAC Symposium, IAMAP Meeting, Hamburg, August 1981.

Herr, F.L. and W.R. Barger, 1978: Molecular hydrogen in the near-surface atmosphere and dissolved in waters of the tropical North Atlantic. J. Geophys. Res., 83, 6199-6205.

Hicks, B.B. and P.S. Liss, 1976: Transfer of SO_2 and other reactive gases across the air-sea interface. Tellus, 28, 348-354.

Higbie, R., 1935: The rate of absorption of a pure gas into a still liquid during short periods of exposure. Trans. Am. Inst. Chem. Engr., 35, 365-373.

Hill, F.B., V.P. Aneja and R.M. Felder, 1978: A technique for measurement of biogenic sulfur emission fluxes. J. Environ. Sci. Health A13, 199-225.

Hoover, T.E. and D.C. Berkshire, 1969: Effects of hydration on carbon dioxide exchange across an air-water interface. J. Geophys. Res., 74, 456-464.

Huhnerfoss, H., W. Walter and G. Kruspe, 1977: On the variability of surface tension with mean wind speed. J. Phys. Oceanogr., 7, 567-571.

Hunter, K.A. and P.S. Liss, 1977: The input of organic material to the oceans: air-sea interactions and the organic chemical composition of the sea surface. Mar. Chem., 5, 361-379.

Hunter, K.A. and P.S. Liss, 1981: Organic sea surface films. In: Marine Organic Chemistry (E.K. Duursma and R. Dawson, eds.) Elsevier, 259-298.

Hunter-Smith, R.J., P.W. Balls and P.S. Liss, 1983: Henry's law constants and the air-sea exchange of various low molecular weight halocarbon gases. Tellus, in press.

Ingvorsen, K. and B.B. Jorgensen, 1979: Combined measurement of oxygen and sulfide in water samples. Limnol. Oceanogr., 24, 390-393.

Jahne, B. and K.O. Munnich, 1980: Momentum induced gas exchange through a smooth water surface, models and experimental results from linear and circular wind-water-tunnels. In: Berichte SFB 94, Universität Hamburg, 55-62.

Jahne, B., K.O. Munnich and U. Siegenthaler, 1979: Measurements of gas exchange and momentum transfer in a circular wind-water-tunnel. Tellus, 31, 321-329.

Jarvis, N.L., C.O. Timmons and W.A. Zisman, 1962: The effect of monomolecular films on the surface temperature of water. In: Retardation of Evaporation by Monolayers: Transport Processes (V.K. La Mer ed.) Academic, 41-58.

Johnson, K.S., R.M. Pytkowicz and C.S. Wong, 1979: Biological production and the exchange of oxygen and carbon dioxide across the sea surface in Stuart Channel, British Columbia. Limnol. Oceanogr., 24, 474-482.

Jones, E.P. and S.D. Smith, 1977: A first measurement of sea-air CO_2 flux by eddy correlation. J. Geophys. Res., 82, 5990-5992.

Kabel, R.L., 1976: Atmospheric impact on nutrient budgets. J. Great Lakes Res., 2, 114-126.

Kanwisher, J., 1963: On the exchange of gases between the atmosphere and the sea. Deep-Sea Res., 10, 195-207.

Kerman, B.R., 1983: A model of gas transfer at the air-sea interface by breaking waves. J. Geophys. Res., submitted.

Kromer, B. and W. Roether, 1983: Field measurements of air-sea gas exchange by the radon deficit method during JASIN 1978 and FGGE 1979. "Meteor" Forsch.-Ergebnisse A, in press.

La Mer, V.K., 1962: Preface. In Retardation of Evaporation by Monolayers: Transport Processes. Academic, xiii-xiv.

La Mer, V.K. and T.W. Healy, 1965: Evaporation of water: its retardation by monolayers. Science, 148, 36-42.

Lee, Y.N. and S.E. Schwartz, 1981a: Reaction kinetics of nitrogen dioxide with liquid water at low partial pressures. J. Phys. Chem., 85, 840-848.

Lee, Y.N. and S.E. Schwartz, 1981b: Evaluation of the rate of uptake of nitrogen dioxide by atmospheric and surface liquid water. J. Geophys. Res., 86, 11,971-11,983.

Lenschow, D.H., R. Pearson and B.B. Stankov, 1982: Measurements of ozone vertical flux to ocean and forest. J. Geophys. Res., 87, 8833-8837.

Lion .W. and J.O. Leckie, 1981: The biogeochemistry of the air-sea interface. Ann. Rev. Earth Planet. Sci., 9, 449-486.

Liss, P.S., 1971: Exchange of SO_2 between the atmosphere and natural waters. Nature, 233, 327-329.

Liss, P.S., 1973: Processes of gas exchange across an air-water interface. Deep-Sea Res., 20, 221-238.

Liss, P.S., 1975: Chemistry of the sea surface microlayer. In: Chemical Oceanography (J.P. Riley and G. Skirrow, eds.), Vol. 2, Academic Press, 193-243.

Liss, P.S., 1977: Effect of surface films on gas exchange across the air-sea interface. Rapp. P.-v. Reun. Cons. int. Explor. Mer., 171, 120-124.

Liss, P.S. and P.G. Slater, 1974: Flux of gases across the air-sea interface. Nature, 247, 181-184.

Liss, P.S. and P.G. Slater, 1976: Mechanism and rate of gas transfer across the air-sea interface. In: Atmosphere-Surface Exchange of Particulate and Gaseous Pollutants (1974), (R.J. Engelmann and G.A. Sehmel, eds.), ERDA, 354-366.

Liss, P.S. and F.N. Martinelli, 1978: The effect of oil films on the transfer of oxygen and water vapour across an air-water interface. Thalassia Jugoslavica, 14, 215-220.

Liss, P.S., P.W. Balls, F.N. Martinelli and M. Coantic, 1981:
    The effect of evaporation and condensation on gas transfer
    across an air-water interface. Oceanologica Acta, 4,
    129-138.
Logan, J.A., B.M. McElroy, S.C. Wofsy, and M.J. Prather, 1979:
    Oxidation of CS_2 and COS: sources for atmospheric SO_2.
    Nature, 281, 185-188.
Lovelock, J.E., 1974: CS_2 and the natural sulphur cycle. Nature,
    248, 625-626.
Lovelock, J.E., R.J. Maggs and R.A. Rasmussen, 1972: Atmospheric
    dimethyl sulphide and the natural sulphur cycle. Nature, 237,
    452-453.
Lovelock, J.E., R.J. Maggs and R.J. Wade, 1973: Halogenated
    hydrocarbons in and over the Atlantic. Nature, 241, 194-196.
Lowe, D.C. and U. Schmidt, 1981: Formaldehyde in marine air.
    Abstracts ROAC Symposium, IAMAP Meeting, Hamburg, August 1981.
Lumby, J.R. and A.R. Folkard, 1956: Variation in the surface
    tension of sea water in situ. Bull. Inst. Oceanogr. Monaco,
    No. 1080, 19 pp.
MacIntyre, F., 1971: Enhancement of gas transfer by interfacial
    ripples. Phys. Fluids, 14, 1596-1604.
MacIntyre, F., 1974a: Chemical fractionation and sea-surface
    microlayer processes. In: The Sea (E.D. Goldberg, ed.),
    Wiley, Vol. 5, 245-299.
MacIntyre, F., 1974b: The top millimeter of the ocean. Scientific
    American, May 1974, 62-77.
Mackay, D. and A.T.K. Yuen, 1981: Transfer rates of gaseous
    pollutants between the atmosphere and natural waters. In:
    Atmospheric Pollutants in Natural Waters, (S.J. Eisenreich,
    ed.), Ann Arbor Science, 55-65.
Mancy, K.H. and D.A. Okun, 1965: The effects of surface active
    agents on aeration. J. Wat. Pollut. Control Fed., 37,
    212-227.
Maroulis, P.J., A.L. Torres, A.B. Goldberg and A.R. Bandy, 1980:
    Atmospheric SO_2 measurements on project Gametag. J. Geophys.
    Res., 85, 7345-7349.
Martinelli, F.N., 1979: The effect of surface films on gas ex-
    change across the air-sea interface. Thesis, University of
    East Anglia, 320 pp.
Merlivat, L., 1980: Study of gas exchange in a wind tunnel. In:
    Berichte SFB 94, Universität Hamburg, 49-53.
Merlivat, L. and M. Coantic, 1975: Study of mass transfer at the
    air-water interface by an isotopic method. J. Geophys. Res.,
    80, 3455-3464.
Merlivat, L. and L. Memery, 1983: Gas exchange across an air-
    water interface: Experimental results and modeling of bubble
    contribution to transfer. J. Geophys. Res., 88, 707-724.
Meszaros, E., 1978: Concentration of sulfur compounds in remote
    continental and oceanic areas. Atmos. Environ., 12, 699-705.
Miyake, Y. and S. Tsunogai, 1963: Evaporation of iodine from the
    ocean. J. Geophys. Res., 68, 3989-3994.

Nguyen, B.C., A. Gaudry, B. Bonsang and G. Lambert, 1978:
    Reevaluation of the role of dimethyl sulphide in the sulphur
    budget. Nature, 275, 637-639.
Ockelmann, G., F.X. Meixner and H.W. Georgii, 1981: SO_2 measure-
    ments over the Arctic and Atlantic Oceans. Abstracts ROAC
    Symposium, IAMAP Meeting, Hamburg, August 1981.
Peng, T.H., T. Takahashi and W.S. Broecker, 1974: Surface radon
    measurements in the North Pacific Ocean Station Papa. J.
    Geophys. Res., 79, 1772-1780.
Peng, T.H., W.S. Broecker, G.G. Mathieu, Y.H. Li and A.E. Bainbridge,
    1979: Radon evasion rates in the Atlantic and Pacific Oceans
    as determined during the GEOSECS Program. J. Geophys. Res.
    84, 2471-2486.
Petermann, J., 1976: Der einfluss der oberflächenspannung wässriger
    systeme auf die kinetik des gasaustausches. Thesis,
    Universität Hamburg, 276 pp.
Pond, S., G.T. Phelps and J.E. Paquin, 1971: Measurements of the
    turbulent fluxes of momentum, moisture and sensible heat over
    the ocean. J. Atmos. Sci., 28, 901-917.
Pytkowicz, R.M., 1964: Oxygen exchange rates off the Oregon coast.
    Deep-Sea Res., 11, 381-389.
Quinn, J.A. and N.C. Otto, 1971: Carbon dioxide exchange at the
    air-sea interface: flux augmentation by chemical reaction.
    J. Geophys. Res., 76, 1539-1549.
Rasmussen, R.A., M.A.K. Khalil and S.D. Hoyt, 1982a: Atmospheric
    carbonyl sulfide (OCS): Techniques for measurement in air
    and water. Chemosphere, 11, 869-875.
Rasmussen, R.A., M.A.K. Khalil, R. Gunawardena and S.D. Hoyt,
    1982b: Atmospheric methyl iodide (CH_3I). J. Geophys. Res.,
    87, 3086-3090.
Rasmussen, R.A., D. Pierotti, J. Krasnec and B. Halter, 1976:
    Trip report on the cruise of the Alpha Helix research vessel
    from San Diego, California to San Martin, Peru, March 5 to
    20, 1976.
Rasmussen, R.A., L.E. Rasmussen, M.A.K. Khalil and R.W. Dalluge,
    1980: Concentration distribution of methyl chloride in the
    atmosphere. J. Geophys. Res., 85, 7350-7356.
Redfield, A.C., 1948: The exchange of oxygen across the sea
    surface. J. Mar. Res., 7, 347-361.
Reichardt, H., 1951: Vollständige darstellung der turbulenten
    geschwindigkeitsverteilung in glatten leitungen. Z. angew.
    Math. Mech., 31, 208-219.
Rudolph, J. and D.H. Ehhalt, 1981: Measurements of C_2-C_5 hydro-
    carbons over the North Atlantic. J. Geophys. Res., 86,
    11,959-11,964.
Schmidt, U., 1974: Molecular hydrogen in the atmosphere. Tellus,
    26, 78-90.
Schmidt, U. and G. Kulessa, 1981: The oceans as a source of
    atmospheric hydrogen (H_2). Abstracts ROAC Symposium, IAMAP
    Meeting, Hamburg, August 1981.

Schooley, A.H., 1969:   Evaporation in the laboratory and at sea.
    J. Mar. Res., 27, 335-338.
Scranton, M.I. and P.G. Brewer, 1977:   Occurrence of methane in
    the near-surface waters of the western subtropical North
    Atlantic. Deep-Sea Res., 24, 127-138.
Seiler, W., 1974:   The cycle of atmospheric CO. Tellus, 26, 116-
    135.
Seiler, W., 1981:   The ocean as a source and sink for atmospheric
    trace gases. Abstracts ROAC Symposium, IAMAP Meeting, Hamburg,
    August 1981.
Setser, P.J., J.L. Bullister, E.C. Frank, N.L. Guinasso and D.R.
    Schink, 1982:   Relationships between reduced gases, nutrients,
    and fluorescence in surface waters off Baja California. Deep-
    Sea Res., 29, 1203-1215.
Sheppard, J.C., H. Westberg, J.F. Hopper, K. Ganesan and P.
    Zimmerman, 1982:   Inventory of global methane sources and
    their production rates. J. Geophys. Res., 87, 1305-1312.
Sieburth, J.McN. and J.T. Conover, 1965:   Slicks associated with
    trichodesmium blooms in the Sargasso Sea. Nature, 205, 830-
    831.
Siegenthaler, U. and H. Oeschger, 1978:   Predicting future atmos-
    pheric carbon dioxide levels. Science, 199, 388-395.
Singh, H.B., L.J. Salas, H. Shigeishi and E. Scribner, 1979:
    Atmospheric halocarbons, hydrocarbons, and sulfur hexafluoride:
    global distributions, sources and sinks. Science, 203, 899-
    903.
Slater, P.G., 1974:   The exchange of gases across an air-water
    interface. Thesis, University of East Anglia, 205 pp.
Slinn, W.G.N., 1983:   A Potpourri of deposition and resuspension
    topics. In: Proc. 4th. Int. Conf. on Precipitation Scaveng-
    ing, Dry Deposition and Resuspension (H.R. Pruppacher, A.G.
    Semonin, and W.G.N. Slinn, eds.), Pergamon Press, in press.
Slinn, W.G.N., L. Hasse, B.B. Hicks, A.W. Hogan, D. Lal, P.S. Liss,
    K.O. Munnich, G.A. Sehmel and O. Vittori, 1978:   Some aspects
    of the transfer of atmospheric trace constituents past the
    air-sea interface. Atmos. Environ., 12, 2055-2087.
Sturdy, G. and W.H. Fischer, 1966:   Surface tension of slick
    patches near kelp beds. Nature, 211, 951-952.
Sugiura, Y., E.R. Ibert and D.W. Hood, 1963:   Mass transfer of
    carbon dioxide across sea surfaces. J. Mar. Res., 21, 11-24.
Swinnerton, J.W. and R.A. Lamontagne, 1974:   Oceanic distribution
    of low-molecular-weight hydrocarbons. Baseline measurements.
    Environ. Sci. Tech., 8, 657-663.
Sze, N.D. and M.K.W. Ko, 1979: CS_2 and COS in the stratospheric
    sulphur budget. Nature, 280, 308-310.
Thompson, A.M., 1980:   Wet and dry removal of tropospheric
    formaldehyde at a coastal site. Tellus, 32, 376-383.
Thorpe, S.A., 1982:   On the clouds of bubbles formed by breaking
    wind-waves in deep water, and their role in air-sea gas trans-
    fer. Phil. Trans. R. Soc. Lond. A, 304, 155-210.

Tiefenau, H. and P. Fabian, 1972: The specific ozone destruction
    of the ocean surface and its dependence on horizontal wind
    velocity from profile measurement. Arch. Met. Geoph. Biokl.
    Ser. A, 21, 399-412.
Tsunogai, S. and N. Tanaka, 1980: Flux of oxygen across the air-
    sea interface as determined by the analysis of dissolved com-
    ponents in sea water. Geochem. J., 14, 227-234.
Turco, R.P., R.C. Whitten, O.B. Toon, J.B. Pollack and P. Hammill,
    1980: OCS, stratospheric aerosols and climate, Nature, 283,
    283-286.
Wangersky, P.J., 1976: The surface film as a physical environment.
    Ann. Rev. Ecol. Syst., 7, 161-176.
Watson, A.J., J.E. Lovelock and D.H. Stedman, 1980: The problem
    of atmospheric methyl chloride. In: Proc. NATO ASI on
    Atmospheric Ozone, (A.C. Aikin, ed.),U.S. Federal Aviation
    Administration, 365-372.
Webb, E.K., G.I. Pearman and R. Leuning, 1980: Correction of
    flux measurements for density effects due to heat and water
    vapour transfer. Quart. J. Roy. Met. Soc., 106, 85-100.
Weiss, R.F., 1981: Nitrous oxide and methane in ocean surface
    waters. Abstracts ROAC Symposium IAMAP Meeting, Hamburg,
    August 1981.
Wesley, M.L., D.R. Cook, R.L. Hart and R.M. Williams, 1982: Air-
    sea exchange of CO_2 and evidence for enhanced upward fluxes.
    J. Geophys. Res., 87, 8827-8832.
Whitman, W.G., 1923: The two-film theory of gas absorption.
    Chem. metall. Engng., 29, 146-148.
Wilson, D.R., J.W. Swinnerton and R.A. Lamontagne, 1970: Pro-
    duction of carbon monoxide and gaseous hydrocarbons in sea-
    water: relation to dissolved organic carbon. Science, 168,
    1577-1579.
Wyman, J., P.F. Scholander, G.A. Edwards and I. Irving, 1952:
    On the stability of gas bubbles in sea water. J. Mar. Res.,
    11, 47-62.
Yaglom, A.M. and B.A. Kader, 1974: Heat and mass transfer between
    a rough wall and turbulent fluid flow at high Reynolds and
    Peclet numbers. J. Fluid Mech., 62, 601-623.
Zafiriou, O.C., 1975: Reaction of methyl halides with seawater
    and marine aerosols. J. Mar. Res., 33, 75-81.
Zafiriou, O.C. and M. McFarland, 1980: Determination of trace
    levels of nitric oxide in aqueous solution. Anal. Chem., 52,
    1662-1667.
Zafiriou, O.C., J. Alford, M. Herrerra, E. Peltzer, A.M. Thompson,
    and R.B. Gagosian, 1980a: Formaldehyde in remote marine air
    and rain: flux measurements and estimates. Geophys. Res.
    Lett., 7, 341-344.
Zafiriou, O.C., M. McFarland and R.H. Bromund, 1980b: Nitric
    oxide in seawater. Science, 207, 637-639.
Zeininger, 1975: Loslichkeiten von Frigen 12 und Frigen 11 in
    meerwasser. Internal Report Dr. Z/Mu-5957 III.1 Farb. Hochst.
    AG. West Germany.

AIR-TO-SEA TRANSFER OF PARTICLES

W. George N. Slinn

Pacific Northwest Laboratory
Richland, WA 99352 U.S.A.

1. INTRODUCTION

When starting to read a chapter such as this, I have difficulty discerning where the author wants to lead. Usually, there
are some generalities (this chapter describes some physics of wet
and dry removal -- for chemists), some restrictions (there is
insufficient space, here, to cover all the details), and some
hazy outline of the approach (because of these space limitations,
a deductive approach will be taken). Then, the author usually
attempts to draw a road map with words. However, for me, the
phrase "then, in the next section" has a definite mesmerizing
quality. Some of the fault, I think, lies with   editors and
publishers who refuse to let the author list section and subsection headings. After all, that is what the author is trying to
do; and for me, it's so much easier to review the list, later,
when I'm totally lost. Fortunately in the present case, the
editors have been more receptive to rational arguments(!), and
therefore I invite the reader to glance at the Table of Contents,
now, to gain a glimpse of the chosen route.

But even if the route is known, there are questions about
the mode of travel. I was advised to "teach to chemists" and
"minimize the math", and have tried. But after finishing the
writing, I find that (on occasion) a few symbols, formulae, and
"curvey signs" have crept into crevices between words and lines.
I'm expecting (and have already received) some criticism from
various chemists (apparently best not to name), but this time I'm
ready with some responses. My first response is a story (origin
lost) about the person who was probably the U.S.'s top chemist,
Gibbs. As I recall the story (vaguely!), at the turn of the

299

century, or so, he was attending a faculty meeting (at which he
seldom spoke), dealing with the perennial question about forcing
students to learn other languages. It was suggested that at
least one "foreign-language requirement" be replaced by a requirement in mathematics, but the suggestion was being attacked since
the subject was languages. Whereupon Gibbs rose to make one of
his few remarks: "Mathematics is a language!" I would vote with
Gibbs -- but not that math be considered a foreign language: it's
the language of science. As Galileo wrote in 1623: "Philosophy
is written in this grand book -- I mean the universe -- which
stands continually open to our gaze, but it cannot be understood
unless one first learns to comprehend the language and interpret
the characters in which it is written. It is written in the language of mathematics ...... without which it is humanly impossible
to understand a single word of it; without these, one is wander-
ing about it in a dark labyrinth."

And I can't help wondering if there wouldn't be more light
in the labyrinth, and more-reasonable paths found, if the only
language we humans used was math -- plus maybe a little humming,
whistling, laughing, and their necessary complements. As the
philosopher/theologian Alan Watts wrote (in his little book
Cloud-hidden, Whereabouts Unknown, 1968): "Even the fascinating
investigations of science have become a pest since ... scientists
have taken themselves seriously, like theologians, humming and
hawing, bumbling and rumbling, about whether so-and-so's hypoth-
eses are really, truly, and absolutely sound -- which is often
all that they are." Therefore, for this chapter, I propose to
extend the Chinese proverb "one picture is worth more than ten
thousand words" to "one formula can be worth more than ten
thousand pictures!", and get on with developing the formulae.

2. RESIDENCE TIMES

There are two main (and only slightly-different) reasons for
wanting to know particle fluxes to the oceans. One reason is to
understand just how much we humans are fouling the oceans (and,
as appropriate, do something about it) since the oceans influence
so much: precipitation, the earth's radiation budget, the concentrations of most atmospheric gases (e.g., oxygen, carbon dioxide,
etc.). The other main reason is to describe the "tropospheric
residence time" of particles, in turn, in part, so blame can be
distributed appropriately (e.g., how much of Europe's airborne
lead and sulfate have crossed the Atlantic from N. America?). Of
course, answers to these two types of questions are related (the
fraction remaining is unity minus the fraction removed!), but the
analyses have differences: given the local air concentration of
particles, not too much meteorology is needed to estimate the
local fluxes to the ocean; in contrast, to predict the local air

concentrations (and similarly, to predict the residence time) is
overwhelmingly a meteorological problem. But even if the thrust
is to estimate residence times, still it is necessary to know the
fluxes -- so this is a reasonable place to begin, after I make a
few comments about "particles".

2.1   Aerosol Particles

In this volume, the three preceeding chapters have dealt
with air-sea exchange of gases. This and the next two chapters
emphasize particles. And though this would be a logical place to
describe atmospheric aerosols, my space allotment does not permit
an adequate description: whole books are devoted to aerosol particles (e.g., Twomey, 1977). Yet I would like to emphasize two
points.

The first point is to advise caution about "particle concentration". Sometimes, "concentration" means total number of particles per unit volume, and sometimes, total mass (of all elements
or compounds, or just one). Moreover, especially in this chapter,
"concentration" can mean concentration as a function of particle
"size" (e.g., particle radius, a) and this may be given as number,
area, or volume (or mass) of particles per unit volume and per
unit increment of particle radius (or diameter, D). Interrelations among various "size distributions" are shown in Table 1.
Particular attention is called to the distributions using increments of the logarithm of the particle diameter (where the base
of the log may be 10 or e). The advantage of these distributions
is that when, say, the number distribution of particles is plotted
on log-log paper, then $(\Delta N/\Delta \ln D) \Delta \log D = N/2.3$ is not distorted
by the use of the logarithmic paper. Thus, a quick glance at
such a graph gives a correct picture of the number of particles
of each size increment; in contrast, $(\Delta N/\Delta D)\Delta \ln D = D^{-1} N$,
which distorts the graph.

The second point to emphasize is the "structure" of particle-
size "spectra". Thirty years (or so) ago, when most particle-size
spectra were plotted as number distributions, the Junge distribu-
tion, $n(a) \sim a^{-\gamma}$ with $\gamma \sim 4$, was frequently found, especially for
particles with radii $a \gtrsim 0.1 \ \mu m$. Subsequently, particle spectra
have been "fit" better with one or more lognormal distributions --
and the "upper tail" of a lognormal distribution can usually be
fit with a "power-law" distribution of the Junge type. About a
decade ago, Whitby and coworkers emphasized a "bimodal" particle
distribution, with an "accumulation mode" in the $0.1-1 \ \mu m$ range
("accumulation" because it is on these particles that most trace
gases and very small particles accumulate; e.g., during pollution
episodes), and a "coarse mode" of larger particles. This bimodal
distribution is especially evident if particle volume (or mass)
distributions are plotted, because such plots heavily weight the

large particles. In contrast, if log-area distributions are
plotted, as in Figure 1, then the particle distribution is seen
to be multimodal, with the number and "strength" of each mode

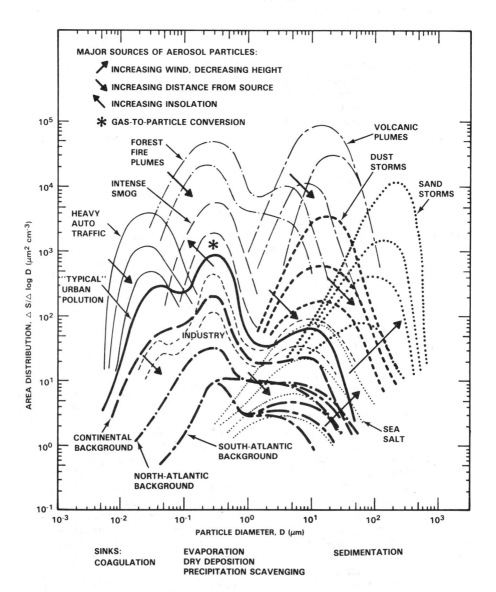

Figure 1. Major sources of atmospheric aerosol particles,
"typical" surface-area distributions of particles in surface-
level air, and indications of the major sinks. (After Slinn,
1975.)

Table 1. Interrelations among various particle "size" distributions. (From Slinn, 1975.)

"Size" Distributions	Usual symbols for cumulative distribution functions	Usual symbols for distribution functions*	Convenient units	Alternative symbols for distribution functions*	Derivation from the number distribution
Number	N	$\Delta N/\Delta D$ †	Number $cm^{-3}\,\mu m^{-1}$	$n(D)$	$n(D)$
Log number	--	$\Delta N/\Delta lnD$**	Number cm^{-3}	$n(lnD)$	$Dn(D)$
Length	L	$\Delta L/\Delta D$	$\mu m\ cm^{-3}\,\mu m^{-1}$	$\ell(D)$	$Dn(D)$
Log length	--	$\Delta L/\Delta lnD$	$\mu m\ cm^{-3}$	$\ell(lnD)$	$D^2 n(D)$
Area	S, A	$\Delta S/\Delta D$	$\mu m^2\ cm^{-3}\,\mu m^{-1}$	$s(D)$	$\pi D^2 n(D)$
Log area	--	$\Delta S/\Delta lnD$	$\mu m^2\ cm^{-3}$	$s(lnD)$	$\pi D^3 n(D)$
Volume	V	$\Delta V/\Delta D$	$\mu m^3 cm^{-3}\,\mu m^{-1}$	$v(D)$	$(\pi/6)D^3 n(D)$
Log volume	--	$\Delta V/\Delta lnD$	$\mu m^3\ cm^{-3}$	$v(lnD)$	$(\pi/6)D^4 n(D)$
Mass	M	$\Delta M/\Delta D$	$\mu g\ m^3\,\mu m^1$	$m(D)$	$\rho_p v(D)$ †
Log mass	--	$\Delta M/\Delta lnD$	$\mu g\ m^{-3}$	$m(lnD)$	$\rho_p v(lnD)$

* To be consistent with mathematical statistics these would be called density functions.
† The characteristic size of an individual particle, D, is not easily identified especially when different measurement techniques are used (e.g. light scattering, aerodynamic behavior of particles, their mobility, etc.). In theoretical studies the particles are usually treated as spheres and D is taken to be the particle diameter.
** Many authors use base-10 rather than base-e logarithms. To convert between the two, use $\Delta lnD = (ln10)\Delta logD = 2.3\ \Delta logD$.
† ρ_p is the average mass density for an individual aerosol particle.

depending on many factors, such as those shown in the figure. More of this later; now, I want to get on with the task of estimating fluxes.

2.2 Wet and Dry Deposition Velocities

My goal, here, is to show the framework for the subsequent description of wet and dry fluxes of particles to the oceans. For particles (in contrast to the case for gases) it is reasonable to expect that the dry deposition flux, F_d, is proportional to the concentration of particles in air, C_a, near the air/sea interface:

$$F_d \propto C_{a,h} \ , \tag{1}$$

where the subscript h, on C_a, identifies a convenient reference height (e.g., the height of the air intake to the sampler). It is assumed, thereby, that the particles are "well mixed" beneath this reference height (see Section 6.3 for comments about the constant-flux assumption). Also, it is noted that the air concentration of particles, C_a, may have any of many different dimensions (e.g., number of particles of a given size class and

per unit volume, mass per unit volume, etc.). In (1), if the dry
flux, F_d, has units of, say, kg $m^{-2}s^{-1}$, and if C_a has "consistent"
units of kg m^{-3}, then the proportionality constant must have
units of m s^{-1}, and is known as the dry deposition velocity,
v_d,

$$v_d = -F_d/C_{a,h} \quad , \tag{2}$$

where the negative sign has been introduced so that v_d will be
positive when the flux is down (i.e., positive flux will be taken
to be in the positive z-direction, viz., up).

Some points about (2) are worth noticing. (i) Since $C_{a,h}$
normally depends on the reference height, h (even if the particles are well-mixed below h), but F_d may not (and will not,
for a "constant-flux layer"), then v_d will depend on h, and
should be written as $v_{d,h}$. However, the dependence of v_d on
height is usually quite weak, e.g., as ln (h/z_0), where z_0 is
the roughness height. Consequently, even if different investigators use substantially different reference heights, the different v_d's don't differ much (compared to other uncertainties, to
be described). For example, if z_0 = 1 cm, h_1 = 1 m, and h_2 =
10 m, then $v_{d,10 \text{ m}}/v_{d,1 \text{ m}} \simeq \ell n \ 10^2/\ell n \ 10^3$ = 2/3 -- and in this
area of study, we are not yet to the state-of-accuracy to worry
about factors of two or three (in some cases, see later, the
uncertainty is 2 or 3 orders-of-magnitude!). (ii) If the flux is
dictated by gravitational settling, then $-F_d \simeq v_g C_a$, and therefore $v_d = v_g$, the gravitational settling speed. For example, for a
particle of radius a = 10 μm and density ρ_p = 1 g cm^{-3}, then $v_g \simeq$
1 cm s^{-1}; whereas for a = 1 μm and ρ_p = 1 g cm^{-3}, $v_g \simeq 10^{-2}$ cm
s^{-1}. (iii) But if there is only gravitational settling (i.e.,
$v_d = v_g$, independent of height, and if C_a varies with height,
then we can't have a constant flux layer, i.e., $-F_d = v_g C_a(z) =$
$-F_d(z)$. More about this later. (iv) In some cases, the net flux
of particles (e.g., sea-salt particles) is from (not to) the
ocean. Then v_d could become negative, but there are other mathematical ways to describe this case of "resuspension" (e.g., in
terms of a resuspension velocity; see Section 6.2). (v) For
gases, it needn't be that F_d is related linearly to C_a; in
fact, F_d may be essentially independent of C_a and dominated by
the concentration of the gas in the sea, C_s. But the case of
air/sea exchange of gases is described in other chapters; here
the thrust will be to describe v_d for particles for which (2)
is adequate, if resuspension is temporarily ignored. We seek
descriptions for v_d in terms of particle characteristics, and
atmospheric and surface conditions; e.g., Figure 2 shows a partial
answer.

Just as the dry flux can be written in terms of a dry deposition velocity, the wet flux can be written in terms of a wet

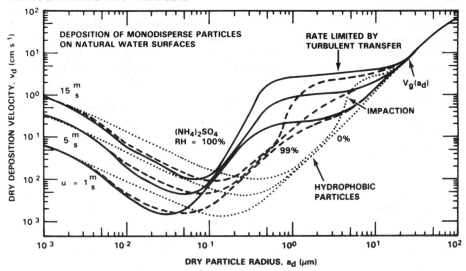

Figure 2. Theoretical predictions of the influence of wind speed and of particle growth by water-vapor condensation, for particle deposition to natural waters. (From Slinn and Slinn, 1980.)

deposition velocity, v_w. Thus, associated with precipitation (e.g., snow), the magnitude of the flux of particles can first be written as

$$F_{w,h} = \frac{\text{"amount"}}{\text{area-time}} = \frac{\text{amount in precip.}}{\text{volume of precip.}} \times$$

$$\frac{\text{volume of precip.}}{\text{area-time}} = C_{p,h} p_h \quad , \tag{3}$$

where C_p is the concentration of particles in the precipitation (e.g., in units of kg m^{-3} = g ℓ^{-1}), and where p is the volume flux of precipitation (e.g., in m^3 m^{-2} s^{-1} = m s^{-1}). Then, if this wet flux is normalized by the particle concentration in the air at some reference height h, a wet deposition velocity results:

$$v_{w,h} = F_{w,h} / C_{a,h} = (C_{p,h}/C_{a,h}) \, p_h = (s_r p)_h \quad , \tag{4}$$

where the ratio of reference-height concentrations is known as the scavenging ratio, $s_{r,h} = (C_p/C_a)_h$.

Some points about (4) are worth noticing. (i) It is a nuisance to be formal, keeping F_w as positive up, and introducing negative signs; therefore, throughout, F_w will mean the magnitude of the wet flux. Also, it may seem (and may be) a bit pedantic to identify so many quantities with the reference height, h.

(Usually the rain rate doesn't change much between the sampler
height and the ocean surface!) However, later we'll need the
precipitation rate inside clouds, and this definitely does change
with height (e.g., it must be zero at cloud top). (ii) There is
even less justification in the case of wet flux (than dry) to
assume that F_w is related linearly to $C_{a,h}$, even for particles.
Thus, the particles can be collected by precipitation aloft, in
the clouds, and there the particle concentration may be essentially unrelated to $C_{a,h}$, unless the atmosphere is well mixed.
(iii) In fact, it is frequently the case over the oceans that a
linear relation between F_w and C_a at h applies better for gases
than particles: frequently, equilibrium conditions prevail, and
for rain (subscript r) $C_{r,h} = sC_{a,h}$, where s is the solubility
coefficient for the gas, and therefore $s_{r,h} = s$, $v_{w,h} = sp_h$,
and the wet flux of the gas at height h, $F_{w,h} = sp_hC_{a,h}$ is thus
related to $C_{a,h}$, linearly. (iv) For particles, the wet deposition velocity can be large. To demonstrate this, suppose $s_{r,h} =$
3.6×10^5. (See Figure 3, which illustrates some data for scavenging ratios over land. These scavenging ratios are dimensionless density ratios: mass/unit volume in rain ÷ mass/unit volume
in air.) Then, if $p = 1$ cm hr^{-1},

$$v_w = ps_r = 3.6 \times 10^5 \times 1 \text{ cm}/(3600 \text{ s}) = 10^2 \text{ cm s}^{-1} \quad , \quad (5)$$

which is very much larger than typical dry deposition velocities
(compare v_d values in Figure 2). However, it doesn't rain all
the time, and if we use (instead of 1 cm hr^{-1}) an annual-average
precip. rate of 100 cm yr^{-1}, then for a scavenging ratio of 3 ×
10^5, the annual average wet deposition velocity would be

$$\langle v_w \rangle = \langle ps_r \rangle = 3 \times 10^5 \times 10^2 \text{ cm}/(3 \times 10^7 \text{ s}) = 1 \text{ cm s}^{-1} \quad , \quad (6)$$

which suggests that the long-term-average wet and dry fluxes of
particles to the ocean are of roughly comparable magnitudes. But
I don't think so.

     It will require most of this chapter to explain why I think
wet removal dominates the residence time of most atmospheric particles, and similarly, why wet deposition is more important than
dry deposition for particle deposition to the oceans, at least if
···. Those "···'s" represent a host of "weasel words", concerned with particle size and chemical composition, precipitation
amounts, distribution of sources, etc. However, for some inland
lakes and similar bodies of salt water, over which the low-
altitude pollution concentration is large, then wet and dry deposition fluxes may be nearly the same. This is suggested by
Figure 4, which shows the theoretical dry-deposition velocity
from Figure 2 (solid curve), and measured wet-depositon velocities for a number of elements (with the assumption of an annual

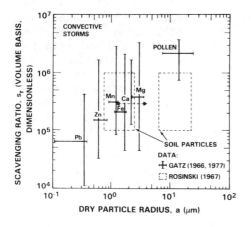

Figure 3. Some scavenging ratios for particles scavenged by convective storms. (From Slinn, 1978.)

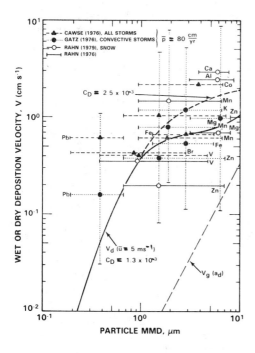

Figure 4. An indication of the relative importance of wet and dry deposition to inland waters. (From Slinn and Slinn, 1981.)

precipitation of 80 cm). It is true that this comparison raises many questions (How accurate is the dry-deposition theory? Why is the comparison between theory for monodisperse (= single-size) particles and data for polydisperse aerosols? Why are there differences in the data for Pb and SO_4^{2-} obtained by different investigators? etc.), and these questions will be addressed later. For now, though, I would like just to leave it as a quali-tative expectation that for inland waters, wet and dry deposition can be of roughly-similar magnitude. Moreover, I want to leave the impres-sion that there are substan-tial uncertainties about deposition to both lakes and oceans. Perhaps after another 30 years of research in this field (e.g., since the pioneering work by Chamberlain, 1953), then the relative importance of wet and dry removal will be less uncertain.

In this chapter, my goals are to indicate some of the knowns and unknowns about wet and dry fluxes of particles to the oceans. First, I'll focus on wet removal, then dry. The description of wet deposi-tion will take a little longer, because I'll be introducing some topics in aerosol physics common to both removal processes. Otherwise, though, the two processes are quite dissim-ilar. For wet removal, the major task is to understand

how particles become associated with precipitation; understanding
how they then get to the ocean (via precipitation) is relatively
simple. In contrast, for dry deposition, the difficulty is to
understand how the particles move through the atmosphere, especially through the last "deposition layer," next to the surface.
But before getting to this, I would like to assume we know the
answers for v_d and v_w and show how these answers could be used
to estimate particle residence times in the troposphere. I will
avoid the case of residence times in the stratosphere, because
usually this "reservoir" doesn't satisfy the well-mixed condition
required by the theory (e.g., Slinn, 1980), and therefore "stratospheric residence times" are usually just the times required for
particles to mix to the troposphere.

2.3 A Simple Model for τ_w and τ_d

To estimate the residence (or turnover) times of particles
in the troposphere, I will start with a simple and crude model.
I'll take the dry flux to be $\overline{v}_d \, \overline{C}_a$ and the wet flux to be $\overline{v}_w \, \overline{C}_a$,
where the overbars represent some average values, accounting for
the particles' vertical distribution. (For now, I don't want to
define that carefully!) Then for the "box model" sketched in
Figure 5, with box height = h and width = w, a steady-state mass
balance is

INFLOW MASS = (MASS/UNIT VOLUME) X [VOLUME INFLOW]

$$= [\overline{C}_a \; \overline{u} \; \Delta t \; h \; w]_x \tag{7}$$

$$= \text{OUTFLOW} = [\overline{C}_a \; \overline{u} \; \Delta t \; h \; w]_{x+\Delta x} + (\overline{v}_d + \overline{v}_w) \, \Delta t \, \Delta x \; w \; \overline{C}_a \, .$$

Cancelling Δt and w, dividing by Δx and letting $\Delta x \to 0$, and
assuming u and h are constants (but see Section 6.5), we get

$$\overline{u} \, \frac{d}{dx} \, \overline{C}_a = -\left\{ \frac{\overline{v}_d}{h} + \frac{\overline{v}_w}{h} \right\} \overline{C}_a \, , \tag{8}$$

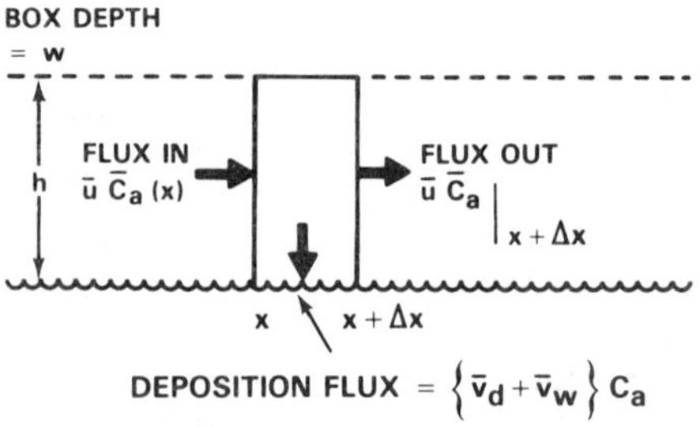

Figure 5. The simple box model used to obtain Eq. 7.

which has the solution (letting $t = x/\bar{u}$)

$$\bar{C}_a(t) = \bar{C}_a(0) \exp\left\{-t/\bar{\tau}\right\}, \tag{9}$$

where the (average, e-fold) "residence" or "turnover" time is

$$\frac{1}{\bar{\tau}} = \frac{1}{\bar{\tau}_d} + \frac{1}{\bar{\tau}_w} + \cdots = \frac{\bar{v}_d}{h} + \frac{\bar{v}_w}{h} + \cdots , \tag{10}$$

in which the "..." has been added, in case there are other (first-order) removal processes, such as chemical reactions or radioactive decay. Instead of viewing $\bar{\tau}$ as a residence time, it might be better to see that $\bar{\tau}^{-1}$ is an overall removal rate, when there are more-than-one removal processes.

As an example of (10), if $\bar{v}_w = 0.5$ cm s^{-1}, $\bar{v}_d = 0.1$ cm s^{-1}, and h = 3 km, then $\bar{\tau}_d = 3 \times 10^5$ cm/0.1 cm s$^{-1} \simeq 35$ days, $\bar{\tau}_w = 3 \times 10^5$ cm/ 0.5 cm s$^{-1} \simeq 7$ days, and

$$\bar{\tau} = \frac{\bar{\tau}_d \bar{\tau}_w}{\bar{\tau}_w + \bar{\tau}_d} \simeq 6 \text{ days} . \tag{11}$$

If this little analysis and example are anywhere near reasonable (and I think they are for particles such as Pb and SO_4^{2-}), then it suggests that wet removal has dominant influence on the tropospheric residence time of particles (except for very large particles, e.g., a $\gtrsim 10$ μm, for which gravitational settling dominates). Therefore, it seems reasonable to try, now, to improve on our estimate for τ_w. This task will be the thrust of the rest of Section 2, and Sections 3 and 4 as well.

2.4 Scavenging Ratios and τ_w

Already, τ_w (wet removal's contribution to a pollutant's tropospheric residence time) has been estimated in terms of scavenging ratios s_r (or in terms of the wet deposition velocity $v_w = ps_r$), but look at that again. As an analogy, suppose you wanted to estimate the average tropospheric residence time of humans! One way is as follows. If the world's population had come to a steady-state value of 4×10^9 people, with a birth rate (= death rate, for steady state) of 80×10^6 people per year, then the average "residence time" of people would be

$$\bar{\tau} = \frac{\text{amount}}{\text{inflow}} = \frac{\text{amount}}{\text{outflow}} = \frac{4 \times 10^9 \text{ people}}{80 \times 10^6 \text{ people/yr}} = 50 \text{ years} . \tag{12}$$

Note that there would be a distribution of residence times about
this mean value, and if you think about it, you could see how
different "residence times" could be defined (see Bolin and Rodhe,
1973). For example, the average "residence time" for humans may
be about 50 years, but the average age of (living) people may be
only about 25 years.

If the same method is used to estimate the residence time of
sulfur in the troposphere, then

$$\bar{\tau}_s = \frac{\text{amount}}{\text{inflow}} \simeq \frac{(1.4-2.0)\text{Tg}}{100 \text{ Tg/yr}} \simeq 1 \text{ week} \quad , \tag{13}$$

and for water vapor in temperate latitudes (Junge, 1963)

$$\bar{\tau}_{wv} = \frac{\text{amount above unit area}}{\text{outflow per unit area}} \simeq \frac{1.8 \text{ g cm}^{-2}}{90 \text{ g cm}^{-2} \text{ yr}^{-1}} \simeq 1 \text{ week} \tag{14}$$

(which suggests, but doesn't prove, that the atmospheric water
cycle has a strong influence on the atmospheric cycling of
sulfur).

Now use this same method to estimate wet removal's contribution to the residence time of any trace constituent in the
troposphere:

$$\bar{\tau}_w = \frac{\text{amount above unit area}}{\text{outflow per unit area}} = \frac{\int \bar{C}_a dz}{\bar{p}_o \bar{C}_{p,o}} = \frac{\int \overline{(C_a/C_{a,o})} dz}{\bar{s}_{r,o}\bar{p}_o} = \frac{\bar{h}_w}{\bar{v}_{w,o}} \quad , \tag{15}$$

where the integral is over the height of the troposphere, and
subscript zero identifies "surface-level" values. The result (15)
for $\bar{\tau}_w$ is similar to the result in (10), but now we have a
better understanding for what was meant, then, by "accounting for
the particles' vertical distribution." Here, we have $\bar{v}_{w,o} = (\bar{p}\bar{s}_r)_o$, and $\bar{h}_w$ is the mean height of the particles during precipitation. However, we must go farther if we are to understand
more about wet deposition velocities and scavenging ratios.

2.5 Scavenging Rates and τ_w

Scavenging ratios ($s_r = C_p/C_a$) can be measured, and they
can be understood in terms of (and derived from) scavenging rates.
For particles, say of a given size class a to a + Δa, the scavenging rate, $\Psi(a)$, is the instantaneous fraction of the particles (by
number, mass, or other characteristic) removed per unit time by
precipitation

$$\Psi(a;\vec{r},t) = \lim_{\Delta t \to 0} \frac{-\Delta C_a/C_a}{\Delta t} \tag{16}$$

where the negative sign is used because ΔC_a is negative, and we want Ψ to be positive. If we ignore spatial variations in C_a (so that the change, as we move with the pollution, arises only because of the time variation), then (16) is $\partial C_a/\partial t = -\Psi C_a$; and if Ψ were independent of time (steady precipitation), then

$$C_a(a;t) = C_a(a;o) \exp\left\{-\Psi t\right\} , \tag{17}$$

i.e., similar to any "first-order removal process." If Ψ were time dependent, (17) would have $\exp\left\{-\int \Psi dt\right\}$.

It might be worthwhile to comment on whether or not scavenging is a first-order process; i.e., on the assumption that the removal rate, $\partial C_a/\partial t = -\Psi C_a$, is first order in C_a (or that Ψ is independent of C_a). I presume no one would expect $\partial C_a/\partial t$ to vary with C_a more "strongly" than C_a^1. (Why, for example, would C_a^2 be expected? The number of particles captured by a raindrop per unit time could be expected to be proportional to the number of particles present, but why on C_a^2?) But what about on $C_a^{1/2}$, or even independent of C_a? In other words, could a hydrometeor (= raindrop, snowflake, hailstone, etc.) become "saturated" with particles, "refusing" to accept more? Presumably that could occur (it does occur for gases when, say, a raindrop becomes saturated with the gas -- see Peters, this volume), but I will ignore the possibilities (snow scavenging during dust storms or volcanic emissions?) for particles. That is, I will assume that hydrometeors do not saturate with particles, and that particle scavenging is a first-order process.

Then, returning to the main thrust, what is the relationship between scavenging rates, ratios, and τ_w? That's rather simple to see, at least when the precipitation is steady. In terms of the scavenging ratio, the wet flux of particles to the surface was written as

$$F_{w,o} = P_o C_{p,o} = P_o [C_p/C_{a,o}]C_{a,o} = [ps_r C_a]_o = [v_w C_a]_o . \tag{18}$$

Meanwhile, in terms of scavenging rates, the wet flux of particles to the surface is the sum of the removal from all heights aloft (if it is assumed that what is removed, aloft, does reach the earth's surface). Per unit volume, the removal per unit time, $-\partial C_a/\partial t$, is ΨC_a. From a volume $\Delta x \Delta y \Delta z$, the removal during Δt is therefore $\Psi C_a \Delta x \Delta y \Delta z \Delta t$. Consequently, the wet flux (divide by $\Delta x \Delta y \Delta t$) is the sum of the removal from all heights:

$$F_{w,o} = \int \psi C_a dz = \int \psi (C_a/C_{a,o}) dz \; C_{a,o} \quad , \tag{19}$$

in which the integration range can be taken to be from zero to
infinity (since, presumably, $\psi \to 0$ at large z; there isn't much
rain at infinity!). Thus, by equating (18) and (19), and thereby
assuming that conditions are steady, we see the relationship
between v_w (or s_r) and ψ:

$$v_{w,o} = (ps_r)_o = \int_o^\infty \psi [C_a/C_{a,o}] dz \quad . \tag{20}$$

Also, from (15), where we had $\bar{\tau}_w = \bar{h}_w/\bar{v}_{w,o} = \int (C_a/C_{a,o}) dz/(ps_r)_o$,
we see that

$$\bar{\tau}_w = \int C_a dz \Big/ \int \psi C_a dz \equiv \langle \psi \rangle^{-1} \quad , \tag{21}$$

which is a result that could have been expected: wet removal's
contribution to the residence time, $\bar{\tau}_w$, is just the inverse of
the removal rate (appropriately averaged over all heights, and
properly accounting for times when it's not raining; i.e., when
$\psi = 0$). I'll show an example later, after ψ has been related to
the precipitation rate.

2.6 Scavenging Efficiencies and τ_w

Already, we have two formulations for τ_w: in terms of
scavenging rates, $\tau_w = \langle \psi \rangle^{-1}$, and in terms of scavenging ratios,
$\bar{\tau}_w = \bar{h}_w/\bar{v}_{w,o} = \bar{h}_w/(\bar{p}\bar{s}_r)_o$. Yet it is useful to consider another:
this one in terms of what I call the scavenging efficiency, ε.
By scavenging efficiency, I mean the fraction of the "pollution"
processed by a storm that is scavenged by it, i.e., the fraction
ε, between 0 and 1, that is deposited on the earth's surface.
The scavenging efficiency is thus an "integral property" of the
storm (and will depend on the pollutant), where the integral is
over the duration and spatial extent of the storm.

Later, in Section 4.3, I'll show some quantitative relations
among scavenging efficiencies, rates, and ratios; for now, let me
just mention an obvious, qualitative relationship. If the (continuously changing!) inflow and outflow areas of a storm (A_i and
A_o), and the durations that the pollution is exposed to the precipitation were known, then

$$\varepsilon = \frac{\text{Amt. Out in Precip.}}{\text{Amt. In}} = \frac{[\iint v_w C_a \, dA \, dt]_{out}}{[\iint \text{Influx} \, dA \, dt]_{in}} \quad . \tag{22}$$

Also, ε can be related to a time integral of the volume-average
scavenging rate; but it is difficult, because quantities must be

suitably time lagged. It might also be useful to mention that the
concept of a scavenging efficiency does not mean that the (instantaneous) scavenging process is not first-order in the particle's
concentration. (Nor does it mean that it is first order!)
Instead, this concept recognizes that the removal process continues only for a finite duration. Also, this duration might be
(and usually is) much shorter than the duration of the precipitation: a cumulonimbus cloud, for example, can continue to precipitate for many hours, but a specific particle entrained by the
storm can "find itself" blown out of the anvil, or dumped with the
rain, in about 10^3 seconds.

For now, then, let me gloss over these complications and pre-
tend ε is known. Then how can ε be used to estimate τ_w? The
essence of the answer is very simple: if, on average, a fraction
$\bar{\varepsilon}$ of the pollution is scavenged by each storm, and if the pollution (or associated air mass) experiences storms with an average
frequency $\bar{\gamma}$ (in a Lagrangian sense; i.e., moving with the air),
then the answer to be derived is that $\bar{\tau}_w = (\bar{\varepsilon}\,\bar{\gamma}\,)^{-1}$. For example,
if $\bar{\varepsilon} = 1/2$, and if in winter $\bar{\gamma}_w = (2.5 \text{ days})^{-1}$, and in summer
$\bar{\gamma}_s = (5 \text{ days})^{-1}$, then

$$\bar{\tau}_w = (\bar{\varepsilon}\,\bar{\gamma}\,)^{-1} = \begin{cases} 2 \ (2.5 \text{ days}) = 5 \text{ days, winter} \\ 2 \ (5 \text{ days}) = 10 \text{ days, summer} \end{cases} . \tag{23}$$

Now, with the answer given, let me turn to the derivation! (But
I trust that everyone realizes that the numerical values just used
were "fudged" - I do not know the "correct" answer for $\bar{\tau}_w$, and
as far as I know, no one else does either!)

2.7 Scavenging as a Stochastic Process

Perhaps the dominant feature of precipitation scavenging is
that it is a stochastic process (i.e., random in time), not deter-
ministic. Thus, the relation used in the previous section, $\bar{\tau}_w = (\bar{\varepsilon}\,\bar{\gamma}\,)^{-1}$, gives the average residence time, and what is sought in
this section is to determine more of the statistics of the pro-
cess, beyond just its average. Of course, a case can be made
that scavenging is deterministic, just as a similar case can be
made that coin tossing is deterministic (if all forces on a
flipped coin and the initial conditions were known exactly, then
there would be no doubt about whether "heads" or "tails" would be
the outcome), but essentially never would the details be known
with sufficient accuracy. Besides, even if they were known, it
is questionable that we would want them: usually, "all" that we
want to know are a few "statistics" of the stochastic process,
such as its mean and variance. These can be determined from the
probability density function, f, for the outcome, say x, where
f(x)dx gives the probability that the outcome is between x and
x + dx.

To derive $\bar{\tau}_w = (\bar{\varepsilon}\,\bar{\nu})^{-1}$ and other results, suppose that, initially, the amount q_0 of some pollutant was released to the atmosphere, and suppose that this pollution is removed from the atmosphere only by precipitation (ignore dry deposition). To estimate the amount $q(t)$ still airborne at time t, suppose that the random fraction $\underset{\sim}{\varepsilon}_j$ of the pollution ($0 \le \underset{\sim}{\varepsilon}_j \le 1$), entering storm j, is scavenged by it. (The tilde underneath a variable, as in $\underset{\sim}{\varepsilon}_j$, will be used to identify it as a random variable.) Then, as sketched in Figure 6, the amount remaining airborne after encountering n storms would be the random variable

$$\underset{\sim}{q}_n(t)/q_0 = \prod_{j=1}^{\underset{\sim}{n}} (1 - \underset{\sim}{\varepsilon}_j) = \prod_{j=1}^{\underset{\sim}{n}} \underset{\sim}{\delta}_j \quad . \tag{24}$$

In (24), note that $\underset{\sim}{n}$ is a random variable, that $(1-\underset{\sim}{\varepsilon}_j) = \underset{\sim}{\delta}_j$ is the precipitation inefficiency of storm j (i.e., the fraction not scavenged), and that we are implicitly assuming that the duration of any storm is short compared with the time t (or with the average time $\bar{\tau}_w$); i.e., (24) does not describe $q(t)$ <u>during</u> any storm.

To obtain statistics of $q_n(t)$ from (24), there are a number of ways to proceed (all of them approximations!), but I do not have the space here to describe any in detail. Let me just sketch the results, and refer interested readers to any good text on

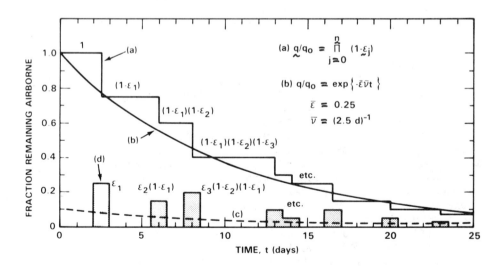

Figure 6. A schematic indication of the influence of scavenging by storms on the amount of some trace constituent remaining airborne. (From Slinn et al., 1978.)

probability and statistics (e.g., Papoulis, 1965). For what follows, it is easier to work with the logarithm of (24), since then
we will have a random variable, $\ln(q_n/q_0)$, which is the sum
(rather than the product) of random variables:

$$\ln(\underset{\sim}{q}_n(t)/q_0) = \sum_{j=1}^{n} \ln (1-\underset{\sim}{\varepsilon}_j) = \sum_{j=1}^{n} \ln \underset{\sim}{\delta}_j \quad . \tag{25}$$

Now, by refering to a statistics text, we could find the
following: (i) If n were deterministic (not a random variable),
say $\bar{n}$, and if the random variables $\underset{\sim}{\varepsilon}_j$ are uncorrelated (i.e.,
$\underset{\sim}{\varepsilon}_j$ is unrelated to $\underset{\sim}{\varepsilon}_{j+1}$, for example), then the mean (or
expected value, E) of the sum is just the sum of the means, and
the variance (σ^2) is the sum of the variances:

$$E\left\{\ln(\underset{\sim}{q}_n/q_0)\right\} = \ln(\bar{q}_n/q_0) = \sum_{j=1}^{\bar{n}} \ln \bar{\delta}_j = n \ln \bar{\delta} = \ln \bar{\delta}^{\,n} \tag{26}$$

$$\sigma^2_{\ln(\underset{\sim}{q}_n/q_0)} = \sum_{j=1}^{\bar{n}} \sigma^2_{\ln \delta_j} = n \,\sigma^2_{\ln \bar{\delta}} \quad , \tag{27}$$

where, for the last equality in (26) and (27), it has been
assumed there is a common mean value and variance of $(1-\underset{\sim}{\varepsilon}_j)$
for all j. (Why not?!) From (26), defining $\bar{n} = \bar{\jmath}t$, where $\bar{\jmath}$ is
the average frequency of storms as experienced by the pollution,
and taking $\ln \bar{\delta} \equiv \ln(1-\bar{\varepsilon}) \simeq -\bar{\varepsilon}$ for small $\bar{\varepsilon}$, then

$$\ln(\bar{q}_n/q_0) \simeq \bar{n}\bar{\varepsilon} = -\bar{\varepsilon}\bar{\jmath}t, \text{ or } \bar{q}(t) = q_0 \exp(-\bar{\varepsilon}\bar{\jmath}t) \quad , \tag{28}$$

i.e., a mean wet-removal time scale can be defined via $\bar{\tau}_w = (\bar{\varepsilon}\bar{\jmath})^{-1}$, which is the answer used in (23). Note, however, that
this answer is an approximation, valid only if $\varepsilon^2 \ll \varepsilon$. [Inci-
dentally, for those familiar with Junge's result, and controver-
sies about it (Gibbs and Slinn, 1973; Baker et al., 1979), (26)
and (27) yield for the coefficient of deviation (the standard
deviation divided by the mean):

$$(cd)_{\ln \underset{\sim}{q}_n/q_0} = \frac{\sqrt{\bar{n}}\,\sigma_{\ln \delta}}{\bar{n} \ln \bar{\delta}} \sim \frac{\sigma_{\ln \delta}}{\sqrt{\bar{n}}\,\bar{\varepsilon}} \sim \left[\frac{\bar{\tau}_w}{t}\right]^{1/2} \tag{29}$$

in which I've assumed that $\sigma^2_{\ln \delta} \sim \varepsilon$. The result (29) is not the
same as Junge's result.]

(ii) On the other hand, if the number of storms, $\underset{\sim}{n}$, is a random variable (as it is), then we need another theorem (e.g., Papoulis, 1965, p. 248). This one states that if all the random variables $(\ln \underset{\sim}{\delta}_j)$ have the same expected value $(\ln \bar{\delta})$, and if they are uncorrelated with the same variance $(\sigma^2_{\ln \delta})$, then

$$E\left\{\ln(q_{\underset{\sim}{n}}/q_0)\right\} = E\left\{\underset{\sim}{n}\right\} \ln \bar{\delta} = \bar{n} \ln \bar{\delta} \ , \qquad (30)$$

i.e., the same result as in (26), when $\underset{\sim}{n}$ was deterministic. But for the second moment, we get the different result:

$$E\left\{\ln^2(q_{\underset{\sim}{n}}/q_0)\right\} = E\left\{\underset{\sim}{n}^2\right\} \ln^2 \bar{\delta} + E\left\{\underset{\sim}{n}\right\}\sigma^2_{\ln \delta} \ . \qquad (31)$$

As a special case of (31), if precipitation events form a series of Poisson impulses (short bursts distributed randomly in time, such as for radioactive decay or "shot noise" in electronics), then the variance of n is equal to its mean (i.e., $E\{\underset{\sim}{n}^2\}$ − $E^2\{\underset{\sim}{n}\} = E\{\underset{\sim}{n}\} = \bar{n}$) and some algebra leads to a result similar to (29):

$$(cd)_{\ln \, \underset{\sim}{q}_n/q_0} = \frac{\bar{n}^{-1/2}\left[\ln^2 \bar{\delta} + \sigma^2_{\ln \delta}\right]^{1/2}}{\bar{n} \ln \bar{\delta}} \sim \frac{1}{\sqrt{\bar{\jmath}t}} \ . \qquad (32)$$

(iii) But the main result I would like to show you, in this view of precipitation scavenging as a stochastic process, is this. With use of the central limit theorem (that, under conditions easily satisfied, the <u>sum</u> of a "sufficiently large" number of random variables, regardless of their distribution, has a Gaussian or normal distribution), then (25) yields that the probability density function (pdf) of q/q_0 (or of $q_0/\underset{\sim}{q}$) has the lognormal distribution:

$$f(q_0/\underset{\sim}{q}) = \left[(q/q_0) \, \sigma_T \sqrt{2\pi}\,\right]^{-1} \exp\left\{-[\ln(q_0/q) + \mu_T]^2/2\sigma^2_T\right\} \ , \qquad (33)$$

where I have used $\ln(q_0/q) > 0$ in the exponent, and where $\mu_T = \bar{n} \ln \bar{\delta}$ and $\sigma^2_T = \bar{n}\sigma_\varepsilon^2$, in which $\bar{\delta} = (1-\bar{\varepsilon})$ and σ_ε^2 is the variance of $\ln(1-\varepsilon)$. From taking moments of (33), we can get (e.g., see Aitchison and Brown, 1976)

$$(cd)_{q_0/\underset{\sim}{q}} = \left[\exp(\sigma^2_T)-1\right]^{1/2} \simeq \sigma_T \quad (\text{if } \sigma_T \ll 1) \ . \qquad (34)$$

This is the same result as in (29) and (32) if we look now at the inverse of (34).

But of more significance is this: Eq. (33) yields that there
will be substantial variability in the amount of pollution remaining airborne; i.e., it is a "long-tailed" lognormal distribution.
Thus, in this model, the mean falls exponentially with time (see
Eq. 28), but at any time downwind, the variation about the mean
has a lognormal distribution; i.e., quite a large probability for
a large fraction still airborne.

2.8 Summary

Let me use a paragraph to show physicists a quick way to
derive the results of this first section. Starting from the continuity equation

$$\frac{\partial C_a}{\partial t} = - \nabla \cdot \vec{\text{Flux}} + \text{Gain} - \text{Loss} \quad,$$

take one term in the loss per unit volume to be ΨC_a, ignore the
gain term, integrate over the troposphere (ignoring any flux
through the tropopause), use Gauss's theorem, and divide by the
total amount of pollution present, q. Then

$$- \frac{1}{q} \frac{\partial q}{\partial t} = \frac{\oint \vec{F} \cdot \vec{ds}}{\int C_a \, dV} + \frac{\int \Psi C_a \, dV}{\int C_a \, dV} + \cdots \quad,$$

or, with term-by-term identification, the overall removal rate is
given by

$$\frac{1}{\tau} = \frac{1}{\tau_d} + \frac{1}{\tau_w} + \cdots \quad.$$

Plus there was the other point: precipitation scavenging is a
stochastic process, with large (lognormal) variability. But
enough of this general formulation! It's time to replace some of
the symbols by numbers.

3. SCAVENGING BY PRECIPITATION

The question now addressed is this: given the precipitation
(rate, type, hydrometeor size distribution), and given properties
of the particles (especially their sizes and chemical compositions), then what will be the rate at which the particles are
scavenged? In the next section, entitled scavenging by storms,
there will be some description of sources of the precipitation.
Here, let me start simply, with the precipitation given, and consider what may appear to be a simple problem, namely:

3.1   Rain Scavenging of Bugs

At the recent Dahlem Conference on Atmospheric Chemistry, a
chemist from Harwell (Stuart Penkett) asked some penetrating questions (which I didn't answer well): "Why is it that, when it's
raining, bugs don't seem to be scavenged? That is, they don't
seem to mind the rain; there seems to be as many flying around,
during and after the rain, as before. Is it just because of the
spaces between the drops?"

Now that I've had more time to think, I'd like to try to
answer those questions -- and similar ones for other "particles".
First, how much space is there between drops? Or, stated differently: during rain, how much of the sky is "covered" by drops?
Alternatively, we could ask: if drops retained their shapes after
they hit the ground (as does hail), then how much of the ground
would be covered by drops?

Presumably the answer depends on the total rainfall P =
$\int p \, dt$, measured, say, in units of cm or $cm^3 \, cm^{-2}$. To get a
rough answer, suppose all drops had radius $\bar{R}$. Then to obtain a
volume of rain per unit area, P, will require n raindrops per
unit area, where

$$n = \frac{\text{volume rain/unit area}}{\text{volume of each drop}} = \frac{P}{4/3 \, \pi \, \bar{R}^3} \quad \frac{cm^3/cm^2}{cm^3} \quad . \tag{35}$$

Meanwhile, the cross-sectional area of each (spherical) drop is
πR^2. Therefore, the fraction of the ground or sky covered by
drops (area/unit area) is

$$f = \pi \bar{R}^2 \, n = \pi \bar{R}^2 \, P \Big/ \left(\frac{4}{3} \pi \bar{R}^3 \right) \simeq P/\bar{R} \quad , \tag{36}$$

where I've used $4/3 \simeq 1$ (this analysis can't support greater
accuracy!). Alternatively, using the rainfall rate (rather than
the rainfall amount), we can say that the fraction of the sky
(area/area) covered per unit time is

$$\dot{f} \simeq p/\bar{R} \; [cm^2 \, cm^{-2} \, s^{-1}] \quad . \tag{37}$$

Let's try to evaluate (37). If $p = 1$ mm hr^{-1} (light rain),
and if $R = 0.1$ mm, then $f = 1$ mm $hr^{-1}/0.1$ mm $= 10$ $cm^2 \, cm^{-2}$
$hr^{-1} \div 60$ min $hr^{-1} = 1/6$ $cm^2 \, cm^{-2} \, min^{-1}$. As an other example,
if $p = 1$ cm hr^{-1} and $R = 1$ mm, then again we get $f = 1/6$ $cm^2 \, cm^{-2}$
min^{-1}, which is rather strange. It would be expected that the
ground is covered more rapidly during heavy rain than during light
rain; perhaps those estimates for drop sizes are unreasonable.

For steady rains, Mason (1971) gives, for the mass-mean drop size, $R_m = 0.35$ mm $[p/1$ mm hr$^{-1}]^{1/4}$. If we take $\bar{R} = R_m$, then (37) becomes $f \simeq 3\ p^{3/4}$ (cm^2 cm^{-2} hr^{-1}), with p in mm hr^{-1}. That seems more reasonable, and for $p = 1$ and 10 mm hr^{-1}, $f \simeq 1/20$ cm^2 cm^{-2} min^{-1} and $f \simeq 1/4$ min^{-1}, respectively. But glossing over these details, we see that in about 10 minutes, the sky should be completely "covered" by drops (assuming that not too many drops fall in each other's path). Consequently, unless bugs aren't collected by drops (bounce off?), or it's just-as-hard to swat them with drops as it is with your hand ("dart off"!), then they should be scavenged in about 10 minutes. Maybe Stuart didn't stay out in the rain long enough!

But it seems likely that bugs and other "particles" are not collected with 100% efficiency, and that this result for the "scavenging rate", $p/\bar{R}$, (area covered/unit area and time) will need to be multiplied by some "collection efficiency". That is what I will turn to now, but to examine it carefully would take much longer than a single lecture. In total, it's a very difficult problem, and as yet, not completely solved. And in case you become lost as I skim over the details, let me write the answer we'll get after another 10 pages or so: the scavenging rate is just the fractional area covered per unit time (i.e., the result just obtained, $p/\bar{R}$), multiplied by a collection efficiency, E, and by a (generally unknown!) numerical factor, $c \simeq 1$. Thus the answer for the scavenging rate will be $\Psi = cpE/\bar{R}$.

3.2 Scavenging Rates and the Collection Efficiency

First, how does the collection efficiency enter into the mathematical description of scavenging? Consider first the case of rain scavenging, either within or beneath a cloud. And, as sketched in Figure 7, consider a raindrop of "radius" R (large drops aren't spherical!) falling at terminal velocity $v_t(R)$. Also, suppose that the aerosol particles are spherical, that their gravitational settling speed is v_g, and that the number density of particles of radii a ("a-particles") is $n(a;\vec{r},t)$. To an observer sitting on the raindrop, the undeflected flow is toward the drop at speed v_t, and (at large distances away) the particles appear to approach at speed (v_t-v_g). If the particles were very massive, so massive that they wouldn't be "blown aside" by the airflow about the drop, and if every collision resulted in coalescence (no "bounce off"), then during a time interval Δt, all particles within the volume $(v_t-v_g)\Delta t\ \pi(R+a)^2$ would be captured by the drop. But particles are not infinitely massive, some will be "blown aside", and some that hit will bounce. Therefore, we write for the number of a-particles actually captured during Δt by a single drop: $n(v_t-v_g)\Delta t\ \pi(R+a)^2\ E'(a,R)$, where E' is the collection efficiency.

There are thus two concepts in the idea of the collection efficiency: first collision, then retention. If these two are identified separately, with E the collision efficiency and η the retention efficiency, then $E' = \eta E$. One interpretation of the collision efficiency is that $\pi(R+a)^2 E$ is the area, ahead of the drop, within which a particle will eventually collide with the drop. If $E = 1$, then all particles within the path of the drop will collide with it (and if η also is unity, then these colliding particles will be collected).

But what is the scavenging rate? If $n(v_t-v_g)\Delta t \pi(R+a)^2 \eta E$ is the number of a-particles collected by a single drop, and if there are $N(R;\vec{r},t)dR$ (noninteracting) drops per unit volume and of radii R to R+dR ("R-drops"), then the number of particles collected per unit volume by all R-drops is $NdR\, n(v_t-v_g)\Delta t$ $\pi(R+a)^2 \eta E$. Finally, then, the decrease in the number of a-particles during Δt is found by adding the collection by drops of all sizes:

$$\Delta n = - n\Delta t \int_0^\infty N\ (v_t-v_g)\ \pi(R+a)^2 \eta E\ dR\ .$$

Meanwhile, the scavenging rate, Ψ, was defined via $\partial n/\partial t = -\Psi n$. Consequently, the rain scavenging rate is

$$\Psi_r(a;\vec{r},t) = \int_0^\infty [v_t(R)-v_g(a)]\ \pi(R+a)^2 \eta\ E(a,R)\ N(R;\vec{r},t)\ dR\ . \qquad (38)$$

If the particles (and drops) are not spherical, then it is easy to modify this result appropriately. Similarly, the snow scavenging rate is

$$\Psi_s(a;\vec{r},t) = \int_0^\infty [v_t-v_g]\ A_x\, \eta\, \ell(a,\lambda)\ N(\ell;\vec{r},t)\ d\ell\ , \qquad (39)$$

where A_x is the cross-sectional area of the hydrometeor (plus the size of the particle, just as for the case of drops where we used $A_x = \pi(R+a)^2$). Also, please notice that I have explicitly identified the dependence of the scavenging rates on position and time ($\vec{r}$ and t), so that these scavenging rates can be used for either below- or in-cloud scavenging.

I would agree with the skeptic who says that, as yet, we haven't done much. The formulation does account for the hydrometeor size distribution, but mainly it just transfers our old ignorance about the scavenging rate into new ignorance about the collection efficiency! That's true, but we can now try to determine the collection efficiency. Unfortunately, however, there is insufficient space, here, to examine the collection efficiency in the needed detail. I could easily spend five lectures on this

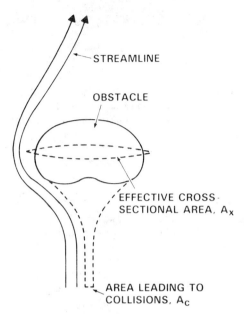

STREAMLINE

OBSTACLE

EFFECTIVE CROSS-
SECTIONAL AREA, A_x

AREA LEADING TO
COLLISIONS, A_c

RELATIVE SPEED, $(V_t - V_g)$
COLLISION EFFICIENCY $E = A_c/A_x$
RETENTION EFFICIENCY $= \eta$
COLLECTION EFFICIENCY $= \eta E$

Figure 7. Definitions of the
various efficiencies.

topic, alone. Thus, my problem (in a few pages, to give
you some idea about collection efficiencies) is indeed
formidable. To help me,
please try to just "go with
the flow" for awhile: try
to get the general picture;
don't get "hung up" on
details -- especially the
many details that will be
missing! What I plan to do
is sketch some ideas about
collection via Brownian
diffusion, interception, and
impaction. (No, I'm not
going to define those words
yet: go with the flow!).

3.3 Collection by Diffusion

First, look at a very
simple case: collection of
particles by a stationary
raindrop! Assume that all
particles that collide with
the drop, stick to it (retention efficiency, $\eta = 1$). The
particles are assumed to move
in the air via Brownian dif-
fusion, with flux $\vec{F} = -\mathfrak{D} \nabla n$,
where $\mathfrak{D}$ is their Brownian diffusivity. Assume that steady conditions prevail, and that n is subject to the boundary conditions
$n(r \rightarrow \infty) = n_0$, a constant, and at the drop, $n(r=R) = 0$ (perfect
sink; i.e., $\eta = 1$). For a steady state to prevail, the total number of particles, flowing past any (mathematical) sphere at radius
r, must be a constant (otherwise there would be a buildup at some
r). Therefore, with F_r the radial component of the flux,

$$F_r \, 4\pi r^2 = -\mathfrak{D}\frac{\partial n}{\partial r} \, 4\pi r^2 = \text{Const.} \tag{40}$$

Consequently (check!), $n = n_0 [1-R/r]$, and the magnitude of the
flux to the drop is

$$\left. |\vec{F}| \right|_{r=R} = \left. |\mathfrak{D} \nabla n| \right|_{r=R} = \mathfrak{D} n_0/R . \tag{41}$$

If this flux is nondimensionalized with the only available dimensional quantities in this problem ($\mathfrak{D}$, n_0, and R), then the

nondimensional mass flux, called the Sherwood number, is

$$Sh = |\vec{F}|_{r=R} \Big/ (\mathfrak{D}n_o/R) = 1 \quad . \tag{42}$$

Incidentally, chemical engineers usually use the diameter of the drop, rather than its radius, for nondimensionalizing the flux; therefore, in this case, they get the numerical value 2 for their Sherwood number (cf. Peters, this volume).

For a moving drop, the gradient of particles near the drop's surface becomes steeper, and therefore the flux is larger than for a stationary drop. For a stationary drop, it can be seen from the previous paragraph that the gradient is about n_o/R. (In fact, from (41), it is seen that the gradient is exactly n_o/R at $r = R$.) When the drop is "ventilated", particles are brought closer to the "upwind" side of the drop, and the flux there is increased over $\mathfrak{D}n_o/R$. Simultaneously, there's a decrease downwind. But the flux can decrease only to zero (assuming no resuspension), whereas on the upwind side, the flux can increase substantially more than twice the stagnant-drop value. Consequently, as you know whenever you feel cooler in a wind, mass (and heat) transfer increase with wind speed. But it can't depend just on wind speed, because otherwise the answer would depend on the units used to measure the speed (m s^{-1}, mph, furlongs per fortnight!); it must depend on dimensionless quantities, constructed from the wind speed and any other variables of the problem, such as the size of the obstacle (say R, say in cm), the fluid density (ρ, g cm^{-3}), its viscosity (ν, cm^2 s^{-1}) and, for the particles, on their diffusivity ($\mathfrak{D}$, cm^2 s^{-1}).

From these dimensional quantities, the following dimensionless groups can be formed.
- The Reynolds number, Re = Rv_t/ν, the ratio of the speed with which the air moves past the drop (v_t) to the "speed" at which the "signal" diffuses out into the stream, ν/R (cm^2 s^{-1}/ cm = cm s^{-1}) "telling" the fluid that there's an obstacle present; and therefore, for Re $\gg$ 1, the "ventilated region" would be very tight around the drop, gradients would be large in the resulting, thin, boundary layer, and the mass transfer would therefore be expected to increase with increasing Reynolds number.

Figure 8 qualitatively indicates how flow fields change with Reynolds number. Figure 8a suggests the case of Re $\ll$ 1 (e.g., Stokes flow about a sphere); in this case the viscous "boundary layer", "containing" the signal that a body is present, is very much larger than the size of the obstacle. Figure 8b shows the case of a thin boundary layer, Re $\gg$ 1. Figure 8c shows a hypothetical flow field (called potential flow) for Re $\rightarrow \infty$ (or better,

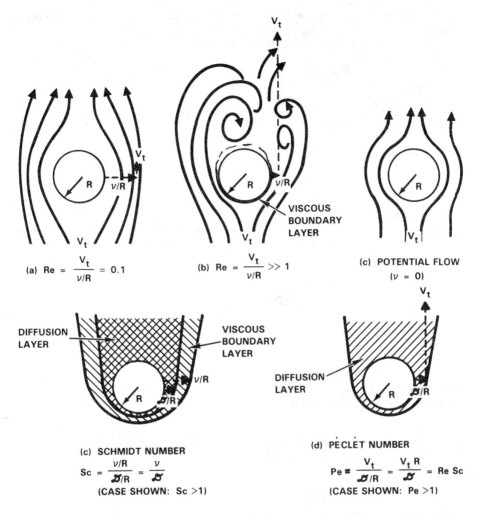

Figure 8. Sketches to convey meanings of the Reynolds, Schmidt, and Peclet numbers.

for an _inviscid_ fluid). Such a flow field does not exist, but on the upstream side of any body at high Re number, and outside the boundary layer, it does provide a fairly adequate (and simple!) description of the flow. This "potential flow field" (so named because a function called the velocity potential can then be defined) will be used in the next subsection.

- The Schmidt number, Sc = ϑ/δ, the ratio of momentum's (or vorticity's) diffusivity, to that for the particles. "Vorticity" is a measure of how much the fluid literally "curls up". (In

fact, the vorticity $\equiv$ curl $\vec{v}$). At the surface of a body, a continuum fluid must stop; away from the body, $\vec{v} \neq 0$; therefore, a
body is a "source" of vorticity for the fluid. This vorticity
diffuses into the fluid, while the particles diffuse toward the
body, and if $Sc \gg 1$, the velocity or vorticity boundary layer
($\sim \vec{v}$) would be thicker than an even-thinner "particle-diffusion
layer" ($\sim \mathcal{D}$), the particle gradients would be large, and the flux
of particles would therefore be larger, increasing with Sc (see
Figure 8d).

- The Peclet number, $Pe = ReSc = Rv_t/\mathcal{D}$, like the Reynolds
number, the ratio of the transport velocity to a diffusion
"velocity" $\mathcal{D}/R$, but in this case for particles (Figure 8e).

Now I do not have time to derive the resulting mass transfer
to a moving drop, but I hope the result seems reasonable. In any
event, the results are mainly based on correlations of data. Of
these correlations (see Peters, this volume) one is the Frössling
correlation:

$$Sh = 1 + 0.4 \ Re^{1/2} \ Sc^{1/3} \ , \tag{43}$$

which is the area-average, nondimensional mass flux. In (43),
note that for $Re = 0$, we regain the result for a stationary drop,
(42), $Sh = 1$. Also, note that Sh increases with Re and Sc
(because of "tighter" boundary layers). With the flux known, the
contribution to the collection efficiency (= collision efficiency
when $\eta = 1$) from Brownian diffusion is then

$$E_B = \frac{\text{Rate of Particle Collection}}{\text{Rate of Particles Approaching}} = \frac{(\mathcal{D}n_o/R) \ Sh \ (4 \pi R^2)}{\pi(R+a)^2 \ n_o \ (v_t - v_g)} \ , \tag{44}$$

and if we take $a \ll R$ and $v_g \ll v_t$ (which are sufficiently accurate for the submicron particles whose Brownian diffusion is significant), then

$$E_B \simeq \frac{4\mathcal{D}}{Rv_t} \ Sh = \frac{4}{Pe} \ [1 + 0.4 \ Re^{1/2} \ Sc^{1/3}] \ . \tag{45}$$

This result will be used later. It (e.g., the experimental constant 0.4 -- when Re is based on the radius!) is probably accurate
to within 20%, although Peters (this volume) shows a correlation
for high Sc number with the constant 0.4 replaced by $0.95/\sqrt{2} = 0.67$.

3.4 Impaction, Interception, Etc.

Particles can be collected by mechanisms other than by
Brownian diffusion (whereby particles "random walk" across the
streamlines to hit the hydrometeor). One of these other mechanisms is interception, which is simple to describe qualitatively,
but not quantitatively. Conceptually, the idea behind interception is that even if the particles do not "random walk", and even
if they are massless (i.e., exactly follow the streamlines around
the hydrometeor), still the particles can be collected because of
their finite size. Thus, as a first approximation, if particles
were coming towards a drop within a distance of one particle
radius, a, from the centerline streamline, then they would be
expected to intercept the drop, and the collection efficiency
(= collision efficiency if η = 1) would be

$$E = \frac{\text{collection rate}}{\text{onflow rate}} = \frac{\eta\, a^2\, n_o\, (v_t - v_g)}{\eta (R+a)^2\, n_o\, (v_t - v_g)} \simeq \left(\frac{a}{R}\right)^2 \equiv I^2 \, . \qquad (46)$$

However, what if the streamlines converge near the drop (as they
do for nonviscous flow) or even diverge (for viscous, boundary-
layer flow)? Here is where I must skip the details, and refer
you to texts such as the one by Fuchs (1964). For the case of
potential (nonviscous) flow about a sphere, Fuchs gives

$$E_{IN} = 3\, a/R = 3I \qquad (47)$$

(and note that for a = 1 μm and R = 1 mm, then 3I = 3 x 10^{-3} is
very much larger than I^2 = 10^{-6}).

But it will be useful to spend a little more time on the
"intertial" or "impaction" contribution to collection, because
the same concepts will be used in sections 5 and 6, dealing with
dry deposition. Impaction depends on the particle's mass, or
inertia, or (better) on the particle's Stokes number (or impaction parameter), which will take awhile to describe. To do this,
start with Newton's equation of motion ($\vec{a} = \vec{F}/m$) for a particle
in a fluid. Take the forces to be from gravity and from the
fluid, for which it is assumed that the difference between the
particle's velocity, $\vec{v}_p$, and the fluid's velocity, $\vec{v}_f$, is
small enough so that the drag on the particle is related linearly
to the velocity difference ($\vec{v}_p - \vec{v}_f$). That is, we assume there
is Stokes (or "creeping") flow about the particle (but not necessarily about any obstacle in the flow, such as a raindrop!). Then
Newton's law is

$$\frac{d\vec{v}_p}{dt} = -\frac{1}{\tau_s} (\vec{v}_p - \vec{v}_f) + \vec{g} \quad , \tag{48}$$

where τ_s^{-1} (units of s^{-1}) is the drag coefficient per unit
mass. τ_s is known as the particle stopping time. For a spherical particle of density ρ_p and radius a (larger than the mean
free path for air molecules), Stokes' result for the drag can be
used to find that

$$\tau_s = \frac{2}{9} \frac{a^2}{\nu} \frac{\rho_p}{\rho_f} \quad , \tag{49}$$

where ρ_f and ν are the density and coefficient of viscosity for
the fluid.

Why τ_s is known as the particle stopping or relaxation time
can be easily seen from (48) when $\vec{v}_f = 0 = \vec{g}$. Then, if the particle's initial velocity is $\vec{v}_{p,o}$, (48) gives

$$\vec{v}_p = \vec{v}_{p,o} \exp(-t/\tau_s) \quad , \tag{50}$$

and hence τ_s is the e-fold time constant for the particle to skid
to a stop [which therefore could have been expected to depend on
particle size and density, and on the density and viscosity of
the fluid, and the only grouping of these variables that has
dimensions of time, see (49), is $a^2 \rho_p/(\nu \rho_f)$]. Also from (48),
a convenient expression for τ_s can be found by looking at the case
of a particle falling at a steady rate ($d\vec{v}_p/dt = 0$), through a
stationary fluid ($\vec{v}_f = 0$), with a balance between gravitational
and drag forces. Then $\vec{v}_p$ is the gravitational settling speed,
and directly from (48) we get

$$\tau_s = v_g/g \quad . \tag{51}$$

For example, for a 10 μm, unit density particle, with $v_g \simeq 1$ cm
s^{-1}, then $\tau_s \simeq 1$ cm s$^{-1} \div 10^3$ cm s$^{-2} \simeq 10^{-3}$ s = 1 ms. In other
words, small particles "skid to a stop" very rapidly; for a 1 μm
particle, with $v_g \simeq 10^{-2}$ cm s^{-1}, $\tau_s \simeq 10$ μs! Table 2 lists τ_s and
$\mathcal{D}$ (the Brownian diffusivity) for a range of sizes of particles
with $\rho_p = 1$ g cm^{-3}.

Particle inertia enters the collection efficiency problem via
the Stokes number, S. To see this, we nondimensionalize the equation of motion for a particle imbedded in a flow about an obstacle.
Let the characteristic time scale of the flow field be τ_f. τ_f
can be a characteristic dimension of the obstacle (e.g., drop
radius, R) divided by a characteristic speed for the flow (e.g.,

Table 2. Radius (a in μm), diffusivity ($\mathcal{D}$ in cm^2 s^{-1}), and stopping time (τ_s in s) for unit density particles (Fuchs, 1964).

a	10^{-3}	10^{-2}	10^{-1}	10^0	10^1
$\mathcal{D}$	1.28×10^{-2}	1.35×10^{-4}	2.21×10^{-6}	1.27×10^{-7}	1.38×10^{-8}
τ_s	1.33×10^{-9}	1.40×10^{-8}	2.28×10^{-7}	1.31×10^{-5}	1.23×10^{-3}

v_t or (v_t-v_g)). Then, with g = 0 and t' = t/τ_f, (48) becomes

$$\frac{d\vec{v}'_p}{dt'} = - \frac{R/v_t}{\tau_s} (\vec{v}'_p - \vec{v}'_f) \equiv -\frac{1}{S} (\vec{v}'_p - \vec{v}'_f) \quad , \qquad (52)$$

in which the velocities have been nondimensionalized with the characteristic upstream speed, and the Stokes number

$$S = \tau_s/\tau_f \qquad (53)$$

is the ratio of the two characteristic times of the problem: particle stopping time $\div$ time scale of the flow. It is therefore seen from (52) that the Stokes number contains all the physical parameters of the problem, in one dimensionless group. It is the natural unit for measuring particle mass. For example, rather than say "small particles" we should say "small Stokes numbers", for then we mean the stopping time is small relative to something (namely, relative to the time scale of the flow yield). That τ_s is 10 μs, for example, is really of no significance, unless we then also say that the characteristic time of the flow field is 1 μs or 100 μs or ....

Two limiting cases are worth examining. One is for very massive particles (better, S $\gg$ 1): they will not respond to changes in flow direction (since $\tau_s \gg \tau_f$), and therefore they will collide with the obstacle. Therefore, for large S, all particles in the path of, say, a raindrop will collide with it. That is, the "impaction" collision efficiency satisfies $E_{IM}(S \rightarrow \infty) = 1$. In fact, it can be shown (Slinn, 1974) that for large S and potential flow about a sphere

$$E_{IM} = 1 - (1/S) + \mathcal{O}(1/S^2) \qquad (54)$$

(read $\mathcal{O}(1/S^2)$ as "of-the-order-of $1/S^2$"). For example, for the case of "bug scavenging" (guessing that v_g for a dead bug is

about 5 cm s^{-1} (!), and using the approximation for raindrops that
v_t = (8000 s^{-1}) R), we get

$$S = \frac{\tau_s}{\tau_f} = \frac{v_g/g}{R/v_t} = \frac{v_g v_t}{g\,R} \simeq \frac{(5\ \text{cm s}^{-1})(8000\ \text{s}^{-1})}{10^3\ \text{cm s}^{-2}} = 40 \quad , \quad (55)$$

and therefore from (54),

$$E_{IM} = 1 - (1/40) + \mathcal{O}(1/40^2) \doteq 0.97 \quad ; \tag{56}$$

i.e., 97% of (dead!) bugs in the paths of raindrops will collide.
Thus, the drop/bug collison efficiency is large, and unless
Stuart's observations were inadequate, we must again conclude
that either the bugs "bounce off" or "dart off"!

   Incidentally, it is useful to notice that if the terminal
velocity of drops can be adequately approximated by v_t =
(8000 s^{-1}) R (and, from Figure 9, this is seen to be accurate
to better than about 20% for drops with 0.1 $\lesssim$ R $\lesssim$ 1 mm), then
τ_f = R/v_t = 0.12 ms, independent of drop size. In other words,
then the Stokes number, S = τ_s/τ_f -- and therefore the contribution of particle inertia to the collision efficiency -- is independent of drop size. We'll use this result, later, in Section
3.6. By the way, notice that, at the top of Figure 9, approximate
Reynolds numbers for the drops are given.

   The other limiting case worth examining is for very small
Stokes numbers. Then the particle "almost exactly" follows the
fluid (since $\tau_s \ll \tau_f$). Therefore, let us look at particle
motion where the fluid changes most dramatically; i.e., along the
upstream axis of symmetry, near the stagnation point. The case
of potential (inviscid) flow was analyzed by Langmuir. If the
z-axis is taken along the axis of symmetry with z = 0 the stagnation point, then nearby, v_f = -cz, where c is a numerical constant depending on the geometry of the obstacle (e.g., from Fuchs,
c = 3 for a sphere, 2 for a circular cylinder, and 4/π for a circular disk). Then, with v_p = dz/dt = $\dot{z}$, (52) becomes

$$S\,\ddot{z} + \dot{z} + cz = 0 \quad . \tag{57}$$

This equation is the same as for damped, oscillatory motion --
and the particle would execute simple harmonic motion about z = 0
(in the flow field v_f = -cz) if there were no drag and if there
wasn't a body there, at z = 0, to stop the swing! But the main
point is this: if the "damping" is too large (i.e., if 1/S is
too large, or S too small), then the particle will never (in
finite time) "swing" even up to z = 0 (i.e., would never be collected by the obstacle, except, e.g., by interception or other

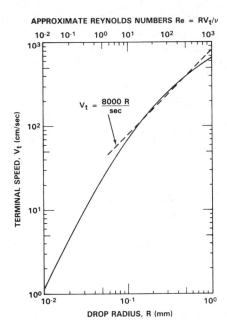

Figure 9. Approximate fit for a raindrop's terminal
speed as a function of its size (radius at bottom;
approximate Reynolds number at top). (From Slinn,
1983a.)

processes). As you may recall from math, conditions for this
"critical damping" can be found by seeking the roots of the "characteristic equation", found from trying, for the solution to (57),
$z = A \exp(\lambda t)$. This gives $S\lambda^2 + \lambda + c = o$ and the roots

$$\lambda_{1,2} = [-1 \pm (1-4cS)^{1/2}]/2S \quad, \tag{58}$$

from which it is seen that critical damping (a nonoscillatory
solution) occurs for $4cS < 1$. In other words, there will be no
inertial contribution to collection if the Stokes number is
smaller than a "critical Stokes number" $S* = 1/(4c)$. Thus, near
the stagnation point, for potential flow about a sphere, circular
cylinder, and disk, then $E_{IM}(S < S*) = 0$ where $S* = 1/12$, $1/8$, and
$\pi/16$, respectively.

It is not easy to obtain the critical Stokes number for other
flows. The results obtained in the previous paragraph were for
"potential flows", and for near the upstream stagnation point.
However, since potential flows are known to be fairly accurate,
upstream of obstacles and for flows with large Reynolds number

(e.g., large speeds), then generally we can expect that if a particle won't hit the obstacle (if $S < S*$) for potential flow, then
it will be even-less-likely to hit in slower, viscous flows. This
expectation is born out in numerical studies, results from which
are illustrated in Figure 10.

3.5 Collision Efficiency Summary

      In this brief introduction to the collection efficiency prob-
lem, I have essentially totally ignored the retention-efficiency
problem, and will continue to do so. For submicron particles,
there are (as far as I know) no available data suggesting that
the retention efficiency, η, is other than unity. But for particles larger than about 10 μm, when the ratio of particle surface
area (and therefore the van der Waals or dipole/dipole force) to
particle volume (and therefore particle mass) becomes smaller,
there are data that show $\eta < 1$, and "bounce-off" is significant
(e.g., Slinn, 1982a). Here, though, I will take $\eta = 1$, and
therefore set the collection efficiency equal to the collision
efficiency.

      Figures 11 and 12 are summaries for particle/raindrop and
particle/ice-crystal collision efficiencies. At the top of Figure 11 is a proposed expression for the collision efficiency,
which contains: (i) a contribution from Brownian diffusion, as
we had earlier,

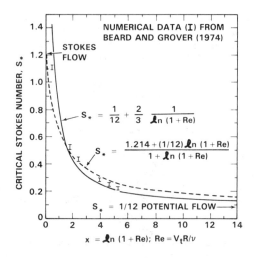

Figure 10. Approximate fits for the critical Stokes
numbers, for particle/drop collisions, as a function of
drop Reynolds number. (From Slinn, 1977a.)

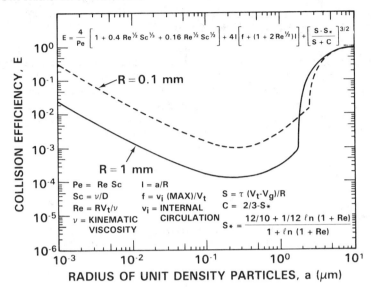

Figure 11. An approximate relation for the drop/particle collision efficiency. (From Slinn, 1983a.)

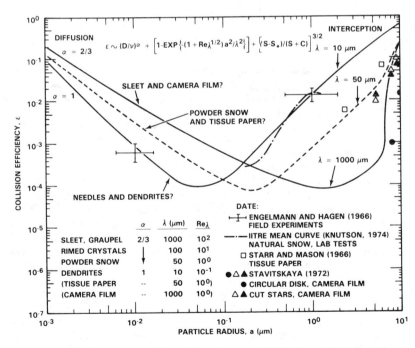

Figure 12. Suggestions for the ice-crystal/particle collision efficiency. (From Slinn, 1977b, where references to the data sources can be found.)

$$E_B = \frac{4}{Pe} [1 + 0.4 \, Re^{1/2} \, Sc^{1/3}] \quad , \tag{59}$$

plus an attempt to account for internal circulation in raindrops;
(ii) a contribution from interception, more complicated than the
3I of (47) (and not at all definite!); and (iii) a contribution
from impaction, which vanishes for S < S* (S* as given in Figure 11
is from fitting the numerical results shown in Figure 10), and
which, for large S, has the limiting form:

$$E_{IM} = 1 - \frac{3}{2S} [S_* + C] + \mathcal{O}(1/S^2) \quad , \tag{60}$$

which is the same as (54) when C = 2/3 - S*. Notice, also, that
E_{IM} is essentially the same for the two drop radii for which E
is illustrated. Figure 12 shows a similar expression for $\mathcal{E}$ for
ice crystals, but it is even more uncertain than the case for
drops. I've shown some data for the ice-crystal case (and could
have shown data for drops, but the figure would have become quite
cluttered), but much more data are needed, both for drops and ice-
crystals, before these expressions for the collection efficiencies
can be trusted to within a factor of two.

Actually, that last sentence must be amplified. What I
meant was: we probably know diffusion and impaction (and maybe
interception) to within a factor of two, but there may be other
processes acting (such as particle growth, electrophoresis, thermophoresis, etc.) and then the resulting, actual collection efficiency can be orders-of-magnitude different from the collision
efficiencies shown in Figures 11 and 12. Figure 13 is illustra-
tive. The bottom curve in this figure, labelled t = 0 and R =
0.5 mm was obtained from the expression written on Figure 11. In
contrast, and in some places three-orders-of-magnitude larger
than this theoretical/laboratory-data curve, the dashed curve
shows the average of seven sets of field data obtained by Radke
et al. (1980). Elsewhere (Slinn, 1977b), I have suggested that
the larger-than-expected, field-measured E's, for particles larger
than about 0.5 μm radius, was probably caused by particle growth
by water-vapor condensation; therefore, the particles were larger
(larger Stokes number) when scavenged, wet, than when sampled,
dry. However, the matter is not yet settled, and there is even
more uncertainty about the cause of the large E for smaller particles: as is suggested on the LHS of Figure 13, Brownian diffusion of these particles to plume droplets (or other large
"particles" with $E_{IM} \simeq 1$) is a factor of about 10^2 too slow to
mesh with the field data -- which, itself, may not be reliable
(Slinn, 1983b). In summary, it's clear there's still work to
do. But there's fun, too, and in that vein, let me finish what I
wanted to say about the scavenging of bugs.

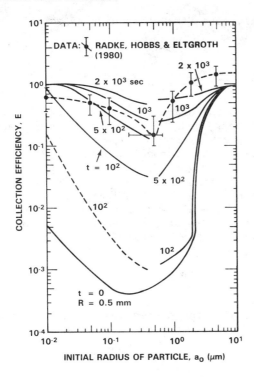

Figure 13. Comparison of theory with field data for the drop/particle collection efficiency. (From Slinn, 1983a.)

3.6 Approximate Scavenging Rates

In case you've forgotten, the reason we started all this collection-efficiency business was because our simple model for bug scavenging didn't work. In that simple model, we had that the fraction of the sky area "covered" by drops, per unit time, was $p/\bar{R}$, and from this saw that the sky should be completely covered in about 10 minutes. It therefore seemed, if Dr. Penkett's observations are correct, that the bug/drop collection efficiency is less than unity, and we therefore introduced collision and retention efficiencies into the formalism. The result for the rain scavenging rate was

$$\psi_r(a;\vec{r},t) = \int_0^\infty N(R;\vec{r},t)\,\pi(R+a)^2\,(v_t-v_g)\,\eta\,E(a,R)\,dR \quad . \tag{61}$$

What I want to do, now, is develop some approximations to this rain scavenging rate, ψ_r (and similarly for the snow scavenging rate, ψ_s) and then compare these approximate relations with data.

Toward this goal, note first that usually $a \ll R$ and $v_g \ll v_t$. Therefore,

$$\psi_r \simeq \int_0^\infty N(R;\vec{r},t)\ \pi R^2\ v_t\ \eta\ E(a,R)\ dR \quad . \tag{62}$$

Also, we have seen that there are substantial uncertainties in the collection efficiency, and therefore, the overall accuracy of our predictions will probably not be impaired if we just evaluate E using a single, mean radius for all drops, say $\bar{R}$. This approximation should be especially good for inertial impaction, because we saw that E_{IM} was essentially independent of R. (And impaction usually dominates when we seek the total **mass** scavenged, since for spherical particles, mass $\sim a^3$, and therefore large particles contribute substantially more mass to the total amount scavenged). Therefore, we make the further approximation

$$\psi_r \simeq \eta\ \bar{E}(a,\bar{R}) \int_0^\infty N(R)\ \pi R^2\ v_t\ dR \quad . \tag{63}$$

Now compare (63) with the expression for the volume flux of rain (i.e., the precipitation rate; e.g., in $cm^3\ cm^{-2}\ s^{-1}$ or $cm\ s^{-1}$):

$$p(\vec{r},t) = \int_0^\infty N(R;\vec{r},t)\ \frac{4}{3}\pi R^3\ v_t\ dR \quad . \tag{64}$$

In both (63) and (64), approximate raindrop-number distributions, N(R), could be introduced, and years ago I followed that procedure. But I became quite discouraged, because there are so many number distributions available (Marshall-Palmer, Best, Khrgian-Mazin, gamma, lognormal, etc. -- all different, all fairly good fits for **some** data, but poor for other storms, and usually poor for the entire duration of even a single storm!), and because there appears to be no way to predict which N(R) would be best for a future storm! Instead, look again at how "close" are the integrands of (63) and (64) -- they differ only by an R -- and think again about the uncertainties in the entire formalism (to refresh your memory, look again at Figure 13 and the three orders-of-magnitude discrepancy between the theoretical E and E derived from field data!).

I therefore have suggested that, until this whole field of endeavor gets to the state where we begin arguing about factors of two or three (a state that may never arise, unless the atmosphere becomes laminar!), we can simply place an extra R in the integrand of (63) and divide by some mean radius, $\bar{R}$, outside the integral -- with $\bar{R}$ chosen to best fit the data for ψ_r. Then the integral is just p, and the proposed approximation is

$$\Psi_r(a,\bar{R};\vec{r},t) \simeq \eta\, \frac{\bar{E}(a,\bar{R})}{\bar{R}}\, \frac{3}{4} \int_0^\infty N(R)\, \frac{4}{3}\pi R^3\, v_t\, dR$$

$$\simeq c'\eta\, \bar{E}(a,\bar{R})\, p(\vec{r},t)/\bar{R} \simeq c\,\bar{E}\,p/\bar{R} \qquad\qquad (65)$$

where c' is a numerical factor near unity (which can be combined
with the retention efficiency via $c = c'\eta$, since η is another
unknown "numerical factor", though it presumably depends on a and
R).

    If a similar series of approximations are made for snow scav-
enging, there results (with a little more difficulty)

$$\Psi_s(a,\bar{\lambda};\vec{r},t) \simeq \gamma\, \bar{\xi}(a,\bar{\lambda})\, p(\vec{r},t)/\bar{D} \qquad\qquad (66)$$

where γ is another "fudge factor" near unity, p is the precipitation rate (rain water equivalent), $\bar{\lambda}$ is a characteristic "capture
length" used in the collision efficiency (see Figure 12), and
where D has different values for different crystal types: 140 μm
for graupel particles, 40 μm for ice needles, 27 μm for rimed
plates and stellar dendrites, 10 μm for powder snow and spatial
dendrites, and 3.8 μm for plane dendrites. With these very small
numerical values for $\bar{D}$ in (66), it might be thought that the snow
scavenging rate would then be very large. Compare these $\bar{D}$ values
with typical $\bar{R}$ values that would be used in (65). That is true
(since snow flakes generally have much more surface area, than do
drops, to collect particles). However, snow flakes usually fall
more slowly than drops, therefore the Stokes number is smaller,
and therefore so is the impaction contribution to the collision
efficiency (see the RHS of Figure 12).

    How good are these approximations? Generally, we don't know.
Dana (1970) performed a series of experiments, releasing tracer
particles from a tower and sampling rain (from frontal storms) for
the mass of tracer scavenged. He concluded that

$$\Psi_r = \bar{E}\,(1.6\text{ mm}^{-1})\, p \quad, \qquad\qquad (67)$$

which is the same as (65) if $c/\bar{R} = 1.6$ mm^{-1}, e.g., c = 1/2 and
$\bar{R} = 0.3$ mm. For convective storms, the average of the results
given by Vali (1977) is

$$\Psi_r/\bar{E} = 2.5 \times 10^{-4}\text{ s}^{-1}\,(p/1\text{ mm hr}^{-1})^{0.8} \quad, \qquad\qquad (68)$$

which is close to what is expected from (65) if we use c = 1/3
and use for R the expression for the mass-mean raindrop radius as
given by Mason (1971) (but for frontal storms)

$$R_m = 0.35 \text{ mm } (p/1 \text{ mm hr}^{-1})^{1/4} \quad .$$ (69)

And in Figure 14 are shown some data for tracer particles scavenged by convective storms, as measured by Burtsev et al. (1970),
along with the expression (65) with the rainfall-rate dependence
of $\bar{R} = R_m$ as in (69) and values for c and $\bar{E}$ as shown. Generally, then, the rainfall-rate-dependence of Ψ_r, as given by the
approximation (65), seems fairly well established, but more data
are needed before firm conclusions can be drawn.

For snow scavenging, the data base is even weaker. For Fig-
ure 15, for the dashed curves, I used $\bar{\ell} = 1$ and $\gamma = \pi/4$, and to
"fit" the data, I used $\gamma\bar{\ell} = 7.3 \times 10^{-4}$ for iodine (possibly dis-
solved in plume droplets) scavenged by ice needles, and $\gamma\bar{\ell} = 1.8 \times 10^{-4}$ for silver iodide particles scavenged by various crystal
types, usually powdered snow or spatial dendrites. This value for
$\gamma\bar{\ell}$ is also $\gamma = 1/3$ and $\bar{\ell}$ as given in the plot of collision efficiencies shown previously in Figure 12.

It is clear that a major problem is in specifying the
collision/collection efficiencies. Not only are there uncertainties arising from the possibilities of particle growth,
electrical influences on collection, thermophoresis (particles
driven by a temperature gradient), and Stefan flow (particles
swept along with, e.g., water
vapor), but there are major
difficulties associated with
experimental techniques usually used. In particular,
usually the precipitation is
sampled for total mass scavenged, and therefore, unless
a single-particle-size
("monodisperse") aerosol is
used, the presence of a few
large particles will contribute overwhelmingly to the
total mass scavenged. This
is illustrated in Figure 16,
where the "polydispersity"
of the aerosol distribution
is characterized by the geometric standard deviation,
σ_g, of an assumed lognormal
distribution. ($\sigma_g = 2$ is a

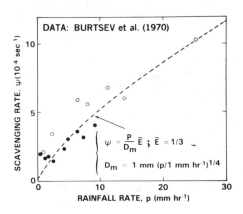

Figure 14. Comparison of data
with the approximate relation for
the rain scavenging rate. (From
Slinn, 1977b.)

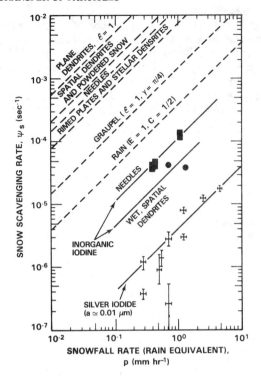

Figure 15. Approximate relations for the snow scavenging rate (for the dashed curves, = 1) and a comparison with (the extremely sparse!) data. (From Slinn, 1983a.)

typical value for an aerosol generated with readily available methods; $\sigma_g = 1$ is for a monodisperse aerosol.) These curves, generated using essentially the $E(a)$ curves shown in Figure 11 and a lognormal drop distribution) show that if $\sigma_g > 2$, then the large E values (for the few large particles present) result in a mass-average collision efficiency of essentially unity, if the geometric mean particle radius, a_g, is greater than about 1 μm. The lower half of Figure 16 shows the n-averaged scavenging rate (n = 3 for the mass average, at least for spherical particles); i.e.,

$$\langle^n\psi_r\rangle = \frac{\int_0^\infty a^n\, f(a) \int_0^\infty N(R)\, \pi(R+a)^2(v_t - v_g)\eta E(a,R)\, dR\, da}{\int_0^\infty a^n\, f(a)\, da} \tag{70}$$

where $f(a)da$ is the number of a-particles per unit volume.

To close, I can't resist the temptation to say (even though a reviewer wrote "please resist!") that this whole business still contains a lot of bugs! But I hope that, otherwise, the view is relatively clear. As working approximations, I suggest the scavenging rates as given by (65) and (66); and these just "say" that the scavenging rates are the rates at which the sky is covered by hydrometeors (e.g., p/$\bar{R}$), multiplied by (poorly known!) collection efficiencies.

4. SCAVENGING BY STORMS

Earlier, it was assumed that the precipitation "just happened" -- it was assumed known. Here, some attention will be given to where it comes from -- i.e., I want to describe some features of the scavenging by entire storms. Some of you might be thinking: "Why doesn't he just say that earlier was the case of below-cloud scavenging and here, in-cloud scavenging?" In response, there are two main reasons for not distinguishing below- from in-cloud scavenging (please don't use the words "washout" and "rainout"!), and for chemists it might be useful to belabor both.

First, below-cloud scavenging almost never occurs. When I was at Oregon State University, nestled between the Coastal Mountains on the west and the Cascades on the east, we tried for two years to catch a case when rain fell through air trapped in the valley. It might have occurred, but we never caught a single case. The reason for our failure was not only because of ventilation of the valley by horizontal winds during storms, but also because of the vertical winds associated with the rainshafts. Please think about that: there must be downdrafts associated with precipitation. Each hydrometeor "feels" a drag from the air (resulting in a steady terminal-speed, with the drag balancing the gravitational force), and by Newton's third law, therefore the hydrometeors cause an equal and opposite (viz., down) drag on the air. Moreover, as Professor Dingle said so many times about those who imagine air to remain stationary in a "mixed layer", through which rain is supposed to fall: "Where do they think the rain comes from, if not from ascent of the moist, lower-level air?"

And the other reason for <u>not</u> distinguishing between below- and in-cloud scavenging is because the same formalism is appropriate for both. Thus, the approximate relation for the rain scavenging rate, $\Psi_r(a;x,y,z,t) = c\bar{E}(a,\bar{R})\ p(x,y,z,t)/\bar{R}$, can be used for any height, if we take care, especially, for the height-dependence of the rainfall rate, p(z). Also, to those who say: "But what about the importance of 'nucleation scavenging' in a cloud (i.e., particle growth by water vapor condensation)?", I

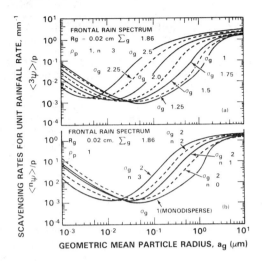

Figure 16. Indications of the dramatic change in scavenging rate
for a polydisperse aerosol (measured by σ_g of an assumed
lognormal particle size distribution) vs. for monodisperse
particles (σ_g = 1), and the dependence of the average
scavenging rate on the order of the average taken (e.g., n = 3
would be for the mass-averaged Ψ for spherical particles). (From
Slinn, 1983a; original from Dana and Hales, 1976.)

respond: "This particle growth must be included in the theory if
we are to adequately describe particle scavenging from anywhere
in the atmosphere, when it's raining (i.e., when the humidity is
high)". Thus, in summary, the concepts are the same for both in-
and below-cloud scavenging (if the latter ever occurs!), and the
total wet flux can be found by integrating the scavenging rate
(which has the same form for all heights) over all heights, both
below and within a cloud. Still, it is now time to move on from
the assumption that the precipitation rate is some parameter that,
somehow or other, someone else found and gave to us. Thus, now
we turn to some aspects of scavenging by storms.

4.1 Bomb Debris Data and Junge's Model

Even before "acid(ic) rain" became a household "word", interest in precipitation scavenging was widespread because of the
"fallout" of radioactivity after atmospheric tests of nuclear
weapons. Figure 17 illustrates the type of data collected during
those days, and shows that, over many orders-of-magnitude variation in the air concentration of (total fission-product) radioactivity, the ratio of concentrations, in rain to in air, was
relatively constant. The resulting best-fit for the (surface-
level) scavenging ratio (± one standard deviation) is

$$s_{r,o} = (C_p/C_a)_o = (1.0 \pm 0.3) \times 10^6 , \qquad (71)$$

where, please notice, the (dimensionless) scavenging ratio is derived using concentrations measured in "volume units" (or densities): here, curies/dm^3, in water to in air. In contrast, some authors use mass-mixing ratios; e.g., curies (or μg or...) per gram of water divided by curies per gram of air. To convert between these two ways that s_r is usually reported, the other author's s_r (10^{-3} smaller than the s_r's here) must be divided by the density of air, which at sea level is about 1.23 x 10^{-3} g cm^{-3}. Notice also in (71) that $s_r \simeq 10^6$; i.e., precipitation concentrates the radioactivity, on a volume bases, by a factor of about a million.

The amazing increase in concentration (by a factor of ~10^6), caused by precipitation, is quite easy to understand. If, in clouds, there are ~10^2 droplets (or ice-crystals) per cm^3, and if about 10^6, 10 μm cloud droplets are needed to create one, 1-mm drop, then the pollution initially distributed in 10^6 cloud droplets (i.e., in 10^6/10^2 cm^{-3} = 10^4 cm^3 of cloud air) is concentrated into a single drop of volume ~10^{-2} cm^3; i.e., concentrated by a (volume) factor of about 10^4 cm^3/10^{-2} cm^3 = 10^6.

Junge (1963) improved on this simple model. To see the outline of his approach, consider Figure 18. If a fraction ε_i of the pollution enters the cloud water, then its concentration there is $\varepsilon_i \langle C_a \rangle$ (say in μg/m^3) or $\varepsilon_i \langle C_a \rangle$/L (in μg/g H$_2$O), where L is the condensed water content of the cloud (typically, 0.1 < L < 10 g m^{-3}), and where $\langle C_a \rangle$ represents an average C_a in the cloud. The advantage of using the mass-mixing ratio is that, if there is no dilution of the cloud water en route to the earth's surface, then as the cloud water coagulates to form precipitation, the mass mixing ratio remains constant; therefore, the concentration in the precipitation will also be $\varepsilon_i \langle C_a \rangle$/L (in μg/g H$_2$O) or $C_p = \rho \times \varepsilon_i \langle C_a \rangle$/L (in μg cm^{-3}), where ρ is the density of water. Consequently, Junge's scavenging ratio is

$$s_{r,o} = \frac{C_{p,o}}{C_{a,o}} = \varepsilon_i \frac{\langle C_a \rangle}{C_{a,o}} \frac{\rho}{L} \delta, \qquad (72)$$

in which I have included on (unknown) factor δ, to account for any possible dilution (or concentration, e.g., by evaporation) as the precip. falls to the earth's surface. Notice from (72) that if $\langle C_a \rangle/C_{a,o} \simeq 1$ (but it would be expected to be >1 for bomb debris initially lodged in the stratosphere -- and possibly >1 for pollutants over the ocean, where dry deposition could reduce $C_{a,o}$ -- and < 1 for pollution over the continents, near low-level sources),

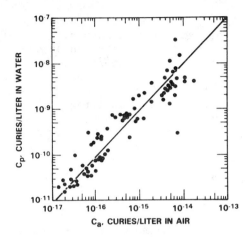

Figure 17. An example of the correlation between concentrations of total fission products in precipitation, C_p, and their concentrations in surface-level air, C_a. (From Slinn, 1983a; original from Gedeonov et al., 1970.)

and if $\delta \approx 1$, then $s_{r,o} \simeq \varepsilon_i \rho/L \simeq \varepsilon_i$ (1 g cm^{-3})/(10^{-6} g cm^{-3}) $\simeq$ $10^6 \varepsilon_i$, which is similar to the correlation for the bomb debris data, given by (71), provided $\varepsilon_i \simeq 1$; i.e., provided that a substantial fraction of this pollution is incorporated into cloud water.

That most particulate mass in a cloud would be incorporated into cloud water is eminently reasonable if we reconsider how clouds form. First, moist (usually, low-level) air becomes saturated: via flow over a topographical barrier (leading to "orographic" precipitation), via buoyant lifting e.g., caused by increased heating of a sunlit hillside (leading to "convective" clouds), via forced lifting by a colder air mass ("cold-front" precipitation), via warm air overriding colder air ("warm-front" precipitation), via radiational cooling ("radiation fog"), etc. But water-vapor saturation, alone, is insufficient to

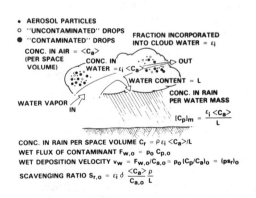

Figure 18. A schematic development of Junge's model for the scavenging ratio. (After Slinn, 1983a.)

cause cloud formation: in ultra-clean laboratory conditions, the
supersaturation can be increased to many hundreds-of-percent
before droplets condense, and droplets can be cooled to -30°C
before they will freeze. In the atmosphere, in contrast, particles are available to act as "nucleation sites", permitting
cloud-droplet condensation at a few hundredths of a percent supersaturation, and permitting ice-crystal formation at a few degrees
below 0°C. Thus, particles become involved in the cloud-droplet
and ice-crystal formation processes, themselves, and therefore, it
is quite reasonable that the fraction of particles incorporated
into the cloud water is large. In Junge's words, he concluded
that the fraction of particle mass that enters cloud water via
this "nucleation scavenging" is: "for continental aerosols of
high concentrations, 0.5; for continental aerosols of low concentration, 0.8; for maritime aerosol and tropospheric aerosols
with a total particle concentration of less than 200 to 300/cm^3,
0.9 to 1.0."

It may be useful to add a few comments about which particles
are most active as "cloud condensation nuclei" (CCN) and as "ice-
freezing nuclei" (IFN). To get a physical picture, lay your hand,
palm down, on the table, and imagine that your knuckles, finger
joints, and nails are water molecules. If your hand is flat and
if you pretend that the table contains more water molecules, then
you see that, say, a single finger joint has a half-plane of
bonds, pulling it down to the table. The corresponding water-
vapor pressure is the saturation vapor pressure over a flat surface of water. Now cup your hand, still with palm down. Now
each water molecule at the water surface (e.g., a knuckle) has
less than a half-plane of bonds, pulling it toward the table.
Therefore, a water molecule could leave a curved surface more
readily than it could from a flat surface, and therefore, to be
in equilibrium, the vapor pressure must be higher over a curved
than over a flat surface. (This is known as the Kelvin effect).
Because of this Kelvin effect, we then see that the larger the
particle, i.e., the flatter the surface, then the more effective
will the particle be as a CCN (provided it is at least "wettable"
by water; i.e., provided that there is some bonding between water
and the particle's molecules).

Moreover, if there is a strong bond between water molecules
and molecules of the "particle" (e.g., a sulfuric-acid droplet),
then the water molecules will be held to the particle even more
strongly than to water, and the saturation vapor pressure can be
even lower. Thus, hygroscopic particles are very active as CCN.
If the particles ionize (e.g., sea-salt particles) then each ion
(e.g., Na$^+$ or Cl$^-$) can "bond" several water molecules in its
neighborhood (within a Debye sphere), again reducing the vapor
pressure needed for saturation. Similarly, since the water-water
bond in ice is greater than in liquid water (it takes heat to

melt ice!), then the saturation vapor pressure is lower over ice
than over water; therefore, droplets evaporate in the neighborhood of ice crystals. To form ice, rather than cool water to
-30°C when it will "spontaneously nucleate", a particle can
nucleate the ice phase at temperatures between about -1° and
-20°C, with the exact temperature depending on particle size and
on the degree to which the particle's crystalline structure promotes the ordering of water's crystalline form. AgI, some clay
particles, and even some organic particles and vapors can nucleate
ice when droplet temperatures are in the range from -2° to -10°C.

Of course, there are other ways for particles to attach to
cloud water besides via nucleation (and, in the case of nucleation scavenging, perhaps we should say that the water attaches to
the particles, rather than vice versa!). Estimates for the rates
of some of these other attachment processes are shown in Figure 19. From this figure, it is suggested that, for most clouds,
even those particles not scavenged by nucleation would become

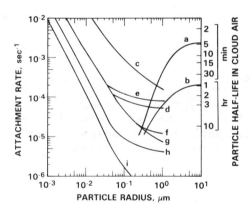

Figure 19. Calculated rates of attachment of aerosol particles
within plumes and clouds. Curves a and b are for inertial
capture by raindrops for rain rates of 10 and 1 mm hr^{-1},
respectively. Curve c represents attachment in an external
electric field with E_0 = 3 x 10^5 volts m^{-1} and particles in
bipolar charge equilibrium. Curve d represents attachment by
turbulent diffusion and convection, and curves e and f show
corrections to d caused by phoretic effects. Curve g is for
Brownian diffusion and convection; curve h represents attachment
to cloud droplets, considering turbulent diffusion without
convection. Curve i represents attachment to raindrops for
Brownian diffusion and convection and a rain rate of 10 mm
hr^{-1}. For curves c to h the droplet concentration was taken to
be 300/cm^3. (From Williams, 1977.)

attached to cloud water, at least within an hour or so after
entering the cloud. Thus, in summary, it is expected that essentially all particle mass (and maybe almost all particles by
number, too) would relatively quickly enter the cloud water;
therefore $\varepsilon_i \simeq 1$.

Yet some limitations on Junge's simple model are evident.
There is, first, the problem of accounting for the rate of attachment of the particles to the cloud water (or v.v.), especially
for convective clouds through which the air may be processed in
~10^3 seconds. A second problem is to account for the time
dependence of the whole process (cloud formation, particle attachment, scavenging, time-dependent precipitation, and cloud dissipation); in contrast, Junge's analysis implicitly assumes the
process is at a steady state. And there remains unanswered the
major question: how much of the pollution is not scavenged by
the cloud, thereby to emerge from a storm, modified, and available for still-longer-range transport? That is, just because the
particles enter the cloud water doesn't mean they reach the
earth's surface: not all clouds precipitate! Therefore, to
examine some of these questions, and to explain more-recent data,
we must dig deeper.

4.2 Particle-Size Dependence of Scavenging Ratios

Figure 20 shows scavenging-ratio data for particles scavenged
by convective storms (Gatz, 1975). The width of the horizontal
bar at each data "point" reflects different mass-mean radii of
the particles for the elements found in six U.S. cities, including St. Louis (solid dot) where the scavenging ratios were measured. Also, the data are precipitation-weighted scavenging
ratios,

$$\bar{s}_r = \sum_i P_i s_{r,i} / \sum_i P_i \quad , \qquad (73)$$

where P_i is the total (measured) precipitation from storm i.
The data represent the summary from two summer seasons, sampling
essentially all convective-storm precipitation. Samples were
collected daily (the open funnels therefore collected some material by dry deposition), and were analyzed for both soluble and
insoluble material. There was found to be essentially no soluble
Fe. It is also noted that arithmetic-mean scavenging ratios would
be larger than the precipitation-weighted values, (73), because
it was found that $s_{r,i}$ was usually larger for lighter rains
(more evaporation?). I will return to this later; for now, I
want to emphasize the particle-size dependence displayed by these
data.

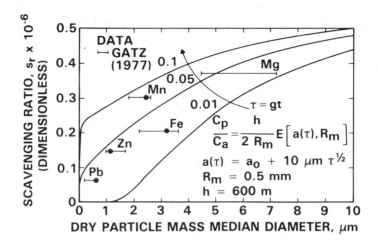

Figure 20. Comparison of predictions of Eq. (76) with rainfall-weighted scavenging ratios measured by Gatz (1977) for various metals scavenged by convective storms. In the evaluation of Eq. (76), different condensational-growth rates, g, were chosen; E was as given in Figure 11, and the characteristic height, h, was chosen solely on the basis of improving the fit to the data. (After Slinn, 1977b.)

What I want to suggest, of course, is that the particle-size dependence shown in Figure 20 reflects the particle-size dependence of the collection efficiency. Toward this suggestion, suppose a cumulonimbus (Cb) cloud is sitting above us, steadily processing moist, polluted air. Then in this (to us, steady; to the particles, unsteady!) process, the flux of particles brought to the ground via rain is the vertical sum of the removal, per unit volume, aloft:

$$F_w = \int_o^\infty \psi_r C_a dz \simeq \int_o^\infty [cp(z)\ \bar{E}(a,\bar{R})/\bar{R}(z)]\ C_a(z)dz \quad , \tag{74}$$

where I have used the approximate expression for the rain scavenging rate, (65). But we also have the wet flux in terms of a scavenging ratio:

$$F_w = p_o C_{p,o} = p_o(C_p/C_a)_o\ C_{a,o} = p_o s_{r,o} C_{a,o} = v_{w,o}\ C_{a,o} \quad . \tag{75}$$

Therefore, for this steady-state storm,

$$s_{r,o} \simeq c \int_o^\infty \bar{E}(a,\bar{R})\ \frac{C_a(z)}{C_{a,o}}\ \frac{p(z)}{p_o}\ \frac{dz}{\bar{R}} \simeq \frac{ch}{\bar{R}}\ \bar{E} \quad , \tag{76}$$

where the last expression represents some (crude) approximation to the integral, which surely is extremely difficult to evaluate. As an example of (76), for the three curves labelled $\tau = gt = 0.01$, 0.05 and 0.1 in Figure 20, I have used c = 1/2, $\bar{R} = R_m = 0.5$ mm, h = 600 m (or it could be said I chose $ch/\bar{R} = 6 \times 10^5$ -- with the choice made to help the fit to the data), and I used the collection efficiency $\bar{E}[a(\tau), R_m]$ as given previously for inertial impaction $E_{IM} \sim 1 - 1/S + \mathcal{O}(1/S^2)$, plus an attempt to account for particle growth.

   To account for particle growth by water-vapor condensation, I used the simple growth relation $a(\tau) = a_0 + (10\,\mu m)\tau^{1/2}$, and plotted only three cases, for three values of the nondimensional time, $\tau = gt$, where g is a growth rate. It is easy to see that condensational growth of particles results in a $\sim t^{1/2}$. From (41), the flux $\sim \mathcal{D}(\rho_v - \rho_{vs})/a$; therefore, $d(4/3\,\pi a^3\rho)/dt \sim [\mathcal{D}(\rho_v - \rho_{vs})/a]\,4\pi a^2$, $\therefore a^2\,da/dt \sim a$, and hence $a^2 \sim t$ or a $\sim t^{1/2}$. Also, the numerical values for $\tau = gt$ were chosen because, for typical supersaturations $(\rho_v - \rho_{vs})$ in clouds, it is common that there is insufficient time in convective clouds for particles (or cloud droplets) to grow much larger than 10 μm. In contrast, for stable frontal storms, there is more time, the particles can become larger (though the supersaturation may be smaller!), and therefore it might be expected that the scavenging ratios (especially for small particles, e.g, Pb), would be larger than for the case of convective storms. Indeed, this is found to be the case, as is illustrated in Figures 21 and 22 (in which, please note, a switch has been made to mass-basis scavenging ratios).

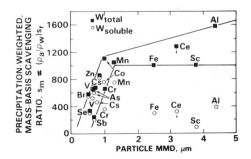

Figure 21. Variations of precipitation-weighted scavenging ratio (on a mass basis) with particle size. Data obtained by Cawse (1974) at Chilton, U.K. during July-December, 1973; much of the data is probably for frontal storms. The data include contributions from both dry and wet deposition. (Adapted from Gatz, 1975.)

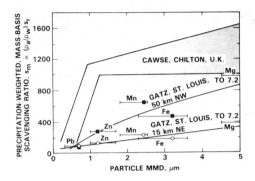

Figure 22. Comparison of rainfall-weighted scavenging ratios (mass basis) for Chilton (monthly samples) and for St. Louis (daily samples). (Adapted from Gatz, 1975.)

But there are many questions and comments about all this, and very little time to respond. For example: (i) If droplets grow only to 10 μm, how can it rain? Reducing much of cloud physics to a few sentences, the response is: there probably are two main processes. One process, for warm clouds, is that "a few" (one in a million!) particles grow to larger sizes (say $\sim$ 50 μm; e.g., on a large, soluble Mg particle -- see Figure 22), then fall relative to the smaller droplets, capturing them, and relatively rapidly (e-fold time constant of about (2.5 min)/L, with the liquid-water content, L, in g m^{-3} -- see Slinn and Gibbs, 1971) becoming rain raindrops. On the other hand, for cold clouds (ice present), once one droplet freezes -- or in some other manner an ice crystal starts -- then neighboring droplets evaporate (ρ_{vs} is less over ice than over water at the same temperature), and then with possible coagulation (or riming), a large snowflake will fall, melting near the freezing level.

(ii) What about nucleation scavenging? Well, if that seems to have been ignored, then I am sorry my point was not clear: nucleation scavenging is what I have been talking about, by accounting for particle growth. It is semantics. It is true, however, that I have ignored the scavenging of that one (one-in-a-million!) particle that grew large enough to start the rain. The scavenging of these few particles will need to be treated as in Junge's model (or ignored!); the theory, here, needs some "scavengers" to do the scavenging (even if they vanish at the top of the cloud, where p(z) $\rightarrow$ 0. However, there is a problem here (A. Pszenny, personal communication) caused by uncertainty in the words "soluble" vs "insoluble". Operationally, "soluble" distinguishes "particles" that pass through a specified filter (Cawse, see Figure 21, used Whatman 41 filter paper, and Gatz used "0.5 μm pore diameter membrane filters"). Consequently, "soluble" can mean "insoluble, but small"! However, these small, insoluble particles may not contribute much mass, and it may therefore be that the error is negligible.

(iii) Note that the scavenging ratio as given by (76) depends on the pollutant's vertical distribution (in $C(z)/C_0$); and for example, the smaller is $C_{a,o}$, then the larger will be s_r. This factor, alone (rather than longer times for particle growth) could explain why Gatz's data from 50 km NW of St. Louis are larger than the 15-km data, and why Cawse's data (for semi-rural sites in the U.K., with presumably still-lower $C_{a,o}$) give still-larger s_r values. These s_r-values (e.g., for Pb) may be still-larger over the mid oceans (e.g., see Buat-Menard, this volume) because of the reduction in $C_{a,o}$ by dry deposition, and because, in relatively-clean air, clouds must "make do" with otherwise unpreferred nuclei.

And the final point I want to make will take a little longer. I want to impress on you how many averages these scavenging ratios represent. First, to obtain the scavenging rate, we take the collection efficiency (itself a consequence of many processes) and average over all drop sizes

$$\Psi_r(a;z,t) = \int_0^\infty dR\, N(R;z,t)\, \pi R^2\, v_t\, E(a,R) \ . \qquad (77)$$

Then, to obtain the scavenging ratio (for a steady storm) we "average" the scavenging rate over all heights

$$s_r(a;t) = \frac{1}{\rho_o C_{a,o}} \int_0^\infty dz\, C_a(z) \int_0^\infty dR\, N\, \pi R^2\, v_t\, E \ . \qquad (78)$$

Next, to obtain the mass-average scavenging ratio, we average s over the mass-distribution, say $m(a)$:

$$\langle{}^3 s_r(t)\rangle = \frac{1}{M} \int_0^\infty dm\, m(a) \int_0^\infty dz\, C_a(z) \int_0^\infty dR\, N\, \pi R^2\, v_t E \Big/ (\rho_o C_{a,o}) \ , \qquad (79)$$

where M is the total mass ($= \int m(a)\, da$). And then Gatz has reported a time-averaged s_r (averaged over each convective storm) weighted by the amount of precipitation

$$\bar{s}_r = \sum_i P_i \frac{1}{T} \int_0^T \langle{}^3 s_r(t)\rangle\, dt \Big/ \sum_i P_i \ , \qquad (80)$$

and Cawse has, in addition, summed over all storm types! Consequently, with so many averages taken, perhaps the most surprising result is that any variations remain at all!

If a few of these averages are removed, then some of the variations re-emerge. For example, in Figure 23, the vertical

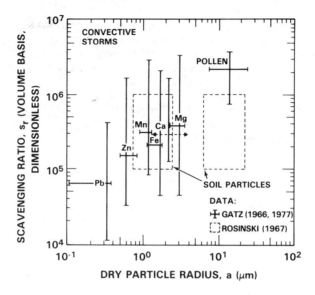

Figure 23. An indication of variabilities in
scavenging-ratio data. (From Slinn, 1978.)

extent of each "error bar" extends over the range in which 5% to
95% of Gatz's s_r-values were measured, and the "boxes" suggest
the range of s_r-values determined by Rosinski (1967) for soil
particles. Also, Figure 24 shows some dependence of the scavenging ratios on rainfall amount. This dependence is quite weak, as
might be expected from $s_r = ch\bar{E}/\bar{R}$ with $\bar{R} \sim p^{1/4}$ (although this
is the rainfall rate, p, not the rainfall amount, P, and therefore
the consequences of evaporation are not included). Figure 25 suggests a stronger dependence of s_r on P, and different for different elements, possibly because of different trajectories during
different storms from the different sources of the different elements. And if these differences are not enough for you, I offer
Figure 26, which speaks for itself! Yet I am not trying to dis-
courage either the use or the measurement of scavenging ratios.
They are useful, especially if all available information is used;
and they would be more useful, if more information (e.g., on the
pollutant's vertical distribution, particle-size distributions,
precipitation rates, storm characteristics, etc.) would be measured and reported along with the values for the scavenging
ratios.

4.3 Scavenging and Precipitation Efficiencies

But once again there is insufficient space to go into these
and so many other details. Instead, I must push ahead; this time
into a major new area, through which a narrow path can be found

leading back to where we started, looking for estimates for τ_w in terms of scavenging efficiencies. And a warning in advance: as tangled as the underbrush of formulae may seem, it would be much more dense if we did not rather brutally hack out many of the complications that in reality exist.

First, a crude estimate for the scavenging <u>efficiency</u>; i.e., the fraction of the total amount of some pollutant, ingested by a storm, that is scavenged by it:

$$\varepsilon_s = \frac{\text{Total amount of the pollutant scavenged}}{\text{Total amount ingested}} . \qquad (81)$$

To estimate this scavenging efficiency, accurately, averages would need to be taken over the entire region of the precipitation and for its duration:

$$\text{Pollutant out} = \int_{Out} dA \int_{Out} dt \, p_o \, C_{p,o} = \int dA \int dt \, (ps_r C_a)_o . \qquad (82)$$

Let us pretend that can be done, and write the answer as

$$\text{Pollutant Out} = [\langle \bar{s}_r \rangle \langle \bar{C}_a \rangle P]_o , \qquad (83)$$

where P is the total volume of precipitation from the storm, and $\langle \bar{s}_r \rangle$ and $\langle \bar{C}_a \rangle$ represent "suitable" time- and area-average values. Similarly, take for the pollutant inflow

$$\text{Pollutant In} = \int_{In} dA \int_{In} dt \, (C_{a,i} \, \bar{u}_i) = \langle \bar{C}_{a,i} \rangle V_{in} , \qquad (84)$$

where V_{in} is the volume of inflow air. Then

$$\varepsilon_s = \frac{\text{Out}}{\text{In}} = \frac{\langle \bar{s}_{r,o} \rangle \langle \bar{C}_{a,o} \rangle P_o}{\langle \bar{C}_{a,i} \rangle V_{in}} \equiv \langle \bar{s}_{r,o} \rangle \frac{\langle \bar{C}_{a,o} \rangle}{\langle \bar{C}_{a,i} \rangle} \frac{P_o}{V_{in}} \frac{\rho_w}{\langle \bar{\rho}_{wv,i} \rangle} \frac{\langle \bar{\rho}_{wv,i} \rangle}{\rho_w} \qquad (85)$$

where ρ_w is the density of water from the storm, and $\bar{\rho}_{wv,i}$ is the average density of the water vapor (absolute humidity) entering the storm. But

$$(\rho_w P_o)/\langle \bar{\rho}_{wv,i} \rangle V_{in} \equiv \varepsilon_p \qquad (86)$$

is the fraction of the mass of inflowing water, $\langle \bar{\rho}_{wv} \rangle V$, that emerges from the storm as precipitation. I will call this the precipitation efficiency, ε_p. (Be careful: as will be seen

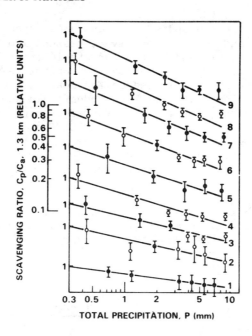

Figure 24. Monthly-average scavenging ratios for a number of
radionuclides as a function of total precipitation. The
radionuclides are: 1-^{125}Sb, 2-^{89}Sr, 3-^{106}Ru, 4-^{137}Cs, 5-^{140}Ba,
6-^{90}Sr, 7-^{54}Mn, 8-^{144}Ce, 9-^{95}Zr. The curves are displaced
vertically to improve clarity. (Original data taken from a
number of A.E.R.E. Harwell reports; graph redrawn from Makhon'ko,
Avramenk, and Makhon'ko, 1970.)

later, there is a host of different "precipitation efficiencies"
used by different authors!) Also, we can define a scavenging
ratio for water vapor, s_{wv}, just as for any constituent

$$\bar{s}_{wv} = \rho_w / \langle \bar{\rho}_{wv,i} \rangle \ . \tag{87}$$

Consequently, and crudely, the scavenging <u>efficiency</u> is related
to the precipitation efficiency via

$$\varepsilon_s = \frac{\langle \bar{s}_{r,o} \rangle}{\langle \bar{s}_{wv} \rangle} \frac{\langle \bar{c}_{a,o} \rangle}{\langle \bar{c}_{a,i} \rangle} \varepsilon_p \ . \tag{88}$$

Look at some features of this result, (88).
(i) $\langle c_{a,o} \rangle / \langle c_{a,i} \rangle$ is the ratio of the average, surface-level air
concentration of the pollution (basically, in the downdraft) to

its concentration in the updraft. This ratio depends on the
storm's dynamics and would generally be measured to be near unity
for a convective storm, and larger for stable frontal storms. It
would usually be measured near unity for convective storms because
a fairly long time is usually needed for most instruments to measure $C_{a,o}$, and during that time, the updraft air would probably
be sampled. [However, for gases scavenged by convective storms,
care is needed: for gases, with $s_{r,o} = [C_p/C_a]_o = s$ (where s is
the solubility coefficient), the $C_{a,o}$, here, is the gas concentration in the downdraft air, which can be from the mid-troposphere;
therefore, it would be quite inconsistent to use $s_{r,o} = s$ and
assume $C_{a,o} \simeq C_{a,i}$.]

(ii) $\langle s_{r,o} \rangle / \langle s_{wv} \rangle$ is the ratio of the scavenging ratio for the
pollutant of interest, to s_r for water vapor. A typical value
for s_{wv} is easily estimated; e.g., for rain

$$s_{wv} = \frac{\text{concentration of water in precip.}}{\text{concentration of water vapor in air}}$$

$$\simeq \frac{1 \text{ g cm}^{-3}}{(\sim10 \text{ g/kg air})(1.23 \times 10^{-6} \text{ kg air/cm}^3)} \simeq 10^5, \quad (89)$$

which is similar in magnitude to s_r for particles (cf. Figures 20
to 23).

(iii) That s_r for particles is similar to s_{wv} could have been
expected, since precipitation, itself, usually depends on the collection (coagulation/riming) of cloud droplets (or ice crystals)
by a single, larger hydrometeor. Thus, crudely,

$$\frac{s_r}{s_{wv}} = \frac{\int dz \int da_c \; c \; \bar{E}(a_c,\bar{R}) \; p(z) \; [C(a_c,z)/C_{a,o}]/\bar{R}}{\int dz \int da_w \; c \; \bar{E}(a_w,\bar{R}) \; p(z) \; [\rho_w(a,z)/\rho_{wv,i}]/\bar{R}} \quad (90)$$

where $E(a_c)$ is the collection efficiency of (the droplets containing) the chemical of interest, $E(a_w)$ is the collection efficiency of the cloud water (e.g., all droplets), and there are
integrals over appropriate mass distributions. If all integrals
are crudely approximated, and the condensed water content of the
cloud is identified as the integral over the water mass distribution, then (90) yields

$$\frac{s_r}{s_{wv}} \simeq \frac{\bar{E}(\bar{a}_c,\bar{R})}{\bar{E}(\bar{a}_w,\bar{R})} \; \frac{\rho_{wv,i}}{L} \; \frac{\langle C_a \rangle}{C_{a,o}}, \quad (91)$$

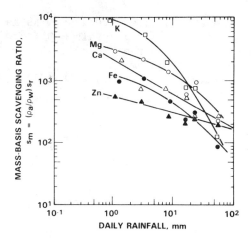

Figure 25. Variations of scavenging ratios with amounts of daily precipitations. (From Gatz, 1975.)

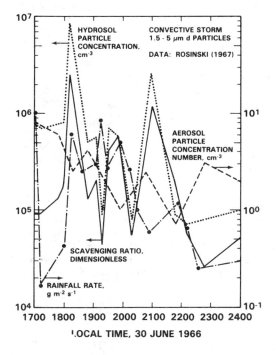

Figure 26. Variations in C_p (·····), C_a(---), p(-·-·-·), and s_r (——) during a single convective storm. Data from Rosinski, 1967; figure from Slinn, 1978.)

where the overbars represent "suitable" averages. As an example
of (91), if $\langle C_a \rangle \simeq C_{a,o}$, if $\rho_{wv,i} = 5$ g/kg air $\simeq 5$ g m^{-3}, and if
the average liquid water content of the cloud is ~ 4 g m^{-3}, then
$s_r = 1.2\{\overline{E}(a_c)/\overline{E}(a_w)\}\,s_{wv}$, suggesting that if $s_{wv} \sim 10^5$, cf.
(89), then $s_r \simeq 10^5\,\overline{E}(a_c)/\overline{E}(a_w)$. [In fact, with (87), (91)
becomes

$$s_r \simeq \frac{\rho_w}{L}\,\frac{\langle C_a \rangle}{C_{a,o}}\,\frac{\overline{E}(\overline{a}_c, \overline{R})}{\overline{E}(a_w, \overline{R})} \; , \tag{92}$$

which is essentially the same as Junge's model, (72), but with
his ε_i replaced by the ratio of collection efficiencies.]

(iv) But the major point of this exercise was to relate the scav-
enging efficiency, ε_s, to the precipitation efficiency ε_p. From
(88), if $\langle \overline{s}_r \rangle \simeq \langle \overline{s}_{wv} \rangle$ and $\langle \overline{C}_{a,o} \rangle \simeq \langle C_{a,i} \rangle$, then $\varepsilon_s \simeq \varepsilon_p$; i.e., the
efficiency with which a storm removes particles is expected to be
similar to its efficiency for removing its condensed water. For
gases, with $s_{r,o} = s$, the solubility coefficient, then $\varepsilon_s \simeq s \cdot \varepsilon_p/\langle s_{wv} \rangle$; i.e., the scavenging efficiency is (typically) lower
than the precipitation efficiency, by the ratio of the solubility
of the gas in water to the "solubility" of water vapor in water
($\sim 10^5$). However, some gases at low concentrations (e.g., H_2SO_4
and HNO_3) are more soluble in water than water vapor; therefore, a storm (which may scavenge only 10%, say, of its water
vapor) can at the same time scavenge essentially 100% of the
ingested H_2SO_4 and HNO_3.

For particles, the result $\varepsilon_s \simeq \varepsilon_p$ was to be expected. But
is this result of any value? What is the precipitation effi-
ciency? That is another difficult question.

4.4 Storm Efficiencies

For the title of this section, I did not use "precipitation
efficiency" because experimentalists have defined so many different ones -- even though there are so few data sets available!
There is, for example, the cloud-water removal efficiency

$$\varepsilon_{cw} = \frac{\text{water out}}{\text{water available}} = \frac{\text{precip. out}}{\text{max. condensed}}$$

$$= \frac{(\text{vapor in}) - (\text{vapor out})}{(\text{vapor in}) - (\text{vapor at top})} \; , \tag{93}$$

where by "top" is meant the level in the storm where most water
is condensed. Data are also available for a reduced, cloud-water
removal efficiency, ε'_{cw}, used as an estimate of the increase
in a frontal storm's ε_{cw} caused by an orographic barrier. Data

are also available for ε_{cw} for individual rainbands in a frontal storm, and I will identify these by ε_{cw}^{RB} (RB for rain band) to distinguish them from ε_{cw}^{OR} for isolated orographic clouds, and from $(\varepsilon_{cw}^{OR})'$ for the enhancement of ε_{cw} in a frontal storm caused by an orgraphic barrier. Then there is the precipitation efficiency used in the previous section, ε_p = water out ÷ vapor in, and a "net" precipitation efficiency that uses, in the denominator, only the <u>net</u> vapor flux into the updraft.

To illustrate three of these efficiencies, consider Figure 27, which shows the principal branches of the water (W), vapor (V), and air (A) circulations in a single thunderstorm (20 km wide) buried in a squall line (a line of thunderstorms, common in the NE U.S. during summer, running ahead of a cold front, plowing its way through moist, Gulf Coast air). From the data for this cumulonimbus cloud (Cb) we can get

$$\varepsilon_p^{Cb} = \frac{\text{Precip. Out}}{\text{Vapor In}} = \frac{4.7}{8.8} = 0.53$$

$$\varepsilon_{cw}^{Cb} = \frac{\text{Precip. Out}}{\text{Water Available}} = \frac{4.7}{8.8 + 0.7 - 0.6} = 0.53$$

$$(\varepsilon_p^{Cb})_{net} = \frac{\text{Precip. Out}}{\text{Net Vapor In}} = \frac{4.7}{(8.8 - 4.3)} = 1.04 \quad .$$

That this third efficiency is greater than 100% reflects inadequacies in the definitions (for example, the 0.7 kton/s entrained in the downdraft has been ignored, and no account has been taken of time lags between inflow and outflow), but more significantly, it is just a trivial indication of the enormous difficulties in conducting such studies. Please, for a moment, stop to think how you would conduct one of these field studies: How many rain gauges would you use? How many aircraft? Who would you "con" into flying them?!

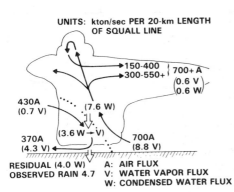

UNITS: kton/sec PER 20-km LENGTH OF SQUALL LINE

RESIDUAL (4.0 W)
OBSERVED RAIN 4.7

A: AIR FLUX
V: WATER VAPOR FLUX
W: CONDENSED WATER FLUX

Figure 27. Principal branches of the fluxes of air (A), water vapor (V), and water (W), corresponding to a single thunderstorm in a squall line. (From Newton, 1966.)

The early studies of $(\varepsilon_{cw}^{OR})'$, the increase in a frontal storm's ε_{cw} caused by an orographic barrier, were relatively crude (e.g., the water, remaining as vapor above the mountain crest, was deduced from a cloud model rather than measured).

However, because of the reduced expense of the studies, the
investigators (Elliott and Hovind, 1964) could investigate a
substantial number of cases. In contrast, Dirks (1972) and
Marwitz (1974) used aircraft to obtain data for isolated orographic clouds, and each investigator studied one case. Recently,
Hobbs et al. (1980) have used radar and aircraft to deduce ε_{cw}
for individual rainbands within frontal storms, e.g., see Figure 28. A summary of all these results (which are, to my knowledge, all that are available) is shown in Table 3.

For isolated convective storms, more data are available.
Marwitz (1972) has brought some order to the data by plotting
ε_p^{Cb}, not (ε_p^{Cb})$_{net}$, as a function of wind-speed shear. This
shear is the wind speed at cloud top (say 25 m s^{-1}), less the
wind speed at cloud base (say 5 m s^{-1}), divided by cloud height
(say 10 km, and then for this case the wind-speed shear would be
2×10^{-3} s^{-1}). The results, as summarized by Foote and Fankhauser
(1972), are shown in Figure 29. It can be expected that ε_p^{Cb}
would return to a value near zero for no wind shear, since without shear, a Cb's downdraft will suppress its updraft; therefore
the Cb would soon dissipate. On the other hand, for large wind
shear, Marwitz suggests the strong winds aloft will cause the precipitation to fall outside the saturated cloud environment (causing evaporation), and/or will blow the ice off the Cb's anvil
(leading to "orphaned anvils"!).

As far as I know, that summarizes all available data for
these "storm efficiencies." Then where are we? If you will look

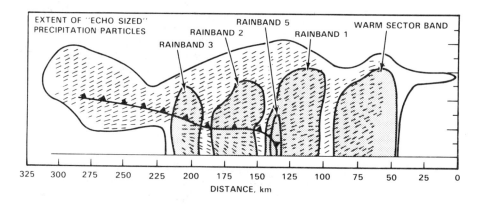

Figure 28. Rainbands within a frontal storm. The cold front is
identified by ◣◣ ; the detailed horizontal wind field (shown by
the short lines) indicate wind direction only; precipitation
efficiencies are given in Table 3. (From Hobbs et al., 1980.)

Table 3. Cloud-Water-Removal Efficienices Obtained by Elliott and Hovind (1964), Dirks (1972), Marwitz (1974) and Hobbs et al. (1980)

Notes	Data
San Gabriel Mts. ($\sim$8000'), 31 cases, stable storms $(\varepsilon_{cw}^{OR})' =$	26%
, 8 cases, unstable " =	27%
Santa Ynez Mts. ($\sim$4000'), 21 cases, stable " =	17%
, 22 cases, unstable " =	26%
Medicine Bow Mts. ($\sim$3500'), Cld. Top	
$T = -23$ C, $\bar{u} = 12$ ms^{-1} $\varepsilon_{cw}^{OR} =$	65%
$T = -19$ C, $\bar{u} = 22$ ms^{-1} " =	55%
$T = -35$ C, $\bar{u} = 20$ ms^{-1} " =	25%
San Juan Mts. ($\sim$4500'), $T = -28$ C, $\bar{u} = 30$ ms^{-1} " =	62%
Warm Sector Rainband, Precip = 100 - 123 kg s^{-1} m^{-1} $\varepsilon_{cw}^{RB} =$	40-50%
Wide cold-frontal band #1, Precip = 40-50 kg s^{-1} m^{-1} " =	80-100%
Wide cold-frontal band #2, Precip = 6 kg s^{-1} m^{-1} " =	20%
Narrow cold band #5, Precip = 11-18 kg s^{-1} m^{-1} " =	30-50%

again at the data, perhaps you will appreciate the standard line: "The beauty of working with efficiencies is that you can always specify them within a factor of two: 1/2 ± 1/2!" Perhaps we can do better, 1/2 ± 1/4, but I'm not too sure: it is understandable why meteorologists conduct field studies on mature storms giving lots of precipitation, but what about all those clouds and storms that barely drop enough rain to wet the dust? I think I'll stick with 1/2 ± 1/2, until more data become available. And therefore the fraction of particles not scavenged by storms, approximately $(1-\varepsilon_p)$, is estimated to be 1/2 ± 1/2, if that's any help!

4.5 Summary for τ_w

In summary, I started this chapter with a simple model that gave for the average tropospheric residence time of particles

$$\frac{1}{\bar{\tau}} = \frac{1}{\bar{\tau}_w} + \frac{1}{\bar{\tau}_d} + \frac{1}{\bar{\tau}_{ch}} + \frac{1}{\bar{\tau}_{ph}} + \cdots \qquad (94)$$

where τ_w, τ_d, τ_{ch}, τ_{ph} ... are the residence times if only wet, dry, chemical, physical, ... first-order-removal processes were acting alone. For example, for ^{85}Kr, the only significant removal process is radioactive decay, and $\bar{\tau} \simeq \tau_{ph} \simeq 10$ years. There are more elegant ways to derive (94) than with the simple box model used here, but the result is the same (e.g., Slinn et al., 1978; Slinn, 1980).

So far, the ingredients to estimate $\bar{\tau}_w$ by three different (but related) methods have been examined. Thus, (i), in terms of scavenging <u>ratios</u>:

$$\bar{\tau}_w = \frac{\text{Inventory}}{\text{Outflow}} = \frac{\int C(z)/C_o \, dz}{(\overline{ps}_r)_o} = \frac{\bar{h}_w}{\bar{v}_w} \quad ; \tag{95}$$

and if $\bar{h}_w = 2$ km, $\bar{s}_{r,0} = 10^5$ and $\bar{p}_0 = 100$ cm/yr, then

$$\bar{\tau}_w = \frac{2 \times 10^5 \text{ cm}}{10^5 \times 100 \text{ cm}} \times 50 \text{ weeks} = 1 \text{ week} \quad . \tag{96}$$

(ii) In terms of scavenging <u>rates</u>:

$$\bar{\tau}_w = \langle \bar{\Psi} \rangle^{-1} = \langle c\bar{E}\bar{p}/\bar{R} \rangle^{-1} \quad ; \tag{97}$$

and if $c = 0.5$, $\bar{E} = 0.1$, $\bar{R} = 1$ mm, and $\bar{p} = 100$ cm/yr, then

$$\bar{\tau}_w = \frac{(0.1 \text{ cm})(50 \text{ weeks})}{(0.5)(0.1)(100 \text{ cm})} = 1 \text{ week} \quad . \tag{98}$$

And (iii), in terms of scavenging efficiencies: $\tau_w = (\bar{\varepsilon}\bar{\nu})^{-1}$, where $\bar{\varepsilon}$ is the average fraction scavenged per storm, and $\bar{\nu}$ is the average frequency of storms experienced by the pollution; and if $\bar{\varepsilon} = 1/2$ and $\bar{\nu} = (3.5 \text{ days})^{-1}$, then $\bar{\tau}_w = 1$ week.

Maybe that appears to be a nice neat little summary, with all three methods giving $\bar{\tau}_w \sim 1$ week, but there is one little complication: the parameters were "fudged" to give $\bar{\tau}_w = 1$ week! Moreover, I don't know what $\bar{\tau}_w$ should be, and as far as I know, no one else knows the "correct" answer, either!

Earlier, I showed some of the uncertainties in the scavenging ratios [and therefore in $\bar{\tau}_w = \bar{h}_w/(\overline{ps}_r)_o$], and in the collection efficiencies (and therefore in $\bar{\tau}_w = \langle c\bar{p}\bar{E}/\bar{R} \rangle^{-1}$). Here, let me mention some of the uncertainties in the estimate $\bar{\tau}_w = (\bar{\varepsilon}\bar{\nu})^{-1}$. First, recall that this $\bar{\tau}_w$ is not the e-fold residence time unless $\bar{\varepsilon}$ is small; "in reality" (see Eqs. 26 and 30),

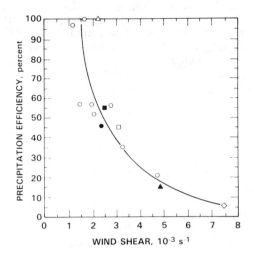

Figure 29. Precipitation efficiencies, ε_p^{Cb}, for cumulonimbus
storms as a function of wind shear. (Figure after Foote and
Fankhauser, 1973, in turn adapted from Marwitz, 1972, where
original sources of the data can be found.)

$$\bar{q}(t)/q_0 = (1-\bar{\varepsilon})^{\bar{\jmath}t} = \exp\left\{\bar{\jmath}t \ln(1-\bar{\varepsilon})\right\} = \exp\left\{-t/\bar{\tau}_w^*\right\} \qquad (99)$$

where $\bar{\tau}_w^* = [\bar{\jmath}|\ln(1-\bar{\varepsilon})|]^{-1}$. But this is a small source of error
compared to our problems with estimating $\bar{\varepsilon}$ ($1/2 \pm 1/2!$), and $\bar{\jmath}$.
And what I'd like to do, to finish this "Chautauqua" on scaveng-
ing, is show you some aspects of where this field of study is, in
estimating something as seemingly simple as $\bar{\jmath}$ (the mean frequency
at which an "air parcel" experiences storms).

Start simply. If storms scavenged all ingested particles,
then obviously the average τ_w would be the average time until
the particles encounter a storm. If the particles are released
at an arbitrary time, if we neglect the possibility that it's
raining during the time of the release [at most locations in temperate latitudes, it rains for less than 10% of the time (believe
it or not!)], then obviously $\bar{\tau}_w$ would be one half the average
duration of dry periods -- whatever that means! That is, be careful: there is the average time between storms as recorded by an
observer fixed on the earth (known as the "Eulerian" time), and
there is the average time between storms as experienced by an
observer moving with the air (the "Lagrangian" time); and this
Lagrangian time will presumably depend on whether the Lagrangian
observer "starts" each observation near the earth's surface or
near the tropopause! Darn -- and we were going to start simply!

Well, start with the Eulerian time. As you might expect, many people have looked at weather-station data to determine the expected duration of dry days. For example, Weiss (1964) looked at 50 years of data from Kansas City, Missouri, and found that the probability of a sequence of n dry days (with less than 0.01 inches of rain) was approximately

$$P_0 (1-P_0)^{n-1}$$

where, for example, $P_0 = 0.192$ in January, and 0.221 in July. As a specific example, in January and July 1975 in Kansas City, the expected (Eulerian) duration of dry periods were 96 and 106 hours, respectively. Therefore, if these Eulerian times could be used to estimate $\bar{\nu}$, and if the first storm removed all particles ingested, then $(\tau_w)_{Eul} \simeq 50$ hours (where I've divided by a factor of two because it is assumed that the particles are released at any time during any dry period).

But the particles don't just sit there in Kansas City, waiting for precipitation to scavenge them! They move with the air, and therefore the question is: what is the average duration of Lagrangian times between storms? We looked at that (Vickers and Slinn, 1979) by following (low-level!) trajectories from Kansas City, starting every 6 hours, and stopping when the trajectory "hit" recorded precipitation. (In two cases in January, and four in July, 1975, precipitation was not encountered before the trajectories left the U.S., and for these cases, the time until encountering precipitation was estimated more subjectively, based on the location of weather systems in Canada and over the Atlantic.) The results for these low-level Lagrangian times, for the average durations until the particles would hit a storm, starting from an arbitrary time during a dry period, were 38 hours in January and 54 hours in July. Hamrud et al. (1981) performed a similar but more extensive study for Northern Europe, but again using only low-level (850 mb) trajectories. Their average results for the average time from an arbitrary moment in a dry period until it ends with $p \geqslant 0.1$ mm hr^{-1}, for 36 stations, were: 61 hours, summer semester (April-September); and 60 hours, winter. Thus, as a first estimate for Northern Europe, $\bar{\tau}_w \simeq 2.5$ days, assuming all particles are scavenged by the first strom (i.e., $\varepsilon_s = 1$).

But ε_s is almost never as large as unity (cf., Figure 29 for Cb's), and that is where the trouble begins -- and where we are today. A Cb, for example, can leave more than half of the ingested particles near the tropopause (or in the stratosphere!); then, how long until those particles encounter a storm? I don't know! (But I expect that the answer will be such that, then, $\bar{\tau}_w \simeq (\bar{\varepsilon}\bar{\nu})^{-1} \simeq 1$ week!).

And, in my opinion, it is quite improper to use a scavenging rate of, say, 1 hr^{-1} (or $cp\bar{E}/\bar{R}$), and multiply it by an (Eulerian!) duration of precipitation of 3 or 6 or 24 hours. This gives $\epsilon_s \simeq 1$. The error is in not recognizing that even for frontal storms with their rainbands, particles do not remain long in precipitation-producing regions: even in clouds, air moves!

Originally, I had planned to devote a full lecture to these and other aspects of cloud physics, but there isn't time. Now it's time to turn to dry deposition. But for those interested, what I would have said about cloud physics, I have written elsewhere (Slinn and Hales, 1983). And for those interested in pursuing estimates for τ_w, they may want to start by reading the (difficult!) recent publication by Rodhe and Grandell (1981), who review the many models developed to estimate τ_w, and develop more. One sentence, in particular, caught my attention: "To be honest, it should be pointed out that the uncertainties in the estimate of (the various parameters) are so large that differences between the various models in many cases are of limited significance."

5. DRY DEPOSITION OF PARTICLES

What is the flux of particles to the ocean if there is no precipitation; i.e., what is the dry deposition flux, F_d, or the dry deposition velocity, v_d? Truthfully, we must admit that this, too, is unknown. But some information and ideas are available, and the goal here is to show some of them. I plan to start with relatively easy topics (I said relatively easy!), for which some solutions are beginning to become available. In Section 5, solutions will be emphasized; in Section 6, problems.

5.1 The Dry Flux

First, the flux. If a group of molecules, particles, or people of concentration C (I'm going to drop the subscript "a" on C, since all concentrations will be in air) were all traveling at velocity $\vec{v}^*$, then the (dry) flux $\vec{F} = \vec{v}^* C$. But even for a crowd of people, except soldiers marching in step, what do we mean by $\vec{v}^*$? We can define an average velocity, locally, but that doesn't mean everyone is moving with this average velocity (except in the case of soldiers). For molecules (and particles), it is customary to define a local $\vec{v}$, averaged over a small region and time duration, but large enough to contain many molecules and long enough for many intermolecular collisions (a typical time between collisions is the mean distance between molecules, λ , divided by their mean thermal speed, $\bar{c}$; i.e., $\lambda/\bar{c} \simeq (0.1 \ \mu m)/(10^4 \ cm \ s^{-1})$ 10^{-9} s at STP). Because of this averaging, and because particles and molecules don't march like soldiers (except in a beam), there

can be a resulting diffusional drift velocity, relative to the
mean motion. This diffusional drift (Brownian or molecular diffusion) "exists" because of our decision to define an average
velocity. The only thing random about a random walk is our definition of what is straight!

But what we want is the flux. If we knew the velocity, $\vec{v}_i$,
of every particle, then the number-flux would be the sum over all
particles of $\vec{v}_i$. If we use a local-average $\vec{v}$, then the local
flux $\vec{F}$ is $\vec{v}$ C plus the flux associated with the "random walk."
However, if there is no concentration gradient, then the random
walk would result in no net flux (just as many "particles" would
randomly walk from A to B as from B to A). Thus, there is a diffusional flux associated with the random walk only if, in addition, there is a concentration gradient, ∇C. This gradient
doesn't "drive" the drift (as so many authors suggest): the
"drift" exists regardless of the gradient, but there is a resulting "diffusional" flux if there is a concentration gradient.
Phenomenologically (Fick's law), the resulting diffusional flux
is $-\mathscr{D} \nabla C$; i.e., down the gradient and proportional to it, where $\mathscr{D}$
is the molecular diffusivity (for gases) or Brownian diffusivity
(for particles). It is obvious from checking the dimensions,
that $\mathscr{D}$ has dimensions of L^2/T^{-1}; e.g., $cm^2 s^{-1}$.

It will be useful to have reviewed a simple model for this
diffusional flux. Recall from the kinetic theory of gases that
the positive-z flux of molecules, past a plane at height z, can
be estimated to be $\bar{c}$ C(z - λ) = $\bar{c}[C(z) + (\partial C/\partial z)(-\lambda) + ...]$,
where (again) λ is the mean-free path and $\bar{c}$ is the thermal speed,
and the ... represent higher order terms in this Taylor series.
The negative-z flux, similarly, is $\bar{c}$ C(z + λ) = $c[C(z) +
(\partial C/\partial z)(\lambda) + ...]$, where it is assumed that the temperature and
therefore $\bar{c}$ is independent of z. The net, positive-z flux is
then the difference, $2\bar{c}\lambda \partial C/\partial z$, and this is the z-component of
$-\mathscr{D} \nabla C$, when the diffusivity $\mathscr{D} = 2\bar{c}\lambda$. For example, for low-
molecular-weight gases, with $\lambda \simeq 0.1 \, \mu m$ and $\bar{c} \simeq 10^4 \, cm \, s^{-1}$,
then $\mathscr{D} \simeq 0.1 \, cm^2 \, s^{-1}$.

We have, then, that the flux is

$$\vec{F} = \vec{v} \, C - \mathscr{D} \nabla C \quad ; \tag{100}$$

but especially for particles, what is this local, average $\vec{v}$? It
can be expected that the major component of $\vec{v}$ is the local fluid
velocity, $\vec{v}_f$, but be careful: the particles need not exactly
follow $\vec{v}_f$. Certainly the particles can move relative to the
fluid because of gravitational settling, $v_g(-\hat{k})$, where $\hat{k}$ is a
unit vector in the positive-z direction. (This v_g is assumed
to contain a correction for buoyancy, which is usually negligible
in air.) Also, though, particles may have some other slip veloc-

ity, $\vec{v}_s$, relative to the fluid, and derived from any of a number
of causes: failure of the particles to exactly follow rapid
changes in $\vec{v}_f$ (which of course will depend on the particle's
Stokes number!), drift because of electrical effects, and other
"phoretic effects," such as thermophoresis and diffusiophoresis/
Stefan flow. [Fundamentally, Stefan flow arises because of our
use of the mean velocity $\vec{v}_f$ for the fluid, when it is a mixture
of more than one gas (e.g., "air" molecules and water vapor).]
In summary, then, we can write the (dry) flux of particles, schematically, as

$$\vec{F} = [\vec{v}_f + \vec{v}_s + v_g(-\hat{k})]C - \mathcal{D} \nabla C \quad , \tag{101}$$

or for the z-component (with $\vec{v}_f = (u,v,w)$, etc.),

$$F_z = [w_f + w_s - v_g]C - \mathcal{D} \, \partial C/\partial z \quad . \tag{102}$$

I trust that (102) looks simple enough; unfortunately, though,
there is more.

5.2 Turbulence

The trouble is the atmosphere is usually turbulent, and what
we usually want is some long-term average of the flux (e.g., averaged over a time period long enough to collect a sample that can
be chemically analyzed). That is, what we usually want is the
z-component of the flux, averaged over a duration T, usually
dictated by "instrument" response time:

$$\tilde{F}_z = \frac{1}{T} \int_0^T \left\{ [w_f + w_s - v_g]C - \mathcal{D} \frac{\partial C}{\partial z} \right\} dt \quad . \tag{103}$$

Let's work on this average for awhile. If we break all terms
into their T-mean values (identified with a tilde) plus variations
about these averages (identified with primes), e.g., $C = \tilde{C} + C'$,
if we then perform the average dictated by (103), if we notice,
for example, that $\widetilde{w'} = 0 = \widetilde{C'}$ (since w' and C' were already just
the fluctuations about mean values), and if we gloss over statis-
tical difficulties that arise if the turbulence is not stationary
and if the variations have different frequency distributions, then
(103) becomes

$$\tilde{F}_z = \widetilde{w_f'C'} + \widetilde{w_s'C'} + (\tilde{w}_s - v_g)\tilde{C} - \mathcal{D} \frac{\partial \tilde{C}}{\partial z} \quad , \tag{104}$$

where I've also taken $\tilde{w}_f = 0$ (i.e., no mean vertical-motion of
the air, just fluctuations), and assumed that gravity and the diffusivity don't fluctuate! You may think that the result (104)

was simple to obtain, but then you should be surprised how many
authors have fouled it up, and neglected $\widetilde{w_s'C'}$.

"Rigorously", that's about as far as we can go, but others
have gone farther. In particular, it is common to introduce, in
analogy with molecular diffusion, a "turbulent diffusion" via the
"gradient-flux ansatz" ("ansatz" = assumption):

$$\widetilde{w_f'C'} = -K_z \frac{\partial \widetilde{C}}{\partial z} \ , \tag{105}$$

where K_z is the (zz-component of the second-order tensor) "turbulent diffusivity". Similarly, and equally justifiably (unjusti-
fiably!), we could assume

$$\widetilde{w_s'C'} = -\mathcal{K}_p \frac{\partial \widetilde{C}}{\partial z} \ , \tag{106}$$

where $\mathcal{K}_p$ is a "particle diffusivity", which would depend on particle characteristics. If these are used in (104), it becomes

$$\widetilde{F}_z = (\widetilde{w}_s - v_g)\widetilde{C} - [K_z + \mathcal{K}_p + \mathcal{D}] \frac{\partial \widetilde{C}}{\partial z} \ . \tag{107}$$

Is the use of these "other diffusivities" justified? The
answer isn't known. Of course, the response can be: it's just a
definition, e.g., $K_z = -\widetilde{w_f'C'}/(\partial \widetilde{C}/\partial z)$, so certainly it's justified! But can it be justified by its usefulness? The concepts
would be useful (very useful, to replace the unknown C' with C!)
if K (and $\mathcal{K}_p$) were independent of C' and $\widetilde{C}$; i.e., were properties only of the fluid. Is that reasonable? Probably yes -- in
the bulk of the fluid, far from the interface. To see this, follow an argument similar to the one used in the kinetic theory of
gases (see a few paragraphs earlier): net flux up = (flux up from
z - ℓ) - (flux down from z + ℓ), where ℓ is some "turbulent mixing
length", = u*$\widetilde{C}$ at z - ℓ less u*$\widetilde{C}$ at z + ℓ, where u* is some
"mixing speed",

$$= [u*\widetilde{C} - \frac{\partial}{\partial z}(u*\widetilde{C})\ell + \ldots] - [u*\widetilde{C} + \frac{\partial}{\partial z}(u*\widetilde{C})\ell + \ldots]$$

$$= 2[u*\ell \, \partial \widetilde{C}/\partial z + (1/3!) \, u* \, \ell^3 \, \partial^3 \widetilde{C}/\partial z^3 + \ldots] \ ,$$

assuming u* is independent of z. (By the way, note that terms
containing even-order derivatives cancel.) From this, we get
K = 2u*ℓ.

But what are u* and ℓ? It might be expected that u* could be
related to turbulent fluctuations in the velocity (e.g. to $\widetilde{w'^2}$

or to the friction velocity, u*). But what is ℓ? It might be
expected to be roughly the "size" of the "turbulent eddies" (whatever that means!), and these in turn might be expected to be
roughly equal to the height above the interface. In fact, if we
take $K = \varkappa$ u*z, with $\varkappa = 0.4$ = von Karman's constant and u* =
the friction velocity (see later), then this is the standard
expression for K in a boundary layer, uninfluenced by buoyancy
(i.e., "neutral stability").

But immediately we can see troubles with this gradient-flux
ansatz, especially near the interface, exactly where we most want
help (since we want the flux <u>at</u> the interface). The trouble is:
if we use $K = \varkappa$ u*z (i.e., $\ell \sim z$), and if, as is usually the
case, $C \sim \ell n\ z$ (to see this, integrate Flux = const = $K\ \partial C/\partial z$),
then when these are used in the series expansion, it is seen that
the series is teetering on the verge of divergence. Try it: if
you use $\ell = z$, you'll see that the nth term has $(n-1)!\ z^{-n}$ multiplied by $z^n/n!$, i.e., you'll get a sequence of terms with $1/n$,
and the series diverges. However, if you use $\ell = \varkappa z/2$, then the
series converges (a result pointed out to me by Hasse). It might
then be thought: "definitely use the smaller ℓ!", but why not use
$\ell = z$ and u* $= \varkappa$ u*/2, which again leads to a divergence series?

But there is more, here, beyond manipulating factors of $(\varkappa/2)$
to force a series to converge   (when the entire expansion is a
crude approximation). Let me mention three other factors. One
relates to the entire concept of using (local) derivatives. Much
of physics (and chemistry) is "local and linear", by which I mean
that what is going on near a point, x_0, can be approximated by
what is occurring at the point, plus a linear correction: $f(x) =$
$f(x) + f'(x_0)(x-x_0)$, where the (first-order) derivative, f', is
evaluated at the point x_0. This is the essence of Ohm's law,
Fick's law, Fourier's heat law, and many similar "laws" -- including this K-theory of turbulence -- and these lead to descriptions
of processes in terms of differential equations, whose essence is
to relate quantities, locally (e.g., how the local current is
related to the local gradient of the electrostatic potential).
But nature is obliged to be neither local nor linear! Ohm's law,
etc., can be inadequate if the gradients are large -- and the concentration gradient is usually largest near the air-sea interface,
exactly where we most want to describe what is going on. Moreover, beyond failure of linearity, there can be not only <u>failure</u>
of "localness", it may be fundamentally inappropriate. Thus,
whenever conditions at a point depend not only on local condition,
but on conditions elsewhere, then local derivatives (and resulting differential equations) will be inadequate: the processes
must be described in terms of integral equations. For example,
integral equations are needed to describe light propagation
through a medium (especially near its boundaries), to describe
neutron transport and the velocity of gas molecules, and to describe the behavior of any two systems with distributed masses (or

other relevant parameters) that are coupled (e.g., a membrane generating sound waves in air). And therefore my first point is that
it is likely to be fundamentally inappropriate to seek a differential (as opposed to an integral) description of turbulence:
the essence of turbulence, especially near boundaries, seems to
be that local conditions are dominated by the fluid's behavior
elsewhere, at distances comparable to the local height.

I will try to be briefer and more positive, presenting my
other two points. One is to recall a definition of a good theory:
a theory that continues to be adequate beyond its range of applicability! In that sense, the Navier-Stokes equations of fluid
mechanics represent a surprisingly good theory: they "shouldn't
be used," for example, to describe conditions inside a shock wave
in air (where gradients are large over distances of-the-order-of
the mean-free path); yet when they are, they describe conditions
fairly well. Similarly, K-theory may be a "good theory". In
fact, moving on to my final point, the Navier-Stokes equations
(with their first-order derivatives) give a better description of
conditions inside a shock wave than do the second-order Burnett
equations for fluids, which use a more complex way to "close" the
equations (rather than using Stokes' linear relation between
stress and rate of strain). This is a point that should be appreciated by all those working with "second-order closure models" of
turbulence: it is likely that the first-order, K-theory represents the first term in an asymptotic expansion (just as with the
Navier-Stokes equations), and a characteristic feature of asymptotic expansions is that the description can deteriorate if more
terms are taken. Defenders of second-order closure models may
point to cases (e.g., in a forest canopy) where there is flux up
(as opposed to down) the gradient; however, it can be argued that,
in these cases, there has been a poor definition of mean quantities or the omission of some terms (e.g., the pressure gradient)
in the equations (e.g., see Slinn, 1982a). Thus, not only may
K-theory be a "good theory", it may get worse if you try to
improve it!

Thereby, you might think I would support efforts to keep
K-theory, and just improve on the description of K. But then it
becomes somewhat of a game. The game is usually played with the
momentum: the wind profile $\tilde{u}(z)$ is measured, the vertical flux
of horizontal momentum, $\widehat{u'w'}$, is measured, and the game is to find
a $K(z)$ such that the measured $\widehat{u'w'}$ is $K(z)$ multiplied by the measured $\partial \tilde{u}/\partial z$. As a result, proposed K's have had some incredible
forms, beyond the simple $\varkappa u_* z$. For example, I say the game has
gone too far when Davies (1966) suggests:

$$K/\nu = z_+^4 - z_+^{0.08} \ [1000 \{ 2.5 \times 10^7 /Re \}^{z+/(400 + z_+)}]^{-1} \qquad (108)$$

where ν is the kinematic viscosity and $z_+ = z/(\nu/u*)$. A similar "game" has been played with the "particle eddy diffusivity"; e.g., Sehmel (1971) has tried to fit some of his measurements of particle flux using

$$(K + \chi_p)/\nu = 0.531 \exp\{-0.033\ u_*\}\ z_+^{2.6}\ S_+^{1.2} \tag{109}$$

where $S_s = \tau_s u_*^2/\nu$ is the Stokes number based on the fluid's length scale, ν^+/u_*, and velocity, u_*, near a smooth surface.

Now, I don't mind a good game (even if it's one of disguise!) but I do complain about these K's: I think they are inadequate for the case of particle deposition -- even in forms such as (109), but especially for K's such as (108). It's one thing to measure $\widetilde{u'w'}$ and $\partial\tilde{u}/\partial z$ and then find a K to fit, but it's quite another to treat this K, "derived" from momentum transfer, as if it's also appropriate for mass transfer. Let me put it this way. I don't expect anyone will argue with the following statement: "The average mass flux is the mass flux during average conditions." I expect no argument because the statement is meaningless! -- until "average conditions" are defined. And if they are defined as the conditions that give average flux, then I still expect no argument -- because then the statement is a useless tautology. But I do argue with the statement "the average mass flux is the mass flux during average conditions -- where 'average conditions' are defined with an 'average' K; i.e., derived to fit the average momentum transfer, not mass transfer." My simple response is: "Prove it!" Prove, for example, that one, large, violent "eddy" doesn't deposit essentially all the mass!

And I see hints that something's wrong. For example, years ago, von Karman and Prandtl analyzed mass transfer using (essentially) $K = \chi u_* z$ and Reynolds' earlier suggestion that momentum and mass transfer were analogous; the result (essentially) was that the mass-transfer velocity should vary as Sc^{-1}. (Recall that $Sc = \nu/\mathfrak{D}$.) As more data became available for larger Schmidt numbers, the mass transfer for turbulent flows was seen to be correlated better with $Sc^{-2/3}$, which is what theory suggested -- but for laminar flows! But eventually someone saw they could "derive" this, even for turbulent flows, if they assumed $K \sim z^3$ as $z \to 0$, rather than $K \sim z$. But more data, with larger Sc (especially for liquids, since even gases in liquids have $\mathfrak{D} \sim 10^{-5}$ cm^2 s^{-1}) suggested $Sc^{-1/2}$. Two theories then emerged. One was Higbies' eddy-replacement model, which accounts for time variations and yields a transfer velocity as an (error) function of $\mathfrak{D}^{1/2}$, and the other went back to K-theory: with $K \sim z^n$, $(n > 1)$, then the transfer velocity $Sc^{-(1-1/n)}$, [actually, (King, 1966), multiply this by $(n/\pi) \sin (\pi/n)$], and so all bets (past present, and future!) are covered.

But what happened to reality? I think the reality is that
we will need to obtain the mass transfer "eddy-by-eddy," and then
average over all eddies to get the average mass transfer. Stated
differently: the average mass transfer appears to be <u>not</u> the mass
transfer by an average eddy (defined via an average K). And the
problem with this field, and the reason why these notes will be
incomplete, is that we don't yet know the mass transfer "eddy-by-
eddy." Later, I'll sketch some preliminary ideas; for now, let
me just mention that it <u>does</u> make a difference. For example, for
1 μm particles in air, with Sc $\simeq 10^6$, then the ratio of Sc^{-1} to
$Sc^{-2/3}$ to $Sc^{-1/2}$ is as $10^{-6}:10^{-4}:10^{-3}$. The question, then, is
whether or not we can "hone down" the uncertainty to a little less
than a factor of 10^3!

5.3 A Simple Two-Layer Model

If I immediately tried to jump to a better model, I think
even the physicists would holler "slow down!"; therefore, I will
start with a simple model (Slinn and Slinn, 1980) even though it
is poor. After showing its outline, I'll then show why some of
its details are inadequate. In Section 6, I'll show that even
the outline is all fouled up! With that advertisement, even the
chemists might prefer if I did jump directly to a better model.
But the trouble with that is: I don't know how to!

In this simple model, the atmosphere below a convenient
reference height (e.g., 10 m) is conceptually divided into two
layers; see Figure 30. Geochemists, especially, have this pro-
pensity to introduce various layers, because they have the hope
that, in each, there will be
fewer variables and processes
with which to contend. Thus,
here it is hoped that in the
upper layer, turbulence dominates the transfer; and that
in the lower layer, turbulence is negligible. The
height of this lower layer
(which I call the "deposition" layer), next to the
interface, is not yet
defined -- except with the
idea that above it, transfer
by turbulence can no longer
be ignored.

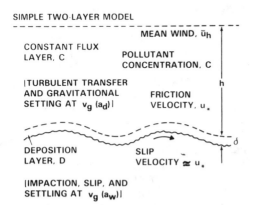

Before continuing with
the model, it may be useful
to spend some time on transfer by turbulence. Let me

Figure 30. Schematic of the two-
layer model described in the text.
(From Slinn and Slinn, 1980.)

approach it this way. Away from the interface, one "thing" that's transferred vertically by turbulence is the horizontal momentum, $\rho_a \tilde{u}$. I hope that makes sense. To see it, suppose a jet engine was blasting out some horizontal momentum, $\rho_a \tilde{u}$, into a turbulent atmosphere. The farther you go downwind from the engine, the farther off the jet axis you would find some of the jet's horizontal momentum. Thus, the horizontal momentum can be transported vertically. This vertical flux of horizontal momentum is called the shear stress, τ, (it is a stress = force per unit area; check its dimensions!) and meteorologists customarily report its measurement either in terms of the "friction velocity" u_*, where $\tau \equiv \rho_a u_*^2$, or in terms of the drag coefficient, C_D, where $C_{D,h} = u_*^2/\tilde{u}_h^2$. For example, as you'll see more in the lectures by Hasse (this volume), over the sea and with a reference height h = 10 m, $C_D \simeq 1.3 \times 10^{-3}$ (dimensionless). Then, by the way, if $\tilde{u}_h = 5$ m s^{-1}, $u_* = C_D^{1/2} \tilde{u} = 18$ cm s^{-1}. But here, while considering vertical transport by turbulence, what I want to define is Chamberlain's deposition velocity (or "transfer velocity") for momentum, by turbulence:

$$v_{d,m} = \frac{\text{Flux}}{\text{Conc.}} = \frac{\tau}{\rho_a \tilde{u}_h} = \frac{\rho_a u_*^2}{\rho_a \tilde{u}_h} = \frac{u_*^2}{\tilde{u}_h} = \frac{u_*^2}{\tilde{u}_h^2} \tilde{u}_h = C_{D,h} \tilde{u}_h \ . \tag{110}$$

For example, with $C_{D,h} = 1.3 \times 10^{-3}$ and $\tilde{u}_h = 5$ m s^{-1}, this deposition velocity/transfer velocity by turbulence is 0.65 cm s^{-1}.

Now, back to the two-layer model. The question is: is there a region (our upper layer) where turbulent transfer dominates? Well: what can compete (in the upper layer) with this turbulent transfer of $C_D u$ (whose magnitude is about 1 cm s^{-1})? Obviously, there is gravity, at least for "large" particles (if a $\gtrsim 10$ μm, $v_g \gtrsim 1$ cm s^{-1}). What other "slip velocity" besides gravity? I don't know of any. Stefan flow, whereby particles in otherwise stationary air move with, say, a water-vapor flux, is normally not large enough. Thus, if the water vapor flux is $\dot{m}''$, then the the mean motion of the mixture, the Stefan velocity of the particles, is

$$v_s = \frac{0 \, \rho_a + (\dot{m}''/\rho_v) \, \rho_v}{\rho_a + \rho_v} \simeq \frac{\dot{m}''}{\rho_a} \ , \tag{111}$$

and even if $\dot{m}'' = 1$ g cm^{-2}hr^{-1} (1 cm hr^{-1} is a lot of evaporation!), then v_s is only about 0.2 cm s^{-1}. The thermophoretic drift velocity would be even smaller (Slinn and Hales, 1971), and so would electric drift, I expect. As for Brownian diffusion, a characteristic transfer velocity is $\mathcal{D}/L$, where L is a characteristic length scale, and even for low-molecular-weight gases in

air (with a large $\mathcal{D}$ of ~ 0.1 cm^{2}s^{-1}), with L ~ 1 m, $\mathcal{D}/$L $\sim 10^{-3}$ cm
s^{-1} and therefore, transfer by Brownian or molecular diffusion is
negligible (when L is large!). Consequently, for the upper
"constant-flux" (or "C") layer, I'll assume the dominant terms in
the transfer velocity to be governed by gravitational settling
and turbulence:

$$\beta\, C_D \widetilde{u} \,, \quad v_g(a_d) \,, \quad \text{and maybe } \dot{m}''/\rho_a \,, \tag{112}$$

where I've introduced a "fudge factor", β, in front of $C_D\widetilde{u}$, for
reasons to be explained later, and I've identified the particle
radius, a, with the subscript d (for dry) -- later to distinguish
a wet-particle radius, a_w, if the particles grow by water-vapor
condensation "near" the interface. [What is special near the
interface? What is this business about particles growing by condensation near the interface? Does he think it's humid only
there? There's a continuous gradient of the humidity....] Yes,
I know. But what do you want: a full-blown model right at the
start? Just "go-with-the-flow" for a while. Worrying about the
height at which the particles actually grow is one of the least
of our problems -- we know how to solve that one: the particles
grow continuously, as the humidity increases.

What is special near the interface is that turbulent transfer
diminishes. Why? Because, e.g., for momentum transfer, $\widetilde{u_f'w_f'}$
must vanish at the interface, defined to be where $w_f' = 0$ (i.e.,
it is assumed that the air doesn't flow into the water). At a
solid surface, also u_f' is zero at the interface (the "no-slip
condition" of continuum fluid mechanics). But water can slip (be
moved by the stress from the air), and therefore, u' needn't be
zero at the air-sea interface, only w'. True, there can be a w'
near the surface (e.g., with air flow over the waves), but exactly
at the interface (except for air engulfed by breaking waves and
released by bursting bubbles) $w_f' = 0$. Also, though, w' (the z-
component of the particle's slip velocity relative to the air)
needn't be zero at the interface; for example, a "large" particle
(large Stokes number!) will not stop so quickly as the air, and
thereby can be impacted on the water. So we can expect that the
transfer velocity, across the "deposition layer", should depend
on the Stokes number, where the "deposition layer" thickness is,
so far, undefined -- except with the idea that, there, turbulent
transfer (e.g., $\widetilde{u_f'w_f'}$) is negligible.

But then: how does the momentum finally get to the surface,
if not by turbulence? The answer, of course, is by molecular
diffusion or "viscosity": mathematically, the stress is $\tau =$
$-\rho_a\widetilde{u'w'} + \mu\, \partial\widetilde{u}/\partial z$ (where the dynamic viscosity, $\mu = \rho_a\nu$), and
if $\widetilde{u'w'}$ (the "Reynolds stress") becomes small, all that is left

is the molecular term, which can be large if the velocity gradient, $\partial \tilde{u}/\partial z$, is sufficiently large. Thus, if we approximate the gradient $\partial \tilde{u}/\partial z$ by $u*/\delta$ (what $\tilde{u}$ would you use?), then we see that there must be a "sublayer", next to the interface, where viscous drag dominates and whose thickness, δ, must be given approximately by

$$\tau = \rho_a u_*^2 \approx \mu u_*/\delta \implies \delta \approx \mu/(\rho_a u_*) = \nu/u_* \quad , \tag{113}$$

which could have been guessed, because there is no other combination of appropriate variables, there, which would give us a variable with the dimensions of length. For example, if $u* = 10$ cm s^{-1} and $\nu = 0.1$ cm^2 s^{-1}, then $\delta \approx 10^{-2}$ cm $= 0.1$ mm $= 100 \,\mu$m. This now gives an idea of the height of this "deposition layer." Also, it gives us a way to define a Stokes number, $S = \tau_s/\tau_f$, the ratio of the particle stopping time to a characteristic "time scale" of the fluid motion, τ_f. With δ as in (113) and a velocity scale $u*$ (what else?), it seems reasonable to define a Stokes number, here, via

$$S = \tau_s/\tau_f = \tau_s/(\delta/u_*) = \tau_s u_*^2/\nu \quad . \tag{114}$$

Fine: there's a "deposition layer" (height somewhere around δ), in which turbulent transfer is negligible, and maybe the particles can "skid across this layer" via $\widetilde{w_s C}'$, though this depends on the Stokes number. How else can the particles get across the layer?

I'm going to save the majority of my response to that question until the next subsection. (A summary of the response is: I really don't know, and I don't think anyone else does, either!) However, for now, let's at least assume there's gravitational settling, $v_g(a_w)$ (with a wet particle radius!), and look at the possibility of Brownian diffusion. Again, the characteristic "diffusion velocity" is $\mathfrak{D}/L$, but now for the length scale, L, rather than using 1 m (as for the upper layer) use δ given by (113). Then

$$\mathfrak{D}/L = (\mathfrak{D}/\nu)u_* = u_* Sc^{-1} \quad , \tag{115}$$

where again, $Sc = \nu/\mathfrak{D}$ is the Schmidt number, and for $Sc \approx 1$ (low-molecular-weight gases), this "diffusion velocity" is $\sim u* \sim 10$ cm s^{-1}! Of course, for (say) 1 μm particles, with $Sc \approx 10^6$, then this velocity is very much smaller, but clearly this transfer by Brownian diffusion, across the deposition layer, is "something to watch for." If we add on the Stefan flow from (111) (and call the water-vapor mass flux, $\dot{m}''$, positive if there's evaporation), then the flux through this "deposition layer" would seem to have dominant terms for the "transfer velocity" (positive, down) through the deposition layer

$$v_g(a_w) \; , \quad -\dot{m}''/\rho_a \; , \quad f_1(u_*,S) \;\; \text{and} \;\; f_2(u_*,Sc) \; , \tag{116}$$

where f_1 and f_2 are two unknown functions of u_*, the Stokes number, and the Schmidt number.

Rigorously, that's about as far as we can go. (Rigorously?)! Now a number of crude tricks, and my goal is to quickly get an "answer" and then criticize it. Earlier, in (104), the dry deposition flux was found to be

$$-\widetilde{F}_z = -\widetilde{w_f'C'} - \widetilde{w_s'C'} + \mathcal{D}\frac{\partial \widetilde{C}}{\partial z} + (v_g - \widetilde{w}_s)\widetilde{C} \; . \tag{117}$$

Then, in (107), the possibility was entertained that the flux might be written in the form

$$-\widetilde{F}_z = [K_z + \mathcal{K}_p + \mathcal{D}]\frac{\partial \widetilde{C}}{\partial z} + (v_g - \widetilde{w}_s)\widetilde{C} \; , \tag{118}$$

and I won't repeat the serious concerns about this form. Now, the even cruder forms are temporarily proposed:

$$-\widetilde{F}_z(z=h) = k_C'(\widetilde{C}_h - \widetilde{C}_\delta) + [v_g(a_d) - \widetilde{w}_s]\widetilde{C}_h \tag{119}$$

$$-\widetilde{F}_z(z=\delta) = k_D'(\widetilde{C}_\delta - \widetilde{C}_i) + [v_g(a_w) - \widetilde{w}_s]\widetilde{C}_\delta \; , \tag{120}$$

in which the "transfer velocities" k_C and k_D are not yet defined. Also, to get it all over-with in one shot, let's make the "constant flux assumption,"

$$\widetilde{F}_z(a=h) = \widetilde{F}_z(z=\delta) \; , \tag{121}$$

use the "perfect sink condition," $\widetilde{C}_i = \widetilde{C}(z = z_i) = 0$, and define the deposition velocity via

$$-\widetilde{F}_z(z=h) = \widetilde{v}_{d,h}(\widetilde{C}_h - \widetilde{C}_i) \; . \tag{122}$$

If (121) and (120) are used to eliminate $\widetilde{C}_{\xi}$ in (119), and the result is used in (122), then a little algebra yields the result

$$\frac{1}{\widetilde{v}_{d,h}} = \frac{1}{k_C} + \frac{1}{k_D} - \frac{v_g(a_d)}{k_C k_D} \tag{123}$$

where

$$k_C = v_g(a_d) + k_C' \quad \text{and} \quad k_D = v_g(a_w) + k_D' \tag{124}$$

(and, if desired, and correcting an error in Slinn and Slinn (1980), $w_s = \dot{m}''/\int_a$ could be subtracted from the two v_g's, wherever they appear).

    Enough: let's back up! If you'll look again at (119), surely you'll question its validity. Even if we knew what k_C was, how can we justify setting a part of the flux equal to $k_C(C_h - C_\xi)$? Really, from (117), the dominant term (neglecting particle slip and Brownian diffusion at z = h) is $\widehat{w_f'C'}$. Presumably k_C is related to w_f', but why is C' expected to be ($\widehat{C}_h$ - C_ξ)? There is no adequate answer to that question: it's all guesswork, and in that vein, guess that

$$k_C' = \beta \, \widetilde{v}_{d,m} = \beta \, C_D \, \widetilde{u}_h \quad , \tag{125}$$

as in (112), with β a fudge factor and $v_{d,m} = C_D\widetilde{u}$, the transfer velocity for momentum. Similar complaints (and more -- see later) can be made about (120), and the guess (see 116) that

$$k_D' = f_1(u_*,S) + f_2(u_*,Sc) \quad . \tag{126}$$

    But we can attempt to help this formalism be realistic if we "tie it down" at all available loose ends. For example, (123) is not too bad in that, if there is no turbulence (i.e., u_*, f_1 and f_2, and k_C and k_D all zero), then we get

$$\frac{1}{\widetilde{v}_d} = \frac{1}{v_g(a_d) - \widetilde{w}_s} + \frac{1}{v_g(a_w) - \widetilde{w}_s}$$
$$- \frac{v_g(a_d) - \widetilde{w}_s}{[v_g(a_d) - \widetilde{w}_s][v_g(a_w) - \widetilde{w}_s]} \quad , \tag{127}$$

or $v_d = v_g(a_d) - \widetilde{w}_s = v_g(a_d) - \dot{m}''/\rho_a$, as desired. Also, for momentum (with $v_g = 0 = \widetilde{w}_s$, the Stokes number, $S = 0 = f_1$, and $Sc = 1$), then

$$\frac{1}{\tilde{v}_{d,m}} = \frac{1}{C_D\tilde{u}} + \frac{1}{f_2(u_*,Sc=1)} \quad , \tag{128}$$

and this will be as desired (i.e., $v_{d,m} = C_D\tilde{u}$) if we take

$$f_2(u_*,Sc=1) = \frac{\beta}{(\beta-1)} \, C_D\tilde{u} \quad . \tag{129}$$

The remaining questions, then, are what are the functional dependencies of f_1 on the Stokes number (and on u_*) and of f_2 on Sc? I will go into those (unanswered!) questions in the next subsection. For now, let me just write the expressions used by Slinn and Slinn (1980):

$$k'_D = f_1 + f_2 = \frac{\beta}{\beta-1} \, C_D\tilde{u}_h[Sc^{-1/2} + 10^{-3/S}] \tag{130}$$

with $S = \tau_s/\tau_f = \tau_s u_*^2/\gamma$ and their choice for $(\beta-1)/\beta$, based on fitting data for particle deposition to smooth, solid surfaces (Slinn, 1976), was 0.4; i.e., $\beta = 5/3$. A resulting plot of the deposition velocity, (123), was shown earlier as Figure 2.

Figure 31 shows results from a recent extension of this theory (Williams, 1982) that attempts to include effects caused by breaking waves. These effects are buried in two parameters: $\alpha = 1.7 \times 10^{-6} \, \tilde{u}_{10}^{3.75}$ (the fraction of the ocean's surface that is broken with waves) and k_{bs} (a particle-size-independent transfer velocity to the broken surface). Williams states: "the broken surface transfer coefficient, k_{bs}, is governed by processes such as scavenging by impaction and coagulation with spray droplets,

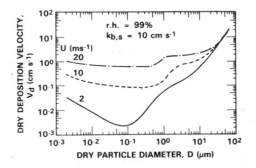

Figure 31. Predicted dry deposition velocities for particles, including an assumed particle-size-independent transfer velocity, k_{bs}, to water's "broken surface." (From Williams, 1982.)

which are likely to be quite efficient." However, as stated in
Slinn and Slinn (1980) and reviewed in Section 6.2, the efficiency
of spray scavenging will be strongly dependent on particle size
(better, on the particle's Stokes, interception, and Schmidt numbers), and therefore Williams' use of a large, particle-size-
independent value for k_{bs} seems to be highly unrealistic. For
Figure 31, k_{bs} was taken to be 10 cm s^{-1}. Yet Williams'
result may be reasonable, but for a different reason: breaking
waves engulf air (later leading to bubbles), and the resulting
transfer velocity will be essentially independent of particle
size, except for those bubbles that do not dissolve in the ocean
and later return some of the particles to the atmosphere, when
the bubbles burst at the air-sea interface. But I do not know of
any estimates of the magnitude of the resulting k_{bs}, and don't
have time to try to present an estimate here. However, I expect
it's small and therefore expect that Figure 31 is unrealistic.
Nevertheless, as a minimum, a comparison of Figures 2 and 31 gives
an idea of some of the uncertainties in the theory; more on this
in Section 6.

There are insufficient (and, I will argue, inadequate) data
to test the theory. Figure 32 shows a comparison of the theory
(the solid line is Eq. 123) with measured dry deposition velocities for a number of substances. However, all of these measurements are for a distribution of particle sizes (hence a few large
particles can make an overwhelmingly large contribution to the
total deposition), and all except for three data points (for
vanadium, phosphate, and sulfate) are for deposition to other
than water surfaces. Yet, for the theoretical curve, it was
assumed that all particles grew (similar to $SO_{\bar{4}}$ particles) to
their equilibrium size at 100% humidity in the deposition layer,
and the theory is for monodisperse particles; i.e., not an average over a particle-size distribution, though that certainly
should be done for this comparison. Also, note that the theoret-
ical curve was evaluated for $\tilde{u}_h = 5$ ms^{-1} and $C_D = 1.3 \times 10^{-3}$; the
heavy dashed curve shows the consequences if $C_D = 2.5 \times 10^{-3}$. The
horizontal width of each "error bar" was obtained from Rahn's
(1976) compilation of the large distributions of measured, mass-
mean diameters. In Section 6, I will describe details supporting
my contention that none of the data sets shown in Figure 32 provide an adequate test of the theory; more significantly, it is my
opinion that none of the data sets can be relied upon to indicate
true values for particle deposition to the oceans.

Figure 33 shows what are (as far as I know) all available
particle deposition data from water/wind-tunnel studies. On the
left-hand side of the figure are three curves, labeled with u*
values of 11, 44, and 117 cm s^{-1}; these are plots of (130), but
using Sc$^{-0.6}$ rather than Sc$^{-1/2}$. On the right-hand side of the

figure are three curves (all only for the case u* = 44 cm s^{-1}, and all are just plots of (130), but with v_g added); these three curves show the dramatic influence of particle growth on the theoretical predictions. Shown are the cases $a_w = a_d$; $a_w = 2a_d$, and a_w as for $(NH_4)_2SO_4$ particles in equilibrium with a relative humidity of 99% in the deposition layer (see Figure 34). It might be thought that the data suggest the theory has merit, but a critical assessment of the data (Slinn and Slinn, 1981) suggests that pecularities of the water/wind-tunnel used by Sehmel and Sutter (1974) may have had overwhelmingly dominant influences on the reported results. In particular, for the high-wind-speed case [and note that for $\tilde{u}(z=10$ cm$) = 13.8$ m s^{-1}, then for the ocean, extrapolating via a logarithmic wind-speed profile, $\tilde{u}(z=10$ m$)$ 30 m s^{-1}], the wave height in the wind tunnel was only about 17 cm! This combination of high wind and small waves results in a large Stokes number, increasing the impaction on waves. In contrast, for the oceans, with much larger waves at large wind speeds, the Stokes number ($S = \tau_s/\tau_f = \tau_s v_f/L$) is much smaller, except for impaction on capillary waves, a topic that needs further study. These and other problems (C measured at only 13 cm; only a short time available for growth of the water-soluble uranine particles) suggest that available data neither prove nor disprove the theory, though presumably it is clear why I think the data suggest that particle growth should be included in future, improved models.

5.4 Some Small-Scale Problems

There are many inadequacies with the theory sketched in the previous subsection. Some of these inadequacies will be described in Section 6 where I'll complain mostly about the overall framework. Here, I'll ignore these "overall" (or large-scale problems); and focus on some "small-scale" problems -- in particular, some difficulties with specifying the Stokes- and Schmidt-number dependencies for the transfer velocity across the deposition layer. But some words of warning. This subsection has caused me a lot of difficulties: you try describing, in an introductory manner, topics that are beyond the state-of-the-science in your speciality! To overcome these difficulties, I've relied on the wisdom of the comic-strip character Charlie Brown: "No problem is so big you can⁻t run away from it!" That is, my solution to the problem is not to try to solve it! Thus the warning: this subsection is for physicists (they have been patient long enough!); chemists just "go with the flow."

The heart of the problem is our lack of knowledge about the air flow near even a solid surface. To illustrate some of the evolution in concepts about such flows, consider the evolution in

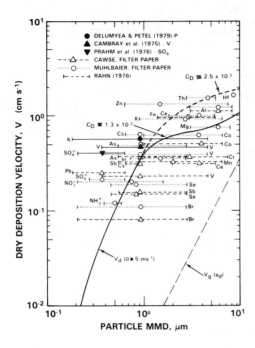

Figure 32. Comparisons of measured, mass-mean
deposition velocities with predictions for monodisperse
particles as given by Eq. 123. (From Slinn and Slinn,
1981, where references to the data sources can be
found.)

proposed Schmidt number dependencies of the transfer velocity
across the layer next to a solid surface.
(i) As I mentioned earlier, the early (~1910-20) concepts were of
a stagnant film next to the interface (many still use this concept!), and the resulting deposition velocity, illustrated with
either Prandtl's

$$\tilde{v}_{d,h} = C_{D,h}\tilde{u}_h/[1 + 5\ C_{D,h}^{1/2}\ (Sc - 1)] \tag{131}$$

or von Karman's

$$\tilde{v}_{d,h} = C_{D,h}\tilde{u}_h\Big/\Big[1 + 5\ C_{D,h}^{1/2}\Big\{(Sc - 1) + \ln\frac{(1 + 5\ Sc)}{6}\Big\}\Big] \tag{132}$$

basically has a Sc^{-1} dependence. This Sc^{-1} dependence can easily
be seen by taking $-\tilde{F}_z = \mathcal{D}\,\partial C/\partial z \sim \mathcal{D}C/\delta_D$, where δ_D is the "film
thickness" of the "diffusion layer", and therefore

$$\tilde{v}_d = - \tilde{F}_z/\tilde{C} \sim \mathcal{D}/\delta_D \sim (\mathcal{D}/\vartheta)(\vartheta/\delta_D) \sim (\vartheta/\delta_D)Sc^{-1} \ . \qquad (133)$$

(ii) Next, there was reliance on Prandtl's boundary layer models,
which indicated that the thickness, δ_D, itself, depends on Sc.
It is easy to see (e.g., Appendix C of Slinn et al., 1978) that,
for laminar flows, $\delta_D \sim \delta_v Sc^{-1/3}$, where δ_v is the thickness of
the viscous layer ($\sim Re^{-1/2}$), and therefore, identifying ϑ/δ_v as
essentially the deposition velocity for momentum, then (133)
becomes

$$\tilde{v}_d \sim (\vartheta/\delta_v)Sc^{-2/3} \sim C_D \tilde{u} \, Sc^{-2/3} \ . \qquad (134)$$

However, this proposed dependence on $Sc^{-2/3}$ has theoretical basis
only for laminar flows, unless one is convinced that there is a
theoretical basis for K-theory and for taking $K \sim z^3$.

(iii) I will not go into the theory leading to the prediction $v_d \sim$
$Sc^{-0.6}$ (basically it is for stagnation point flow), nor into
K-theories.

(iv) But I should mention the Higbie-Danckwerts "surface-renewal"
or "penetration" theories, which try to account for time-dependent
mass transfer to "eddies" (see later in this subsection), when
they penetrate to the surface, and then are "cast-off" from the
surface, to be renewed by another eddy. This "intermittancy" is
analyzed in many engineering texts (e.g., Danckwerts, 1970) and
yields a $Sc^{-1/2}$ dependence for v_d.

In summary, in the absence of constraining data, we have a prolif-
eration of predictions: $v_d \sim Sc^{-1}, Sc^{-2/3}, Sc^{-0.6}, Sc^{-1/2}, \ldots$!

A similar evolution can be traced for proposed Stokes number
dependencies of v_d, even for deposition to solid surfaces. These
I will mention very briefly; I've given more details in Hicks and
Slinn (1983). Friedlander and Johnstone (1957) used K-theory
(with $K \sim z^3$) plus an assumption that particles could be "thrown"
toward the surface, by turbulence, a distance related to the
particle stopping time. Their result was

$$\tilde{v}_d = C_D \tilde{u} \left[1 + C_D^{1/2} \left\{ \frac{1525}{(0.9S)^2} - 50.6 \right\} \right]^{-1} \qquad (135)$$

where $S = \tau_s u_*^2/\vartheta$ is the Stokes number as defined earlier.
Refinements to this model were made by a number of authors.
Other authors introduced particle diffusivities (Sehmel 1971,
1973; Illori 1971), and new models were conceived (Owen, 1969;
Caparaloni et al., 1975; Slinn 1976, 1977b). In particular,

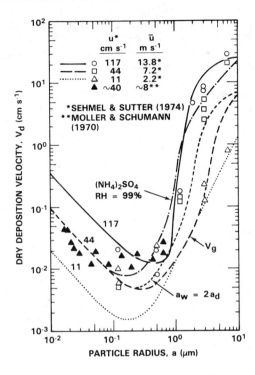

Figure 33. Comparisons of modeled deposition velocities with those measured in water-wind tunnels. (From Slinn and Slinn, 1981.)

to obtain (130), I used Owen's idea that the particles would be convected across the deposition layer (it's not a stagnant film!), relied on the picture of turbulent bursts obtained by Joukowski in the first decade of this century (Figure 35), and chose the particular functional form of $10^{-3/S}$ because of its fit with data obtained by Liu and Agarwal (1974), and because it has the right general form for the "collection efficiency" for particles in a jet impinging on a flat plate. Namely, for large Stokes number, it tends toward unity [$10^{-3/S} = e^{-3(\ln 10)/S} \simeq 1 - 3(\ln 10)/S + \mathcal{O}(1/S^2)$], and for small Stokes numbers, $10^{-3/S}$ rapidly approaches zero, mimicking a critical Stokes number (at $S \simeq 1$). However, rather than dwelling on the past, I'd like to point toward the future (or, as a minimum, do a little handwaving at it!).

I think that future studies will yield a "gentler function" than $10^{-3/S}$ for the Stokes number dependence of v_d: maybe similar to (135) (which, after all, is just a fit to data), or maybe similar to what I showed earlier for the particle/drop inertial-collision term (E $\simeq$ [(S - S*)/(S + C)]$^{2/3}$), or maybe similar to what I used in a recent publication (Slinn, 1982a; in which this equation has a typographical error): E $\simeq$ S^2/(1 + S^2). Soon I'll explain why I have this expectation. As for the question about the Schmidt number dependence (whether v_d depends on Sc^{-1}, Sc$^{-2/3}$, Sc$^{-0.6}$, Sc$^{-1/2}$, ...), I think the answer will be "Yes!" -- and now I'd better explain that (or at least give a few more hints!).

        As I said earlier, the heart of the problem is our lack of knowledge about the air flow near even a solid surface. But more information is becoming available and as sketched in Figure 36, the details are complicated. For example, Joukouski's bursts (Figure 35) may be caused when the "low-speed streaks" are accelerated by the surrounding flow, until their Reynolds number (Re = vz/ν) reaches a value of about 100, the critical value for shear instability (e.g., with v = u*, then when z $\approx$ 100 ν/u* -- i.e., 100 "viscous sublayer" heights -- the shear layer will break into instability). Perhaps the "low-speed streaks", themselves, represent the startup of the flow after a burst, but perhaps they are caused when disturbances from the outer flow reach the wall. If the bursts are dictated only by conditions near the wall, then only the variables there (i.e., u* and ν) can enter into the description of, say, the time period between burst, T_B (e.g., T_B = Cν/u$_*^2$). However, recent studies (e.g., Cantwell, 1981) have shown that data for the bursting period are correlated better with a mixture of both "inner" and "outer" variables (specifically, u*, ν, and $\tilde{u}$), so it now appears that we have the chicken-and-egg question of which comes first, the turbulence or the bursts!

        But, setting aside the question(s) about the cause(s) of the streaks and bursts, we are faced with the question: given that the streaks and bursts exist, how are particles transported to the underlying surface? I think the answer is that the particles get across in all possible ways -- and that it will take us a decade or two to answer the question quantitatively. Thus, I expect that the particles diffuse across the high- and low-speed streak areas (yielding a dependence of v_d on Sc^{-1}, just as for a stagnant film), plus they diffuse across the "startup" boundary layer, when a new streak forms (yielding a dependence on Sc$^{-2/3}$, just as for a laminar boundary layer), plus they diffuse from the stagnation points at the bottoms of the bursts and of the roll vortices (yielding Sc$^{-0.6}$), plus they diffuse from the unsteady bursts (yielding Sc$^{-0.5}$)! In other words, I expect that the f_1 of (130) will be something like

$$\gamma C_D \bar{u} \left[a_1 Sc^{-1} + a_2 Sc^{-2/3} + a_3 Sc^{-0.6} + a_4 Sc^{-1/2} \right]$$

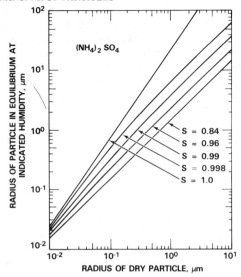

Figure 34. Equilibrium size of $(NH_4)_2SO_4$ particles for the indicated humidities, s. (From Fitzgerald, 1975.)

where the different a's give the fractions of the total area "covered by" the different features. And though these fractions may be relatively fixed for a specific flow and surface (and then change if $\tilde{u}$ changes!), yet for different particle sizes (e.g., as particle size and Schmidt number increase), the deposition velocity will smoothly shift from, say, a dependence on Sc^{-1} to a dependence on $Sc^{-1/2}$ (since $Sc^{-1/2} \gg Sc^{-1}$ for, say 0.1 μm particles). That is what I meant by "Yes!"

As for inertial impaction, we may be able to get by with a simple functional dependence of v_d on the Stokes number, S, because S must be large (larger than some critical Stokes number, S_*) for impaction to occur, and $S = \tau_s/\tau_f$ will be largest where the flow is most violent (small $\tau_f = L/v_f$; i.e., small L and/or large v_f), e.g., where the bursts occur. Yet, as I said, I expect a "more gentle" function than $10^{-3/S}$. The thought behind using a "more gentle" function is that the result represents the average over all bursts that impact particles on the surface, and just as these bursts must possess a continuum of τ_f values, then there would be a continuum of S_* values over which we would need to average. However, in case that sounds simple, I pose the questions: What if the surface has a characteristic roughness (e.g., capillary waves)? Then would most bursts occur just "downwind" of these waves, and have a characteristic τ_f? Also, since a water surface is not immovable, how will this influence the bursts, low-speed streaks, etc.? And besides, in (130), what

coefficient should be associated with f_2 (and f_1!)? Should we
somehow weight f_1 with the fraction of momentum transported by
viscous drag (typically ~ 0.5), and f_2 with the fraction transported by pressure (or form) drag? But it's not easy: there is
impaction even on smooth surfaces (no form drag). And what about
interception? Have fun!

6. MORE PROBLEMS WITH DRY DEPOSITION

The thrust of the previous section was to say: "For particle dry deposition to the oceans, I don't know what the deposition velocity is, and I don't think anyone else does, either --
there doesn't seem to be any good data." Now, for this last section, I want to show you other problems, besides those associated
with transport across the deposition layer. And the thrust is to
say: "It appears to me that we're not likely to get any good data
if we continue the way we've been going." To reach that conclusion about present experimental methods, I will need to show you
some of the theoretical bases for these methods. Consequently, I
will need to use a fair amount of mathematics. I'm sorry about
that, but see no other way.

6.1 The Continuity Equation and Boundary Conditions

The theoretical basis for both wet and dry removal studies
is the mathematical statement of conservation of mass, also known
as the continuity equation (because it is assumed that a continuous concentration field can be defined; i.e., it is assumed
that no holes or voids develop in the continuum). To obtain the
continuity equation, we can examine how the particle concentration changes within an infinitesimal fixed volume of the atmosphere (or, similarly, within the ocean). The number of particles
within the volume, $C \, \Delta x \, \Delta y \, \Delta z$, can increase because of an inflow
at x (i.e., during Δt, there is an increase of $F_x(x) \, \Delta y \, \Delta z \, \Delta t$,
where $F_x(x)$ is the x-component of the flux evaluated at x), and
decrease because of outflow at $x + \Delta x$ (i.e., $F_x(x + \Delta x) \, \Delta y \, \Delta z \, \Delta t$).
Similarly for the other faces of the volume. Consequently, dur-
ing Δt, the net increase in the number of particles within, viz.,
$[C(t + \Delta t) - C(t)] \, \Delta x \, \Delta y \, \Delta z$, can be found. If we divide the
result by $\Delta x \, \Delta y \, \Delta z \, \Delta t$, then in the limit as these tend to zero,
we obtain the continuity equation

$$\frac{\partial C}{\partial t} = -\nabla \cdot \vec{F} + G - \mathcal{L} \tag{136}$$

where the divergence of the flux, $\nabla \cdot \vec{F} = \partial F_x/\partial x + \partial F_y/\partial y + \partial F_z/\partial z$,
and where G and $\mathcal{L}$ are any other gain and loss rates per unit vol-
ume caused, for example, by gas-to-particle conversion (gain), or

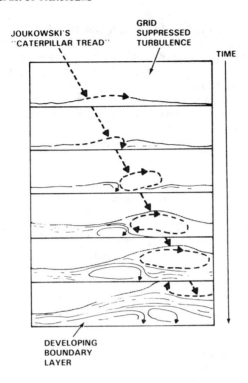

Figure 35. An indication of the generation of "eddies" at a
fluid-solid interface, also indicating where particles could be
convected across the viscous sublayer to the surface. The figure
was redrawn from motion picture frames obtained by Joukowski in
the early 1900's. (From Slinn, 1977a.)

precipitation scavenging (for which we had the volumetric loss
rate $\mathcal{L} = \Psi C$). Incidentally, sometimes C in (136) appears as a
mass-mixing ratio, and then our C is replaced by $\rho_a C$.

     For any partial differential equation (p.d.e.), such as
(136), of course we need boundary conditions (b.c.'s) to identify
the desired function (from an infinite number of possibilities!)
that solves a particular problem. Well, there is a general theory
about b.c.'s for p.d.e.'s, and we could now refer to, say, the
table on page 706 of Morse and Feshbach (1953) to see the types
of b.c.'s needed for this "parabolic" p.d.e. (136). Fortunately,
though, the theory is rather obvious, and to see what b.c.'s are
needed, we can just "use our heads."

     Generally, there are two types of problems that (in princi-
ple!) can be solved. One is of the following form. If C were

known "initially" (e.g., off the east coast of North America),
then (136) will give C everywhere (downwind), say over the
Atlantic and in the troposphere, if we know the flux to the
stratosphere (one boundary condition) and the flux to the ocean
(the other). I hope that seems totally reasonable, physically;
it also happens to be the same requirement of the general theory:
there is an "unique stable solution in the positive direction if
a parabolic equation has Dirichlet (value) or Neumann (slope)
boundary conditions specified on an open boundary."

   Throughout this chapter, we have generally been ignoring the
flux to the stratosphere (assuming it to be zero, except for a
few comments about the influence of deep storms), and have been
focusing on the flux to the ocean. The numerical modelers would
like us to give them this flux in terms of a deposition velocity,
so that then, all they need do is multiply v_d by the concentration
at their lowest "grid level" (e.g., 10 m), giving them the flux to
the ocean. This was the "game" being "played" in the last section, and I must admit that I'm rather tired of playing it. It's
all artificial, derived from the incompetence of numerical modelers! In contrast, with more competence, we could dispense with
the entire idea of a deposition velocity!

   To see this, consider the other obvious type of b.c.'s for
the p.d.e. (136): specify the value (Dirichlet conditions) rather
than the slope or flux (Neumann conditions) at the air-sea interface. Please think about it for a moment, physically: if you
were given the particle concentration over the North Atlantic, at
the air/sea interface (plus, say, no flux to the stratosphere),
then don't you think that (in principle!) you could find the air
concentration? It's obvious, isn't it?

   Moreover, the concentration at the interface is also fairly
obvious, at least for the case of no resuspension (see later for
the case with resuspension). To see the value for C at the interface $C(z = z_i) \equiv C_i$ consider this. For the case of no resuspension, then when a particle enters the water, it is (as far as
the air is "concerned") as if the particles were not there: they
just disappear! That is, for no resuspension, the boundary condition is $C_i = 0$. And this b.c. (plus, say, no flux to the
stratosphere) is sufficient to solve the problem posed. With
$C_i = 0$, then (in general) there is some flux to the sea, and if
we wanted to, we could divide this flux by C at some reference
height, giving a deposition velocity -- for the incompetents who
want one!

   Then, is the problem solved? No! But it's not because the
boundary condition isn't known; it's because details about the
air motions are so complicated, causing us great difficulties in

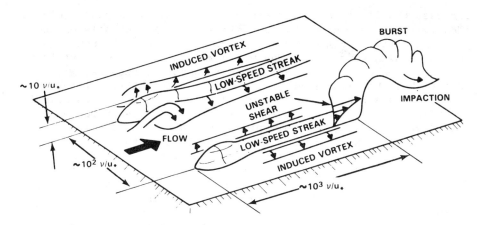

Figure 36. A qualitative sketch of the emerging picture of turbulence near a solid interface.

specifying the flux (for (136), not for v_d!). I'll get back to this, later; first, let me mention a few things about resuspension and about the gain and loss terms in (136).

6.2 Resuspension and Spray Scavenging

If there is resuspension of particles from the sea to the air (and certainly there is -- e.g., sea salt), then the lower b.c. for the equation for C in air is not $C(z = z_i) = 0$. Then what? One way to "handle" resuspension is in terms of a resuspension velocity (the resuspension flux divided by a concentration, e.g., of particles in the sea, near the interface, or at some "convenient" reference depth), but this is the same ruse used by the air modelers: dump the problem on someone else! Another way (the "correct" way!) is obvious: take the flux across the interface to be continuous (the net flux out of the air is the net flux into the sea) and then solve another continuity equation for the concentration of particles in the sea, with an appropriate b.c. at the sea floor, and with appropriate gain and loss terms. True, this would be very difficult, if done right, because we would need (among many other details) to describe mathematically the physics of particles being shot out from the sea, back into the air (e.g., from bursting bubbles). Instead, if we are most interested in the particle concentrations in the air, here is a method I recently proposed that, with judicious use of data, might be quite simple to apply.

The proposed method is to continue to use the "perfect sink" b.c. ($C_i = 0$), as if there were no resuspension, and then account for resuspension in the volumetric gain term, G. A sketch of the idea is shown in Figure 37. In this method, the details of the particle motion (as the particles are, say, shot out from bursting bubbles) are ignored. Why would we want the details of these motions, if what we're after is C? We want the results, i.e., we want to know how many particles reach what height (and this, to be used to distribute the particle "source strength" with height, can be estimated if we know the number of particles produced with "exit" velocity, v_{ex}, by multiplying v_{ex} by the particle stopping time); but once we have this gain term, G, then the subsequent motion of the particles will be dictated by the air motions. That is, turbulence, for example, will quickly "erase" a particle's "memory", of whether it came from the local sea surface or, say, from a continent. I hope that someone will find this proposal useful; it would seem to be relatively easy to apply, and I expect that sufficient data are already available to specify G(z) as a function of sea state (see the chapters by Blanchard and by Buat-Menard, this volume).

Incidentally, while I'm talking about resuspension, let me mention something about resuspension and the interfacial b.c. for gases. For most gases (except those, e.g., SO_2, that are very reactive in sea water), their "resuspension" back into the atmosphere cannot be ignored. Moreover,＼ gas molecules "skid to a stop" so rapidly (their Stokes number is so small) that the gain term is essentially a delta function at the interface. This permits the assumption that gases are "essentially" (not perfectly, just essentially!) in equilibrium at the interface; i.e., $C_{s,i} = sC_{a,i}$, where s is the (Ostwald/Henry's law, equilibrium) solubility coefficient. This boundary condition permits the problem to be solved (in principle!). However, obviously the b.c. is not quite correct because if equilibrium actually did prevail, then (by definition) the flux would be exactly zero (which is not what we want if we think some deposition is occurring!). But the error/inconsistency

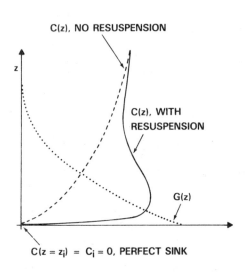

C(z), NO RESUSPENSION

C(z), WITH RESUSPENSION

G(z)

$C(z = z_i) = C_i = 0$, PERFECT SINK

Figure 37. Schematic of the proposed method to account for resuspension with a gain term, G(z).

is usually negligible. This can be seen by comparing the typical deposition flux, $F = v_d C_a$, with $v_d \sim 1$ cm s^{-1}, against only the air-to-sea part of the flux at equilibrium, which from kinetic theory is $1/4 \bar{c} C_a$, where $\bar{c}$ is the mean thermal speed of the molecules, about 3×10^4 cm s^{-1}. Thus, the "amount" that the system is off equilibrium is only $\sim (1$ cm s$^{-1})/(3 \times 10^4$ cm s$^{-1})$, i.e., one part in 30 thousand, and we're no-where-near worrying about that level of accuracy. Thus, for gases, we take the equilibrium b.c., $\tilde{C}_{s,i} = s\tilde{C}_{a,i}$, and use it to evaluate the relatively-miniscule, nonequilibrium flux.

And on another matter, while I'm talking about the gain term, let me also mention the loss term, $\mathcal{L}$. In particular, I want to respond to questions, raised by some, about the potential influence on "dry" deposition, caused by sea spray. Actually, I would suggest we not try to account for sea spray by manipulating it into a term in an expression for v_d, but instead, account for it in the loss term (just as I suggested that resuspension be included in G). In particular, I suggest that for "spray scavenging" we use the same approximate expression as for rain scavenging

$$\mathcal{L} = \psi_r C = [c\bar{E}(a,\bar{R})p(z)/\bar{R}]\ C \quad , \tag{137}$$

where now $p(z)$ is the height-dependent "spray rate" and $\bar{R}$ is a "mean" droplet size. However, if we want to estimate the relative importance of this spray scavenging, compared with other removal mechanisms, then it is convenient to look at a "spray deposition velocity" (what do you want to call it?!):

$$v_{sd} = \frac{\text{spray flux}}{\text{conc}} = \frac{\int \psi_r C dz}{C} \simeq \frac{\bar{p}\bar{E}h_s}{2\bar{R}} \quad . \tag{138}$$

For example, if $\bar{p} = 1$ mm hr^{-1}, $\bar{R} \doteq 0.2$ mm, $\bar{E} = 0.5$, and $h_s = 1$ m, then

$$v_{sd} = \frac{0.1\ \text{cm}}{3600\ \text{s}} \times \frac{0.5}{0.04\ \text{cm}} \times 100\ \text{cm} = 3.5 \times 10^{-2}\ \frac{\text{cm}}{\text{s}} \quad , \tag{139}$$

which is comparable to gravitational settling of 1 μm particles, but that was a large value for the collection efficiency, $\bar{E}$, and I expect that $\bar{p} = 1$ mm hr^{-1} is quite a high value for a typical "spray rate." Of course, it could be larger: if $\bar{R} = 1$ mm, $\bar{E} = 1$, $h = 10$ m, and $\bar{p} = 1$ m/hr (hurricane!), then

$$v_{sd} = \frac{100\ \text{cm}}{3600\ \text{s}} \times \frac{1}{0.2\ \text{cm}} \times 10^3\ \text{cm} = 10^2\ \frac{\text{cm}}{\text{s}} \quad , \tag{140}$$

which is enormous. However, hurricanes are not frequent, and the
point missed by Williams (1982) is that the collection efficiency
is particle-size dependent. Moreover, returning to the problems
I want to address, even if v_{sd} could become as large as 10^2 cm
s^{-1}, we should then question if turbulence could deliver particles fast enough to be scavenged by the spray. That is, it seems
likely that the removal would then be rate-limited by turbulent
transfer through the atmosphere. This leads me back to questions
about resistance models and their basis: the assumption of constant flux.

6.3   The Constant-Flux Condition

     With the goal of attempting to strengthen the theoretical
foundations (and thereby, eventually, the data base) for dry deposition, I would like to return to examine the foundations of the
resistance model used in Section 5.3, and used by so many authors.
The question posed is: When are these resistance models valid?
The quick answer to that question is: The flux must be independent of height, since this is the condition used in the resistance-
model formulation (e.g., see Eq. 121). However, that answer is
not very useful, except that it leads to the rephrased question:
When is it an "adequate" approximation to assume that the flux is
independent of height?

     To answer the last question, look again at the continuity
equation

$$\partial C/\partial t = -\nabla \cdot F + G - \mathcal{L} \quad .$$

(141)

For now, let's ignore the (volumetric) gain and loss terms, and
then perform the usual averaging over "rapid" fluctuations. Thus,
let $C = \tilde{C} + C'$, etc., and in the horizontal flux, $(u \, \hat{i} + v \, \hat{j}) \, C$,
let's assume there's no "cross-wind" variation (i.e., take
$\partial/\partial y = 0$). Then (141) becomes

$$\frac{\partial \tilde{C}}{\partial t} + \frac{\partial}{\partial x} \left[\tilde{u}\tilde{C} + \widetilde{u'C'} \right] = -\frac{\partial}{\partial z} [\tilde{F}_z] \quad .$$

(142)

In this form, the continuity equation is called the convective-
diffusion equation (mostly by chemical engineers) or the advection-
diffusion equation (mostly by meteorologists). Now, average (142)
over some height h (below which we hope to have the z-component
of the flux independent of height), using the definition for the
mean concentration over the height h:

$$\langle \tilde{C} \rangle = \frac{1}{h} \int_o^h \tilde{C} dz \quad ,$$

(143)

and similarly for other mean values. The result is

$$\frac{\partial \langle \widetilde{C} \rangle}{\partial t} + \frac{\partial}{\partial x} [\langle \widetilde{u}\widetilde{C} \rangle + \langle \widetilde{u'C'} \rangle] = - \frac{1}{h} [\widetilde{F}_{z,h} - \widetilde{F}_{z,i}] \quad , \qquad (144)$$

where I've identified the constant, resulting from integrating the RHS of (142), as $F_z(z = z_i) = F_{z,i}$ because if $h \to 0$, then (with the b.c. $C(z_i) = 0$) the LHS becomes zero, and therefore, so must the RHS. Now, if the flux is independent of height (in a "constant-flux layer" of height h), then the RHS is always zero in this layer. Therefore, the question: When is the LHS of (144) zero (or, at least, small enough to be negligible)?

To answer the last question, nondimensionalize (144) (because we need some "standard" against which to judge "small enough"). Nondimensionalize concentrations with $\widetilde{C}_h$, velocities with some characteristic speed U, lengths with some L (see later for what L might be), and time with L/U. Then, using superscript zero to identify nondimensional variables (I'm running out of symbols!), (144) becomes

$$\frac{hU}{L} \left\{ \frac{\partial \langle \widetilde{C} \rangle^0}{\partial t^0} + \frac{\partial}{\partial x^0} [\langle \widetilde{u}\widetilde{C} \rangle^0 + \langle \widetilde{u'C'} \rangle^0] \right\} = - \frac{1}{\widetilde{C}_h} [\widetilde{F}_{z,h} - \widetilde{F}_{z,i}] \quad . \quad (145)$$

But the RHS of (145), proposed to be zero in a constant flux layer, is otherwise the difference between "deposition velocities": (flux ÷ conc) at z = h and at $z = z_i$. Therefore, when we want the LHS of (145) to be "small", we mean it is to be small relative to the deposition velocity. To emphasize this, let me rewrite (145), nondimensionalizing the RHS with, say, $\sigma C_D U$, where $C_D U$ is essentially the deposition velocity for momentum, and $\sigma < 1$, would contain all the dependencies on Stokes number, Schmidt number, etc. Let's forget about the case of dry deposition of "boulders" (with large gravitational settling). Thus, for the LHS of (144) or (145) to be "small", then

$$\frac{hU}{L\sigma C_D U} \left\{ \cdots \right\} = - \left[\frac{\widetilde{F}_{z,h} - \widetilde{F}_{z,i}}{\widetilde{C}_h} \right]^0 \ll 1 \quad , \qquad (146)$$

where I didn't rewrite the terms in the $\left\{ \cdots \right\}$ of (145).

So where are we? If everything has been nondimensionalized reasonably (and complaints could be lodged that we've overestimated the x-diffusion, $\widetilde{u'C'}$, by using $U\widetilde{C}_h$, but an overestimate won't hurt this analysis), then all terms in the { } are of-the-order-of unity. Consequently, the condition for the flux to be

"essentially" independent of height, so that the RHS of (146) is
negligibly "small" (re unity), is that

$$h/(\sigma C_D L) \ll 1 \quad \text{or} \quad h \ll \sigma C_D L \ . \tag{147}$$

Fine, but what is L? If you would look again at (142), you
can see that what we have done, basically, is compare the divergence of the horizontal flux $\partial(uC)/\partial x$, plus time variations,
with the divergence of the vertical flux; therefore, L is the
characteristic length scale for horizontal changes. Consequently,
our result is that, for example, if you are $L = 10^4$ m = 10 km
downwind of the source of particles, and if their deposition
velocity is as large as for momentum (!)($\sigma = 1$; $v_d \simeq C_D U$), then
for $C_D \sim 10^{-3}$, the vertical flux can be treated as essentially
constant over a height, adjacent to the surface, h $\ll \sigma C_D L = 10^{-3}$
(10^4 m) = 10 meters! Look at that result: 10 km's downwind,
$v_d = v_{d,m}$ and a constant-flux layer very much smaller than ten
meters! Now it's true that this analysis has been fairly crude
(for improvements see Slinn, 1983b), and it turns out that the
condition can be weakened some (if K-theory is used, and taking
K = constant). But the essence remains; and for experimentalists
to be on the safe side, they should satisfy L $\gg h/(\sigma C_D)$. And
if they want a 10-m constant-flux layer for particles, with $v_d \sim$
$10^{-2} C_D U$ (i.e., $\sigma = 10^{-2}$; see Figure 2), then they should be
downwind from the source a distance

$$L \gg h/(\sigma C_D) \sim 10 \text{ m}/ 10^{-5} \sim 10^3 \text{ km} \ ! \tag{148}$$

This means that experimentalists must do their constant-flux measurements in mid-ocean, and arrange to have present no other
sources, e.g., from the sea!

And people wonder why "eddy-correlation" measurements of particle (and gaseous) fluxes (i.e., measurements of $\widehat{w'C'}$), measured
on shorelines, piers (or in fields, with C_D larger, but only a
few tens of kilometers from sources) are giving strange results!
The least they could do is measure the fluxes at more than one
height to see that it's not constant, and to see that what they're
measuring is not deposition (and equally-large resuspension!), but
variations in the horizontal flux. (But usually their measurement
sensitivity is not adequate to detect a v_d of, say, 10^{-2} cm s^{-1}
anyway -- and then to take the difference of two small numbers ... !) Thus, as far as I'm concerned, all reported eddy-
flux measurements attempting to measure v_d values expected to
be smaller than their instrument sensitivity, and failing to
satisfy the condition L $\gg h/(\sigma C_D)$ have given useless results,
nothing to do with v_d. Later, I'll return to problems (and maybe

a solution!) concerned with measuring particle deposition velocities, but first I want to return to the theory -- to show you
that it has severe problems of its own.

6.4 Resistance Models

Suppose we do satisfy the condition that the flux is independent of height. Then can we use resistance models? The answer
is yes -- but! And the but is: resistance models for particles
don't have the same form as those for gases; in fact, they don't
have a "resistance-analogy" form at all (flux ~ driving force/
resistance), and it's therefore rather ridiculous to call them
"resistance models." Moreover, I think the "theory" is essentially useless, for reasons I'll explain as soon as I review it.

To do this, start from (104), and assume constant flux:

$$\text{Const} = -\widetilde{F}_z = \widetilde{v}_{d,h}\widetilde{C}_h = -\widetilde{w_f'C'} - \widetilde{w_s'C'} + \mathcal{D}\,\partial\widetilde{C}/\partial z + (v_g - \widetilde{w}_s)\widetilde{C} \;. \qquad (149)$$

Also, suppose (only for a little while, to show you what others
do!) that it's acceptable to use turbulent and particle diffusivities as in (105)-(107). Then it is easy to integrate

$$\left[K(z) + \mathcal{H}_p + \mathcal{D}\right]\frac{\partial\widetilde{C}}{\partial z} + \left[v_g - \widetilde{w}_s\right]\widetilde{C} = \text{const} = \widetilde{v}_{d,h}\widetilde{C}_h \;. \qquad (150)$$

The result (Slinn and Slinn, 1981) can be written in the form

$$\frac{1}{\widetilde{v}_{d,h}} \equiv \frac{\widetilde{C}_h}{-\widetilde{F}_z} = \frac{1 - \exp\{-(v_g - \widetilde{w}_s)\Sigma' r_i\}}{(v_g - \widetilde{w}_s)} \;, \qquad (151)$$

where the sum of the "resistances" is

$$\Sigma r_i = \int_0^{C_h} [-\widetilde{F}_z - (v_g - \widetilde{w}_s)\widetilde{C}]^{-1}\, d\widetilde{C} \;, \qquad (152)$$

and individual, i'th-layer resistances can be identified by breaking the total integral into sums of integrals over individual
layers. For example, if there is a layer in which turbulent
transfer dominates, then with

$$[-\widetilde{F}_z - (v_g - \widetilde{w}_s)\widetilde{C}]^{-1} \simeq [K\, d\widetilde{C}/dz]^{-1} \simeq [\kappa u_* z\, d\widetilde{C}/dz]^{-1} \;, \qquad (153)$$

$$r_i = \int_{z_i}^{z_{i+1}} \frac{dz}{\kappa u_* z} = \frac{1}{\kappa u_*} \ln \left[\frac{z_{i+1}}{z_i} \right] = \frac{\tilde{u}(z_{i+1}) - \tilde{u}(z_i)}{u_*^2} \quad (154)$$

in which, consistent with this $K(z)$, a logarithmic wind profile, $\tilde{u}(z) = (u_*/\kappa)\ln(z/z_o)$ has been assumed.

But the points I want to make are in the other direction from standard results such as (154). One point is that, when gravity and other slip velocities are present, then the general result, (151), is not of the form (used by so many) that the "overall resistance $(\tilde{v}_d^{-1})$ is the sum of the resistances in individual layers." This frequently assumed form [e.g., essentially used here for (119) and (120)] follows from (151) only if $(v_g - \tilde{w}_s)$, say ϵ, is negligibly small re $(\sum r_i)^{-1}$:

$$\frac{1}{\tilde{v}_{d,h}} = \frac{1}{\epsilon} [1 - \exp\{-\epsilon \sum r_i\}] = \frac{1}{\epsilon} \left[1 - \{1 - (\epsilon \sum r_i) + \cdots \right.$$

$$= \sum r_i \left\{ 1 + \mathcal{O}(\epsilon \sum r_i) \right\} . \quad (155)$$

Otherwise, for particles whose gravitational settling cannot be ignored, it is quite incorrect to take $v_d^{-1} = \sum r_i$.

But of more significance is this. Even if a practical case can be found when a constant-flux layer exists, there seems to be little practical advantage to this "resistance-model" formulation, because it contains no new science. It is a somewhat-circuitous construction that, stripped to its frame, amounts to this: (i) assume flux = constant; (ii) assume special forms for $\widehat{w_f'C'}$ and $\widehat{w_s'C'}$; then, (iii) $\tilde{C}$ is related linearly to this constant flux and can be found from (150), and therefore, (iv) a consistent $\tilde{v}_d$ that ensures flux = constant is given by Flux/$\tilde{C}(z)$, in which the flux cancels. In other words, the flux was never found, and cannot be found unless in (152) we know the flux! To break this circuitous development, new science must be introduced, e.g., to say that the flux in one layer is as given in (153). True, it can be argued that this development, e.g., (150), yields a convenient framework into which the additional science can be placed; but it is a constraining framework, constrained, for example, by the assumptions about forms for $\widehat{w_f'C'}$ and the need to specify the layer heights, z_i, as in (154). In contrast, it seems preferable to relate particle dry deposition to momentum deposition, $C_D\tilde{u}$, and try to generalize this form for the case of particles, e.g.,

$$\tilde{v}_d = C_{D,h}\tilde{u}_h [\alpha f_1(Sc) + \beta f_2(S) + \gamma f_3(I)] + \cdots + (v_g - \tilde{w}_s) , \quad (156)$$

where α, β, ... are also to account for the fraction of the momentum "deposited" via viscous versus form drag; and I think that, for a long time to come, heavy reliance will need to be placed on data. And in that vein, let me now turn to mention other ways that good data might be obtained, especially since I have serious doubts that, to date, we have any data to which can be applied the adjective "good."

6.5 Horizontal and Vertical Concentration Gradients

Some attempts to deduce v_d for particles (Lodge, 1978; Delumyea and Petel, 1979) have been based on measurements of particle-concentration decreases as the air travelled across water bodies. The theoretical basis for this method has been the simple "box model" used in Section 2.3, using a "mixed layer" of fixed height H:

$$\frac{d}{dx}\langle\tilde{C}\rangle = -\frac{\tilde{v}_d}{\tilde{u}H}\langle\tilde{C}\rangle \;\Rightarrow\; \langle\tilde{C}\rangle = \langle\tilde{C}_o\rangle \exp\left\{-x/(\tilde{u}H/\tilde{v}_d)\right\} \quad , \qquad (157)$$

which gives the e-fold dry-deposition length scale $\lambda_d = \tilde{u}H/\tilde{v}_d$. For example, if $\tilde{v}_d$ is as large as $C_D\tilde{u}$ and if H = 1 km, then $\lambda_d = 10^3$ km.

But reality isn't that simple! In reality, the mixed layer changes in height, e.g., during the course of the day (as the solar radiation changes), or as the air moves across the water (a distance greater than 10^3 km!) and the "information" (that the air is now flowing over water) diffuses to greater heights. Therefore, if H is recognized as a potential variable, a revised box-model is easily seen to yield, instead of (157),

$$\frac{d\langle\tilde{C}\rangle}{dx} = -\left[\frac{\tilde{v}_d}{\tilde{u}} + \frac{dH}{dx}\right]\frac{\langle\tilde{C}\rangle}{H} \quad . \qquad (158)$$

Look at the magnitude of the new term. If $\tilde{v}_d/\tilde{u}$ is as large as for momentum (i.e., $\tilde{v}_d/\tilde{u} = C_D \simeq 10^{-3}$), then if dH/dx is larger than 10^{-3}, the changing mixed-layer height will dominate the change in $\langle\tilde{C}\rangle$. But a dH/dx of 10^{-3} is only 1 m per kilometer (and if $\tilde{v}_d \simeq 10^{-2}C_D\tilde{u}$ then we're talking about a dH/dx of 1 m per 100 km!). You can see, therefore, why I am extremely skeptical that horizontal concentration-gradient methods can be used to obtain good data for v_d.

Another technique is to measure vertical concentration gradients, although I don't know if anyone has tried this, except over land. The idea behind this method is to measure the vertical gradient [or C(z)], where turbulence dominates the vertical transport, and take the vertical flux to be given by

$$-\tilde{F}_z = K\partial\tilde{C}/\partial z = \tilde{v}_{d,h}\tilde{C}_h \quad . \tag{159}$$

An independent "measurement" is made of K (e.g., by measuring the velocity profile and the momentum flux $\widehat{u'w'}$, and using $-\widehat{u'w'} = \rho_a K\partial\tilde{u}/\partial z$). But this assumes that the particle flux is independent of height, and I already commented on the near impossibility of satisfying that condition if $\tilde{v}_d \ll C_D\tilde{u}$. Now maybe it is understandable why, whenever someone tells me about a new measurement of $\tilde{v}_d$ by the concentration-gradient method, I immediately begin to wish they'd back off!

6.6 Actual Deposition and Mass-Mean Values

There is an obvious way to deduce v_d, via measurements of actual deposition, and though I think that this is the only reasonable way to proceed at the present time, still the method has a number of very serious limitations. One limitation is caused by the type of sampler used: for example, how is particle deposition to a bucket or to a (charged?) polyethylene sheet, or to ... related to particle deposition to the oceans (or other environmental surfaces)? If gravitational settling dominates, then the answer is obvious; but if it does, why is so much effort expended? Is there some doubt about the results obtained by Newton and Stokes?

Another potentially-serious problem is that, if what is wanted is the mass deposited, then deposition of a few large particles of a polydisperse aerosol can overwhelm the deposition. Thus, if the particles are spheres of radius a, mass $\sim a^3$, and even with just gravitational settling with $v_g \sim a^2$ (Stokes), then the mass deposited $\sim a^5$. That is, compared to a 1 μm particle, a single 10 μm particle will result in the deposition of 10^5 more mass! Put differently, if there is only one 10 μm particle for every 10^4, 1 μm particles, still the single 10 μm particle will dominate the total mass deposited. And if inertial impaction is important, then v_d can be even more strongly dependent on particle radius (e.g., $v_d \sim 10^{-3/S}$, with $S \sim a^2$). To overcome such problems, if ever we are to determine the particle-size dependence of v_d, I see no way to proceed (until better techniques are developed for releasing large quantities of monodisperse aerosols), other than by counting (not "weighing") individual particles deposited (e.g., in the sea). I wish someone would accept the challenge and perform such experiments.

Although it was probably obvious that a few-large particles can dominate the dry-depositon mass flux for a polydisperse aerosol, I have a reason for wanting to restate it, this time analytically. Thus, suppose $v_d = ca^2$, and suppose that the particle number distribution, $f(a)$, is lognormal, with geometric mean radius a_g and standard deviation σ_g. Then the mass-mean deposition velocity is just

$$\langle^3 v_d \rangle = \int_0^\infty ca^2 \frac{4}{3}\pi a^3 \rho_p f(a)da \bigg/ \int_0^\infty \frac{4}{3}\pi a^3 \rho_p f(a)da$$

$$= \exp\left\{8(\ln\sigma_g)^2\right\}v_d(a_g) = \exp\left\{2(\ln\sigma_g)^2\right\}v_d(\text{mmr}) \ , \quad (160)$$

where I've used available integrals of the lognormal distribution (e.g., Aitchison and Brown, 1969), and "mmr" stands for mass-mean radius. As an example of (160), if $v_d(\text{mmr}) = 0.1$ cm s^{-1} and $\sigma_g = 3$, then the mass-mean v_d, $\langle^3 v_d \rangle = 1.1$ cm s^{-1}. In other words, the few large particles, in the "long-tail" of the lognormal, overwhelm the value for the mass-mean v_d.

But in spite of our desire to know $v_d(a)$, let's not lose sight of the obvious. What is usually desired is the mass-mean v_d (e.g., in studies of the long-range transport of air pollution and in studies of the global budgets of atmospheric trace constituents). However, there's another obvious point that should not be overlooked (but almost invariably is): since v_d depends on particle size, then the particle size distribution must shift with time (boulders fall out first!), and therefore, the mass-mean v_d will also shift with time. Thus, if someone asks you for the mass-mean v_d for, say Pb, I think the best answer is to say "Right!" -- not a British expeditious "Right!" -- slur it as they do in the Southern United States: "Ri-i-i-ght!"

Or answer them in Gibbs' language -- with some simple mathematics. Thus, suppose that, initially, the particle number distribution were the simple Junge distribution (which must fail for large and small particles!)

$$f(a;x=0) = c_1(a_0/a)^3 \ , \qquad (161)$$

and suppose that $v_d = c_2 a^1$. Then, if a simple box model (fixed mixing-height H) is used, the <u>number</u> of particles of a specific size would decrease with downwind distance, x, via

$$f(a;x) = f(a;0) \exp\left\{-x/\lambda_d\right\} \ , \quad \lambda_d = \bar{u}H/\bar{v}_d(a) \ . \quad (162)$$

Then the mass-mean v_d changes with x,

$$\langle ^3\tilde{v}_d(a)\rangle = \frac{\displaystyle\int_o^\infty c_2 a\, \frac{4}{3}\, \pi a^3 \rho_p c_1 (a_0/a)^3 \exp\left\{\frac{-c_2 ax}{\tilde{u}H}\right\} da}{\displaystyle\int_o^\infty \frac{4}{3}\, \pi a^3 \rho_p c_1 (a_0/a)^3 \exp\left\{\frac{-c_2 ax}{\tilde{u}H}\right\} da} = \frac{\tilde{u}H}{x} \qquad (163)$$

(big boulders do fall out first!); or, alternatively, the total
mass remaining airborne is

$$M(x) = \int_o^\infty \frac{4}{3}\, \pi a^3 \rho_p f(a;x) da = M(o)\left[\tilde{u}H/\{\tilde{v}_d(a_0)x\}\right] . \qquad (164)$$

This last result shows [for the assumed $f(a;o)$ and $v_d(a)$] that
the mass falls <u>not</u> exponentially with downwind distance, but much
more slowly: as x^{-1}. This result brings us back to the general
question posed at the start of these lectures: What is the tropospheric residence time of particles? In fact, with the result
(164), we must be careful about the meaning of the words "residence time": with $M(x)$ as in (164), there is no "e-fold time";
the airborne mass does not decrease exponentially with time.

6.7 Residence Times, Revisited

It would be disappointing to travel so far and return with
less than you started, only if you didn't enjoy the trip. And
besides, it's not that bad -- (162) does have an e-fold residence
time (or length scale); it is only in (164), when we averaged over
the mass-distribution, that we obtained a "linear-fold" time.
Also, that result was for dry deposition, not wet disposition,
which leads back to the question: What is the relative importance
of wet and dry removal?

That's a difficult question. The relative importance of wet
and dry removal depends sensitively on meteorological properties,
and on characteristics of the "pollutant." But, if the air concentrations for wet and dry removal are comparable (or better,
if $\bar{h}_w \simeq \bar{h}_d$), if the measured scavenging ratios are appropriate,
if the dry deposition velocites to the ocean were as shown in
Figure 2, (and if .....), then one way to compare the two removal
processes is to compare the two deposition velocities, v_d and
$v_w = ps_r$, appropriately accounting for wet and dry periods.

Rodhe (1980) pointed out that this is rarely done "appropriately"; and I agree with him that simply using the annual-average
precipitation rate to find an annual-average v_w, $\langle v_w \rangle$, is a poor
approximation near a pollution source. As far as I can see, the
basic reason for this is that there will be more pollution available for wet removal, downwind of a source, after a dry period

(and with no dry deposition), than if it rains continuously at the annual-average rate. However, that depends on the identification of a particular source; and if, instead, we think of the case of a uniform "smear" of pollution (e.g., over the oceans), then to compare wet and dry removal, I think we can simply compare $\langle v_d \rangle$ and $\langle v_w \rangle = \langle ps_r \rangle$, where $\langle p \rangle$ is the annual-average precipitation rate.

Such a comparison was already shown in Figure 4, using $\langle p \rangle = 80$ cm yr^{-1}, the scavenging ratios for rain from Gatz and Cawse, for snow from Rahn (1979), and the theoretical v_d curve [v_d(a)!] from Figure 2. My conclusion (especially if I were to use a mass-average v_d or, as a first approximation, use v_d(a) values for each element, using the largest of its mmd's reviewed by Rahn, 1976), is that wet and dry removal can be of roughly similar magnitude, with "outriders" such as Ca and Al scavenged by snow, probably caused as much by "anomalous" vertical distributions, during specific measurement periods, as by anything else.

But there are a lot of other things "elsing." For example, that simple comparison overlooks the ideas, examined in the last subsection, that v_d shifts to smaller values as the aerosol "ages;" i.e., as the large particles are removed. Thus, the mass-mean v_d, $\langle {}^3v_d \rangle$, for Pb may be 1 cm s^{-1} off the east coast of North America, but then only 0.1 cm s^{-1} over the mid-Atlantic, and 0.01 cm s^{-1} in the Antarctic, whereas Pb's v_w (or s) may hold relatively constant. Also, the simple model of dry deposition from a well-mixed layer of height h_d, leading to $\bar{\tau}_d = \bar{h}_d / \bar{v}_d$, overlooks the very real and important "trapping" of pollution aloft, essentially out of contact with the surface layer (except for gravitational settling). This trapping is why dust from the Sahara, for example, can reach Barbados (e.g., Prospero and Nees, 1977): the hot air from the desert, blown in the trade winds, will ride over the (relatively cool) maritime air, all the way across the Atlantic; and similarly for pollution traveling from North America to Europe, especially during the summer (warm air from the continent traveling over relatively-cool, stable air over the Atlantic).

And there is more. Many times dry deposition (i.e., down) is less important than dry "ascension" (up!). Thus, as sketched in Figure 38, pollution will diffuse vertically after its release. If K-theory can be used, then after time t the pollution will diffuse to a characteristic height $(Kt)^{1/2}$; that is, there is a characteristic "ascension velocity" $v_a = 1/2(K/t)^{1/2}$. For example, if $K = 4 \times 10^5$ cm s^{-1}, then after 10^5s (~1 day) down-wind, the ascension velocity $v_a \sim 1$ cm s^{-1}, essentially independent of particle size (except for v_g), and as large as (or larger than) typical deposition velocities. If this idea of ascension is coupled with trapping above a (colder) layer, and with $\langle {}^3v_d \rangle$

decreasing with time (and therefore that the mass-average τ_d is
not an e-fold time scale), then it is easy to see why I <u>expect</u>
that dry deposition to the oceans, except near their shores, is
<u>not</u> so important as wet deposition. Pushed to the limit, as I am
on time, I would say that I expect that for most atmospheric particles whose global-scale distribution is of interest, then $\bar{\tau}_d \sim$
1 month and $\bar{\tau}_w \sim$ 1 week.

But that again ignores the particle-size dependence of the
removal processes. To estimate the particle-size-dependent resi-
dence times, assume that the height scales $\bar{h}_d$ and $\bar{h}_w$, in $\bar{\tau}_d =$
$\bar{h}_d/\bar{v}_d(a)$ and $\bar{\tau}_w = \bar{h}_w/\bar{v}_w(a)$, are the same and have a value such
that $\bar{\tau}^{-1} = \bar{\tau}_d^{-1} + \bar{\tau}_w^{-1}$ is about a week, for particles that are
active as cloud condensation nuclei (CCN). Then $\bar{\tau}_d(a)$ and $\bar{\tau}_w(a)$
can be found simply by plotting $\bar{v}_d(a)$ and $\bar{v}_w(a)$, and then by looking at the mirror image (see Figure 39). These plots are crude:
the dashed $\bar{v}_d(a)$ curves suggest we don't know the influence of
particle growth, breaking waves, etc., and the dashed $\bar{v}_w(a)$
curves suggest that, in clean air over the oceans, clouds may be
forced to use smaller CCN than those used over the continents.
In spite of these uncertainties, though, I think the general picture is correct; it shows that wet removal dominates the tropospheric residence time of most particles, except for "boulders"
(a $\gtrsim 10$ μm), dominated by gravity, and for very small particles
(a $\lesssim 10^{-2}$ μm), which coagulate with larger particles (Slinn,
1983b).

But also, please remember that those are expected <u>average</u>
values! For precipitation scavenging, please remember the analysis in Section 2.7, leading to the "long-tailed" lognormal distribution. Further, that analysis was not complete (it will be
at least a decade before it can be done decently). We used $\bar{\varepsilon}$ for
the average fraction of the pollution scavenged by an "average
storm," and $\bar{\vartheta}$, the average time (in a Lagrangian sense) between
"average storms." But, as you can see while travelling on commer-
cial aircraft, skimming through clouds at $\sim$10 km or with Cb's
towering above you, some storms can penetrate into the stratosphere, carrying pollution there, and then leading (presumably)
to a very long time before the pollution next encounters a storm.
This may be the cause, for example, of the recent findings by
Barrie et al. (1981) of (U.S.) ragweed pollen in the Arctic, during the spring, before (after!) the (summertime) ragweed-pollen
season in the U.S. Thus, as sketched in Figure 40, the pollen
may have been carried to the stratosphere by a deep summer storm,
traveled in the jet stream for awhile, diffused north (and south)
during the winter, and then descended during the spring with the
general subsidence in the Arctic. In this case, for particles as
large as ragweed pollen ($\sim$10 μm), we're talking about a residence
time of about 9 months! I could say more, but it would be just
more babble.

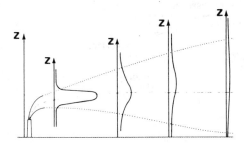

Figure 38. A schematic to illustrate the concept of dry "ascension". (From Slinn, 1982b.)

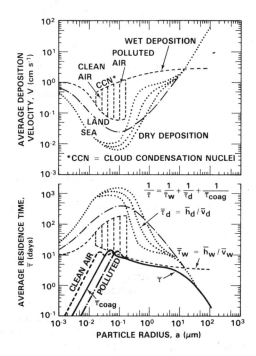

Figure 39. A qualitative indication of how the tropospheric residence time of particles depends on their size; coagulation was estimated as in Junge, 1963. (From Slinn, 1983b.)

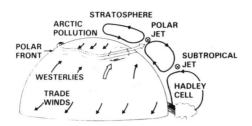

Figure 40. Indication of one way for pollution to reach the Arctic: via injection into the stratosphere by intense, mid-latitude storms. (From Slinn, 1982b.)

ACKNOWLEDGMENTS

I wish to thank a number of attendees at the NATO ASI for their helpful comments on an earlier version of these notes: T. Bidleman, D. Gillette, J. Pankow, L. Peters, A. Pszenny, J. Schmidt, and especially L. Hasse and P. Liss. My studies on air-sea exchange have been funded by U.S. taxpayers via the U.S. Department of Energy, and my productivity would be less if it were not for the substantial administrative help from R. W. Beadle, D. S. Ballantine, and C. E. Elderkin.

GENERAL REFERENCES

Cole, F. W., 1975: Introduction to Meteorology. Wiley, New York, 495 pp.

Fuchs, N. A., 1964: The Mechanics of Aerosols. Pergamon, New York, 408 pp.

Junge, C. E., 1963: Air Chemistry and Radioactivity. Academic Press, New York, 382 pp.

Morse, P. M. and H. Feshbach, 1953: Methods of Theoretical Physics. McGraw-Hill, New York, 1978 pp.

Papoulis, A., 1965: Probability, Random Variables, and Stochastic Processes. McGraw-Hill, New York, 583 pp.

Pruppacher, H. R., R. G. Semonin, and W. G. N. Slinn (eds.), 1983: Precipitation Scavenging, Dry Deposition, and Resuspension (1982). Elsevier, New York, in press.

REFERENCES

Aitchison, J. and J. A. C. Brown, 1976: The Lognormal Distribution. Cambridge Univ. Press, New York, 176 pp.

Baker, M. B., H. Harrison, J. Vinelli and K. B. Erickson, 1979: Simple stochastic models for the sources and sinks of two aerosol types. Tellus, 31, 39-51.

Barrie, L. A., R. M. Hoff and S. M. Daggupaty, 1981: The influ-
    ence of mid-latitudinal pollution sources on haze in the
    Canadian Arctic. Atmos. Env., 15, 1407-1420.
Beard, K. V. and S. N. Grover, 1974: Numerical collision effi-
    ciencies for small raindrops colliding with micron size
    particles. J. Atmos. Sci., 31, 543-550.
Bolin, B. and H. Rodhe, 1973: A note on the concepts of age
    distribution and transit time in natural reservoirs.
    Tellus, 25, 58-62.
Burtsev, I. I., L. V. Burtseva, and S. G. Malakhov, 1970: Washout
    characteristics of the ^{32}P aerosol injected into a cloud.
    Atmospheric Scavenging of Radioisotopes (B. Styra, Ch. A.
    Garbalyauskas, and V. Yu Luyanas, eds.), available as
    TT-69-55099 from the Nat. Tech. Info. Center, Springfield,
    VA, 242-250.
Cantwell, B. J., 1981: Organized motion in turbulent flow. Ann.
    Rev. Fluid Mech., 13, 457-515.
Caporaloni, M., F. Tampieri, F. Trombeti, and O. Vittori, 1975:
    Transfer of particles in nonisotropic air turbulence.
    J. Atmos. Sci., 32, 565-569.
Cawse, P. A., 1974: A Survey of Atmospheric Trace Elements in
    the United Kingdom, AERE Harwell Rept. R 7669, HMSO, London.
Chamberlain, A. C., 1953: Aspects of Travel and Deposition of
    Aerosols and Vapor Clouds, AERE Harwell Rept. R 1261, HMSO,
    London.
Dana, M. T., 1970: Scavenging of soluble dye particles by rain.
    Precipitation Scavenging (1970) (R. J. Engelmann and
    W. G. N. Slinn, coords.), U.S. DOE Tech. Info. Center, Oak
    Ridge, TN and available as CONF-700601 from NTIS,
    Springfield, VA., 137-147.
Dana, M. T. and J. M. Hales, 1976: Statistical aspects of the
    washout of polydisperse aerosols. Atmos. Env., 10, 45-50.
Danckwerts, P. V., 1970: Gas-Liquid Reactions. McGraw-Hill, New
    York, 276 pp.
Davies, C. N., 1966: Deposition from moving aerosols. Aerosol
    Science (C. N. Davies, ed.), Academic Press, New York,
    393-446.
Delumyea, R. and R. L. Petel, 1979: Deposition velocity of
    phosphorus-containing particles over Southern Lake Huron,
    April-October, 1975. Atmos. Env., 13, 287-294.
Dirks, R. A., 1972: The natural efficiency of orographic precipi-
    tation. Preprints Third Nat. Conf. on Weather Modification,
    Rapid City, S.D., Am. Meteorol. Soc., Boston, Mass.,
    pp 96-99.
Elliott, R. D. and E. L. Hovind, 1964: The water balance of
    orographic clouds. J. Appl. Meteorol., 3, 235-239.
Fitzgerald, J. W., 1975: Approximate formulas for the equilibrium
    size of an aerosol particle as a function of its dry size
    and composition and the ambient relative humidity. J. Appl.
    Meteorol., 14, 1044-1049.

Foote, G. B. and J. C. Fankhauser, 1973: Airflow and moisture budget beneath a Northeast Colorado hailstorm. J. Appl. Meteorol., 12, 1330-1353.

Friedlander, S. K. and H. F. Johnstone, 1957: Deposition of suspended particles from turbulent gas streams. Ind. and Engr. Chem., 49, 1151-1159.

Gatz, D. F., 1966: Deposition of Atmospheric Particulate Matter by Convective Storms, PhD Thesis and USAEC Rept. COO-1407-6, University of Michigan, Ann Arbor, MI.

Gatz, D. F., 1975: Pollutant aerosol deposition into Southern Lake Michigan. J. Water, Air and Soil Poll., 5, 239-251.

Gatz, D. F., 1976: Wet deposition estimates using scavenging ratios. Proc. First Speciality Conf. on Atmospheric Contributions to the Chemistry of Lake Water, 28 Sept.-1 Oct. 1975, Int. Assoc. for Great Lakes Research, 21-29.

Gatz, D. F., 1977: Scavenging ratio measurements in METROMEX. Precipitation Scavenging (1974) (R. G. Semonin and R. W. Beadle, coords.), U.S. DOE Tech. Info. Center, Oak Ridge, TN, and available as CONF-741003 from NTIS, Springfield, VA, 71-87.

Gedeonov, L. I., Z. G. Gritchenko, F. M. Flegontov and M. I. Zhilkina, 1970: Coefficient of radioactive-aerosol concentration in atmospheric precipitation. Ibid., Burtsev et al. (1970), 150-156.

Gibbs, A. G. and W. G. N. Slinn, 1973: Fluctuations in trace gas concentrations in the troposphere. J. Geophys. Res., 78, 574-576.

Hamrud, M., H. Rodhe, and J. Grandell, 1981: A numerical comparison between Lagrangian and Eulerian statistics. Tellus, 33, 235-241.

Hicks, B. B. and W. G. N. Slinn, 1973: Surface fluxes of small particles. Fine Particles in the Atmosphere (A. P. Altshuller, ed.), Ann Arbor Science Publishers, Ann Arbor, MI., in press.

Hobbs, P. V., et al., 1980: The mesoscale and microscale structure and organization of clouds and precipitation in midlatitude cyclones. I: A case study of a cold front. J. Atmos. Sci., 37, 568-596.

Illori, T. A., 1971: Turbulent deposition of aerosol particles inside pipes. PhD Thesis, U. of Minnesota, Minneapolis, 146 pp.

Junge, C. E., 1974: Residence time and variability of tropospheric trace gases. Tellus, 26, 477-488.

King, C. J., 1966: Turbulent liquid phase mass transfer at a free gas-liquid interface. Ind. and Eng. Chem. Fund., 5, 1-8.

Liu, B. Y. H. and J. K. Agarwal, 1974: Experimental observation of aerosol deposition in turbulent flow. Aerosol Sci., 5, 145-155.

Lodge, J. P., Jr., 1978: An estimate of deposition velocities over water. Atmos. Env., 12, 973-974.

Makhon'ko, K. P., A. S. Avramenko and E. P. Makhon'ko, 1970: Washout of radioactive isotopes and chemical compounds from the atmosphere. Ibid. Burtsev et al. (1970), 174-184.

Marwitz, J. D., 1972: Precipitation efficiency of thunderstorms on the High Plains. Ibid. Dirks (1972), 245-247.

Marwitz, J. D., 1974: An airflow case study over the San Juan Mountains of Colorado. J. Appl. Meteorol., 13, 450-458.

Mason, B. J., 1971: The Physics of Clouds, 2nd Ed. Clarendon Press, Oxford, 659 pp.

Moller, U. and G. Schumann, 1970: Mechanisms of transport from the atmosphere to the earth's surface. J. Geophys. Res., 75, 3013-3019.

Muhlbaier, J., 1978: The Chemistry of Precipitation Near the Chalk Point Power Plant. PhD Thesis, U. of Maryland, College Park, MD, 276 pp.

Newton, C. W., 1966: Circulations in a large sheared cumulonimbus. Tellus, 18, 699-713.

Owen, P. R., 1969: Pneumatic transport. J. Fluid Meh., 29, 407-432.

Prospero, J. M. and R. T. Nees, 1977: Dust concentration in the atmosphere of the Equatorial N. Atlantic: possible relationship to the Sahelian drought. Science, 196, 1196-1198.

Radke, L. F., M. W. Eltgroth and P. V. Hobbs, 1980: Precipitation scavenging of aerosol particles. J. Appl. Meteorol., 19, 715-722.

Rahn, K. A., 1976: The Chemical Composition of the Atmospheric Aerosol. Tech. Rept. of the Graduate School of Oceanography, U. of Rhode Island, Kingston, R.I., 265 pp.

Rahn, K. A., 1979: Personal communications.

Rodhe, H., 1980: Estimate of wet deposition of pollutants around a point source. Atmos. Env., 14, 1197-1199.

Rodhe, H. and J. Grandell, 1981: Estimates of characteristic times for precipitation scavenging. J. Atmos. Sci., 38, 370-386.

Rosinski, J., 1967: Insoluble particles in hail and rain. J. Appl. Meteorol., 6, 1066-1074.

Sehmel, G. A., 1971: Particle diffusivities and deposition velocities over a horizontal smooth surface. J. Colloid Interface Sci., 37, 891-906.

Sehmel, G. A., 1973: Particle eddy diffusivities and deposition velocities for isothermal flow and smooth surfaces. Aerosol Sci., 4, 125-138.

Sehmel, G. A. and S. L. Sutter, 1974: Particle deposition rates on a water surface as a function of particle diameter and air velocity. J. Rechs. Atmos., III, 911-918.

Slinn, S. A. and W. G. N. Slinn, 1980: Predictions for particle deposition on natural waters. Atmos. Env., 14, 1013-1016.

Slinn, S. A. and W. G. N. Slinn, 1981: Modeling of atmospheric particulate deposition to natural waters. Atmospheric Pollutants in Natural Waters (S. J. Eisenreich, ed.), Ann Arbor Sc. Pub., Ann Arbor, MI, 23-53.

Slinn, W. G. N., 1974: Analytical investigations of inertial
    deposition of small aerosol particles from laminar flows
    onto large obstacles, Parts A and B. PNL Ann. Rept. 1973 to
    USAEC-DBER, BNWL-1850, Pt 3, Battelle-Northwest, Richland,
    WA; also available from NITS, Springfield, VA, 121-132.
Slinn, W. G. N., 1975: Atmospheric aerosol particles in surface-
    level air. Atmos. Env., 9, 763-764.
Slinn, W. G. N., 1976: Dry deposition and resuspension of aerosol
    particles - a new look at some old problems. Atmosphere-
    Surface Exchange of Particulate and Gaseous Pollutants (1974)
    (R. J. Englemann and G. A. Sehmel, coords.), U.S. DOE Tech.
    Info. Center, Oak Ridge, TN; also available as CONF 740921
    from NTIS, Springfield, VA, 1-40.
Slinn, W. G. N., 1977a: Precipitation scavenging: some problems,
    approximate solutions, and suggestions for future research.
    Ibid. Gatz (1977), 1-45.
Slinn, W. G. N., 1977b: Some approximations for the wet and dry
    removal of particles and gases from the atmosphere. J.
    Water, Air, and Soil Poll., 7, 513-543.
Slinn, W. G. N., 1978: Some comments on parameterizations for
    resuspension and for wet and dry deposition of particles and
    gases for use in radiation dose calculations, Nucl. Safety,
    19, 205-219.
Slinn, W. G. N. 1980: Relationships between removal processes
    and residence times for atmospheric pollutants. AIChE
    Symposium Series, 196, Vol. 76, 185-203.
Slinn, W. G. N., 1982a: Predictions for particle deposition to
    vegetative canopies. Atmos. Env., 16, 1785-1794.
Slinn, W. G. N., 1982b: Estimates for the long-range transport
    of air pollution. J. Water, Air, and Soil Poll., 18, 45-64.
Slinn, W. G. N., 1983a: Precipitation Scavenging. Chapt. 11 of
    Atmospheric Sciences and Power Production (D. Randerson,
    ed.), U.S. DOE Tech. Info. Center, Oak Ridge, TN, in press.
Slinn, W. G. N., 1983b: A potpourri of deposition and resuspen-
    sion topics. Precipitation Scavenging, Dry Deposition, and
    Resuspension (1982), Vol. II (H. R. Pruppacher, R. G.
    Semonin, and W. G. N. Slinn, eds.), Elsevier, New York, in
    press.
Slinn, W. G. N. and A. G. Gibbs, 1971: The stochastic growth of
    a rain droplet. J. Atmos. Sci., 28, 973-982.
Slinn, W. G. N. and J. M. Hales, 1971: A reevaluation of the
    role of thermophoresis as a mechanism of in- and below-cloud
    scavenging. J Atmos. Sci., 28, 1465-1471.
Slinn, W. G. N., L. Hasse, B. B. Hicks, A. W. Hogan, D. Lal,
    P. S. Liss, K. O. Munnich, G. A. Sehmel and O. Vittori,
    1978: Some aspects of the transfer of atmospheric trace
    constituents past the air-sea interface. Atmos. Env., 12,
    2055-2087.
Slinn, W. G. N. and J. M. Hales, 1983: Wet Removal of Atmospheric
    Particles. Ibid. Hicks and Slinn (1983), in press.

Twomey, S., 1977: Atmospheric Aerosols. Elsevier, New York,
    302 pp.
Vali, G., 1977: Washout in High Plains Thunderstorms. Ibid. Gatz
    (1977), 494-502.
Vickers, D. and W. G. N. Slinn, 1979: Estimates for wet and dry
    removals' contributions to the residence time for atmospheric
    pollutants in the Eastern United States. SR-0980-6, May
    1979, Air Resources Center, Oregon State Univ., Corvallis,
    OR, U.S.A.; also available from NTIS, Springfield, VA.
Weiss, L. L., 1964: Sequences of wet or dry days described by a
    Mankov chain probability model. Monthly Weather Rev., 92,
    169-176.
Williams, A. L., 1977: Analysis of in-cloud scavenging effi-
    ciencies. Ibid. Gatz (1977), 258-275.
Williams, R. M., 1982: A model for the dry deposition of par-
    ticles to natural water surfaces. Atmos. Env., 16,
    1933-1938.

THE PRODUCTION, DISTRIBUTION, AND BACTERIAL ENRICHMENT
OF THE SEA-SALT AEROSOL

Duncan C. Blanchard

Atmospheric Sciences Research Center
State University of New York at Albany
Albany, N. Y. 12222 U.S.A.

1.   INTRODUCTION

It is often difficult to go back through time to find who
did the pioneer work in a given area of research. This is not
the case for the sea-salt aerosol. Throughout the early years of
this century a few papers on the subject appeared in the literature, but they contained little quantitative information. But
all this changed in the late 1940's with the appearance of the
first of several pioneer papers by Woodcock. His first paper
(Woodcock and Gifford, 1949) described in detail the technique of
obtaining the sea-salt particle size distribution. In a later
paper Woodcock (1952) presented his ideas on the role of the
giant sea-salt particles in the formation of raindrops, and in
1953 he published a paper that for the first time showed the salt
particle distribution as a function of wind speed. Though
numerous investigators have followed in Woodcock's footsteps, his
1953 paper is still relevant and extensively quoted today.

The early interest on the role of the sea-salt aerosol in
the formation of rain (Woodcock, 1952) has expanded to include
the role of salt particles in the evaporation of water from the
sea (Lai and Shemdin, 1974), the reduction of visibility
(Schacher et al., 1981), the harmful effects of sea salt on
ships' turbines (Ruskin et al., 1978) and the enrichment of
heavy metals (Duce and Hoffman, 1976; Weisel, 1981), radioactivity (Fraizier et al., 1977), viruses (Baylor et al., 1977a),
bacteria (Blanchard, 1978), and inert organic material (Hoffman
and Duce, 1977). This chapter is a brief review of the mechanisms of production of the sea-salt aerosol and its distribution
in the marine atmosphere. It also includes recent work on the

P. S. Liss and W. G. N. Slinn (eds.), Air-Sea Exchange of Gases and Particles, 407–454.
Copyright © 1983 by D. Reidel Publishing Company.

enrichment of bacteria in jet and film drops. I will make no
attempt to discuss the salt aerosol over the continents. In all
that follows I assume a quasi steady state between the atmosphere
and sea, with a long air trajectory over warmer water (typical of
the trade winds). Space limitations prevent discussion of many
fine studies, but I would be remiss by not mentioning the pioneering work of Eriksson (1959, 1960), Junge (1963), and Toba
(1966). A bibliography listing 607 papers on all aspects of the
salt aerosol was prepared by Brierly (1970). A recent review of
the sea-salt aerosol, which includes the latest work of our Soviet
colleagues, was done by Podzimek (1980).

Since we recently discussed work on the sea-salt aerosol
(Blanchard and Woodcock, 1980), much of what is here comes from
that paper, with additions and changes reflecting new ideas and
published work since that time.

2.   PRODUCTION OF THE SEA-SALT AEROSOL

More than a century has passed since Beck (1819), Sigerson
(1870), and Aitken (1881) suggested that much of the marine
aerosol was produced by the sea. Little attention was paid to
their suggestions until Jacobs (1937) pointed to the ubiquitous
air bubble as a mechanism for the production of the salt aerosol.
Today it is generally believed that air bubbles bursting at the
surface of the sea are the major source of the salt aerosol. We
begin with bubble-size distributions in the upper few meters of
the sea.

2.1 Bubble-Size Distributions

In 1957, Blanchard and Woodcock published measurements of
bubble distributions in the sea. They considered four sources of
bubble production: (1) raindrops, (2) snowflakes, (3) supersaturation of seawater by temperature change, and (4) whitecaps.
Since the first two require precipitation, they are important
only on an intermittent and local scale. Source 3 requires
rapid temperature changes, which may not be common in the sea.
The last source, whitecaps, is almost certainly the major one for
bubble production. Though not well understood, the breaking of a
wave entrains large amounts of air into the sea to produce a bubble
distribution. Blanchard and Woodcock obtained distributions
about 0.1 m beneath the surface of the water a few seconds after
small waves had broken. Their distribution, with bubbles from
<100 μm to about 500 μm diameter, was heavily weighted toward the
small end. Most of the bubbles were <200 μm. The concentration
of all bubbles was about 10^8 m^{-3}.

Later measurements were made by others (Kolovayev, 1976; Medwin, 1977; Johnson and Cooke, 1979), though not directly beneath a breaking wave. Johnson and Cooke suspended a camera from a float and took pictures of a known volume of water. In measurements obtained during winds from 10 to 13 m s^{-1}, they found the bubble concentration to decrease exponentially by a factor of 30, from 4.8 x 10^5 m^{-3} to 1.6 x 10^4 m^{-3}, as the camera was lowered from a depth of 0.7 to 4 m. Though the concentration underwent a 30-fold decrease, the diameter of the peak concentration did not change; it remained at about 100 µm. Over this same depth range, the maximum observed bubble diameter decreased by less than a factor of three, from 600 to 220 µm.

Measurements of both maximum and minimum bubble sizes must be regarded with some suspicion. The largest bubbles in some breaking waves clearly exceed a centimeter in diameter, as is obvious to anyone who has observed closely a wave breaking at sea. But the concentration of these giant bubbles is extremely low; thus, they would not be seen by Johnson and Cooke's camera. Blanchard and Woodcock overlooked these bubbles, since their bubble trap could not accommodate the large end of the bubble spectrum. At the small end, detection of the bubbles is made difficult both by the problems of simply seeing them, and because surface curvature effects (Blanchard and Woodcock, 1957; Johnson and Cooke, 1979) cause small bubbles to go into solution in less than a minute, even in water 100% saturated with air. Consequently, neither the size of the smallest bubble produced by whitecaps nor its concentration is known, though bubbles <20 or 30 µm have been observed by the above-mentioned investigators. Those who doubt that surface curvature rapidly forces bubbles into solution are quickly made believers when they see a long line of uniformly-sized small bubbles rise a few centimeters from a capillary tip, slow down, and finally disappear into nothingness (Blanchard and Syzdek, 1972a).

Johnson and Cooke's maximum bubble concentration is about 100 times less than that found by Blanchard and Woodcock, but there is no difficulty in reconciling the difference. The former measurements were not made in a breaking wave, where the bubbles are produced, while the latter were. Usually, breaking waves cover at most only a percent or two of the surface of the sea. Some of the smaller bubbles produced by these waves are carried by turbulent mixing and Langmuir circulations, especially in high winds, to depths as great as 10 to 20 m (Kanwisher, 1963; Thorpe, 1982). This mixing, which extends to regions where there are no whitecaps, is accompanied by a drastic decrease in bubble concentration. These ghost-like bubble clouds no doubt are the pseudo steady-state background concentrations found by Johnson and Cooke that exist in the entire surface layers of the

sea when waves are breaking. Superimposed on this are the
relatively short-lived, and much higher, concentrations that
originate "where the action is," in the chaotic entrainment of
air into the sea during the formation of whitecaps. Later I will
show that background concentrations are not nearly as important
in the production of the sea-salt aerosol as are the short-lived
concentrations of bubbles that rise to the surface within perhaps
10 s after they are produced.

In one of his many papers on acoustically-determined bubble
spectra, Medwin (1977) shows that the background concentration
depends on many factors, in addition to the wind: time of day,
season, and the presence or absence of sea slicks. Curiously,
he finds numerous bubbles <200 µm diameter even when no waves are
breaking. As MacIntyre (1978) points out, Medwin's data at the
small end of the bubble spectrum demand a production rate of 300
bubbles m^{-2} s^{-1} of 40 ± 0.5 µm diameter. Why don't they go into
solution? What is the source of these bubbles? Medwin attri-
butes it both to the entrainment of air as the continental
aerosol falls into the sea and to biological processes. Though
I doubt that the former can be of much significance, the latter
may be, especially near the shore where Medwin's data were
obtained. Johnson and Cooke (1981) have presented convincing
evidence that some of Medwin's microbubbles do not go into
solution, because they are stabilized by compressed organic films.
Wu (1981) has reviewed all the data on background bubble spectra.

Laboratory simulations of the production of bubble spectra
by whitecaps have been done by Monahan (1966), who dropped single
charges of water into a tank of seawater, and by Monahan and
Zietlow (1969) who let oppositely-moving wave crests collide.
Since it was difficult in that work to get absolute bubble concentrations, Cipriano (1979) and Cipriano and Blanchard (1981)
designed a "steady-state breaking wave" in which seawater fell
continuously into a tank of seawater from a height of 33 cm at a
rate of 410 cm^3 s^{-1} (Figure 1). The air entrained with the falling water produced a steady-state bubble spectrum, and a droplet-
size distribution in the air above. The former was obtained by
photography and the latter by a variety of particle counters.

These widely different methods to generate bubble
distributions might be expected to produce different shapes or
slopes of the distribution, but Cipriano and Blanchard suspect
there may be a universal slope that is basically independent of
the manner in which the distribution is produced. They suggest
that the quantity of water involved in wave breaking, and the
height from which it falls, controls the absolute but not the
relative distribution of bubble sizes. Figure 2, from Cipriano
et al. (1982), shows bubble concentration as a function of
bubble size in uniform bandwidths of 100 µm. Obtained with the

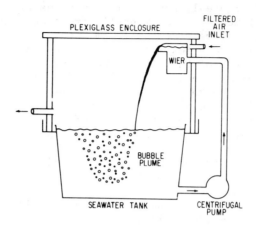

Figure 1.   Simplified scale drawing of model breaking
wave.   Tank is 0.5 m diameter at top.   (From Cipriano
and Blanchard, 1981.)

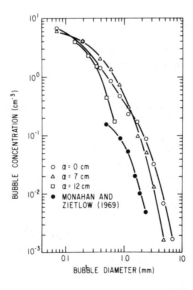

Figure 2.   Model breaking-wave bubble spectra (open
symbols, with 100 μm bandwidth.   Bubble spectrum of
Monahan and Zietlow shows relative concentration only.
(From Cipriano et al., 1982.)

apparatus of Figure 1, bubble distributions are given for three
positions (α) measured horizontally from the center of the upwelling bubble plume. At α = 12 cm, near the edge of the plume,
there are relatively few large bubbles, but at α = 0 and 7 cm,
nearer the center of the plume, the distributions are similar
and extend out to bubbles of nearly 10 mm diameter. The vertical
positioning of the Monahan and Zietlow (1969) distribution is
arbitrary, since the absolute bubble concentration is not known.
However, it is clear that the slope of their distribution is
similar to that found by Cipriano and Blanchard (1981) near the
center of the bubble plume. In both cases the slope β defined by
$dn/dr = Cr^{-\beta}$ is from 3 to 4 for bubbles >1 mm diameter. On the
other hand, we are aware that the "upper tail" of a log normal
distribution can usually be fit with a "power law" distribution,
which probably explains the familiar Junge size distribution of
aerosol particles (see Slinn, this volume).

For α = 7 cm the concentration of bubbles >100 μm was about
20 cm^{-3}; the flux to the surface was nearly 200 cm^{-2} s^{-1}.
Laboratory distributions are interesting, but we need bubble
distributions from breaking waves at sea. The optical technique
of Resch and Avellan (1982) shows promise.

2.2 Bubble Dissolution

Surface curvature effects cause many bubbles produced by
whitecaps to go into solution. Johnson (1979) calculated a
critical depth below which a bubble goes into solution before it
can rise to the surface. For a 100-μm diameter bubble at 0°C,
this critical depth is only 0.3 m.

Both dissolved and particulate organic material attach to
bubbles (Garrett, 1967a; Blanchard, 1975; Hoffman and Duce,
1976; Wallace and Duce, 1978). If the bubbles go into solution,
an organic particle undoubtedly is left (Johnson, 1979). Some
bubbles stop short of dissolution at sizes <10 μm, apparently
stabilized by the organic material (Johnson and Cooke, 1980),
but the correlation between initial bubble and final particle
(bubble?) size is unknown (Garrett, 1981).

Bubbles not reaching critical depth carry the adsorbed
organics to the surface of the sea, where they burst to produce
a sea-salt aerosol. In bubble management, Mother Nature is not
wasteful. Bubbles that burst produce nuclei for cloud formation;
those that do not, produce organic particles for the food chain
of the sea.

2.3 Percentage of Sea Surface Producing Sea-Salt Particles

The higher the wind speed the more numerous the whitecaps.
Many attempts have been made to relate whitecap coverage to wind
speed (Blanchard, 1963; Monahan, 1971; Toba and Chaen, 1973;
Ross and Cardone, 1974; Wu, 1979). The most recent and extensive work is that of Monahan and O'Muircheartaigh (1980). They
find that the relation between the oceanic whitecap-coverage
fraction (W) and the 10-m elevation wind speed (U) is

$$W = 3.84 \times 10^{-6} \, U^{3.41} \quad . \tag{1}$$

This will vary somewhat with the temperature and organic content
of the water, and the thermal stability of the lower atmosphere.

One must be very careful in using any equation of the form
$W = \alpha U^{\lambda}$, where $\lambda > 1$, to obtain an average W from an average $\bar{U}$.
Equation (1) as it stands cannot be used. Over the course of a
week or a month, the positive and negative fluctuations of the
wind U' will be about equal, but the W's produced by positive
fluctuations exceed those produced by the negative. Thus, use
of the mean wind $\bar{U}$ in (1) will give a large underestimate of
mean whitecap coverage $\bar{W}$. We can get an idea of how large this
is by following the argument I used when struggling with a
similar problem in estimating the mean electrical-charge flux
from the sea (Blanchard, 1963). Recognizing that the instantaneous wind $U = \bar{U} + U'$, one can show that, for $\lambda = 3$,

$$\bar{W} = \alpha \bar{U}^3 [1 + 3 \, (\sigma/\bar{U})^2 + \overline{(U')^3}/\bar{U}^3] \tag{2}$$

where σ is the standard deviation of $\bar{U}$. The third term relates
to the skewness of the wind, which, for lack of information,
will be assumed to be zero. The value in brackets is a correction factor that must be applied to (1) to obtain a $\bar{W}$ from $\bar{U}$. I
estimated (Blanchard, 1963) that $\sigma/\bar{U}$ in the North Atlantic is
about 0.8. This produces a correction factor of 3, and for $\lambda = 4$
about 5. For the present case ($\lambda = 3.41$) it is about 4. Let's
apply this to an actual case. The tradewinds near Hawaii
average about 6.2 m s^{-1}. What is the average whitecap coverage?
Using $\bar{U} = 6.2$ in (1), we get a value of about 0.002. That must
be multiplied by 4 to get a time-averaged fractional whitecap
coverage 0.008, or about 1% of the surface of the sea. Here we
assume that the correction factor for Atlantic winds holds for
winds in the Pacific. This probably is not true. It also may
vary with latitude. Since the correction factor is so large, it
is imperative that detailed studies be done with the massive
amount of data available today for winds at sea to get accurate
values of $\sigma/\bar{U}$.

Must we conclude that on average only about 1% of the surface of the sea is active in producing sea-salt particles? Not at all. Kolovayev (1976), Medwin (1977), and Johnson and Cooke (1979) have discovered a background bubble distribution that produces a bubble flux covering the entire surface (100%) of the sea. Is this flux important? Well, at a depth of 0.7 m, all bubbles <140 μm diameter go into solution before reaching the surface (Johnson, 1979). These bubbles account for about 75% of the bubbles (4.8×10^5 m^{-3}) at that depth, at wind speeds of 10 to 13 m s^{-1}. Thus, the ≃25% that rise to the surface are about 10^3 times less in concentration than those found at about 0.1 m in a whitecap produced by much smaller winds (Blanchard and Woodcock, 1957). But, since background bubbles burst over 100% of the surface of the sea, while whitecap bubbles cover only about 1%, we calculate that the background bubbles contribute only about 10% of the whitecap bubbles to salt particle production. This is a rough calculation at best. Additional data on both types of bubbles, and especially on whitecap bubbles, are needed. The relative contribution of the two should be a function of wind speed, the background bubble contribution becoming more important with increasing wind speed. As winds increase, bubbles are mixed deeper into the sea (Thorpe, 1982).

2.4   Ejection of Sea-Salt Particles from the Sea

Though Jacobs (1937) suggested that bubbles no doubt play a major role in the production of the sea-salt aerosol, it was not until 1948, when Woodcock began his bubble studies that a sustained interest in bubble phenomena began. I reviewed the work prior to about 1960 (Blanchard, 1963); in recent years many reviews have appeared on all aspects of the role of bubbles in air-sea exchange of materials (MacIntyre, 1974; Blanchard, 1975; Duce and Hoffman, 1976; Berg and Winchester, 1978). Although it is undoubtedly true that bubbles do not account for all of the sea-salt aerosol, and that droplets torn from the crests of waves do contribute (Lai and Shemdin, 1974; Koga, 1981), it appears that, under conditions usually found at sea, the bubble mechanism predominates.

On the average about 1% of the sea is covered by whitecaps, and in this region there is a flux of bubbles to the surface of at least 3×10^5 m^{-2} s^{-1} (Blanchard and Woodcock, 1957). As mentioned earlier, this value is biased on the low side; so perhaps the flux of 2×10^6 m^{-2} s^{-1} found by Cipriano and Blanchard (1981) in their laboratory studies is closer to the truth. In view of a difference of nearly a factor of 7 in these studies, and because the bubble flux will decrease with increasing whitecap age, it is clear that detailed observations of bubble distributions in breaking waves at sea are needed. I hope that

bubble data from the SEAREX Program (Duce, 1982) will clarify
the situation.

Upon bursting, some of the surface free energy of a bubble
is converted into kinetic energy of a jet of water (Figure 3),
which rises rapidly from the bottom of the collapsing bubble
cavity (Blanchard, 1963; MacIntyre, 1972; Darrozès and Ligneul,
1982). The jet becomes unstable and, depending upon bubble size,
breaks into 1 to 10 drops. Generally, the smaller the bubble,
the more jet drops are produced. The top (first) jet drop is
approximately one-tenth the bubble diameter. The maximum drop
ejection height increases with bubble size, reaching nearly 20 cm
for 2-mm bubbles. For larger bubbles, the ejection height
decreases, and for bubbles >7 or 8 mm no jet drops are produced
(Hayami and Toba, 1958; Blanchard, 1963). As shown in Figure 4,
there is a temperature effect. The ejection height for jet drops
from bubbles >2 mm decreases with increasing water temperature,
but for bubbles <2 mm, the work of Hayami and Toba suggests that
the reverse is true. This is indeed the case. I found the jet
drop height for the smaller bubbles (<0.4 mm) to decrease by
almost a factor of two as the water temperature dropped from
about 25°C to 4°C (Blanchard, 1963). There is no such reversal
for drop size: it increases with temperature for all bubble
sizes. These ejection height and drop size changes are controlled
by the surface tension and viscosity of the water: the surface
tension increases slightly as the temperature decreases from 25°C
to 4°C, while the viscosity increases considerably.

Figure 3. The production of jet drops from a
collapsing bubble cavity.

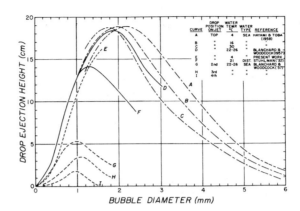

Figure 4. Ejection height of jet drops from bubbles
of an age less than 1 sec. Changes produced by temperature and type of water are shown. (From Blanchard,
1963.)

The changes in salinity normally found at sea play no
significant role in jet-drop dynamics. Stuhlman (1932) found
jet-drop ejection heights from bubbles >1 mm in distilled water
to be far less than that for seawater (Figure 4), but this is not
a direct result of salinity. Rather, it is a result of differences in bubble surface life. Bubbles in seawater have a
surface life of a second or more, while in distilled water they
burst immediately. When bubbles reach the surface, they tend to
overshoot their equilibrium position and rise up partly through
the surface. Surface tension forces take over and force the
bubble back down to an equilibrium position. This takes only 10
ms or less. In seawater the bubble surface life is nearly always
a second or more; thus, bubbles are in an equilibrium position
when they burst, insuring a maximum of surface free energy. But
when bubbles burst immediately, as they do in distilled water,
the bubble is usually above its equilibrium position. The bubble
cavity has a minimum surface area, and thus little energy is
available for jet drop ejection. Occasionally, a bubble in distilled water will remain at the surface for a few seconds before
bursting. When it does, the ejection height of the drops is the
same as in seawater (Blanchard, 1963; Blanchard and Syzdek, 1978).

These results are from lab experiments where bubbles were
produced from capillary tips only 1-2 cm beneath the surface.
Whitecap-induced bubbles are often in the water for many seconds
and may rise a meter or more before reaching the surface. During
this time, adsorption of dissolved organics from the water lowers
the surface free energy, producing a decrease in ejection height

and change in drop size. The decrease of ejection height is a
function of bubble age, size, and the amount of surface-active,
dissolved organic matter in the water. Bubbles of ≃1 mm bursting in seawater from a salt marsh (dissolved organic carbon
about 6 mg/l) showed a decrease in drop ejection height by a
factor of two after only 10 s of bubble rise through the water
(Blanchard and Hoffman, 1978). Bubbles of this size also experience a factor of two decrease in rise speed as they change
from a new or clean bubble to an organic-laden, dirty one
(Detwiler and Blanchard, 1978). Presumably most of the organic
material that enters the atmosphere is carried by drops from
these dirty bubbles (Blanchard, 1975).

Organic films on the surface of the water in laboratory
experiments can be removed and carried into the air aboard the
jet drops (Blanchard, 1963; Bezdek and Carlucci, 1974). Although
organic films (monolayers) can be found on the surface of the sea
(Garrett, 1976b; Hunter and Liss, 1981), they may play little
role in the transfer of organic material to the atmosphere. Tens
of thousands of air bubbles are produced by a whitecap. Their
drag on the water as they rise to the surface produces upwelling,
resulting in an outflow of the water at the surface. This outflow pulls some of the surface monolayer with it, producing a
region momentarily free of organic films. Many of the bubbles
burst in this clean region. However, we know nothing about what
fraction of the total number of bubbles burst before the surface
film closes in on them. Research is needed.

In addition to jet drops, film drops are also produced by
bubble bursting (Figure 5). They arise from the bursting of the
thin film of water that separates the air in the bubble from the
atmosphere. Unlike jet drops, whose numbers usually decrease
with bubble size, the number of film drops increases rapidly with
bubble size (Blanchard, 1963; Day, 1964). Bubbles <0.3 mm do not
appear to produce any film drops, but one of 6 mm can produce a
maximum of about 1000 (Figure 6). The size of film drops can
cover a wide range. The spectrum appears to peak at about 5 μm
diameter, and for drops from larger bubbles has a long tail
extending out beyond 30 μm (Blanchard and Syzdek, 1975, 1982).
At the small end of the spectrum, indirect evidence suggests that
many film drops are <1 μm (Cipriano and Blanchard, 1981). Most
of these data were obtained from single bubbles bursting at a
relatively clean surface. Bubble clusters and organic monolayers,
both on the surface of the bubble and on the surface of the sea,
modify film drop production (Blanchard, 1963; Garrett, 1968;
Paterson and Spillane, 1969; MacIntyre, 1974).

Does the sea produce more film drops than jet drops? The
answer depends in large part upon the size of the bubbles producing
the drops. If, for example, most of the bubbles produced at sea

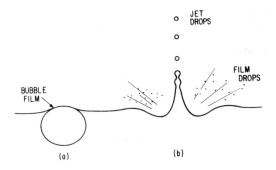

Figure 5. (a) An air bubble at an air-water interface.
(b) The production of jet and film drops from a bursting air bubble.

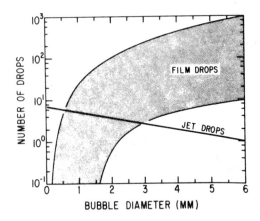

Figure 6. Schematic drawing showing number of film
and jet drops as a function of bubble size. Grey area
illustrates the variability in film-drop number.

are <0.5 mm, it is clear from Figure 6 that more jet drops are
produced. But if most bubbles are >4 mm, the reverse is true;
the film drops predominate. However, both large and small
bubbles are produced by breaking waves. We must know the size
distribution of bubbles to get an answer to our question. The
distributions found by Blanchard and Woodcock (1957) suggest that
jet drops predominate but, as mentioned earlier, their distributions were biased toward the small, jet-drop-producing bubbles.
The bubble distributions found by Cipriano and Blanchard (1981)

in laboratory simulations (Figure 2) are no doubt closer to what
is produced by breaking waves at sea. The volume distributions
of Figure 2 were converted to bubble flux distributions and are
shown in Figure 7. Assuming for simplicity five jet drops per
bubble, one quickly obtains the jet drop flux. The upper bound
(maximum) curve for film drops (Figure 6) was used to get the
film-drop flux. It is clear from a glance at Figure 7 that film
drops predominate. Adding up the contributions of the two types
of drops in the various size ranges gives a jet-drop flux of 9.5
x 10^2 cm^{-2} s^{-1}. The film-drop flux is 6.2 x 10^3 cm^{-2} s^{-1}, about
7 times larger. These drops were produced by a bubble flux of
1.9 x 10^2 cm^{-2} s^{-1}.

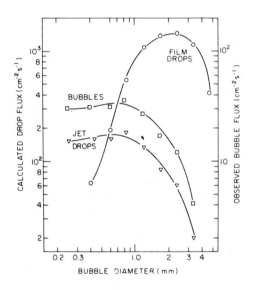

Figure 7. Calculated jet and film drop flux produced
by the bursting of the bubble spectrum observed at
α = 7 cm. Total flux: jet drops, 9.5 x 10^2 cm^{-2} s^{-1};
film drops, 6.2 x 10^3 cm^{-2} s^{-1}; bubbles, 1.9 x 10^2
cm^{-2} s^{-1}. (From Cipriano and Blanchard, 1981.)

    Is this film-drop flux realistic? Not necessarily. Even
if the laboratory bubble-flux distributions are identical to
those produced at sea, we must struggle through a vast wasteland
of ignorance to find how many film drops are produced per bubble
burst. Suppose we want to know how many film drops are produced
by a 6-mm bubble. Figure 6 tells us it is at least 10, but it
might be as many as 1,000. This ambiguity of a factor of 100 is
the same over the entire size range.

We simply don't know what number to choose. Both upper and lower-bound film-drop production curves in Figure 6 are based on my earlier work (Blanchard, 1963). The upper-bound curve was confirmed by Day (1964). He had no lower bound, since all his data fell on my upper-bound curve. The major difference between our data sets is that Day made observations only on bubbles that burst instantly upon arrival at the surface, while my data covered all bubbles, some with zero and others with non-zero surface life. One might be tempted to conclude that bubbles that burst instantly produce a maximum number of film drops, while those that sit on the surface before bursting produce a minimum. Indeed, observations with the unaided eye suggest this, but my data from the thermal gradient diffusion chamber show no clear dependence on bubble surface life.

Even if we did know the bubble-flux distribution to the surface and the number of film drops that each bubble would produce upon bursting, how do we take into account the bubble coalescence that produces the foam that is always observed after each wave breaks? Is this significant? We don't know.

Research is badly needed on the many factors that control film-drop production. If bubble surface life turns out to be important, we need not be overly concerned about the few sea-water bubbles that burst immediately upon arrival at the surface. A zero surface life is usually found only with bubbles in dis-tilled water. In seawater, bubble surface life is generally several seconds, but varies with surface monolayer pressure (Garrett, 1967a), the distance the bubble has risen through the water, and the humidity and speed of the air over the water (Burger and Blanchard, 1982). In the course of several trips to sea, I have made two or three thousand observations with a bubble aging tube (Blanchard, 1963) on bubbles of about 1 mm diameter. Very rarely, perhaps only once in 100 observations, did a bubble burst immediately upon arrival at the surface.

2.5 Seawater in the Atmosphere

A drop of seawater that enters the atmosphere presumably, though not necessarily, has a composition close to that of sea-water, i.e., about 3.5% sea salt by weight, most of which (over 85%) is sodium chloride. Since the equilibrium relative humidity over a seawater surface is 98%, the drop generally begins to evaporate. But the sea salt in the drop does not change. Thus, the salt concentration increases, and the drop's vapor pressure decreases until it is that of the environment. Evaporation ceases; the drop is now a brine drop. If the relative humidity of the atmosphere is less than about 70 to 74%, the drop becomes supersaturated with sea salt (Twomey, 1953), and a phase change may occur to produce a sea-salt particle. However, if the drop

initially is <1 µm, the Kelvin effect (Mason, 1971) dictates
that it can become supersaturated at humidities >74%. The diameter of a dry salt particle as a sphere is one-quarter that of
the parent seawater drop.

Thus, depending on the relative humidity and drop size,
drops produced by the sea exist in the atmosphere as seawater
drops, brine drops, or sea-salt particles. Since salt mass is the
only conservative property, they are all called sea-salt particles, with either the size or the mass of the particles given.
The mass does not include any water that might be associated with
the sea salt.

Experiments years ago suggested that secondary sea-salt
particle production occurs in the atmosphere when seawater drops
change phase to produce "dry" sea-salt particles (Dessens, 1946;
Twomey and McMaster, 1955), but most later experiments failed to
confirm this (Blanchard and Spencer, 1964; Iribarne et al., 1977;
Cipriano, 1979).

3.   DISTRIBUTION OF THE SEA-SALT AEROSOL

Most of our planet is covered with seawater, so it is not
surprising that of all the particulate material cycled through
the atmosphere, the largest component is sea salt. The amount
cycled each year is not known with any degree of certainty.
Estimates vary between 10^9 and 10^{10} tons (Eriksson, 1959, 1960;
Blanchard, 1963; Petrenchuk, 1980). Considering the enormous
complexity of the problem, this difference of a factor of 10 is
understandable. Even larger differences may be found in the
future, unless we decide what altitude we are concerned with.
Do we want to know how much sea salt is cycled through the
lowest meter of the atmosphere? If so, we should expect to find
far more than 10^{10} tons. There, the salt concentrations can be
extremely high (Monahan, 1968), and since the salt-containing
droplets are large, the residence time may be only a matter of
seconds. On the other hand, if our concern is with the salt
cycled through the atmosphere above 5000 m, we no doubt will
find it to be far less than 10^9 tons, with a residence time
measured not in seconds but days. The work of Petrenchuk (1980)
is an attempt to determine the amount of salt cycled through
different altitudes. Future work should go in this direction.

Most papers on the sea-salt aerosol report on mass concentrations. This section begins with what has been learned about
salt concentration and how it varies with wind speed.

3.1   Sea-Salt Concentration as a Function of Wind Speed

Woodcock (1950, 1953) and Woodcock and Gifford (1949) made
the first detailed studies of the distribution of sea-salt particles in the marine atmosphere.   Much of this work was done
windward of the Hawaiian Islands in clear air between the clouds,
and at altitudes of about 600 to 800 m, about that of cloud base.
All of Woodcock's data were obtained from microscope observations
of salt particles collected by exposing small glass slides from
an aircraft.   The isopiestic method (a controlled humidity chamber) plus corrections for collection efficiency were used to get
particle-size distributions.   Comparisons of salt concentrations,
obtained in this way to those obtained by chemical titration of
the chlorides on the glass slides (Woodcock, 1950, 1952) showed
that sea salt, and not other substances was being collected.
The patience and careful attention to detail characteristic of
Woodcock has made his work on sea salt a standard against which
all subsequent work has been compared.

Woodcock's (1953) salt concentrations as a function of wind
speed, redrawn from his original work, are shown in Figure 8.

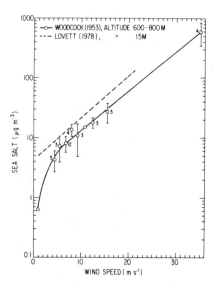

Figure 8.   Sea-salt concentration at two altitudes as
a function of wind speed near the surface of the sea.
The number of observations averaged for each data point
is given (if more than one), as well as standard devia-
tion bars ($\bar{\chi} \pm \sigma$).   (From Blanchard and Woodcock, 1980.)

All concentrations were obtained upwind of Hawaii, with the
exception of those for winds of 35 m s^{-1}, which were obtained
from the top of a lighthouse during a Florida hurricane. Assuming that during hurricane winds the salt concentrations at an
altitude of about 40 m (where Woodcock made his measurements) are
similar to those at 600 m, then it appears that in winds from
about 5 to 35 m s^{-1}, the salt concentration increases exponentially with wind speed. At lower wind speeds, Woodcock's single
observation shows that the salt concentration drops rapidly. This
is hardly surprising, for at wind speeds less than about 3 m s^{-1},
few bubbles are produced by the sea. Not many data sets are
available to confirm Woodcock's observations, but recent aircraft
measurements by Patterson et al. (1980) at altitudes of 250 to
500 m in the marine boundary layer show general agreement.

The variations in salt concentration of a factor or two or
more for a given wind speed (Figure 8) should come as no surprise.
They have been found in many investigations subsequent to
Woodcock's work. There are several reasons why this should be so.
Possibly the most important is that, because the mean sea-salt
particle residence time is a day or two, the salt concentration
is correlated less with the local winds and more with the integrated winds of the air mass in the prior day or so. Variations
are also produced by subsidence caused by large cloud systems in
the vicinity, by differences in the depth of the mixing layer,
and in the number and height of clouds.

The dashed line in Figure 8 is the best-fit line found by
Lovett (1975, 1978) from data obtained at an altitude of 15 m
from three weather ships in the North Atlantic, about 700 km west
of Ireland. Lovett's data are the most numerous of any data set
and cover the largest range of wind speeds, excluding Woodcock's
hurricane observations. In the course of about a year, using
filters with isokinetic sampling heads, Lovett determined salt
concentrations from 1,821 samples (!) collected in winds from 1
to 21 m s^{-1}. His concentrations, like Woodcock's, increase
exponentially with wind speed. They are of course greater than
Woodcock's, since we expect to find the highest concentrations
near the surface of the sea. But unlike Woodcock, Lovett did
not find a rapid decrease in salt at winds <3 m s^{-1}. Possibly
this is due to a higher variability in his winds, as opposed to
the trade winds of Hawaii (which are among the most constant in
the world). In rapidly-varying winds, a high salt concentration
found at low winds might have been produced by higher winds a
day or two before.

The data of Figure 8, plotted in Figure 9 on a log–log plot,
show an interesting aspect of the salt-wind correlation. Keeping
in mind what was just said about Lovett's data at low winds, it
appears that the equilibrium salt concentration at altitudes of

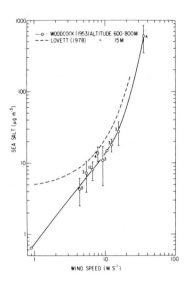

Figure 9.   Sea-salt concentration at two altitudes as
a function of wind speed near the surface of the sea.
(From Blanchard and Woodcock, 1980.)

both 15 and 600-800 m increases with about the first power of the
wind speed for speeds <10 m s^{-1}; but more rapidly after that,
increasing with about the third power at 20 m s^{-1}, and with more
than the fourth power as the winds approach hurricane speed.
Qualitatively this makes sense.   The mixing upward of the salt
(or any aerosol) increases rapidly with wind speed, while the
source area (the whitecaps) increases with more than the cube of
the wind speed.

    The variability of salt concentration found at 600-800 m
altitude is also found near the surface of the sea.   The salt
concentrations shown in Figure 10 were obtained 19 m above the
sea from a tower on the windward shore of the island of Oahu,
Hawaii (Blanchard and Syzdek, 1972b).   The data were derived from
an analysis of the salt collected on both stationary and rotating
platinum wires of 254 µm diameter.   The upper and lower limits
of Woodcock's data are also shown.   At the highest wind speeds,
the decrease in salt concentration with altitude is clear; but
the concentration gradient decreases with decreasing wind speed,
and at winds of about 4 m s^{-1}, the gradient disappears.   This is
contrary to what is shown in Figures 8 and 9, but the apparent
discrepancy, as suggested earlier, may relate to the relative
constancy of the Hawaiian trade winds as opposed to the non-steady

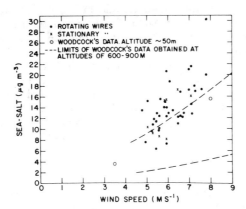

Figure 10. Sea-salt concentrations as a function of
wind speed near the surface of the sea at Oahu, Hawaii.
(From Blanchard and Syzdek, 1972b.)

winds no doubt experienced by Lovett (1978). Later, I will
present numerous observations by Woodcock that leave no doubt
that the vertical concentration gradient decreases with decreasing wind speed.

The collection of salt samples from island stations is done
from towers along the windward shore to avoid problems associated
with high concentrations of salt ejected from the nearby surf
zone (Duce and Woodcock, 1971). Although this insures that the
salt came from the open sea, there is still a question of whether
the island modifies the wind speed from what it is on the open
sea. With an island as small as Oahu, one might think not, since
the winds show little change from night to day. However, there
is evidence that the winds crossing the windward shore (Figure 10)
are about 1 m s^{-1} less than those reported from ships upwind of
Oahu during the 60 days in 1970 when the data were obtained.
Similar results were found by Woodcock (1975) during another
study of winds near Oahu. This would tend to shift the data
points of Figure 10 about 1 m s^{-1} to the right, suggesting that
over the open sea the salt concentration is somewhat less for a
given wind speed than that shown in Figure 10.

In spite of this potential correction to salt samples obtained
from island stations, the samples of Blanchard and Syzdek (1972b),
and those of Hoffman (1971), also obtained from the Oahu tower,
are similar to numerous other studies, some made from ships at sea
and some from island stations. The Oahu tower salt concentrations
range from about 2 to 25 µg m^{-3}, as the wind increases from about
4 to 8 m s^{-1}. Shipboard samples obtained in the North Atlantic

between 10^oN and 25^oN averaged about 12 µg m^{-3}, but varied
between a few tenths and 45 µg m^{-3} (Hoffman et al., 1974).
Detailed studies from two island stations in the tropical North
Atlantic show that the distribution of the sea-salt concentration
is a composite of two log-normal distributions, with one mode
overwhelmingly dominant (Savoie and Prospero, 1977). The average
concentration was about 20 µg m^{-3}, with a standard deviation of
about 7 µg m^{-3} at one station and 11 µg m^{-3} at the other. Lower
concentrations, most <10 µg m^{-3}, were found by Prospero (1979),
who collected 246 samples from ships in many locations on the
World Ocean.

It seems useless to compare in detail the many observations
collected over the World Ocean, since in many cases the wind at
the time of sampling was not reported. But even if the wind
speed is known, other factors, some mentioned earlier, can control
salt concentration. Atmospheric stability, usually not reported
with salt concentration data, exerts a major control. The more
stable the atmosphere, the higher will be the low-level salt
concentration for a given wind speed. Measurements made at a
fixed location on the ocean, such as from island stations as
opposed to ships, are more likely to provide the clues necessary
to determine the importance of the many factors controlling salt
concentrations. In that regard, the excellent and detailed
studies of Savoie and Prospero (1977) extending over many months
are to be applauded.

3.2 Size Distribution of Sea-Salt Particles

Woodcock's sea-salt concentrations of Figure 8 were obtained
from the particle-size distributions. Some of these are shown
in Figure 11 for Beaufort wind forces of 1, 3, 4, 5, 7, and 12,
which represent mean winds at the 10-meter level of 1, 4.4, 6.7,
9.3, 15.5, and 35 m s^{-1}, respectively. These cumulative distributions for altitudes of 600-800 m cover a particle mass range
from 1 to 10^5 µµg, representing drop sizes when they leave the
sea of about 2 to over 80 µm radius, and which produce dry sea-
salt particles of from 0.5 to over 20 µm radius. It is interesting that the number of particles >2 µm radius (as seawater drops)
increases from about 3 x 10^5 m^{-3} to 7.5 x 10^5 m^{-3}, a factor of
only 2.5, while the winds increase from 4.4 to 15.5 m s^{-1}. However, at the large end of the spectrum, the increase is much more
dramatic. The number of particles >40 µm radius increases from
about 0.1 to over 100 m^{-3}, a factor of 10^3.

This could mean that, as the wind increases, relatively more
larger particles are produced by the sea, but a more likely interpretation is that particles at the large end of the spectrum,
with their shorter residence time, respond more quickly to
changes in wind speed. Consequently, the large end of the

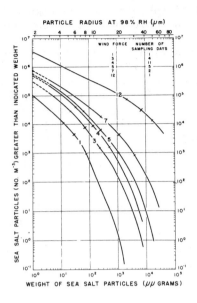

Figure 11. Sea-salt particle cumulative distributions
at altitudes of 600-800 m above the sea as a function
of wind force. (From Blanchard and Woodcock, 1980.)

spectrum is more likely to be in equilibrium with the wind speed
than the small end. The small particles, with long residence
time and sluggish response to changes in wind speed, can be compared (Blanchard, 1963) to a massive fly-wheel whose rate of
rotation remains relatively constant. Recent data by Monahan et
al. (1982) show this quite nicely. They find that the concentration of salt particles of 0.3 μm radius shows very little response
to changes in wind speed, while that of particles of 15 to 17 μm
changes so quickly that it becomes directly proportional to the
source of the particles, the area covered by whitecaps. This
latter finding confirms work by Toba and Chaen (1973).

There is another interesting conclusion that can be derived
from the Woodcock salt-particle distributions. Following
Eriksson (1959, 1960), I used the Woodcock distributions to calculate the sea-salt dry deposition as a function of wind speed
(see Figures 17 and 18 of Blanchard, 1963), and found it to vary
with about the cube of the wind speed. Interestingly, about the
same conclusion can be drawn from the recent experimental work of
McDonald et al. (1982).

Since over the range of wind speeds usually found at sea, 1
to 10 m s^{-1}, the salt concentration increases with barely more

than the first power of the wind speed (Figure 9), it follows
that the ratio of dry deposition to salt concentration varies
with the square of the wind speed. This great increase in
importance of dry deposition relative to salt concentration, as
the wind increases, of course reflects the shift to larger
particles. This can be seen in the Woodcock distributions
(Figure 11). As the wind increases from 1 to 9 m s^{-1}, the mass-
median radius of a seawater drop increases from 5.7 to nearly 12
μm. We are reminded by McDonald et al. (1982) that since the
gravitational settling speed is proportional to the square of
the particle radius and the salt mass to the cube of the radius,
the large particles dominate the dry deposition even though
present in relatively low concentrations.

Woodcock (1972) has extended his particle distributions to
sizes smaller than shown in Figure 11. He finds a total particle
count of 10 to 20 cm^{-3}, the smallest of which have radii of about
0.3 μm as seawater drops. Presumably his smallest particles are
sea salt, though it is possible they are other hygroscopic
particles. This is but a small fraction of the total particle
count of 200 to 300 cm^{-3} found in marine air far from the
continents (Blanchard and Syzdek, 1972b), and raises the question
of whether droplets <0.3 μm produced by the sea contribute
significantly to the total count. In the bubbling experiments of
Cipriano and Blanchard (1981), droplets as small as 0.03 μm
radius were found. Hobbs (1971) doubts that they are significant,
even when compared to the less numerous cloud condensation
nuclei, but Cipriano et al. (1982) believe they are.

The Woodcock distributions (Figure 11) are for altitudes of
600 to 800 m. The only extensive, published data on distributions near the sea is the work of Chaen (1973), who made many
shipboard measurements at an altitude of 6 m. Figures 5 and 6 of
Blanchard and Woodcock (1980) show a comparison of the Woodcock
and Chaen distributions. As expected from earlier work
(Woodcock, 1953), the number concentration is several times less
at the higher than at the lower altitude. At the large end of
the spectrum, where we have particles of 10^4 μμg (40 μm seawater
radius), the decrease with altitude is most rapid. Again, this
is as expected. But, salt-particle distributions to confirm and
extend the Chaen work are needed; and in this regard, I hope the
distributions of Monahan et al. (1982) are published soon.

3.3 The Sea-Salt Inversion

In preparation for an earlier review paper (Blanchard and
Woodcock, 1980) I decided to replot Woodcock's (1962) numerous
observations of salt concentrations to see how they varied with
altitude and wind speed. Woodcock had already done this with
268 observations obtained over many years in numerous parts of
the World Ocean, but with 75 new, unpublished observations by

Woodcock a new analysis seemed necessary. With more data,
smaller altitude intervals could be used, especially at lower
elevations where the observations were more numerous. This
would allow better resolution than in the earlier Woodcock graph.
It should be understood that when clouds were present over the
sea (and they usually were), Woodcock obtained his salt samples
not beneath the clouds (and certainly not in the clouds), but in
the clear air between the clouds. He avoided clouds as much as
possible; the fine mist that often falls from even the small,
non-precipitating cumulus clouds makes it difficult to sample for
sea salt just below cloud base. Surprisingly, a salt inversion
appeared in the Woodcock data. This is seen in Figure 12 where
the average salt concentration, and the standard deviation
($\overline{x} \pm \sigma$),are given as a function of altitude, for average wind
speeds of 1, 3.5, 8, and 14 m s^{-1}.

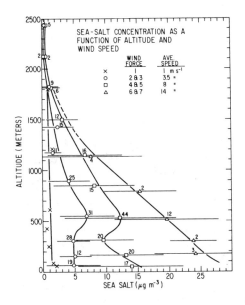

Figure 12. Sea-salt concentration as a function of
altitude and surface wind speed. The number of observations averaged for each data point is given (if more
than one), as well as standard deviation bars ($\overline{x} \pm \sigma$).
A salt inversion appears at the two intermediate wind
speeds. (From Blanchard and Woodcock, 1980.)

    The salt inversion appears between about 300 and 600 m for
winds of 3.5 and 8 m s^{-1}. Though the spread of the data around
the mean is high, the inversion appears to be real, with salt
concentrations between 500 and 600 m nearly 1.4 times that at

300 m. There was no clear evidence of an inversion for winds of 14 m s^{-1}, but with only five observations below 500 m, one cannot draw any conclusions. Above the inversion, the decrease of salt concentration with height for all wind speeds is similar to that observed earlier. The data of Figure 12 were separated into those from both the Atlantic and Pacific and replotted. The salt inversion appeared in both graphs.

These data were replotted on log-log paper, along with data from many other sources (Hoffman, 1971; Monahan, 1968; Blanchard and Syzdek, 1972b; Chaen, 1973; Savoie and Prospero, 1977; Lovett, 1978), to show how the salt load varied from <0.2 m above the sea to >2000 m. The graph, shown in Blanchard and Woodcock (1980) but not here for lack of space, indicates, as predicted by Toba (1965b), that on average a power-law relation exists between salt and altitude, extending from 1 m up to about 300 m. Over this height range, and for winds from 3.5 to 14 m s^{-1}, the average sea-salt concentration ($\overline{\chi}$) can be expressed by

$$\overline{\chi} = 5(6.3 \times 10^{-6} \text{ H})^{(0.21 - 0.39 \log U)} \tag{3}$$

where H is the height in meters, and U the wind speed in meters per second.

The altitude curves relate closely with various regimes or layers in the marine subtropical atmosphere. These layers, shown schematically in Figure 13 along with the salt curve for winds of 8 m s^{-1}, suggest reasons for the height variation of sea salt. Below 0.2 m, salt concentrations are very high. This is the layer of direct influence of large jet drops, most of which rise <0.2 m and return directly to the sea. The layer up to cloud base altitudes at 600 to 800 m is the subcloud layer. Here the air is well mixed; it usually has a dry adiabatic lapse rate, except perhaps for the last 100 m or so, where the air becomes more stable (Bunker et al., 1949). In the cloud layer, extending from cloud base to the temperature inversion at about 2000 m, the temperature lapse rate in the air between the clouds is stable. Here we find a steady decrease of salt with height, reaching negligible values at the temperature inversion.

How do we explain a salt inversion? Years ago Eriksson (1959), using Woodcock's data, found a maximum in the salt concentration at about 500 m but felt that it decreased continually from there down to the sea. Both Eriksson and Toba (1965b) gave different explanations of this apparent surface sink, but they do not apply here, since Figure 12 shows no evidence of a steady decrease of salt below 300 m. A decrease of salt concentration near the surface over large islands and continents is well known (Lodge, 1955; Twomey, 1955; Byers et al., 1957), but this is because salt particles are removed by trees and vegetation

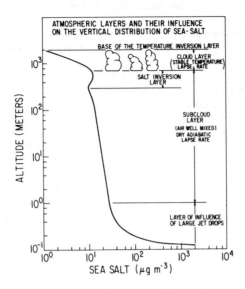

Figure 13. The influence of various atmospheric layers
on the vertical distribution of sea salt. The curve
shows the salt concentration for a wind speed of 8 m
s^{-1}. (From Blanchard and Woodcock, 1980.)

(Toba, 1965a; Tanaka, 1966) rather than being produced there.
Clearly, this is not the case at sea.

A clue to the explanation of the salt inversion is that its
maximum is located just beneath the altitude of cloud base.
There are at least two mechanisms that appear capable of maintaining it. The first is by the fine, mist-like drizzle that
often falls from the clouds (Woodcock and Duce, 1972). This
drizzle contains salt that was carried up from near the sea. The
water evaporates, leaving salt particles in the air. These
particles do not stay beneath the clouds, since, due to wind
shear, the clouds move faster than the subcloud air. A salt
inversion occurs. The second mechanism, not requiring clouds, is
that of an accumulation zone in which high concentrations of
salt, carried up in parcels from near the sea, grow rapidly as
they rise to more humid regions (Woodcock et al., 1963). Some of
the larger salt particles will attain fall speeds that are significant compared with the updraft speed, and so produce an
accumulation or concentration of salt.

3.4 An Attempt to Find the Salt Inversion

    During his numerous salt-sampling flights over the years,
Woodcock was mainly concerned with the role of salt particles in
the formation of rain. He had no need to explore the details of
small vertical gradients, and thus his data for any given flight
cannot be used to detect a salt inversion. When a salt inversion
appeared in the averaging of all Woodcock's data, I decided to
see if it could be detected during a single flight. I went to
Hawaii in October and November of 1981 and made 20 salt-sampling
flights with Woodcock in a single engine aircraft, over the sea
windward of the island of Oahu. About three flights were made
each week, usually late in the morning. On each flight 12 salt
samples were obtained. Sampling altitudes varied, but were
within a range of 30 to 1000 m. Using glass thermometers,
Woodcock noted the wet and dry bulb temperature while I was
collecting each sample.

    Salt was collected on small glass slides (Woodcock, 1952) at
an air-speed of 40 m s^{-1}. A known area of each slide (about 42
mm^2) was washed in a known volume of distilled water. Atomic
adsorption spectroscopy gave the amount of sodium in the sample,
and multiplication by 3.25 gave the amount of sea salt. With
that, the collection efficiency, and the volume of air swept out
by the slide during its exposure from the aircraft (0.05 m^3), one
could calculate the salt concentration. On each flight, one
slide, not exposed, was used as a control.

    This method allowed rapid collection of data. There is,
however, a drawback. Since the particle-size distribution on
each collecting slide was not known, there was no way to make
exact corrections for collection efficiency. Consequently, the
collection efficiency was taken as 100%, and assumed constant
with altitude, resulting in an underestimate of the true salt
concentrations. However, for the purpose of detection of a salt
inversion this should cause little concern, for it is the relative changes in salt concentration with height that are of interest.
In an analysis of his data years ago, Woodcock (personal communication) finds that the collection efficiency, depending upon
a number of factors, generally was between 80 and 95%.

    The salt inversion appeared in 6 or 7 of the 20 salt
soundings, but lack of space prevents detailed discussion. One
of the more interesting salt inversions is shown in Figure 14,
the flight of 2 November 1981. Both salt and temperature soundings are shown, the latter corrected for aerodynamic heating
(0.6°C). The salt inversion is between 350 and 500 m in stable
air near the top of the sub-cloud layer. Salt concentrations
decreased by about a factor of 16 up through the shallow 200-m
thick cloud layer, possibly because of entrainment of cleaner air

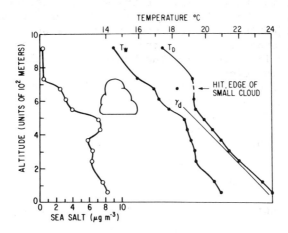

Figure 14. A salt inversion found windward of Oahu,
Hawaii on 2 November 1981.

from above the mixing layer. There were many small clouds, and
we had difficulty completing our 30-s sampling runs without
hitting them. Indeed, we accidentally ran into the edge of a
cloud for a few seconds of the 670-m run, collecting some cloud
water on the thermometers. This caused the dry bulb to partially
act as a wet bulb, lowering the temperature in an otherwise
nearly isothermal cloud layer.

The weather conditions at the time of the flight were
somewhat unusual. A cold front had passed over Oahu from the
north the previous night, bringing widespread and heavy rains.
Surface winds during the flight were estimated to be Beaufort
force 5 (8-11 m s^{-1}) from 025° true. In addition to the many
small clouds, we could see in the distance a few, widely
scattered large cumuli that penetrated far higher. Many of the
other 20 flights were made in "disturbed" weather that was far
different from the steady trades so common in the summer months.

4.   ENRICHMENT OF BACTERIA IN JET AND FILM DROPS

Since bacteria are found in all natural bodies of water, it
should come as no surprise that they are also found in jet and
film drops. Indeed, simple probability considerations suggest
that the concentration of bacteria in the drops should be the
same as that in the water from whence the drops originated.
What is surprising are the findings of many workers in recent
years that the concentrations of bacteria in the drops can be
many hundreds of times greater than in the main body of water.

Interest in the water-to-air transfer of bacteria is not
new. Ever since the pioneering work of Pasteur, it has been
known that microbial aerosols exist in the atmosphere. Early in
this century, experiments by Horrocks (1907) strongly suggested
that one of their sources was the bursting of bubbles. As so
often happens in science, interest lagged and this work was soon
forgotten. Forty years later, Woodcock (1948), unaware of
Horrocks' work, showed that drops from bubbles bursting in
regions of the sea containing high concentrations of plankton
could carry an organic irritant. When inhaled, this aerosol produced coughing and a burning sensation in the respiratory tract.
A few years later Woodcock (1955) suggested not only that bacteria
might get airborne in jet drops from bursting bubbles, but that
the material in the drops might be coming from a very thin layer
at the surface of the sea. Some of my own experiments at that
time in Woodcock's lab proved there was some substance to this
idea, for I found that surface-active monolayers on seawater
could be carried into the air on jet drops. If bacteria were
concentrated at the surface of the sea, Woodcock reasoned that
they should appear in high concentrations in the jet drops.
Curiously, this very important paper was rejected by the first
journal to which it was submitted. The editors said it contained
no new ideas!

Higgins (1964) followed up on Woodcock's suggestions by
collecting the aerosol produced by bubbling water inoculated with
several species of bacteria, including Serratia marcescens (a
rod-shaped bacterium about 0.5 μm dia. and 1 μm long). From an
analysis of the various bacteria in the aerosol, he deduced that
S. marcescens must be concentrated at the surface of the water.
This work strongly suggested, but did not prove, that some
bacteria could be highly enriched in jet drops. Apparently the
first to prove this enrichment were Blanchard and Syzdek (1970).

For those who wish to know about bacteria in the sea, I
recommend the excellent new book by Sieburth (1979). An introduction to the field is given by Sieburth in this volume. He
shows quite convincingly that microorganisms can play a major
role in the biochemistry of the mixed layer of the sea. Bacterial aerosols are discussed in the books by Dimmick and Akers
(1969) and Gregory (1961).

4.1 Jet Drop Enrichment

In all that follows, the bacterial enrichment factor (EF)
is defined as the concentration of bacteria in the jet drop at
the moment of production (before evaporation occurs) divided by
the bacterial concentration in the bulk water from which the jet
drop originated. The term 'concentration factor', as used in
other papers on the subject, is identical to EF. Though some of

the experiments used seawater to prepare the bacterial suspensions, many used freshwater or distilled water. The parameters that control the EF of bacteria in jet drops are not well understood, but four are recognized to be of importance. These are (a) bubble scavenging, (b) drop size, (c) drop position in the jet set, and (d) type of bacteria.

Bubble scavenging. Many investigators have shown that certain species of bacteria can be scavenged by bubbles (Gaudin et al., 1962; Carlucci and Williams, 1965; Grieves and Wang, 1967; Rubin, 1968). The bacteria that attach to a bubble surface (or any water surface) presumably are those that have a sufficient number of hydrophobic sites on the cell wall to make them surface active. The number of bacteria collected by a bubble should be a function of the distance the bubble rises through the water (the bubble rise distance, BRD). Since some of the collected bacteria are transferred to the jet drops, we expect the jet drop EF to be a function of BRD.

This is indeed the case (Blanchard and Syzdek, 1972a; 1974). Figure 15, taken from Blanchard et al. (1981), shows quite clearly the dependence of the EF on the BRD. The EF for the top jet drop increases with BRD for a bubble of 380 μm diameter rising through a suspension of S. marcescens in pond water. The concentration in the bulk water was about 10^6 ml^{-1}. Three experiments are shown. The methods we used to catch a jet drop in flight, to determine both its size and the number of bacteria it carries, are explained in the above-mentioned papers.

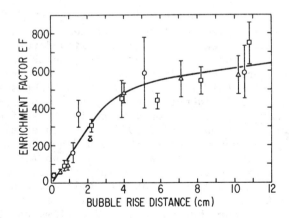

Figure 15. Effect of distance a bubble rises through a suspension of S. marcescens on enrichment factor (EF) for bacteria in the top jet drop. (From Blanchard et al., 1981.)

Initially, the increase of EF with BRD is very rapid; it
increases from near-negligible values to about 400 as BRD increases to only 3 cm. The EF increases more slowly after 3 cm,
attaining only 600 after 10 cm of BRD. Other experiments with
bubbles of equal size showed that an EF of about 1,000 was reached
with a BRD of 25 cm (Blanchard and Syzdek, 1978). The marked
decrease in the rate of increase of EF at about 3 cm, which we
have observed in other experiments, suggests that the bubble
collection efficiency for bacteria decreases at about 3 cm BRD.
We believe this is a result of the bubble changing from a fluid
to a solid sphere. During the first few centimeters of rise, a
bubble of the size used here moves as a fluid sphere (clean
bubble) with a mobile surface (Tedesco and Blanchard, 1979). In
addition to bacteria, it collects surface-active material during
its ascent. Consequently, the mobility of the bubble surface
decreases until enough surfactant has accumulated to render the
surface immobile (Clift et al., 1978). It then rises as a
solid sphere (dirty bubble). The rate of collection of bacteria
by a bubble in the early period of rise, when its surface is
clean, is much higher than the rate later on, when its surface
is dirty (Weber et al., 1982).

Bacteria that reach the surface of the bubble do not do so
by either diffusion or inertial effects. They do it by interception, a collection mechanism that comes into play because the
bacteria have a finite size (Weber, 1981; Slinn, this volume).
Since inertial effects are negligible, the bacteria follow the
fluid streamlines past the bubble. But due to their finite size,
bacteria following streamlines very close to the bubble surface
make contact with the bubble and are scavenged.

Woodcock's (1955) idea that the bacterial EF in jet drops
could be explained by especially-high concentrations of bacteria
at the surface of the water, an idea I believed for a number of
years, probably is not correct. Otherwise, as BRD approaches
zero, EF should approach a finite and significant value. There
is no suggestion of this in Figure 15. This might seem curious
since there is ample evidence that many species of bacteria,
including S. marcescens, can concentrate in the surface microlayer, both at sea (Crow et al., 1975; Sieburth et al., 1976)
and in laboratory suspensions (Norkrans and Sörensson, 1977;
Syzdek, 1982). However, there appears to be no mechanism to
move bacteria from the surface, down into the bursting-bubble
cavity, fast enough to insure that they will catch a ride on the
descending capillary wave (MacIntyre, 1972) that produces the
jet drops. It is true that surface-active films can be transferred
from the surface to the jet drops (Blanchard, 1963; Bezdek and
Carlucci, 1974), but they are far thinner than bacteria and can,
when exerting a surface pressure, move rapidly along the surface.
Syzdek and I once did a series of experiments (unpublished) to

see if the surface microlayer concentrations of S. marcescens
contribute significantly to the jet-drop bacteria. By feeding
the bacterial suspension into the bottom of a 3-cm i.d. tube, we
could control the rate at which the surface overflowed at the
top where the bubbles were bursting. We determined the number
of bacteria in jet drops when bubbles burst at a surface that
had been stagnant for up to 30 min (no overflow), and again when
the surface was overflowing at a rate of 10 ml min⁻¹. There was
no significant difference in the bacteria count. It seems clear
that the jet drop bacteria were coming from bubble scavenging.
Had the bulk surface bacteria played a role, we would have found
more jet-drop bacteria when the bubbles were bursting at an old
and stagnant surface.

    Drop size. In our first experiments (Blanchard and Syzdek,
1970), we used a variety of glass capillaries to produce bubbles
of different sizes at a constant BRD of about one centimeter in
a distilled water suspension of S. marcescens. We determined the
EF for bacteria in the top jet drop, from bubbles ranging from
about 0.25 to about 1.3 mm diameter. Results were given in terms
of EF vs. top-jet-drop size, covering a range of about 25 to 130
μm diameter. A maximum in EF was found at about 60 μm. The
occurrence of a maximum was confirmed by other workers: by
Bezdek and Carlucci (1972), who used seawater suspensions of
S. marinorubra; by Blanchard and Syzdek (1972a), again using
distilled water suspensions of S. marcescens; and by Hejkal et
al. (1980), who used suspensions of several species of marine
bacteria in NaCl solutions.

    The general trend of all the data is shown in the schematic
diagram in Figure 16. The dashed line for the viruses is only a
guess, since no data are available. My reason for placing it to
the left of the curve for bacteria will soon become clear. A
possible explanation for a maximum EF at an intermediate drop
size is suggested by Figure 17, where I have reproduced the jet
drop EF curve of Figure 16, and around it have sketched portions
of air bubbles with bacteria attached to them. Following
MacIntyre (1972), I assume that the thickness of the microlayer
(kR) that is skimmed off the collapsing bubble to produce the
jet drops is proportional to bubble radius (R). For small
bubbles it is probable that kR is less than the size of the
bacterium, and thus many of the bacteria are left behind when
the microlayer is transformed into jet drops. As bubble size
(and kR) becomes smaller, relatively fewer bacteria will get into
the jet drops, and for drops less than about 20 μm (which are
produced from bubbles of about 200 μm), the EF may become less
than one. Experiments need to be done to test this concept for
drops <20 μm. Bubbles of <200 μm that produce such drops are not
easy to make from capillary tips stationary in the water
(Blanchard and Syzdek, 1977), but can be made from tips in a
rotating tank (Blanchard and Syzdek, 1972a).

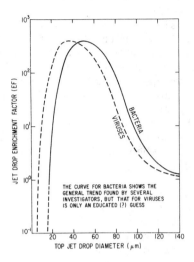

Figure 16. Schematic diagram showing dependence of the top-jet-drop bacterial enrichment factor on drop size.

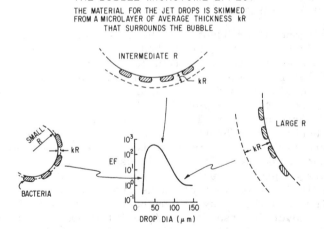

Figure 17. Schematic diagram of a hypothesis to explain the occurrence of the maximum bacterial enrichment factor (EF) shown in Figure 16.

As bubble size increases from small to intermediate size, an optimal thickness kR is reached at which the highly concentrated layer of bacteria attached to the bubble is skimmed off and into the jet drops. This produces a maximum EF. As bubble size continues to increase, kR extends deeper and deeper in the bulk suspension. The drop EF decreases and approaches unity when the layer kR is so thick that the bacterial concentration is essentially that of the bulk suspension. An analogous argument for the origin of jet drop bacteria in the bulk surface microlayer was advanced by Bezdek and Carlucci (1972).

At the end of MacIntyre's (1972) thorough analysis of the details of the water flow in the bursting of a bubble, he says, in reference to some of our bacteria transfer work (Blanchard and Syzdek, 1972), "...the statistics of particle transfer from bubble surface to jet drop might help elucidate flow patterns." He is quite correct in this assertion. The position of the maximum jet drop EF at about 50 μm (Figure 16) can be used to determine the thickness of the layer of water skimmed off the bubble to produce the top jet drop, and to relate this to bubble size. The bubble size at which the maximum EF occurs is about 500 μm diameter (remember that the top jet drop has a diameter about one-tenth that of the bubble). The thickness of the layer giving the maximum EF should be the width of a bacterial cell, about 0.5 μm, assuming that the cells attach to the bubble as shown in Figure 17. Thus, the layer thickness is 0.1% of the diameter of the bubble. Considering the possible sources of error, this agrees quite well with the stimate of 0.05% obtained by MacIntyre from a mass balance of the volume in the drop and that of a spherical shell surrounding the bubble. MacIntyre cautions that his estimate of 0.05% should be accepted "...with the understanding that this is by no means a uniform thickness." He states that the layer thickness increases with distance down the outside of the bubble. If this is true, and if the bacteria scavenged by the bubble, as it rises to the surface, are swept to the lower half of the bubble, which almost certainly is the case, then MacIntyre's estimate of 0.05% for an average thickness would increase to approach the experimentally-determined 0.1%.

The shape of the curve in Figure 16 suggests further experiments. First, since EF decreases so rapidly beyond its maximum value, the thickness of the microlayer skimmed from the bubble probably increases faster than the first power of the bubble radius. Second, instead of bacteria, polystyrene spheres could be used (Quinn et al., 1975). Since the spheres come in a variety of sizes, the drop diameter at which a maximum EF is found should decrease with sphere size. The entire curve in Figure 16 might move to the left with decreasing particle size. This is the rationale behind the placement of the virus curve, since virus particles are smaller than bacteria. Experiments

with the water-to-air transfer of viruses have been done, both
in the laboratory and at sea (Baylor et al., 1977a, 1977b), but
they were not designed to determine the jet drop EF as a function
of bubble size.

   Drop position in the jet set. All that has been said so far
refers only to the first or top drop of the set of jet drops (the
jet set!) produced when a bubble bursts. According to
MacIntyre's (1968, 1972) bubble-microtome hypothesis, the liquid
for the top drop comes from the surface layer of the bubble,
while that for lower drops comes from progressively deeper layers.
Thus, we would expect to find the EF decreasing from the top to
the bottom drop of the jet set. This appears to be true. In
some of our experiments (Blanchard and Syzdek, 1978) a 0.38-mm
diameter bubble rose about 20 cm through a suspension of S.
marcescens in pond water, to burst at the surface and produce 5
drops in the jet set. By a variety of methods, including
electrostatic induction (Blanchard and Syzdek, 1975), we determined the EF of the drops. It was about 1200 for the top drop,
but decreased with drop position, reaching a low value of 8 for
the fifth or lowest drop in the jet set. This marked dependence
of EF on drop position no doubt extends over the entire range of
bubble size, though proof is lacking.

   Type of Bacteria. As mentioned earlier, the efficiency by
which bacteria are collected by air bubbles is not the same for
all species. The collision efficiency can be calculated (Weber,
1981) knowing only the physical parameters of the system (size of
the bubble and bacterial cell, and the viscosity and density of
the water). Whether a cell is collected or not after collision
with a bubble is determined by its sticking or attachment efficiency, a function of the degree of hydrophobicity of the cell
wall. The collection efficiency is the production of the collision
and attachment efficiencies. In the case of S. marcescens the
attachment efficiency appears to be near unity, since the calculated collision efficiency, about 10^{-3}, is about the same as the
observed collection efficiency (Weber et al., 1982).

   It is not necessary to go from one species of bacteria to
another to find large differences in bubble collection efficiency
and drop EF. Under some conditions dramatic differences in jet
drop EF can be found in experiments using only S. marcescens.
Normally the S. marcescens cell produces a pigment, prodigiosin,
that causes the colonies to be blood red in color. However,
environmental stresses such as temperature, nutrients, and UV
radiation, can alter the production of prodigiosin to produce
colonies that range in color from normal blood-red to pink,
orange, and white (Rizki, 1960; Williams, 1973). Colony growth
rates and cell morphology do not change with cell color. We were
surprised to find that, when a bubble is allowed to pass through

a suspension of both red and white S. marcescens cells, the EF
for the top jet drops can be several hundred for the red cells,
but only about unity for the white cells (Blanchard and Syzdek,
1978). We added pink cells to the suspension and found the drop
EF for these cells to be between that for the red and white
cells! The reason for this marked correlation of drop EF with
cell color is not known, though I suspect that the amount of
prodigiosin produced by a cell is correlated with its
hydrophobicity.

4.2 Film Drop Enrichment

    Bubbles larger than about 1 mm most likely produce more film
drops than jet drops (Figure 6). But even if they don't, the jet
drops are so large that they return quickly to the water, leaving
it to the much smaller film drops to carry bacteria into the
atmosphere. Since film drops from these larger bubbles may be
playing a major role in the water-to-air transfer of materials
(Figure 7), the bacterial EF of these drops should be investigated.

    Film drops, unlike jet drops, whose trajectories and narrow
size range are quite predictable, have a wide size range and
trajectories that are unpredictable. Consequently, they cannot
be collected and analyzed for bacteria as easily as are the jet
drops. However, we find that electrostatic induction can be used
to pull film drops upward, to impact randomly over the surface of
an inverted agar plate. After incubation, the number of colonies
tell us the number of viable bacteria carried by the film drops.
It can be shown that it is highly unlikely that more than one cell
is carried per film drop, otherwise the EF would be improbably
high. After exposing a sufficient number of agar plates,
gelatin-coated glass slides can be used to get the film-drop size
distribution. Combining the two data sets will give an average
EF for the film drops. Nothing can be said about the variation
of EF with drop size.

    We have used the apparatus of Figure 18 to determine the EF
for film drops from bubbles of 1.7 mm diameter rising <2 cm
through a suspension of S. marcescens (Blanchard and Syzdek,
1982). Each bubble produced 10 to 20 film drops, which ranged in
size from <2 μm to over 30 μm diameter. Half the drops were
<10 μm. The EF's were between 10 and 20. The only other data we
are aware of are those obtained by Cipriano (1979), who worked
with a seawater suspension of S. marinorubra. His film drop EF's
were around 30. However, he was using an entirely different
method of drop collection that required an inference that film
drops, and not jet drops, were being collected. I believe his
inference was correct, but the design of future experiments of
film drop EF should leave no doubt that film drops are being
collected.

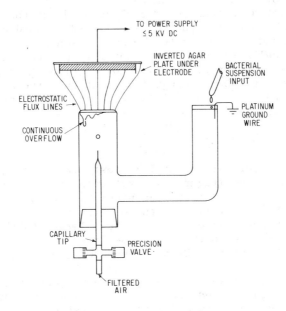

Figure 18. Use of electrostatic induction to pull film drops upward to strike an inverted agar plate. (From Blanchard and Syzdek, 1982.)

4.3 Health Hazards

There are potential health hazards from pathogen-containing aerosols. These aerosols can be produced from bubbles in house-hold toilets (Gerba et al., 1975), wastewater treatment plants (Ledbetter et al., 1973; Hickey and Reist, 1975), and possibly at spray irrigation sites (Bausum et al., 1982). It is probable that some of the pathogens are enriched in jet and film drops. But in spite of numerous studies, it has proved difficult to establish a definitive cause-and-effect relation between these aerosols and the general health of people living near the aerosol source. It is perhaps fortunate for us that airborne bacteria can be rapidly destroyed by ultra-violet radiation. But since the sun does not shine at night, and under certain conditions bacteria can form new cells in airborne particles (Dimmick et al., 1979), the possibility of health hazards must not be dismissed.

There is one pathogen that has been shown to cause disease after being ejected into the atmosphere by bubbling and splashing processes. This is the bacterium Legionella pneumophila, the cause of the respiratory disease, legionellosis, commonly known as the Legionnaires' disease. Although L. pneumophila has been

isolated from many lakes (Fliermans et al., 1979), its
association with the aerosol produced by air-conditioning cooling
towers appears to be the cause of some of the outbreaks of
Legionnaires' disease (Dondero et al., 1980).

For people living along the southeastern shore of the
United States, from Cape Hatteras to Texas and for over 100 km
inland, there is an extraordinarily high percentage (>70%) that
show skin sensitivity to the antigens of Mycobacterium
intracellulare. This bacterium is the cause of a pulmonary
disease whose epidemiology is not known. Gruft et al. (1975)
have suggested that the infection may have been caused by breath-
ing a bacterial aerosol that originated from the bursting of
bubbles in seawater in bays and inlets along the shore. Their
laboratory experiments showed that M. intracellulare survives
well in seawater, and that the bacterial enrichment factors in
jet drops can be several hundred. Subsequent work (Wendt et al.,
1980; Gruft et al., 1981) has shown that M. intracellulare exists
in seawater along the southeastern shore, and it has been
isolated from an aerosol that presumably came from the sea. This,
however, does not prove that a sea-produced, enriched aerosol of
M. intracellulare is causing the infection: there are other
possible sources of the aerosol, including that blown into the
air from the land. Research continues in an attempt to evaluate
the relative importance of sea and land-produced M. intracellulare aerosols.

4.4 Bacteria as Ice Nuclei

The concentration of ice nuclei in the atmosphere is about
a million times less than that of cloud condensation nuclei.
Researches on ice nuclei cover half a century, yet an understanding of the origin of these elusive particles eludes us. Soulage
(1957) appears to have been the first to suggest that some of
the ice nuclei may be bacteria. In a series of thought-provoking
papers, Schnell (1977), Schnell and Vali (1972, 1976), and Vali
et al. (1976) tested many samples of decaying plant and leaf
litter from each of the major climate zones and made the amazing
finding that, regardless of what species of leaf litter they
tested, the ice nucleus spectrum was roughly constant for a given
climate zone but showed large differences from one climate zone
to another. They found the concentration of atmospheric ice
nuclei in any climate zone to correlate highly with the ice
nuclei produced by leaf litter in that climate zone. Some of the
ice nuclei produced from the leaf litter were active at temperatures as warm as -2 to -5°C. The agent producing the ice nuclei
was isolated from the leaf litter and found to be the bacterium
Pseudomonas syringae. Although the bacterial cells themselves
were the ice nuclei, the ice-nucleating ability was not present
in every cell.

Schnell and Vali (1976) found that seawaters rich in plankton have high concentrations of ice nuclei. They suggested that the high ice-nucleus concentrations, found by others in the atmosphere above parts of the ocean, are over regions rich in plankton. The inference is, of course, that many of these ice nuclei came from the sea, presumably from bursting bubbles. From their work with the leaf litter and seawater samples they conclude that "...large proportions of atmospheric ice nuclei are possibly of biogenic origin."

5.   CONCLUSIONS

We know little about bubble-size distributions in breaking waves. Laboratory studies have given insights into the shape of these distributions, but measurements are needed at sea. The lab work suggests that film drops may be more important in the particle flux to the atmosphere than jet drops, but this is based upon certain assumptions regarding film-drop production. These assumptions may not be valid, since the number of film drops from a given-size bubble may vary by a factor of 100. The conditions that control this large variation are not understood and deserve detailed study.

Estimates of particle flux require not only a knowledge of whitecap bubble spectra but also the fractional coverage of the sea by whitecaps. The excellent work of Monahan (1980) and his colleagues has given much information on this coverage.

What is the smallest particle produced by the sea? We don't know, but Cipriano and Blanchard's (1981) laboratory study suggests it is $\leqslant 0.014$ μm diameter as a dry salt particle. Despite many measurements of the atmospheric sea-salt concentration, we know little about the amount of salt cycled through different altitudes. Also, efforts to detect the mechanism of production of the sea-salt inversion should be continued.

Bubble scavenging of bacteria is responsible for high jet-drop enrichment factors, even though bacterial microlayers may exist at the surface of the water. The same may not be true for film drops; they might be enriched by surface microlayers. Film-drop bacteria enrichment may be more important in aerobiology than jet drops, especially since there appears to be a cutoff in the jet-drop enrichment factor for drops less than about 20 μm diameter. The spread of disease by bacteria in airborne droplets has been shown, but the role played by jet and film drop enrichment is unknown.

ACKNOWLEDGEMENTS

    I am grateful to Keith Chave and Virginia Greenberg, both of
the University of Hawaii. Keith provided much-needed lab space
during my search for the sea-salt inversion, and Virginia, a
virtuoso on the flame atomic-adsorption spectrophotometer,
analyzed quickly and accurately nearly 300 sea-salt samples. My
friend and colleague of many scientific adventures, Alfred
Woodcock, who taught me most of what I know about seeing the
unseen in the sea and the air, flew with me on all flights. I
benefited from the suggestions of my colleagues here at the ASRC,
Lawrence Syzdek and Ramon Cipriano, who read the first draft of
this chapter. I especially appreciated the numerous constructive
comments of George Slinn, who found many errors overlooked by the
rest of us. I thank Mary Haley, who did all the typing. This
work was supported by the National Science Foundation under
Grant No. ATM-8015495.

REFERENCES

Aitken, J., 1881: On dust, fogs, and clouds. Trans. Roy. Soc.
    Edinburgh, 30, 337-368.
Bausum, H.T., S.A. Schaub, K.F. Kenyon, and M.J. Small, 1982:
    Comparison of coliphage and bacterial aerosols at a waste-
    water spray irrigation site. Appld. Environ. Microbiol.,
    43, 28-38.
Baylor, E.R., M.B. Baylor, D.C. Blanchard, L.D. Syzdek, and
    C. Appel, 1977a: Virus transfer from surf to wind. Science,
    198, 575-580.
Baylor, E.R., V. Peters, and M.B. Baylor, 1977b: Water-to-air
    transfer of virus. Science, 197, 763-764.
Beck, J.B., 1819: Observations on salt storms and the influence
    of salt and saline air upon animal and vegetable life.
    Amer. J. Sci., 1, 388-397.
Berg, W.W., Jr., and J.W. Winchester, 1978: Aerosol chemistry of
    the marine atmosphere, in: Chemical Oceanography, vol. 7
    (2nd edition), (J.P. Riley and R. Chester, eds.), 173-231,
    Academic Press, New York.
Bezdek, H.F., and A.F. Carlucci, 1972: Surface concentration of
    marine bacteria. Limnol. Oceanogr., 17, 566-569.
Bezdek, H.F., and A.F. Carlucci, 1974: Concentration and
    removal of liquid microlayers from a seawater surface by
    bursting bubbles. Limnol. Oceanogr., 19, 126-132.
Blanchard, D.C., 1963: The electrification of the atmosphere by
    particles from bubbles in the sea. Prog. Oceanog., 1,
    71-202.
Blanchard, D.C., 1975: Bubble scavenging and the water-to-air
    transfer of organic material in the sea, in: Applied
    Chemistry at Protein Interfaces (R. Baier, ed.), Adv. Chem.
    Ser. 145, 360-387.

Blanchard, D.C., 1978: Jet drop enrichment of bacteria, virus, and dissolved organic material. Pure and Appld. Geophys., 116, 302-308.

Blanchard, D.C., and E.J. Hoffman, 1978: Control of jet-drop dynamics by organic material in seawater. J. Geophys. Res., 83, 6187-6191

Blanchard, D.C., and A.T. Spencer, 1964: Condensation nuclei and the crystallization of saline drops. J. Atmos. Sci., 21, 182-186.

Blanchard, D.C., and L.D. Syzdek, 1970: Mechanism for the water-to-air transfer and concentration of bacteria. Science, 170, 626-628.

Blanchard, D.C., and L.D. Syzdek, 1972a: Concentration of bacteria in jet drops from bursting bubbles. J. Geophys. Res., 77, 5087-5099.

Blanchard, D.C., and L.D. Syzdek, 1972b: Variations in Aitken and giant nuclei in marine air. J. Phys. Oceanog., 2, 255-262.

Blanchard, D.C., and L.D. Syzdek, 1974: Importance of bubble scavenging in the water-to-air transfer of organic material and bacteria. J. de Rech. Atmos., 8, 529-540.

Blanchard, D.C., and L.D. Syzdek, 1975: Electrostatic collection of jet and film drops. Limnol. Oceanogr., 20, 762-774.

Blanchard, D.C., and L.D. Syzdek, 1977: Production of air bubbles of a specified size. Chem. Engr. Sci., 32, 1109-1112.

Blanchard, D.C., and L.D. Syzdek, 1978: Seven problems in bubble and jet drop researches. Limnol. Oceanogr., 23, 389-400.

Blanchard, D.C., and L.D. Syzdek, 1982: Water-to-air transfer and enrichment of bacteria in drops from bursting bubbles. Appl. Environ. Microbiol., 43, 1001-1005.

Blanchard, D.C., and A.H. Woodcock, 1957: Bubble formation and modification in the sea and its meteorological significance. Tellus, 9, 145-158.

Blanchard, D.C., and A.H. Woodcock, 1980: The production, concentration and vertical distribution of the sea-salt aerosol. Annals N.Y. Acad. Sci., 338, 330-347.

Blanchard, D.C., L.D. Syzdek, and M.E. Weber, 1981: Bubble scavenging of bacteria in freshwater quickly produces bacterial enrichment in airborne drops. Limnol. Oceanogr., 26, 961-964.

Brierly, W.B., 1970: Bibliography on atmospheric (cyclic) sea-salts. Tech. Rpt. 70-63-ES, U.S. Army Natick Laboratories, Natick, MA, 01760.

Bunker, A.H., B. Haurwitz, J.S. Malkus, and H. Stommel, 1949: Vertical distribution of temperature and humidity over the Caribbean Sea. Pap. Phys. Oceanogr. and Meteor., 11 (1), 82 pp.

Burger, S.R., and D.C. Blanchard, 1982: Bubble lifetime at a water surface. Unpublished manuscript, available on request.

Byers, H.R., J.R. Sievers, and B.J. Tufts, 1957: Distribution in the atmosphere of certain particles capable of serving as condensation nuclei, in: Artificial Stimulation of Rain, 47-72, Pergamon Press, New York.

Carlucci, A.F., and P.M. Williams, 1965: Concentration of bacteria from sea water by bubble scavenging. J. Cons. Perm. Int. Explor. Mer., 30, 28-33.

Chaen, M., 1973: Studies on the production of sea-salt particles on the sea surface. Memoirs of the Faculty of Fisheries, Kagoshima University, vol. 22(2), 49-107.

Cipriano, R., 1979: Bubble and Aerosol Spectra Produced by a Laboratory Simulation of a Breaking Wave. Ph.D. Thesis, State University of New York at Albany.

Cipriano, R.J., and D.C. Blanchard, 1981: Bubble and aerosol spectra produced by a laboratory 'breaking wave.' J. Geophys. Res., 86, 8085-8092.

Cipriano, R.J., D.C. Blanchard, A.W. Hogan, and G.G. Lala, 1982: On the production of Aitken nuclei from breaking waves and their role in the atmosphere. Unpublished manuscript, reprints available.

Clift, R., J.R. Grace, and M.E. Weber, 1978: Bubbles, Drops, and Particles. Academic Press, 380 pp.

Crow, S.A., D.G. Ahearn, W.L. Cook, and A.W. Bourquin, 1975: Densities of bacteria and fungi in coastal surface films as determined by a membrane-adsorption procedure. Limnol. and Oceanogr., 20, 644-646.

Darrozès, J.S., and P. Ligneul, 1982: The production of drops by the bursting of a bubble at an air liquid interface, in: Proc. 2nd Intl. Colloquium on Drops and Bubbles, JPL Pub. 82-7, 157-165, Jet Propulsion Lab., Calif. Inst. Tech., Pasadena, CA.

Day, J.A., 1964: Production of droplets and salt nuclei by the bursting of air bubble films. Quart. J. Roy. Met. Soc., 90, 72-78.

Dessens, H., 1946: Les noyaux de condensation de l'atmosphere. C. R. Acad. Sci. Paris, 223, 915-917.

Detwiler, A., and D.C. Blanchard, 1978: Aging and bursting bubbles in trace-contaminated water. Chem. Engr. Sci., 33, 9-13.

Dimmick, R.L., and A.B. Akers, 1969: An Introduction to Experimental Aerobiology. Wiley-Interscience, 494 pp.

Dimmick, R.L., H. Wolochow, and M.A. Chatigny, 1979: Evidence for more than one division of bacteria within airborne particles. Appl. Environ. Microbiol., 38, 642-643.

Dondero, T.J., Jr., R.C. Rendtorff, G.F. Mallison, R.M. Weeks, J.S. Levy, E.W. Wong, and W. Schaffner, 1980: An outbreak of Legionnaires' Disease associated with a contaminated air-conditioning cooling tower. N. Engl. J. Med., 302, 365-370.

Duce, R.A., 1982: SEAREX: a multi-institutional investigation
    of the sea/air exchange of pollutants and natural sub-
    stances, in: Marine Pollutant Transfer Processes (M.
    Waldichuk, G. Kullenberg, and M. Orren, eds.), Elsevier Pub.
    Co., in press.
Duce, R.A., and E.J. Hoffman, 1976: Chemical fractionation at
    the air/sea interface. Annual Rev. Earth and Planetary
    Sciences, 4, 187-228.
Duce, R.A., and A.H. Woodcock, 1971: Difference in chemical com-
    position of atmospheric sea salt particles produced in the
    surf zone and on the open sea in Hawaii. Tellus, 23, 427-435.
Eriksson, E., 1959: The yearly circulation of chloride and sul-
    fur in nature; meteorological, geochemical and pedological
    implications. Part 1. Tellus, 11, 375-403.
Eriksson, E., 1960: The yearly circulation of chloride and sul-
    sur in nature; meteorological, geochemical and pedological
    implications, Part II. Tellus, 12, 63-109.
Fliermans, C.B., W.B. Cherry, L.H. Orrison, and L. Thacker, 1979:
    Isolation of Legionella pneumophila from nonepidemic-related
    aquatic habitats. Appld. Environ. Microbiol., 37, 1239-1242.
Fraizier, A., M. Masson, and J.C. Guary, 1977: Recherches
    preliminaires sur le role des aerosols dans le transport de
    certains radioelements du milieu marin au milieu terrestre.
    J. Rech. Atmos., 11, 49-60.
Garrett, W.D., 1967a: Stabilization of air bubbles at the air-
    sea interface by surface-active material. Deep-Sea Research,
    14, 661-672.
Garrett, W.D., 1967b: The organic chemical composition of the
    ocean surface. Deep-Sea Research, 14, 221-227.
Garrett, W.D., 1968: The influence of monomolecular surface
    films on the production of condensation nuclei from bubbled
    sea water. J. Geophys. Res., 73, 5145-5150.
Garrett, W.D., 1981: Comment on "Organic particle and aggregate
    formation resulting from the dissolution of bubbles in sea-
    water" (Johnson and Cooke). Limnol. and Oceanogr., 26,
    989-992.
Gaudin, A.M., N.S. Davis, and S.E. Bangs, 1962: Flotation of
    Escherichia coli with sodium chloride. Biotech. and
    Bioengr., 4, 211-222.
Gerba, C.P., C. Wallis, and J.L. Melnick, 1975: Microbiological
    hazards of household toilets: droplet production and the
    fate of residual organisms. Appld. Microbiology, 30,
    229-237.
Gregory, P.H., 1961: The Microbiology of the Atmosphere. Inter-
    science Publishers, Inc., New York, 251 pp.
Grieves, R.B., and S.L. Wang, 1967: Foam separation of bacteria
    with a cationic surfactant. Biotechnology and Bioengr., 9,
    187-194.

Gruft, H., J.O. Falkinham, III, and B.C. Parker, 1981: Recent
    experience in the epidemiology of disease caused by atypical
    mycobacteria. Rev. Infec. Dis., 3, 990-996.
Gruft, H., J. Katz, and D.C. Blanchard, 1975: Postulated source
    of Mycobacterium intracellulare (Battey) infection. Amer.
    J. Epidemiology, 102, 311-318.
Hayami, S., and Y. Toba, 1958: Drop production by bursting of
    air bubbles on the sea surface (1) experiments at still sea
    water surface. J. Ocean. Soc. Japan, 14, 145-150.
Hejkal, T.W., P.A. LaRock, and J.W. Winchester, 1980: Water-to-
    air fractionation of bacteria. Appld. Environ. Microbiology,
    39, 335-338.
Hickey, J.L.S, and P.C. Reist, 1975: Health significance of
    airborne microorganisms from wastewater treatment processes.
    Part II: Health significance and alternatives for action.
    J. Water Poll. Con. Fed., 47, 2758-2773.
Higgins, F.B., Jr., 1964: Bacterial Aerosols from Bursting
    Bubbles. Ph.D. thesis, Georgia Institute of Technology.
Hobbs, P.V., 1971: Simultaneous airborne measurements of cloud
    condensation nuclei and sodium-containing particles over the
    ocean. Quart. J. Roy. Met. Soc., 97, 263-271.
Hoffman, G.L., 1971: Particulate Trace Metals in the Hawaiian
    Marine Atmosphere. Ph.D. thesis, University of Hawaii.
Hoffman, E.J., G.L. Hoffman, and R.A. Duce, 1974: Chemical
    fractionation of alkali and alkaline earth metals in atmos-
    pheric particulate matter over the North Atlantic. J. de
    Rech. Atmos., 8, 675-688.
Hoffman, E.J., and R.A. Duce, 1976: Factors influencing the
    organic carbon content of marine aerosols: a laboratory
    study. J. Geophys. Res., 81, 3667-3670.
Hoffman, E.J., and R.A. Duce, 1977: Organic carbon in marine
    atmospheric particulate matter: concentration and particle
    size distribution. Geophys. Res. Letters, 4, 449-452.
Horrocks, W.H., 1907: Experiments made to determine the conditions
    under which "specific" bacteria derived from sewage may be
    present in the air of ventilating pipes, drains, inspection
    chambers and sewers. Proc. Roy. Soc. London, 79 (Ser. B),
    255-266.
Hunter, K.A., and P.S. Liss, 1981: Organic sea surface films,
    in: Marine Organic Chemistry, Chapter 9, 259-298 (E.K.
    Duursma and R. Dawson, eds.), Elsevier Scientific Pub. Co.,
    522 pp.
Iribarne, J.V., D. Corr, B.Y.H. Liu, and D.Y.H. Pui, 1977: On
    the hypothesis of particle fragmentation during evaporation.
    Atmos. Environ., 11, 639-642.
Jacobs, W.C., 1937: Preliminary report on a study of atmospheric
    chlorides. Mon. Wea. Rev., 65, 147-151.
Johnson, B., 1979: The Rate of Organic Particle Production
    Resulting from Bubble Dissolution in the Ocean. Ph.D. thesis,
    Dalhousie University.

Johnson, B., and R.C. Cooke, 1979: Bubble populations and
    spectra in coastal waters; a photographic approach. J.
    Geophys. Res., 84, 3761-3766.
Johnson, B.D., and R.C. Cooke, 1980: Organic particle and
    aggregate formation resulting from the dissolution of
    bubbles in seawater. Limnol. and Oceanogr., 25, 653-661.
Johnson, B.D., and R.C. Cooke, 1981: Generation of stabilized
    microbubbles in seawater. Science, 213, 209-211.
Junge, C.E., 1963: Air Chemistry and Radioactivity, Academic
    Press, New York, 382 pp.
Kanwisher, J., 1963: On the exchange of gases between the atmos-
    phere and the sea. Deep-Sea Res., 10, 195-207.
Koga, M., 1981: Direct production of droplets from breaking
    wind-waves--its observation by a multicolored overlapping
    exposure photography technique. Tellus, 33, 552-563.
Kolovayev, P.A., 1976: Investigation of the concentration and
    statistical size distribution of wind-produced bubbles in
    the near-surface ocean layer. Oceanology, 15, 659-661.
Lai, R.J., and O.H. Shemdin, 1974: Laboratory study of the
    generation of spray over water. J. Geophys. Res., 79,
    3055-3063.
Ledbetter, J.O., L.M. Hauck, and R. Reynolds, 1973: Health
    hazards from wastewater treatment practices. Environmental
    Letters, 4, 225-232.
Lodge, J.P., 1955: A study of sea-salt particles over Puerto
    Rico, J. Meteor., 12, 493-499.
Lovett, R.F., 1975: The Occurrence of Airborne Sea-Salt and Its
    Meteorological Dependence. M.S. thesis, Heriot-Watt
    University, Edinburgh.
Lovett, R.F., 1978: Quantitative measurement of airborne sea-
    salt in the North Atlantic. Tellus, 30, 358-363.
MacIntyre, F., 1968: Bubbles: a boundary-layer "microtome" for
    micron-thick samples of a liquid surface. J. Phys. Chem.,
    72, 589-592.
MacIntyre, F., 1972: Flow patterns in breaking bubbles. J.
    Geophys. Res., 77, 5211-5228.
MacIntyre, F., 1974: Chemical fractionation and sea-surface
    microlayer processes, in: The Sea (Marine Chemistry), vol.
    5, 245-299 (E.D. Goldberg, ed.), John Wiley and Sons, New
    York.
MacIntyre, F., 1978: Additional problems in bubble and jet drop
    research. Limnol. and Oceanogr., 23, 571-573.
Mason, B.J., 1971: The Physics of Clouds. Oxford University
    Press, 671 pp.
McDonald, R.L., C.K. Unni, and R.A. Duce, 1982: Estimation of
    atmospheric sea salt dry deposition: wind speed and particle
    size dependence. J. Geophys. Res., 87, 1246-1250.
Medwin, H., 1977: In situ acoustic measurements of microbubbles
    at sea. J. Geophys. Res., 82, 971-976.

Monahan, E.C., 1966:   Sea Spray and Its Relationship to Low
    Elevation Wind Speed.   Ph.D. thesis, Mass. Institute of
    Technology, 175 pp.
Monahan, E.C., 1968:   Sea spray as a function of low elevation
    wind speed.   J. Geophys. Res., 73, 1127-1137.
Monahan, E.C., 1971:   Oceanic whitecaps.   J. Phys. Oceanog., 1,
    139-144.
Monahan, E.C., C.W. Fairall, K.L. Davidson, and P.J. Boyle, 1982:
    Observed inter-relationships amongst 10 m-elevation winds,
    oceanic whitecaps, and marine aerosols.   Unpublished ms.
Monahan, E.C., and I. O'Muircheartaigh, 1980:   Optimal power-law
    description of oceanic whitecap coverage dependence on wind
    speed.   J. Phys. Oceanogr., 10, 2094-2099.
Monahan, E.C., and C.R. Zietlow, 1969:   Laboratory comparisons of
    fresh-water and salt-water whitecaps.   J. Geophys. Res., 74,
    6961-6966.
Norkrans, B., and F. Sörensson, 1977:   On the marine lipid sur-
    face microlayer--bacterial accumulation in model systems.
    Botanica Marina, 20, 473-478.
Paterson, M.P., and K.T. Spillane, 1969:   Surface films and the
    production of sea-salt aerosol.   Quart. J. Roy. Met. Soc.,
    95, 526-534.
Patterson, E.M., C.S. Kiang, A.C. Delany, A.F. Wartburg, A.C.D.
    Leslie, and B.J. Huebert, 1980:   Global measurements of
    aerosols in remote continental and marine regions: concen-
    trations, size distributions, and optical properties.   J.
    Geophys. Res., 85, 7361-7376.
Petrenchuk, O.P., 1980:   On the budget of sea salts and sulfur
    in the atmosphere.   J. Geophys. Res., 85, 7439-7444.
Podzimek, J., 1980:   Advances in marine aerosol research.   J.
    Rech. Atmos., 14, 35-61.
Prospero, J.M., 1979:   Mineral and sea salt aerosol concentrations
    in various ocean regions.   J. Geophys. Res., 84, 725-730.
Quinn, J.A., R.A. Steinbrook, and J.L. Anderson, 1975:   Breaking
    bubbles and the water-to-air transport of particulate
    matter.   Chem. Engr. Sci., 30, 1177-1184.
Resch, F., and F. Avellan, 1982:   Size distribution of oceanic
    air bubbles entrained in sea-water by wave-breaking, in:
    Proc. 2nd Intl. Colloquium on Drops and Bubbles, JPL Pub.
    82-7, 182-186, Jet Propulsion Lab., Calif. Inst. Tech.,
    Pasadena, CA.
Rizki, M.T.M., 1960:   Factors influencing pigment production in
    a mutant strain of Serratia marcescens.   J. Bact., 80,
    305-310.
Ross, D.B., and V. Cardone, 1974:   Observations of oceanic white-
    caps and their relation to remote measurements of surface
    wind speed.   J. Geophys. Res., 79, 444-452.
Rubin, A.J., 1968: Microflotation: coagulation and foam separa-
    tion of Aerobacter aerogenes.   Biotechnology and Bioengr.,
    10, 89-98.

Ruskin, R.E., R.K. Jeck, F.K. Lepple, and W.A. Von Wald, 1978:
    Salt aerosol survey at gas turbine inlet aboard USS Spruance.
    NRL Rpt. 3804, Naval Research Laboratory, Washington, DC.
    113 pp.

Savoie, D.L., and J.M. Prospero, 1977: Aerosol concentration
    statistics for the northern tropical Atlantic. J. Geophys.
    Res., 82, 5954-5964.

Schacher, G.E., K.L. Davidson, C.W. Fairall, and D.E. Spiel, 1981:
    Calculation of optical extinction from aerosol spectral
    data. Applied Optics, 20, 3951-3957.

Schnell, R.C., 1977: Ice nuclei in seawater, fog water and
    marine air off the coast of Nova Scotia: summer 1975. J.
    Atmos. Sci., 34, 1299-1305.

Schnell, R.C., and G. Vali, 1972: Atmospheric ice nuclei from
    decomposing vegetation. Nature, 236, 163-165.

Schnell, R.C., and G. Vali, 1976: Biogenic ice nuclei. Part I:
    Terrestrial and marine sources. J. Atmos. Sci., 33, 1554-
    1564.

Sieburth, J. McN., 1979: Sea Microbes. Oxford Univ. Press,
    491 pp.

Sieburth, J. McN., P. Willis, K.M. Johnson, C.M. Burney, D.M.
    Lavoie, K.R. Hinga, D.A. Caron, F.W. French, III, P.W.
    Johnson, and P.G. Davis, 1976: Dissolved organic matter and
    heterotropic microneuston in the surface microlayers of the
    North Atlantic. Science, 194, 1415-1418.

Sigerson, G., 1870: Micro-atmospheric researches. Proc. Roy.
    Irish Acad., 1 (2nd Series, Science), 13-20.

Soulage, G., 1957: Les noyaux de congelation de l'atmosphere.
    Ann. Geophys., 13, 103-134.

Stuhlman, O., 1932: The mechanics of effervescence. Physics,
    2, 457-466.

Syzdek, L.D., 1982: Concentrations of Serratia marcescens in the
    surface microlayer. Limnol. and Oceanogr., 27, 172-177.

Tanaka, M., 1966: On the transport and distribution of giant
    sea-salt particles over land [1] theoretical model. Special
    Contributions, Geophys. Inst., Kyoto Univ., No. 6, 47-57.

Tedesco, R., and D.C. Blanchard, 1979: Dynamics of small bubble
    motion and bursting in freshwater. J. Rech. Atmos., 13,
    215-226.

Thorpe, S.A., 1982: On the clouds of bubbles formed by breaking
    wind-waves in deep water, and their role in air-sea gas
    transfer. Phil. Trans. R. Soc. London, A304, 155-210.

Toba, Y., 1965a: On the giant sea-salt particles in the atmos-
    phere. I. General features of the distribution. Tellus,
    17, 131-145.

Toba, Y., 1965b: On the giant sea-salt particles in the atmos-
    phere. II. Theory of the vertical distribution in the 10-m
    layer over the ocean. Tellus, 17, 365-382.

Toba, Y., 1966: On the giant sea-salt particles in the atmos-
    phere. III. An estimate of the production and distribution
    over the world ocean. Tellus, 18, 132-145.

Toba, Y., and M. Chaen, 1973: Quantitative expression of the breaking of wind waves on the sea surface. Records of Oceanographic Works in Japan, 12, 1-11.

Twomey, S., 1953: The identification of individual hygroscopic particles in the atmosphere by a phase-transition method. J. Appl. Phys., 24, 1099-1102.

Twomey, S., 1955: The distribution of sea-salt nuclei in air over land. J. Met., 12, 81-86.

Twomey, S., and K.N. McMaster, 1955: The production of condensation nuclei by crystallizing salt particles. Tellus, 7, 458-461.

Vali, G., M. Christensen, R.W. Fresh, E.L. Galyan, L.R. Maki, and R.C. Schnell, 1976: Biogenic ice nuclei. Part II: Bacterial sources. J. Atmos. Sci., 33, 1565-1570.

Wallace, G.T., Jr., and R.A. Duce, 1978: Transport of particulate organic matter by bubbles in marine waters. Limnol. and Oceanogr., 23, 1155-1167.

Weber, M.E., 1981: Collision efficiencies for small particles with a spherical collector at intermediate Reynolds numbers. J. Separation Process Technology, 2, 29-33.

Weber, M.E., D.C. Blanchard, and L.D. Syzdek, 1982: The mechanism of scavenging of water-borne bacteria by a rising bubble. Limnol. Oceanogr., in press.

Weisel, C.P., 1981: The Atmospheric Flux of Elements from the Ocean. Ph.D. thesis, the University of Rhode Island.

Wendt, S.L., K.L. George, B.C. Parker, H. Gruft, and J.O. Falkinham, III, 1980: Epidemiology of infection by nontuberculous mycobacteria III. Isolation of potentially pathogenic mycobacteria from aerosols. Am. Rev. Respir. Dis., 122, 259-263.

Williams, R.P., 1973: Biosynthesis of prodigiosin, a secondary metabolite of Serratia marcescens. Appl. Microb., 25, 396-402.

Woodcock, A.H., 1948: Note concerning human respiratory irritation associated with high concentrations of plankton and mass mortality of marine organisms. J. Marine Res., 7, 56-62.

Woodcock, A.H., 1950: Sea salt in a tropical storm. J. Met., 7, 397-401.

Woodcock, A.H., 1952: Atmospheric salt particles and raindrops. J. Met., 9, 200-212.

Woodcock, A.H., 1953: Salt nuclei in marine air as a function of altitude and wind force. J. Met., 10, 362-371.

Woodcock, A.H., 1955: Bursting bubbles and air pollution. Sewage and Industrial Wastes, 27, 1189-1192.

Woodcock, A.H., 1962: Solubles, in: The Sea, Physical Oceanography, vol. 1 (M.N. Hill, ed.), 305-312, Interscience, New York.

Woodcock, A.H., 1972: Smaller salt particles in oceanic air and bubble behavior in the sea. J. Geophys. Res., 77, 5316-5321.

Woodcock, A.H., 1975: Anomalous orographic rains of Hawaii.
    Mon. Wea. Rev., 103, 334-343.
Woodcock, A.H., D.C. Blanchard, and C.G.H. Rooth, 1963: Salt-
    induced convection and clouds. J. Atmos. Sci., 20, 159-169.
Woodcock, A.H., and R.A. Duce, 1972: The "large" salt nuclei
    hypothesis of raindrop growth in Hawaii: further measure-
    ments and discussion. J. Rech. Atmos., 6, 639-649.
Woodcock, A.H., and M.M. Gifford, 1949: Sampling atmospheric
    sea-salt nuclei over the ocean. J. Marine Res., 8, 177-197.
Wu, J., 1979: Oceanic whitecaps and sea state. J. Phys.
    Oceanogr., 9, 1064-1068.
Wu, J., 1981: Bubble populations and spectra in near-surface
    ocean: summary and review of field measurements. J.
    Geophys. Res., 86, 457-463.

PARTICLE GEOCHEMISTRY IN THE ATMOSPHERE AND OCEANS

Patrick Buat-Ménard

Centre des Faibles Radioactivités, Laboratoire mixte
CNRS-CEA, BP N°1
91190 , Gif-sur-Yvette, France

1. INTRODUCTION

The purpose of this chapter is to review our present knowledge of the geochemistry of atmospheric and oceanic particles.
I also want to show that these two cycles of particulate matter
are coupled by the physical, chemical, and biological processes that take place at the air-sea interface and in the surface waters of the ocean. This geochemical coupling is particularly strong for trace metals. Within a few years at most, atmospheric inputs of trace metals are transferred to the deep ocean
by the removal of these elements from the surface layers of the
ocean by settling particles, primarily of biogenic origin. On
the other hand, atmospheric sea-salt particles, enriched in organic matter and trace metals, are continuously recycled between
the oceans and the atmosphere and the fraction transported to
land will ultimately find its way back to the oceans, transported
by rivers. As a consequence, air-to-sea and sea-to-air particulate transfers are important for the understanding of many biogeochemical cycles and their alterations by man.

2. GEOCHEMISTRY OF MARINE AEROSOLS

2.1 Introduction

In this section I do not intend or pretend to present an
exhaustive review of the literature in the field of marine aerosol chemistry. My intention is rather to point out what we know,
what we do not know and what we should know to answer some funda-

P. S. Liss and W. G. N. Slinn (eds.), Air-Sea Exchange of Gases and Particles, 455–532.

mental problems still unsolved. Indeed, for many constituents of
the marine aerosol, sources, sinks and transport processes are
not yet fully understood and in a few cases our knowledge is pri-
mitive at best. Excellent reviews and collections of papers and
reports have been issued during the last ten years, and it is
suggested that for detailed information the following references
should be consulted as a first priority : JGR, 1972 ; JRA, 1974 ;
MacIntyre, 1974b ; Liss, 1975 ; Duce and Hoffman, 1976 ; Rahn,
1976 ; NAS, 1978 ; Berg and Winchester, 1978 ; Duce, 1978, 1981 ;
Coantic, 1980 ; Blanchard and Woodcock, 1980 ; Cicerone, 1981 ;
Lion and Leckie, 1981.; Prospero, 1981.

     To avoid confusion in this chapter, the term "marine aerosol"
will apply to all classes of particles found in the marine atmosphere. From this perspective, one definition to which I think
that everyone would agree is that the marine aerosol is a variable mixture of modified marine and continental source material.
It is a trace constituent of the marine atmosphere, since its
concentration may range generally from a few $\mu g\ m^{-3}$ to a few hundred $\mu g\ m^{-3}$. Most of the mass of the marine aerosol is found in
particles with diameters between 0.1 μm and 50 μm. The tropospheric residence time of such particles spans from a few hours to
a few days. There is,therefore,a fast geochemical coupling between
the continents and the oceans, through the atmospheric transport
of particles from land-to-sea and from sea-to-land.

2.2 Rationale

     These last years have seen the publication of a considerable
amount of new data. More and more substances are being investigated in the context of large scale programs, both national and
international. Although vertical distributions of atmospheric
concentrations are still poorly known (primarily because of sampling difficulties), latitudinal and longitudinal distributions,
based on the analysis of atmospheric samples collected from ships,
coastal areas and remote islands, are beginning to be reasonably
well documented. The present growing interest in this field is
not only a matter of curiosity about the chemical composition of
the marine atmosphere : indeed, there are several reasons, due to
global environmental concerns, why we need an improvement of our
knowledge in this field of research :

     a) The accurate description of global biogeochemical cycles
and their alterations by man. Via atmospheric deposition, the
oceans receive from the continents sedimentary material, nutrients, and other biologically useful elements, as well as increasing amounts of pollutants. At present, the open ocean is probably
more affected by pollution inputs through tropospheric transport
than through riverine transport. We need therefore to know the

magnitude of the anthropogenic input from the atmosphere to the
oceans and how this input affects the composition of the oceans.
As will be emphasized later, since there is a considerable amount
of recycling of trace substances accross the air-sea interface,
the evaluation of the net input of these substances from the continents to the oceans via atmospheric deposition is often extremely difficult.

An appreciable fraction of sea-source-aerosols, consisting
mainly of sea-salt, is transported over land and returned back to
the oceans by rivers. An accurate determination of the "cyclic"
component of elemental concentrations in rivers is of crucial
importance for geochemical mass balances of the ocean system. Such
a determination is also extremely difficult since : i) the composition of sea-source-aerosols is not the same as that of bulk
seawater, and ii) a fraction of this composition may consist of
recycled atmospheric inputs of continental origin.

b) <u>Impact of aerosols on climate</u>. The modification of the
chemistry of marine aerosols by man's activities has a potential
climatic impact (Bach, 1976). Atmospheric particles scatter solar
radiation. They also play a central role in cloud-droplet nucleation processes and as a consequence, may influence the micro-
structure of clouds, the Earth's albedo, and precipitation patterns. Such an influence cannot be estimated until the vertical
distribution of the physical and chemical properties of atmospheric aerosols is accurately known.

c)<u>Heterogeneous reactions in the marine atmosphere</u>. As described in this volume by Peters, marine aerosols act as a source
and a sink for an appreciable number of trace gases (nitrogen
compounds, sulfur compounds, organic compounds, halogenated compounds). The source and sink function for such compounds has to
be precisely investigated to assess properly their atmospheric
cycles.

d) <u>Precipitation chemistry</u>. Since cloud and rain droplets
are efficient scavengers of atmospheric aerosols (see Slinn, this
volume), the chemistry of precipitation obviously depends on that
of the preexisting aerosol. Here again, the knowledge of the
physical and chemical properties of atmospheric aerosols is of
crucial importance. Later in this chapter, I will emphasize the
case of those elements or compounds which are present both in
the particulate form and the gaseous form in the marine atmosphere.
Indeed, for some of these elements or compounds, the situation is
somewhat paradoxical : although the concentration in the gaseous
form dominates the concentration in the particulate form, the
rain content of that element or compound is controlled mainly by
the scavenging of the particulate form.

2.3   The Data Bank

Compared to ten years ago, the quality and quantity of data
available have increased by orders of magnitude. In the early
seventies, most of our knowledge was restricted to major components, such as sea-salt, insoluble continental dust, and their
associated major elements. Since then, many trace elements and
compounds, such as heavy metals and organic compounds, have been
identified and studied. Most of this is the result of improved
sampling and analytical techniques. An example is collection substrates. Filtering substrates are now specifically tested for the
element or compound being studied. For instance, inorganic substrates (such as glass fiber) are commonly used for the study of
particulate organic compounds, whereas organic substrates are used
to sample particulate inorganic compounds. Often, especially for
studies in remote areas, the filter material has to be chemically
processed (solvent extraction, acid-washing) to reduce filter-
blank   levels to a minimum. Collection artifacts, such as adsorption of vapor-phase compounds or loss by volatilization during
sampling, are now well documented in many cases. However, the use
of 'state-of-the-art' sampling and analytical techniques is by itself insufficient to obtain accurate data. Stringent precautions
against local contamination during sampling and sample contamination during handling and analysis have proven to be, for most
trace substances, an absolute necessity, especially for sampling
in remote areas.

2.4   Basic Criteria for Source Identification

Except for a very limited number of compounds, such as synthetic organic pollutants and artificial radionuclides, understanding the composition of the marine aerosol depends first on
our ability to identify sources. Our knowledge about the nature
and intensity of these sources is highly variable : for a given
element, what is now documented is not necessarily what is the
most important. In principle, the source problem can be solved if
we know the chemical composition and size distribution of aerosols
originating from a given source. Unfortunately, this ideal situation is almost never encountered in the marine atmosphere for the
following reasons : i) modification of the source material at
the land-air or sea-air interface, and ii) chemical and physical
modifications during transport in the atmosphere. In this second
case, important factors are sedimentation and coagulation during
transport, processes below, above and within clouds, and gas-to-
particle conversions. It is therefore the end product of all these
processes that we are dealing with when we sample the marine aerosol. Nevertheless, despite the fact that the marine aerosol cannot be considered to be conservative, we have in hand enough tools
so that solving the source problem is not a hopeless effort.

The basic approach, still widely used, is to consider
two of the major sources in terms of mass, the ocean and the surface of the earth's crust, and to characterize the presence of
atmospheric particulate material derived from such sources using
"reference elements". Na is most commonly used as the reference
element for the oceanic source, whereas Al, Fe or Sc are generally
used to characterize the presence of crustal aerosols (JGR, 1972 ;
JRA, 1974 ; Rahn, 1976 ; Duce et al., 1976a ; Buat-Ménard and
Chesselet, 1979).

When using bulk seawater and the composition of the crust as
normalizing factors, the general finding is that many elements in
the marine aerosol are present at much higher concentrations than
can be accounted for by the "unfractionated" dispersion of soil
or seawater. This situation appears to be world wide for many
trace metals (especially heavy metals), organic compounds, P, S,
N, and I. Many studies have focused on the causes of these apparent "enrichments" in marine aerosols because most of the anomalously-enriched elements or compounds are of interest from the
point of view of global pollution. Indeed, the long-range transport of anthropogenic aerosols was first recognized as a potential source for these anomalous elements. Because of their volatility (Bertine and Goldberg, 1971 ; Zoller et al., 1974 ; Duce
et al., 1975) certain heavy metals can apparently be injected
into the atmosphere in the vapor phase by high temperature combustion processes (fossil fuel combustion, incineration of wastes,
metallurgy, etc...). This will lead to their enrichment in the
air, compared to the concentration of more refractory elements.
This emission in the vapor phase is followed by homogeneous or,
more probably, heterogeneous condensation (see Peters, this volume). As a consequence, the enriched elements will occur preferentially in the sub-micrometer fraction of the aerosol, whereas
refractory elements occur on much larger particles. This has been
shown to be the case for metals such as Cu, Zn, Sb, Cd, and Pb
not only in urban air (Rahn, 1976) but also in the marine atmosphere. For example, at Bermuda (Duce et al., 1976a) and over the
Western Mediterranean Sea (Arnold et al., 1982) results of the
analysis of high-volume cascade impactor samples do indicate that
these enriched elements are concentrated in the submicrometer
size-range, whereas oceanic and crustal source elements are concentrated in much larger particles.

It has been realized, however, that such an observation in
the marine atmosphere does not unequivocally indict pollution
sources (Duce et al., 1975). Natural processes also volatilize
elements in the atmosphere. These processes are numerous, but
poorly documented. Among them are high-temperature volatilization
by volcanic activity (Mroz and Zoller, 1975 ; Buat-Ménard and

Arnold, 1978), crustal degassing (Goldberg, 1976), and biological
mobilization through methylation by land plants and marine life
(Wood, 1974 ; Beauford et al., 1977). Hence, by itself, the determination of elemental concentrations as a function of particle
size only suggests the existence of primary and secondary sources
for the marine aerosol. Such an approach, although not sufficient,
is nevertheless a first-priority step in source identification.

2.5 General Considerations of the Composition of Crustal and
     and Oceanic Source Aerosols.

In the marine atmosphere, excess over predicted concentrations may also result when the aerosol source material has a composition different from that which is assumed. Indeed, enrichment
factors are defined relative to a reference element in a specific
source. Generally, for crustal weathering products, and using Al
as the reference element, the enrichment factor, EF_{crust}, for an
element, X, is

$$EF_{crust} = (X/Al)_{air} / (X/Al)_{crust} \qquad (1)$$

where $(X/Al)_{air}$ and $(X/Al)_{crust}$ refer, respectively, to the ratio
of the concentration of element X to that of Al in the atmosphere
and in average crustal material. Likewise, for oceanic source material,

$$EF_{sea} = (X/Na)_{air} / (X/Na)_{sea} \qquad (2)$$

where the composition of bulk seawater is used for the calculation of EF_{sea}. Some caution must,therefore,be exercised in interpreting enrichment factors that are different from unity. This
does not seem to be too serious a problem in the case of soil
derived material, but is most certainly a problem for sea-source
aerosols.

Schutz and Rahn (1982) have studied the elemental composition of several desert soils as a function of particle size,
with special attention paid to the aerosol size-range. Compared
with the composition of bulk crustal rock, the great majority of
the elements (for the r<10 μm fraction of all the samples) have
enrichment factors within a factor 2-3 of unity, Na, Ca, and Sr
have mean enrichment factors <0.5 ; seven elements (Cu, Zn, As,
Sb, I, Au, Ag) have enrichment factors >3 (Table 1). Interestingly, these latter elements are generally enriched in the marine
aerosol as well, but with order-of-magnitude greater enrichments.
These results imply that, relative to the soils studied, the true

enrichment factors of Na, Ca, and Sr in marine aerosols are somewhat greater than values calculated from rock, and that the true
enrichments of Cu, Zn, etc., are less than values calculated from
rock. Nevertheless, it appears from this study that the aerosol-
size range of the desert soils is quite similar in composition to
bulk crustal rock, and that the latter therefore continue to be
an acceptable reference material for calculating aerosol-crust
enrichment factors. This has been found to be the case for Saharan aerosols collected over the tropical North Atlantic (Buat-
Ménard, 1979 ; Glaccum, 1978). Also, as found by Schutz and Rahn
(1982), the constancy of composition within the aerosol size
range of most soils, combined with the similar composition of the
soils studied, supports the concept of a single reference material
for crustal enrichment factor calculations. This work also suggests that differences in bulk compositions of soils will not
necessarily appear in aerosols produced from them, especially
more than 1000 km from the source, since such differences appear
to be concentrated in the coarse fraction of soils. On the other
hand, this argues against using the elemental composition of a
mineral aerosol to deduce the identity of a distant source region.
For this, other characteristics of soils, such as their mineralogical composition, may be more suitable (Glaccum and Prospero,
1980 ; Gaudichet and Buat-Ménard, 1982).

Table 1. Geometric mean enrichment factors for the r < 10 μm
fraction of desert soils from Sahara and United States
(from Schutz and Rahn, 1982)

$EF_{(Al, rock)}$	Element
< 0.5	Sr, Ca, Na
0.5 - 3.0	Rare earths, Si, Hf, Co, Ba, V, Ti, Cs, Zr, Mn, W, Th, Rb, Mg, Ga, Cr, Fe, Cl
3.0 - 10	Cu, Zn, As, Sb, I
> 10	Au, Ag

The situation is far more complex for the chemical composition of sea-source aerosols. This topic will be treated extensively later in this chapter. But it will be useful to keep in mind
that the source region for sea-source aerosols is a very thin
layer at the air/sea interface, the sea-surface microlayer, together with the material, particulate or dissolved, scavenged by

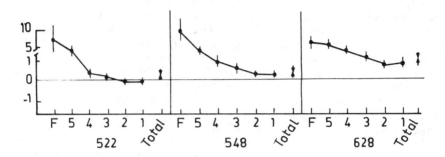

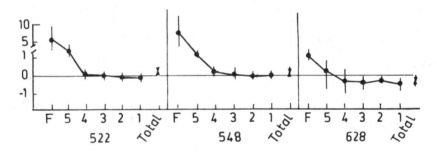

IMPACTOR SAMPLE AND STAGE NO

Figure 1. Top : Total K enrichment, relative to bulk seawater
(E (K, sea) = (K/Na) air / (K/Na)$_{sea}$) - 1), as a function of
particle size in marine aerosols from Bermuda. Equivalent aerodynamic radius cutoffs at 50% collection efficiency for particles
with a density of 1 g cm^{-3} are : Stage 1 = 3.6 μm ; Stage 2 =
1.5 μm ; Stage 3 = 0.75 μm ; Stage 4 = 0.48 μm ; Stage 5 = 0.25 μm ;
Final Filter (F) < 0.25 μm. Bottom : soil corrected K enrichment,
relative to bulk seawater, where the K concentration used is the
difference between the total concentration of K and the concentration of K associated with soil material.
Adapted from Hoffman et al. (1980).

rising bubbles (see Blanchard, this volume). As will be seen, the
now well-documented enrichment of many trace substances in the
sea-surface microlayer (organic compounds, trace metals, P, I, etc.)
suggests strongly that these substances should be enriched, relative to the composition of bulk surface-seawater, on the particles ejected into the atmosphere when bubbles burst at the air/
sea interface. The big question is the exact magnitude of this
enrichment. Although we know reasonably well sea-salt-particle
production processes and rates, this does not necessarily imply
that the sea-source component of a given substance in the marine
aerosol will exhibit the same distribution of its atmospheric concentration, as a function of particle size, as that of sea-salt.
Thus, if we follow the theoretical considerations of MacIntyre
(1972, 1974a), organic or organo-metallic compounds, directly associated with sea-salt particles, can be expected to show a variation of their abundance relative to sea-salt as a function of
sea-salt particle size. Also, mineral and organic particles present at the air-sea interface or at the bubble-seawater interface
can contribute directly to the chemical composition of marine
aerosols, and there is no reason to infer that such particles exhibit the same size distribution as that of sea-salt. Finally, the
emission of some substances from the ocean surface can take place
in the vapor phase, followed by a gas-to-particle conversion process which is known to lead preferentially to sub-micrometer particles. In contrast, most of the sea-salt mass is present in particles larger than 5 μm in diameter. Moreover, because it depends,
among other factors, on the biological activity of surface waters,
the enrichment of trace substances in sea-source aerosols is likely to vary with space and time. For all these reasons, it seems
clear, at least to me, that we cannot identify an oceanic source
for atmospheric particulates if we rely only upon the knowledge
of the chemical composition of the marine aerosol. Together with
concentration data, other criteria must be used that are more
specific for the substance and source being investigated. These
other criteria can include : chemical analysis of individual particles, measurements of the isotopic composition of some elements
and radioisotopes, analysis of specific "source markers" such as
organic compounds or classes of compounds, chemical speciation
studies, and carefully designed laboratory experiments.

Following these general considerations, I will give some
examples that show how the assessment of contributions from various potential sources has been partially or almost totally made.

2.6 Two Examples of Source Differentiation for Marine Aerosols

Potassium in the marine atmosphere. Early findings of K enrichment relative to Na in marine aerosols were often attributed
to chemical fractionation at the air-sea interface (Chesselet et

al., 1972a, b). Hoffman and Duce (1972) and Hoffman et al.(1974a)
concluded from their studies that, in most cases, enrichments
were caused either by contamination or by the presence of K in
the marine aerosol from non-marine sources ; most of the excess
K in marine aerosols, relative to a true sea-salt component, appeared to be derived from soil. Buat-Ménard et al. (1974) and
Morelli et al. (1974) did suggest, however, that K might be enriched to some extent in sea-salt particles produced in biologi-
cally-productive waters, such as upwelling areas. Two recent studies of size separated samples collected at Bermuda have allowed
a better and perhaps final answer to the controversy (Meinert
and Winchester, 1977 ; Hoffman et al., 1980). When the chemical
composition of marine atmospheric particles is corrected for a
soil derived component (Fig.1), K shows no deviation from bulk-
seawater composition for particles with radii greater than 0.5µm.
However, excess K above that expected from either a bulk seawater
or soil source is observed on particles with radii less than
0.5µm. Winkler (1975) also found anomalously high K concentrations
on particles with r<0.2µm over the North Atlantic. Laboratory studies of K enrichment during bubble bursting and results from analysis of in-situ generated sea-salt aerosols in the Sargasso Sea
(Hoffman and Duce,1977a ; Weisel et al., 1982) do not show fractionation of K on any size of particles generated from the sea surface. Therefore the excess K in sub-micrometer marine aerosols
may have a non-marine,non-crustal source. Crozat (1979) has shown
that tropical forests act as a source of sub-micrometer aerosols
rich in K. His results seem to be confirmed by those of Lawson
and Winchester(1979b) in South America and therefore suggest that
long-range transport of K-rich particles of terrestrial vegeta-
tive origin is responsible for the excess K in the sub-micrometer
size fraction of marine aerosols. This could be one explanation
for the results of Buat-Ménard et al. (1974), obtained in the
Gulf of Guinea, where the presence of such K aerosols is more than
likely, but could not be documented when their study was under-
taken.

The various origins of phosphorus in marine aerosols. Phosphorus is an essential nutrient for the growth of marine life. It
is only recently that attention has been paid to the transport of
phosphorus to the oceans via the atmosphere from crustal weathering and pollution (Graham and Duce, 1979 ; Graham et al., 1979 ;
Graham and Duce, 1981, 1982). However, atmospheric phosphorus
also has an oceanic source and phosphorus can be significantly
enriched on sea-salt aerosols, as has been shown by laboratory
and in-situ bubbling studies (Graham and Duce, 1979). The problem, once again, is to discern the levels of non-marine P in marine aerosols, and thereby, the net input of P from the continents
to the oceans. This has been partially solved by measuring its
different chemical forms. In their study, Graham et al. (1979)
were able to show that the enrichment of reactive phosphorus (so-

luble in distilled water, primarily $PO_4^=$, $HPO_4^=$, etc.) on sea-salt
aerosols was only a factor 2-8 above seawater, whereas that of
"organic" phosphorus (determined by the difference between total
phosphorus and reactive phosphorus) was often 100-200, suggesting
that most of the sea-source phosphorus is associated with surface
active organic substances. The results of Graham and Duce (1981)
strongly suggest that, over the eastern Equatorial Pacific, crustal weathering is the primary source of the reactive P, the ocean
is the primary source of organic P, and pollution apparently represents a minor source for both forms in this area. About 60% of
the total P is reactive, whereas in Hawaii and Samoa the reactive
P is 35% and 20% respectively, reflecting the increasing distances from continental regions.

At Bermuda and over the western North Atlantic, the situation appears a little more complex (Graham and Duce, 1982). As in
their study over the Pacific, these authors have used factor analysis and multi-variate regression analysis of sodium and aluminium (oceanic and crustal source indicators), excess vanadium (an
indicator of anthropogenic sources), reactive P and organic P. An
unexpected result is the large amount of P, both reactive and "organic", found to be associated with excess vanadium and therefore
presumably of pollution origin. Air mass trajectories indicated
North America as the major source of excess vanadium as well as
of reactive phosphorus (Figures 2 and 3). Also, during a period
of high aluminium concentration(presumably due to the transport
of Saharan aerosols) a high value of "organic" P was found. This
is compatible with the fact that the Sahara aerosol contains mostly water insoluble phosphates. This complex situation is illustrated by Figure 4, which shows the distribution of Na, Al, excess V, and P as a function of particle size for a cascade impactor sample collected at Bermuda. According to Graham and Duce,
the Na value at stage 4 is anomalously high, compared to other
Na profiles obtained at this site or anywhere else in the marine
atmosphere and is thus probably incorrect. The total and reactive
P distribution shows a peak at stage 2, where material of crustal
or sea-salt origin is usually present. However, there is a significant fraction of P on the small-size fraction, where the bulk
of excess vanadium is also found. This Figure provides the additional evidence for the presence over the North Atlantic of relatively large amounts of P from anthropogenic sources.

I think that this example shows how indirect approaches can
be combined to differentiate rather successfully between various
potential sources of the marine aerosol. In many cases, this represents the state of the art. However, in some other cases, the
source characterization can be almost unequivocal due to the existence of specific tracers and source markers.

2.7 Tracers and Source Markers

Before showing a few examples from recent studies, I want to emphasize that whatever the tracer or the source marker is, that component of the marine aerosol "tells" only about its own behaviour and origin. For any other substance being simultaneously investigated, the conclusions drawn from a comparison with the behaviour of that tracer or source marker must be very carefully examined to avoid misinterpretation and overstatement.

^{210}Pb in the marine atmosphere. ^{210}Pb (22-yr half life) is a radioactive nuclide produced in the air by the decay of gaseous ^{222}Rn (3.8-day half life) emanating from continental soils. Because its source is the Earth's surface rather than the stratosphere, as is the case for many cosmogenic and bomb produced radionuclides, ^{210}Pb is a better analogue of components injected from land sources into the troposphere. Also, as is the case over land, ^{210}Pb in the marine atmosphere is primarily present in the smallest size fraction of the aerosol (Sanak et al., 1981). It is therefore tempting to use this nuclide to provide insights into the atmospheric geochemistry of other land-derived particulate substances of similar sizes. Another argument in favor of the use of this nuclide is that, in the remote marine atmosphere, concentrations of trace substances of continental origin are extremely low; therefore it is extremely difficult to avoid contamination problems, whereas these problems are much less critical for ^{210}Pb. This tracer has been successfully used to study the transport of continental aerosols over remote areas such as the Arctic (AE, 1981; Rahn, 1981) and the Pacific Ocean (Turekian and Cochran, 1981).

There are, however, some limitations to the use of this tracer. The first limitation comes from the fact that the source strength and the concentration of ^{210}Pb over land are rather homogeneous. In contrast, this is not the case for many trace substances of continental origin (e.g. from soil erosion, emission from vegetation, industrial emissions, etc..). Consequently, the trace substance to ^{210}Pb ratio in the marine atmosphere will depend on the value of this ratio in the source region. This ratio may also vary during transport from land to marine locations, if the trace substance and ^{210}Pb are not removed from the atmosphere in similar ways.

To illustrate these considerations, I will use results from recent studies of Pb, mineral dust, and ^{210}Pb over the Pacific (Duce et al., 1980 ; Turekian and Cochran, 1981 ; Settle et al., 1982 ; Settle and Patterson, 1982). Consider first the case of lead. Atmospheric Pb, at least in the northern hemisphere, is primarily anthropogenic (e.g., smelter fumes and gasoline exhausts). Therefore, as is the case for ^{210}Pb, atmospheric Pb is mainly derived from gaseous sources on land and is associated with the

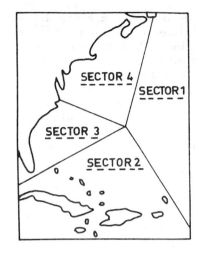

Figure 2. Source sectors for air masses arriving at Bermuda.

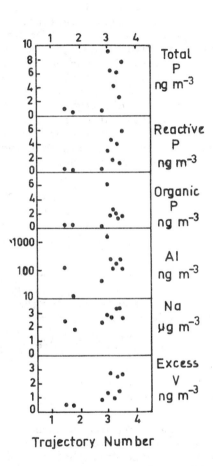

Figure 3. Concentrations of P, Na, Al, and excess V as a function of trajectory number.

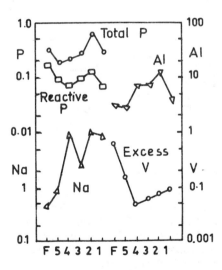

Figure 4. Concentrations of P, Na, Al and excess V (ng m^{-3}) as a function of particle size (see Figure 1) as measured at Bermuda.

Figures 2, 3, and 4 are adapted from Graham and Duce (1982).

smallest size fraction of the marine aerosol. In a given air mass, or during steady meteorological conditions at a given marine site, we might expect the Pb/^{210}Pb ratio to be rather uniform. This seems to be the case. However, in air masses that have passed over different continental areas, this ratio does vary significantly. Consequently, at a given marine location, this ratio may exhibit seasonal variations. This implies that measurements of that Pb/^{210}Pb ratio must be extended over a period of at least a year, to yield meaningful and usable Pb concentration and flux predictions. This can be seen if we look at Table 2. At Enewetak Atoll, in 1979, the Pb/^{210}Pb ratio is somewhat variable during different sampling periods (by a factor of 2). If we now shift from one oceanic region to another, this ratio (both in air and rain samples) varies by almost two orders of magnitude. This ratio is highest at marine locations, such as Bermuda, that can be directly downwind of highly industrialised regions such as the North American Continent (Settle et al., 1982).

Table 2. Pb and ^{210}Pb in Enewetak air, dpm = disintegrations per minute. Pb data are from Settle and Patterson, 1982, and ^{210}Pb data are from Turekian and Cochran, 1981.

Collection dates 1979	Mean Pb ng/m^3	Mean ^{210}Pb dpm/10^{3}m^3	Pb/^{210}Pb ng / dpm
April 22 to May 9	0.23	11.2	20
July 12 to August 10	0.12	3.4	35

The situation appears to be worse when ^{210}Pb is plotted against a compound of continental origin and with a different particle size distribution. Consider the case of eolian transport of mineral dust originating from soil erosion. With sedimentation of the coarse fraction, the particle size distribution shifts towards smaller sizes as the dust moves offshore (Prospero, 1981). Thus, in a given air mass, the dust/^{210}Pb ratio will decrease with distance from the source region. Also, as with lead, the initial ratio will be extremely geographically dependent, being very high in a dust storm and very low where soil erosion is minimal. For example, at Enewetak Atoll (from April to July 1979), the mineral dust concentration (indicated by Al) decreased by almost two orders of magnitude (Duce et al., 1980), whereas the ^{210}Pb concentration decreased only by a factor of 4 (Turekian and Cochran,

1981) ; see Figure 5. There is strong evidence that the high dust
levels in the air in April originated from an Asian soil-source
region, while later in the year the dust was probably coming from
another source region as suggested by mineralogical data (Gaudichet and Buat-Ménard, 1982).On the basis of stable lead isotope
measurements (Settle and Patterson, 1982) this region was possibly the North American Continent. As a crude approximation, Figure 5 can be viewed as a mixing curve between high ^{222}Rn (and
therefore ^{210}Pb), high Al air,and low ^{222}Rn, low Al air, with the
ratio between ^{222}Rn and Al obviously being different for the two
end members (Turekian and Cochran, 1981). Off the coast of North
Africa, I have found that, while the range of Al concentration
was spread over 3 orders-of-magnitude (depending whether the sampling area was inside or outside the Saharan dust plume),fluctu-
ations of such magnitude were not observed for the concentration
of ^{222}Rn (Figure 6).

These considerations imply that we must be very cautious
with the use of such tracers if we want to predict quantitatively
the concentrations of other substances in marine aerosols. On the
other hand, such tracers should be measured in the marine atmosphere, since they are extremely useful for validating models of
atmospheric transport of continental aerosols to the oceans
(Hoang and Servant, 1974 ; Wilkniss et al., 1974 ; Turekian et al.
1977 ; NAS, 1978).

Excess ^{210}Po, a tracer of volcanic aerosols ? Volcanic ema-
nations of volatile compounds (e.g., S, heavy metals, metalloids)
are a   potential source on a global scale for some of the enriched elements in marine aerosols (Mroz and Zoller, 1975 ; Buat-
Ménard and Arnold, 1978 ; Haulet et al., 1977). Although still
very uncertain, global estimates of the magnitude of this source
have been made (Lantzy and Mackenzie, 1978 ; Arnold et al., 1981).
The relative importance in a given location of volcanic versus
pollution sources cannot, however, be directly assessed from
source-strength estimates. For example, since many volcanoes
emit at altitudes above the marine boundary layer, the probability for long-range transport of volcanic aerosols is likely to be
higher than for surface releases of pollution aerosols. Ideally,
an assessment of the relative contributions from these two sources could be made if we had accurate atmospheric transport and
removal models. However, since this is not yet the case, the
use of tracers has to be strongly encouraged. There is evidence
now that the presence of sub-micrometer volcanic aerosols can be
almost unequivocally assessed by the measurements of excess
^{210}Po over ^{210}Pb (^{210}Po/^{210}Pb activity ratio >1), (Lambert et al.
1982). This follows because $\approx$50% of the global ^{210}Po has a volcanic source. Ideally, if we knew the abundance of volatile elements relative to ^{210}Po in volcanic plumes, we could estimate the
volcanic component of their concentration in the aerosols. Such

an approach has been used by Arnold et al. (1982) to evaluate
more accurately the influence of volcanic activity (primarily
from Mt Etna, Sicily) on the heavy metal chemistry of marine aerosols over the western Mediterranean Sea. For some samples, the
volcanic influence has been detected and shown to be significant
for Se, Zn, Cu and Cd.

   Indicators of anthropogenic sources. Tracers of pollution
aerosols are potentially numerous. Non-natural substances and
compounds, such as synthetic organics, can be very valuable as
tracers (Bidleman and Olney, 1974 ; Harvey and Steinhauer, 1974 ;
Atlas and Giam, 1981). In the last 20 years, studies of bomb –
produced atmospheric radionuclides have provided some of the first
insights into long-range atmospheric transport (NAS, 1978). There
is also some evidence that specific tracers can indicate the level of atmospheric pollution. This is certainly the case for stable
lead. In the marine atmosphere, lead aerosol concentrations vary
tremendously, from 1-10 ng m^{-3} in the Northern Hemisphere Westerlies
to values as low as 0.02 ng m^{-3} in the Tropical South Pacific
(Settle and Patterson, 1982).

   Unfortunately, the chemical composition of urban aerosols is
highly variable (Rahn, 1976) so normalization to Pb concentrations
gives only an order-of-magnitude estimate of the contribution of
pollution sources to the total concentration of any given element
in the marine aerosol. However, this approach could be more quantitative, if the source region could be identified. In this context, the measurement of stable lead isotopes appears to be potentially useful. From the results of such measurements, Settle
and Patterson (1982) have shown that, at Enewetak Atoll (in 1979),
Japan was the major source of industrial lead during the dry
season (in Spring) while the United States were the major source
during the wet season (in Summer). Prior to these data, transport
of aerosols from Asia to the Enewetak area had been suggested by
meteorological analyses (Duce et al., 1980), but that was not the
case for the North American source. This is an example of how
specific indicators can help to establish the existence of long-
range atmospheric transport pathways , which might have been overlooked.

   In this context, I should also mention the studies related
to the transport of pollution aerosol to the Arctic (AE, 1981).
The use of the Mn/V ratio has proven to be extremely powerful to
distinguish between North American and Eurasian sources for Arctic
air pollution. The basis of the approach (Rahn, 1981) is that V
is a subtle indicator of mid-latitude pollution sources (combustion of residual oil), while enriched Mn in aerosols comes from
a variety of pollution sources. The ratio of non-crustal Mn to
non-crustal V is 7±4 times greater in Eurasian aerosols than in
North American aerosols. When allowance is made for a decrease in

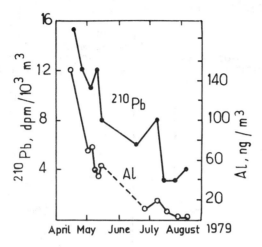

Figure 5. [210]Pb concentrations (Turekian and Cochran, 1981) and Al concentrations (Duce et al., 1980) in air-filter samples collected at Enewetak (11°20'N, 162°20'E) between April and August 1979.

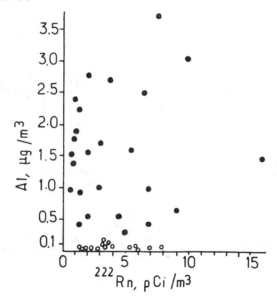

Figure 6. [222]Rn concentrations versus Al concentrations in air-filter samples collected over the Tropical North Atlantic inside the Saharan dust plume (●), and outside the Saharan dust plume (o).
Adapted from Buat-Ménard, 1979.

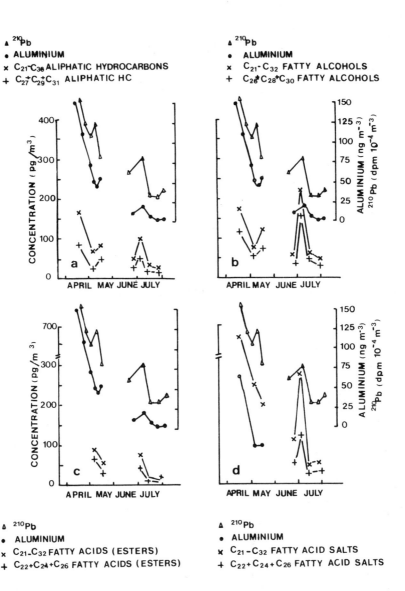

Figure 7. Aliphatic hydrocarbon (a) ; fatty alcohol (b) ;
fatty acid ester (c) ; and fatty acid salt concentrations (d)
in air-filter samples, as a function of time during the Enewetak
field experiment (Gagosian et al., 1982). Aluminium (Duce et al.,
1980) and ^{210}Pb (Turekian and Cochran, 1981) are also plotted.

this ratio during transport, the aerosols of the Norwegian and
North American Arctic are consistent with an Eurasian source and
inconsistent with a North American source. From such studies, it
appears that interelemental ratios (between on the one hand,
elements which are pollution derived but from a single source,
and on the other hand, elements which are pollution derived but
from many pollution sources) can provide useful information
about continental source regions for marine aerosols. This type
of geochemical information provides another basis for the vali-
dation of atmospheric-transport models (NAS, 1978). Yet,we must
also use atmospheric models to plan any sampling strategy, and
therefore we must progress in an iterative manner, step by step.

From what we have seen up to now, one conclusion is that the
continental component of marine aerosols is highly variable in
space and. time. Consequently, the chemistry of marine aerosols may
well be "regional". This must be appreciated before attempting
quantitative estimates of atmospheric fluxes to the ocean or considering aerosol impacts on climate. For example, a single dust storm
over a given oceanic area will contribute to most of the yearly
dust flux to the ocean. On the other hand, background concentrations, occuring most of the time at that same location, probably
have dominant influence on climate.

Organic material of continental origin in marine aerosols. A
major continental source, not yet considered in this review, is the
land biomass. The dearth of data here is evident. Organic aerosols
are produced in two ways : as particles and as the result of gas-
to-particle conversion. In terms of global production of particulate organic carbon, gas-to-particle conversion seems to be the
most important (Duce, 1978 ; Jaenicke, 1978). Potentially, there
is,thus,a net input of organic matter from the continents to the
oceans. The problem, once more, is to unequivocally identify the
presence of that continental material in the marine atmosphere,
since the ocean, by itself, is also a source of organic aerosols.
To try to resolve this problem, two complementary approaches have
been used.

The first approach, often called the "source marker" approach
uses the unique lipid-molecular signatures of various classes of
marine and terrestrial life to ascertain the "source fingerprint"
for the organic materials. Such studies suggest that organic particulate material in the marine atmosphere has significant terrestrial and marine sources (Barger and Garrett, 1970, 1976 ; Quinn
and Wade, 1972 ; Marty et al., 1979 ; Barbier et al., 1981 ; Simoneit, 1977, 1979 ; Eichmann et al., 1979, 1980 ; Gagosian et al.,
1981, 1982). For a given class of organic compound, it is possible
to quantify the respective sources. For example, at Enewetak Atoll,
Gagosian et al. (1981, 1982) identified terrestrial, vascular plant

sources of hydrocarbons, probably originating from wind erosion
of Eurasian soil and direct emission from vegetation (Figure 7).
Also, they found a series of apparent marine-derived lower molecular weight alcohols and fatty acid esters and salts.

The 'source marker' approach is, however, limited to species
that are conservative in the marine atmosphere and that make up
only a small percentage of the total organic carbon in marine aerosols. Most of the total atmospheric particulate organic carbon
in the marine atmosphere is present on particles with radii <0.5μm
(Hoffman and Duce, 1977b : Chesselet et al., 1981), which suggests
that gas-to-particle conversion processes are the major source
for this carbon. Hundreds of single and chain reactions can occur
in the vapor phase organic pool (either land derived or sea derived),and these reactions will ultimately produce small-particle
organic carbon. Using a mass balance approach, Duce (1978) suggested that this small organic carbon might be primarily derived from
continental sources, most likely from the release of hydrocarbons
in the vapor phase by vegetation.

This hypothesis seems to be strongly supported by the results
of Chesselet et al.(1981), which are based on the use of stable
carbon isotopes. The use of stable carbon isotopes is a second
way to approach the problem of identifying marine versus continental sources. For bulk aerosol samples collected at Enewetak
Atoll, and over the Sargasso Sea, they found that the $\delta^{13}C$ values
for total particulate organic carbon in the marine atmosphere were
close to values generally found for continental vegetation, coal
and products of petroleum combustion ($\delta^{13}C$ =-26 $\pm$ 2 ‰). Analysis
of size-separated aerosol samples from Enewetak (Figure 8) confirmed that the organic carbon of continental origin was predominantly present on the small particles. The $\delta^{13}C$ values for the
large-particle organic carbon are similar to $\delta^{13}C$ values for organic carbon of marine origin, -21 $\pm$ 2 ‰ in low and mid-latitude
regions, suggesting that the large-particle organic carbon in the
marine atmosphere is of marine origin. If confirmed by further
studies, the use of carbon isotopes offers an excellent oppor-
tunity for untangling the relative roles of small (net input) and
large (recycled material) particles, and may allow us to estimate
more accurately the net input of organic carbon from the atmosphere to the oceans.

It is relevant here that the size distribution of organic
carbon in marine aerosols is quite similar to the size distribution of some trace metals, which are found to be enriched relative
to a bulk seawater or an average crustal material composition ;
elements such as Cu, Zn, Cd, and Pb are found predominantly on
small particles, but show a small but sizable large-particle component. With respect to that large-particle component (where cor-

respondingly, the organic carbon appears to be derived from the sea-surface), the metal/organic carbon ratio could well be that of the recycled component of marine aerosols. Indeed, there is now some evidence that the enrichment of metals relative to bulk seawater in sea-source aerosols arises from the association of these elements with organic matter, as will be shown in the next section of this chapter.

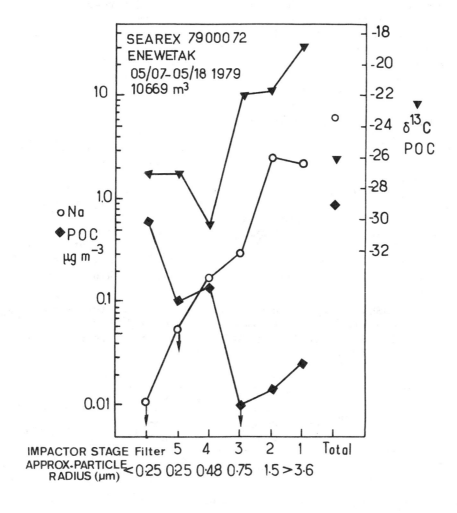

Figure 8. Sodium, organic carbon, and $\delta^{13}C$ as a function of particle size in Enewetak air

$$\delta^{13}C = (\frac{^{13}C/^{12}C \text{ (sample)}}{^{13}C/^{12}C \text{ (standard)}} - 1) \times 1000$$

From data reported in Chesselet et al., 1981.

2.8 Interaction between Trace Gases and Particles in the Marine
    Atmosphere

We have just seen that gas-to-particle conversion processes
may dominate the chemistry of organic carbon in marine aerosols.
In the case of organic carbon, details of these processes are not
known. Together with organic carbon, sulfur and nitrogen compounds
are major components of the so-called background aerosol, which is
mostly of secondary origin. The study of the atmospheric cycles
of S and N receives more and more attention because of increased
anthropogenic perturbations of these cycles, and the potential
impact of such perturbations on climate and the acidity of rain.
Those cycles have been extensively reviewed recently (Granat et
al., 1976 ; Simpson et al., 1977).

What I want to emphasize here is the interaction of the S and
N cycles, which can be seen from data on S and N concentrations as
a function of particle size in marine aerosols. Gravenhorst et al.
(1981) and Savoie and Prospero (1982) have found that, over the
North Atlantic, the maximum nitrate concentration in aerosols is
found on particles with radii of about $1\mu m$, whereas 90% of NH_4^+ is
found on particles with radii less than $0.5\mu m$. For NH_4^+, this distribution would result from the heterogeneous reaction of gaseous
NH_3 with acidic, sulfur-containing particles, mostly with radii
less than $0.5\mu m$. On the other hand, the nitrate distribution would
result from the attachment of HNO_3 or NO_2 to the alkaline sea-salt
particles (see Figure 9 and Table 3). Indeed, the values found for

TABLE 3. Geometric mean, mass-median diameters (MMD)
and total concentrations (Conc.) of sodium and nitrate
in the Tropical North Atlantic marine aerosols (Savoie
and Prospero, 1982), n = number of samples

| Location | n | sodium | | nitrate | | $MMD(NO_3^-)/$ |
		MMD (μm)	Conc. ($\mu g\ m^{-3}$)	MMD (μm)	Conc. ($\mu g\ m^{-3}$)	$MMD(Na^+)$
Miami, Key Biscayne (1981)	11	7.43	2.27	4.00	.975	.536
Miami, Virginia Key (1981)	6	7.14	2.63	4.02	.623	.562
" "(1974)	5	6.55	2.17	3.89	.858	.594
Sal Island (1974)	9	7.04	5.19	4.21	.956	.598
Barbados (1974)	10	--	--	3.6	.390	--

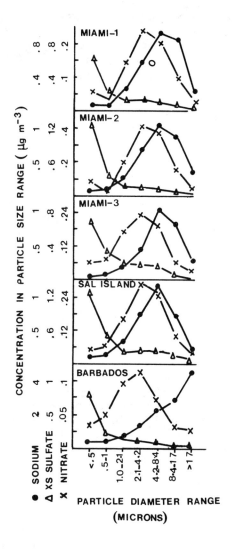

Figure 9. Typical particle size distribution of sodium, non-sea-salt sulfate (xs), and nitrate at several coastal sites.

Adapted from Savoie and Prospero, 1982

the ratios of the mass-median diameters of nitrate to that of
sodium (see Table 3) are close to estimates of surface median diameter to mass median diameter ratios in sea-salt aerosols (0.5 to
0.8), (Walsh et al., 1978). According to Savoie and Prospero
(1982), this indicates that the nitrate mass is roughly distributed as the surface area of the sea-salt aerosols, which should
be expected if aerosol nitrate is produced by the absorption and
subsequent reaction of NO_2, or by dissolution of gaseous HNO_3 on
sea-salt aerosols.

The interaction between trace gases and particles in the marine atmosphere is also important for understanding halogen cycles.
This topic has been thoroughly reviewed by Cicerone (1981). For
halogens, the marine aerosol appears to act both as a source and
a sink for the gas phase, depending on the concentrations of other
trace gases, as well as the aerosol composition (including organics). For example, Rancher and Kritz (1980) have found, during
8 days in the Gulf of Guinea , that there was about twice as much
particulate bromide during night than during day, the opposite for
inorganic, gaseous Br. They suggested a reversible day and night
exchange of Br between the particulate and gas phases. To understand such a process, data on inorganic Br speciation are clearly
needed.

2.9 Conclusions

It seems to me that our understanding of aerosol chemistry
in the marine atmosphere has progressed significantly during the
last few years. Sources and sinks and transport processes can now
be investigated on regional scales. This is necessary because
global scale approaches (based on crude mass-balance calculations)
can only give a crude idea of the various source and sink functions, mean residence times, etc. Residence times may vary from
place to place and from time to time, depending, for example, on the
amount of precipitation.

Still, in many cases, our knowledge is restricted to the
Northern Hemisphere. Obviously, reliable data are needed for the
Southern Hemisphere. At the tropospheric level, interhemispheric
mixing time is long (about 6 months) compared to the residence
time of aerosol particles (a few days). Also, continental and anthropogenic sources are less important in the Southern Hemisphere
than in the Northern Hemisphere. Therefore, the Southern Hemisphere appears the best place for natural background assessments.

A final remark is that most of our knowledge is based on data
obtained either from remote islands or from ships. There is a substantial lack of concentration data as a function of altitude,
especially below and above clouds. Such data are needed for the
understanding of cloud and rain chemistry.

I have left,for the following section,the whole problem of assessing the magnitude of the recycling of particulate material between the surface of the ocean and the marine atmosphere. The main problem is that the surface waters of the ocean are both a source and and a sink for aerosols. Physical, biological, and chemical processes are involved and all influence sea-to-air and air-to-sea particulate transfer. From this perspective, the geochemistry of sea-source aerosols is part of the geochemistry of the surface waters of the ocean.

3. GEOCHEMISTRY OF PARTICLE TRANSFER BETWEEN THE ATMOSPHERE AND THE OCEANS

3.1 Introduction

In the previous section, we have seen some evidence that atmospheric particulate matter derived from the ocean is not just sea-salt. The presence of modified sea-source material in the atmosphere complicates estimates of the net flux of particles to the oceans. In an attempt to understand the chemical enrichment occurring when sea-source aerosols are produced, we shall examine potential processes responsible for that enrichment, especially for organic matter and trace metals. Also, since the air-sea interface is the first place where inputs from the atmosphere are in contact with the marine hydrosphere, we shall examine potential fates. Extensive reviews of chemistry near the surface of the ocean have been published recently (Liss, 1975 ; Duce and Hoffman, 1976 ; Lion and Leckie, 1981). The emphasis here will be placed on the geochemical implications.

3.2 Bubbles and the Chemical Composition of Sea-Source Aerosols

Most sea-salt particles with atmospheric residence times longer than a few minutes are expected to be produced by bubbles breaking at the sea-surface. The direct ejection of spray from breaking waves is most probably a less efficient transport mechanism. Detailed information about bubble production processes and rates, bubble sizes, atmospheric sea-salt particle size and number distributions will be found in the chapter by Blanchard (this volume). When bubbles break at the air-sea interface, they skim off a very thin layer of water to produce atmospheric film and jet droplets. This "microtome effect" for jet drops has been investigated theoretically and experimentally by MacIntyre (1972, 1974b). The material present in the top jet drop is apparently originally spread out over the interior of the bubble surface at a thickness equal to approximately 0.05% of the bubble diameter. Since the diameter of the jet drop is about 10% of the diameter of the bubble producing it, a 100 µm bubble will produce a 10 µm particle

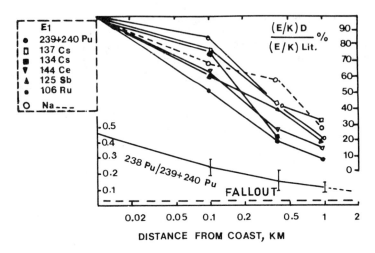

Figure 10.Artificial radionuclides in coastal grass species
of La Hague area (Ecalgrain Bay) as a function of distance
from coast. Potassium-normalized activity distribution in percent of coast value, and plutonium isotopic ratio.

Adapted from Martin et al., 1981.

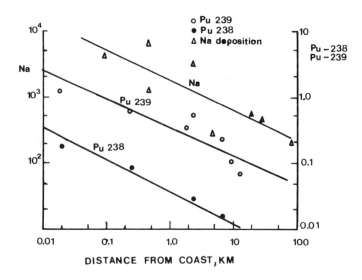

Figure 11. Deposition of Na (Na deposition by rain, µg cm^{-2} yr^{-1}),
Pu-238 and Pu-239 (Pu-238 and Pu-239 in soils, pCi cm^{-2}, fallout
contribution subtracted) versus distance from the sea in West
Cumbria, U.K. (Adapted from Cambray and Eakins, 1980).

composed of material originally present within 0.05µm of the
bubble surface. Since the sea-surface microlayer can be highly
enriched in surface-active organic material and other substances,
these materials can therefore be considerably enriched on jet
drops, compared to their concentration in seawater a few cm below
the surface. These materials can also be present on the bubble
skin itself. The average lifetime of a bubble is longer than a
few seconds, whereas most of the surface chemical phenomena occurring at the surface of a rising bubble are much faster. Not only
surfactants, but also particles can be attached onto rising bubbles ; such particles, together with those present at the air-sea
interface, can also be injected into the air when the bubbles
burst. Together with jet droplets, film droplets are produced when
the bubble film cap or film shatters. Cipriano and Blanchard
(1981) have recently shown that most of the smaller sea-salt particles with significant lifetimes in the atmosphere (i.e., with
radii smaller than about 20 µm) could be derived from film drops.
Very little is known about the composition of film drops, although
they certainly contain material present in the microlayer of the
sea. They may, however, be composed of a much thicker layer of
the water surface (about 2 µm) than the jet droplets.

What, then, is the chemical composition of the surface microlayer ? We have to keep in mind that the term 'microlayer' implies
no particular thickness, and instead, simply refers to the surface
layer that is sampled. Because the various samplers collect different thicknesses of the ocean surface and are based on different
principles, comparison of studies using different types of collec-
tors is difficult. These various systems have been recently reviewed (Garrett and Duce, 1980 ; Lion and Leckie, 1981). The
depths sampled vary over 4 orders of magnitude, from 1 mm to about
100 nm. Thus, to varying degrees, these samplers dilute a monomolecular, surface-layer sample with bulk-marine water. Therefore
sample concentrations may have to be multiplied by a factor of up
to 10^4 to reflect concentrations of a monomolecular layer. This is
the main reason why, data on enrichments in sea-borne aerosols
cannot be fully quantitative. In general, the chemical composition
of the surface microlayer is evidently significantly different
from that of water 10-20 cm or more below the surface : a detailed
report is not possible here (see Lion and Leckie, 1981). The substances concentrated in the microlayer include a wide variety of
organic materials (both natural and anthropogenic, and both particulate and dissolved), nitrogen and phosphate, trace metals, and
marine organisms (Sieburth, this volume). Thus the potential
exists for sea-source aerosols to have a chemical composition significantly different from that of subsurface water.

Perhaps, the best way to investigate the composition of sea-
source aerosols is to use a method based on the bubble microtome
effect itself. Sea-surface samplers using this technique have

been designed either for laboratory experiments or work at sea.
Usually, air is forced through glass frits ; recently, hydro -
phobic, porous membranes have been used (Belot et al., 1982). The
particles ejected into the air by the bursting bubbles are collected by either impaction or filtration. Despite inherent artifacts, such as producing realistic bubble sizes, some consistent
and semi-quantitative results have been obtained. It appears that
the major ions of seawater (alkali and alkaline earth metals, and
Cl and Br) are not significantly enriched in most of the mass of
the sea-salt aerosol (Pattenden et al., 1981 ; Weisel et al.,
1982). This finding is consistent with predictions made for major
ion fractionation at the air-sea interface produced by pure inorganic processes. These various processes or mechanisms have been
reviewed in detail by MacIntyre (1974b) : Gibbs surface adsorption,
the Ludwig-Soret effect, the alteration of the surface water
structure, and electrical double layer effects. On the other hand,
organic carbon, P, I, and trace metals have been found to be enriched by 2 to 5 orders of magnitude relative to subsurface water.

    Especially for metals, sampling contamination causes serious
concern about the reliability of published data. However, results
based on the use of radioactive nuclides strongly argue in favor
of large enrichments of metals in sea-source aerosols. To me, the
most convincing evidence, of the magnitude of the recycling process for trace metals comes from results of atmospheric measurements of radionuclides previously discharged to sea (Fraizier et
al., 1977 ; Cambray and Eakins, 1980 ; Pattenden et al., 1980 ;
Martin et al., 1981). In coastal areas, close to the nuclear
plants at La Hague, France, and Windscale, UK, excess concentrations of radioactive plutonium, americium, antimony, caesium, ru-
thenium, and cerium have been found in aerosols, dry and wet deposition samples, grasses and lichens. These excess concentrations
decrease with increasing distance from the coast, inland (Figures
10 and 11).

    These studies conclude that these excesses most likely come
with sea spray, probably associated with the largest particles.
For example, the results for americium and plutonium can be explained by assuming an enrichment of a factor of about 1000 in sea
spray compared to local waters. The laboratory experiments by
Belot et al. (1982) come to the same level of enrichment of americium in aerosols produced by bursting bubbles. Using a cascade
impactor, they also found that the americium enrichment is the
highest for the largest particles. Therefore, enrichment factors
of 100-10000 which have been found for some metals such as Zn, Cd,
and Pb on artificial bubble-derived aerosols (Pattenden et al.,
1981 ; Piotrowicz et al., 1979 ; Weisel, 1981) could well apply to
the real world. In my own work over the Tropical North Atlantic

(Buat-Ménard, 1979), I have been able, during a five day sampling
period, to collect pairs of aerosol samples at two sampling
heights, 3m and 8m above sea level. The results (Figure 12) showed
increased Na, Cu, Zn, Sb, and Pb concentrations at 3m compared to
8m. Such an increase was not found for crustal elements such as
Al and Fe. The excesses found at 3m were attributed to a sea-
source local effect. For the trace metals, the computed enrichment factors in these sea-source aerosols, relative to normal seawater are given in Table 4. They are in the same range as those
found on artificial bubble-derived aerosols.

As mentioned in the previous section, enriched heavy metals
in the marine aerosol have a small, but sizable, fraction of their
total concentration on large particles. That fraction could be
sea-derived. In the case of lead, more evidence for large enrichments associated with sea spray comes from the combined results
of Settle and Patterson (1982) and Duce (1982) for aerosols collected at Enewetak Atoll, during the SEAREX program. It was found
that the dry deposition velocity of aerosol lead to a flat plate
located about 20m above sea level was 0.7 cm s^{-1}, much greater
than that expected from the mass-median radius of aerosol lead
($\sim$0.2µm). However, $\sim$10% of the lead was contained in particles
larger than 4µm, the size containing 90% of the sea-salt mass.
According to the theoretical predictions of dry deposition velocities (Slinn and Slinn, 1980), the predicted total dry deposition
velocity for lead, when calculated from the measured Pb particle-
size distribution agrees well with that observed. The small percentage of lead on the largest particles controls the Pb deposition. Moreover, the most reasonable explanation for the existence
of the large-particle lead component is that this component is
associated with sea spray. From these results (probably among the
most reliable lead data published to date), one can infer an enrichment factor, relative to local surface seawater, as large as
15,000.

With respect to organic carbon, the results of Chesselet et
al. (1981 ; see Figure 8), previously mentioned, suggest that the
large-particle organic carbon in marine aerosols is derived from
the sea-surface. Based on stable carbon isotope measurements, the
computed enrichment factor, relative to total organic carbon
(dissolved plus particulate) in subsurface seawater, is about 100,
in good agreement with results from laboratory studies by Hoffman
and Duce (1977b). It is interesting to note that metals appear to
be more enriched than organic carbon in sea-source aerosols.

Before we try to get some insight into the processes responsible for the observed enrichments of organic matter, metals, and
other substances in sea-source aerosols, I would like to make
some comments about the geochemical implications of such observations :

1. Measured total dry deposition fluxes of such substances might
be mostly due to recycled fluxes of these substances between the
surface of the ocean and the marine atmosphere. Even if most airborne mass of organic carbon and enriched metals is present on
the smallest particles, the gross deposition may be dominated by
the relatively small, sea-surface-derived fraction, since it typically resides on large particles which have high dry deposition
velocities. The contribution of the recycled component to fluxes
in rain is also open for debate, since it depends on the scavenging efficiency of particles by cloud and rain droplets as a function of particle size. If, as suggested by theoretical and experimental work described in the next section, the scavenging efficiency is higher for large particles, one can speculate that for
such substances, up to 50% of the gross rain fluxes could be from
material recycled from the ocean surface.
2. The fraction of sea-source aerosols deposited on land, mainly
in coastal regions, may contribute significantly to gross river
inputs, not only for sea-salt, but also for the enriched elements
and substances in such aerosols. This is a suggested cause of why
observed fluxes of some heavy metals, transported by rivers, are
higher than theoretical predictions derived from rock weathering
(Martin and Meybeck, 1979).
3. Organic and metallic pollutants present in coastal waters
(from either direct injection in the water or from anthropogenic
inputs from the atmosphere) can be recycled back to land and contribute to coastal atmospheric pollution under sea-breeze conditions. The proper assessments of such contributions should be
made in the near future.

3.3 Enrichment Processes at the Air-Sea and Bubble – Seawater
    Interfaces : Relative Importance of Dissolved and Particulate
    Material in Surface Seawater for Controlling the Chemistry
    of Sea-Derived Aerosols

     There are very few direct observations of genuine sea-deri-
ved material in the atmosphere. Lambert and Jehanno (1980-1981)
have made such observations, using analytical scanning electron
microscopy, on artificial bubble-derived aerosols. They have
found that, together with sea-salt, marine organisms and debris
of such organisms are ejected into the atmosphere by the bubble
bursting process. Bacteria from the sea have also been found to
be concentrated in jet drops from bursting bubbles (see Blanchard
this volume). There is,therefore,evidence that a direct transport of particulate material from surface seawater into the atmosphere does occur. Lambert and Jehanno also found that sea-salt
crystals, formed after dehydration of sea-surface droplets, were
embedded in thin films of organic material, about $0.1\mu m$ thick.
Discrete organic particles were also found. On some of these or-

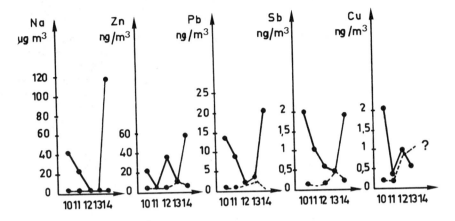

Figure 12. Atmospheric concentrations of sodium, zinc, lead, antimony and copper in 5 consecutive pairs of air filter samples taken at 3 m (——) and 8 m (---) above sea-level in the Tropical North Atlantic

From Buat-Ménard, 1979.

Table 4. Mean enrichment factors relative to subsurface seawater in sea-derived aerosols from the open ocean

Element	(1)		(2)	
	mean	range	mean	range
Pb	5×10^4	$3\times10^4-8\times10^4$	5×10^3	$3\times10^2-2\times10^4$
Zn	9×10^2	$7\times10^2-9\times10^2$	4×10^4	$1\times10^3-3\times10^5$

(1) Buat-Ménard (1979), eastern Tropical North Atlantic
(2) Weisel (1982), Sargasso Sea

ganic films or particles, detectable concentrations of metals
such as Zn have been observed. These observations raise the two
following questions : a) What is the relative importance of dissolved and particulate material in the sea-surface for controlling the chemistry of sea-derived aerosols ? and b) What is the
degree of association of trace metals with organic matter at the
air-sea and bubble-seawater interfaces ? Since it has been shown
that organic matter and metals occur both in the dissolved and
particulate form in the ocean, I will try to address these two
questions simultaneously. This important area of marine chemistry
is still poorly understood and many published data are questionable, so that we must be very cautious about drawing definite conclusions.

The causes of organic-material accumulation at the air-sea
interface will be described here only briefly (for detailed information, see Blanchard, 1975 ; Lion and Leckie, 1981 ; Sieburth,
1982, this volume). Organic enrichment is predictable on the basis
of simple thermodynamic considerations. Most of the material enriched at the sea-surface is either surface active or associated
in some way with surface-active material. Surface active organic
substances have a molecular structure containing both hydrophobic
and hydrophilic portions. Under breaking wave conditions, the primary transport mechanism is by bubbles. Consequently, the chemical
compositions of the oceanic surface microlayer, the surface of
bubbles rising through the water, and the jet and film drops are
closely interrelated. Together with this surface-active material,
mineral particles, and marine organisms such as bacteria and plankton, can be brought to the sea-surface. This process has been quantitatively documented by Wallace and Duce (1978a) for particulate
organic carbon and has been estimated to be the major source for
particulate organic carbon in the microlayer. Moreover, there is
evidence that organic particles can be produced both at the air-
sea interface and when bubbles dissolve. For example, under strong
lateral surface compressional forces, organic surface films may
irreversibly collapse into particles. Also, organic particles may
be produced by photochemical reactions, dissolved organic material
being polymerized and altered into forms that are increasingly insoluble. In addition, when air bubbles dissolve in the sea, adsorbed surface-active organic molecules are compressed, eventually
leading to the formation of organic particles (Johnson and Cooke,
1980). Finally, organic particles at the sea-surface can be supplied by atmospheric input and from oil spills (tar-like particulate residue, plastic particles, soot particles, etc.). The data
summarized here leave little doubt that not only dissolved organic
matter but also particulate organic material can be ejected into
the atmosphere by the bursting bubble process. As a consequence,
the size distribution of marine-derived organic material in aerosols
may not necessarily follow the sea-salt particle size distribution.

There is some evidence that the composition of dissolved
organic material in the microlayer is different from that of the
particulate organic material. For example, Jullien (1982) has shown
that, in the open Mediterranean, sea-surface microlayer proteins
represent less than 1% of the dissolved organic material but 14%
of the particulate organic material. Such differences might permit the identification of these two sources of organic matter in
marine aerosols. In this context, the use of biological "source
markers" may be helpful. Barbier et al. (1981) have found that
over the Tropical North Atlantic, the sterol composition of marine aerosols is similar to the sterol composition of <u>particulate</u>
organic material in the microlayer, which in turn is different
from the sterol composition of <u>dissolved</u> organic material in the
microlayer.

<u>Geochemical behavior of trace metals in the surface waters
of the ocean.</u> We know now that the cycle of trace metals in the
ocean is closely interrelated with the biological cycle. In the
last 10 years, extraordinary progress has been achieved, derived
almost from fanatical attention to problems of sample contamination. A rather coherent picture is now emerging. Insight into the
trace-metal cycle in the open ocean was first provided by studies
of the behavior of artificial and natural radionuclides, primarily brought to ocean surface waters by atmospheric fallout (Lowman et al., 1971). Such studies have enabled time scales of processes to be assessed for trace metals cycling in the ocean. This
cycling involves primarily active or passive scavenging of metals
by particles (Turekian, 1977). Dissolved metals are continuously
extracted by marine plants and animals. The level of metals in
living organisms is often 3 orders-of-magnitude higher than in
seawater. This accumulation of metals appears to be even greater
in biogenic detritus, such as fecal pellets, probably the result
of self-regulation by zooplankton of its metal requirements and
of detoxification processes (Lambert, 1981). It seems that different processes are at work scavenging dissolved metals in seawater : selective accumulation in organic tissues and hard parts,
precipitation of metallic minerals, and surface adsorption onto
solid particles. This latter process might be the most important
geochemically. There is considerable evidence that any particle,
immersed in natural seawater, rapidly loses its surface charge
characteristics and becomes negatively charged. This is consistent
with the formation of a macromolecular organic surface film. The
nature of such organic films is still largely unknown (Hunter and
Liss, 1979 ; Lion and Leckie, 1981). Nevertheless, they should
play a key role in controlling adsorption properties of suspended
particles in the ocean (Balistrieri et al., 1981).

Theoretical calculations, as well as laboratory experiments
using radiotracers, indicate that many trace metals may be com-
plexed by surface-active organic ligands dissolved and/or

adsorbed onto particulate surfaces (Lion and Leckie, 1981).
Hunter (1977) has constructed a chemical model for metal ion
binding in the sea-surface microlayer. He assumed that the sea-
surface is populated by carboxyl (-COOH) groups only and, by
assuming a linear, free-energy relationship between carboxyl-
metal binding ability and the experimentally-determined solubility products of metal carboxylates (Hunter and Liss, 1976), he
has been able to establish a relative binding strength for various metals (Figure 13). This theory has been used to predict the
enhancement to be expected for the metals in the microlayer and
in aerosol samples. The agreement between the predicted enhancements and those measured in the field is generally satisfactory,
except for Hg and Cd (Figure 14).

Another approach involving the use of radiotracers has been
used by Lambert (1981). Natural, dissolved, organic matter is
allowed to adsorb on a mineral matrix of Al_2O_3, suspended in
prefiltered seawater. Radiotracers (^{59}Fe, ^{54}Mn, ^{65}Zn) are added
at concentrations below or equal to the concentrations of the
corresponding stable elements in seawater. These tracers have
been pre-equilibrated with a small quantity of seawater prior to
the experiments. For most metals, an equilibrium between dissolved and adsorbed phases is obtained within 1 hour and lasts for
several days. The main results are : a) The adsorption of metals
onto the organic matter coating the particles occurs at any
depth, but is more important in surface waters. b) The influence
of biology (presence or absence of specific, dissolved, organic
compounds) is well marked for the adsorption of iron. Also, Mn
adsorption is closely dependent on pH and therefore probably on
speciation. It has been observed that the influence of light on
dissolved organic matter can enhance the adsorption of Zn and Mn.
c) The particles are coated with only a small amount of organic
matter, providing a limited number of sites for metal adsorption.
Therefore, the surface area of particles and their sinking rates
are the main features controlling the concentrations of dissolved
metals in surface waters.

All these considerations argue strongly in favor of the accumulation of trace metals at the air-sea interface, associated
with dissolved or adsorbed, surface-organic compounds. Also,
trace metals, dissolved or particulate in the microlayer, can
originate from atmospheric deposition. Wallace et al. (1977) and
Wallace and Duce (1978b) have investigated extensively the relative importance of downward and upward transport for particulate
trace metals in Sargasso Sea waters ; they found that downward
and upward transport to the microlayer were of the same order of
magnitude (Table 5). These studies show that the upward transport results mostly from adsorption and transport by rising bubbles. In North Sea waters, this upward flux was found by Hunter

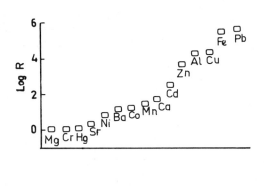

Figure 13. Relative binding strength (R) of -COOH groups for various metals in the sea surface.
From Hunter, 1977.

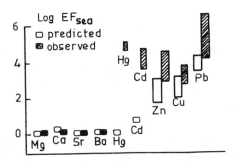

Figure 14. Predicted and observed enrichments of various metals in marine aerosols.
From Hunter, 1977.

(1980) to be even higher than atmospheric deposition, at least for Fe, Cu, Zn, and Cd.

Table 5 . Comparison of bubble transport and atmospheric fluxes of particulate trace metals to the air-sea inter-face (μg m^{-2}s^{-1}), (from Wallace and Duce, 1978b).

Metal	Bubble flux		Atmospheric flux
	Bermuda Coastal	Sargasso Sea	
Al	3.8×10^{-3}	1.1×10^{-3}	1.3×10^{-3}
Fe	3.2×10^{-3}	9.0×10^{-4}	9.5×10^{-4}
Mn	6.6×10^{-5}	1.9×10^{-5}	1.5×10^{-5}
Ni	$5.8 \quad 10^{-5}$	$<1.3 \times 10^{-5}$	1.6×10^{-6}
Cr	5.6×10^{-5}	3.7×10^{-5}	9.9×10^{-6}
Cu	5.6×10^{-5}	1.4×10^{-5}	3.2×10^{-6}
Zn	–	$<5.0 \times 10^{-5}$	8.3×10^{-6}
Pb	3.4×10^{-5}	1.4×10^{-5}*	1.2×10^{-5}
Cd	3.7×10^{-6}	5.8×10^{-7}	6.4×10^{-7}

* Estimated

As is the case for organic matter (Williams, 1967 ; Jullien 1982), particulate trace metals are generally more enriched than dissolved metals at the air-sea interface (Duce et al., 1972 ; Piotrowicz et al., 1972). These studies also indicate that orga-nically bound dissolved metals are more enriched than the inor-ganic fraction. Depending on various factors such as bubble size, the relative importance of jet and film droplets, local biologi-cal productivity, etc., it seems reasonable to predict that the magnitude of the enrichment of metals in sea-derived aerosols will vary with space and time. Moreover , one cannot infer that the ratio of sea-derived trace metals (or organic compounds) to sea-salt will be constant as a function of sea-salt particle size. Furthermore, since atmospheric deposition is also a source for dis-solved and particulate trace metals in seawater (Wallace et al., 1977 ; Buat-Ménard and Chesselet, 1979), the sea-derived component of these elements in marine air is partly a recycled input from the atmosphere. At present, it seems very difficult to make pre-cise, quantitative assessments of the respective importance of the various processes responsible for enriching these elements in sea-derived aerosols. In my opinion, this is an important area for future work, if we want to evaluate accurately the recycled fluxes of metals between the ocean and the atmosphere.

3.4 The Fate of Reactive Elements Entering the Ocean

The biogeochemical processes which we have described tell us about more than just the composition of sea-derived aerosols. As soon as atmospheric particulate matter reaches the surface of the sea, biological processing can start, and therefore, can drive the fate of this material in the ocean. Our present knowledge indicates that we can, to some extent, predict that fate. We will consider here primarily the fate of reactive substances and elements, which includes most atmospheric pollutants which are known to be potentially toxic to marine life (e.g. heavy metals and synthetic organics). While the fate of non-reactive elements in seawater is mostly controlled by water mixing processes, the fate of reactive elements is dependent on the cycle of biogenic particulate matter in the ocean. I will use as an example the fate of atmospheric trace metals.

We are faced here with two types of behavior : "conservative" for those atmospheric particulate elements which are essentially insoluble in seawater (such as those strongly associated with aluminosilicate minerals) and apparently "non-conservative" for those elements which partially or totally dissolve upon entering seawater, or are brought to the ocean in an already soluble form (dissolved in rainwater). The simple model which I will present (Buat-Ménard and Chesselet, 1979) is outlined in Figure 15. This model assumes a 2-box ocean (surface waters and deep waters) and is based principally on the behavior of natural and artificial radionuclides in the ocean.

We have seen how dissolved metals can be scavenged by particles and organisms as soon as they enter the microlayer. Also, in that microlayer, the settling of small particles (less than 10 μm) will be accelerated by aggregation processes. Wallace and Duce (1978a) have shown from bubble-flotation studies that large particles are formed at the air-sea interface, and these aggregates have the ability to accelerate the flux of small particles to sub-surface water. This removal process can be very rapid. It has been shown that residence times of particulate trace metals in the microlayer must be of the order of minutes or less (Hoffman et al., 1974b ; Hunter, 1980).

Studies of artificial metallic radionuclides from nuclear-fallout have shown that, in less than a month, atmospheric inputs are partially found in planktonic organisms and biogenic detritus (Chesselet, 1969). This clearly indicates the involvement of metals, at least once, in the biological cycle. In surface waters, where most of the biological activity takes place, metals can be, like nutrients, recycled many times between the dissolved phase and the particulate phase. At depths greater than about one hun-

dred meters, biogenic detritus and other particles are less frequently recycled so that they leave the surface-water compartment
by sedimentation. Thus, for reactive elements, two removal processes from surface waters to deep waters are at work : water mixing,
and transport by settling particles. As a consequence, the residence time of such elements in the surface-water compartment must
be shorter than that of the water itself (about 20 years). The
presently-available radionuclide data indicate that the residence
times of dissolved and particulate trace metals should range between a few months and 5 years, for a mixed surface layer of about
100 m depth (Nozaki et al., 1976 ; Bacon et al., 1976 ; Broecker
et al., 1973 ; Nozaki and Tsunogai, 1976 ; Pearcy et al., 1977).

    Such short residence times imply that the major removal pro-
cess   of trace metals from surface waters to deep waters is particle controlled rather than due to mixing processes. Recent studies
of the depth distribution of dissolved trace elements, such as Cu,
Ni, Cd, Zn, As, and Se (Boyle et al., 1977 ; Bruland et al., 1978a,b;
Sclater et al., 1976 ; Andreae, 1979 ; Measures and Burton, 1980)
support the conclusion that scavenging by sinking biogenic particles is the major removal process for these elements from the surface-water reservoir to deep waters. As with nutrients, dissolved
concentrations of such elements are lower in surface waters than
in deep waters, where partial or near total dissolution of the
particulate flux occurs with increasing depth. Most of the vertical particulate flux in the ocean is due to the settling of large
biogenic particles (Osterberg et al., 1963 ; McCave, 1975 ; Sackett, 1978 ; Bishop et al., 1978 ; Brewer et al., 1980 ; Honjo,
1980), fecal pellets, calcareous and siliceous hard parts, etc..,
which can reach the deep sea sediments within a month.

    These considerations imply that a relatively fast geochemical
coupling exists between the atmospheric inputs of trace metals
(and other reactive substances) and their transport to the deep
ocean. This rapid particulate transport "short-circuits" water
mixing processes and has therefore a major geochemical implication
which is of great interest to the health of the oceans. Indeed,
for trace metals, atmospheric inputs of recent origin, such as anthropogenic inputs, are partially or totally recorded in the flux
associated with particles sinking from surface waters. Therefore,
induced chemical changes in the deep water column will be first
observed in that particulate flux. Also, for these elements for
which removal in the particulate form is the dominant sink from
surface waters, a simple mass balance between input via the atmosphere and particulate removal from the surface layers can tell us
how important net atmospheric deposition is as a source for these
elements in the surface ocean.

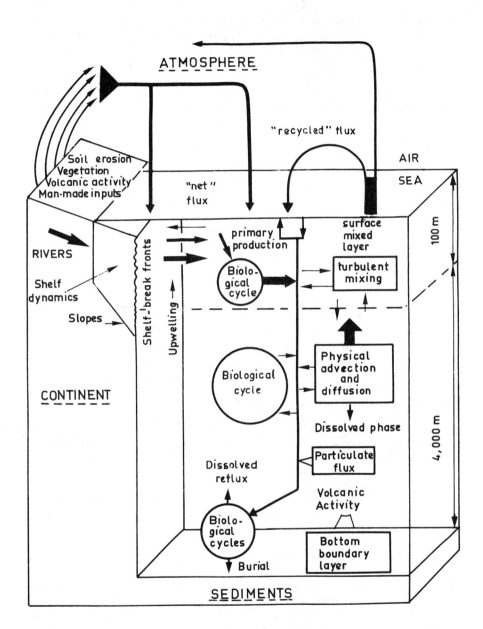

Figure 15. Schematic diagram of major ocean processes for reactive elements.

Once in deep waters, dispersal of atmospheric inputs of trace metals will depend on their internal cycle in that reservoir. As in surface waters, chemically active elements will undergo complex mineralization/solubilization processes, with or without getting involved in the biological cycle which is still at work in deep waters, although it is much slower than in surface waters. Elements involved in the biological cycle will recycle between particulate and soluble phases down to different depths, depending on the chemical behaviour and biological involvement of the element. Also, adsorption and desorption reactions on the surface of particles can occur at every depth, as shown recently for thorium isotopes by Bacon and Anderson (1982). A very important distinction has to be made here between small particles and large particles in the ocean. Large particles (>50 μm), which form only 10% or less of the mass in suspension, account for 90% of the mass which sediments (Chesselet, 1979 ; Bishop et al., 1978). Small particles account for the other 10%, although they represent 90% of the standing stock. Large particles (foraminifera, fecal pellets, aggregates) remain in the water column only a few weeks or days. On the other hand, small particles have a mean residence time of 50 to 100 years in the deep water column (Lambert et al., 1981) and are therefore subject to horizontal advection of the water masses in which they occur. As a consequence it has been suggested that the transit of small particles will induce the largest measurable in-situ chemical changes in the dissolved composition of several elements, whereas large particles will serve as a means of transporting organic matter and inorganic matter directly to the sediment-seawater interface (Lal, 1977, 1980). Moreover, large particles may break or be ingested by pelagic organisms en route, which will also lead to in-situ chemical changes via remobilization of some elements. The relative importance of these various processes is not yet known for most trace elements, with the exception of barium (Dehairs et al., 1980). Finally, at the water-sediment interface, due to a variety of biological and physical processes, some trace metals may return to the water column through advection and diffusion (Lambert, 1981). This is another area for future research.

To finish with this overview of the internal cycle of trace metals in the ocean, we may ask ourselves how important is the geochemical coupling between atmospheric particles and oceanic particles with respect to such elements? Qualitative insight is given in Figure 16, where, for 17 trace metals, deviations from crustal abundance for atmospheric and oceanic particles (from the North Atlantic) are indicated by the values of the enrichment factor relative to the crust, Al being used as the reference element. It can be seen that a similar trend exists for increasing EF_{crust} values in atmospheric particles and in oceanic particles from the deep ocean. The high values in oceanic particles cannot

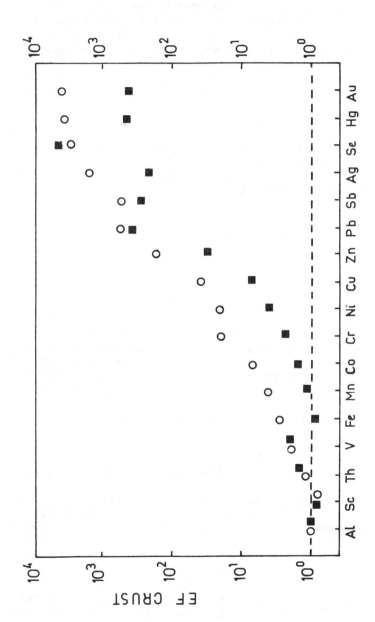

Figure 16. Geometric mean crustal enrichment factors in North Atlantic marine aerosols (squares) and in suspended matter in intermediate and deep waters of the North Atlantic (circles). Elements are ranked by order of increasing EF in oceanic particles.

From Buat-Ménard and Chesselet, 1979

be explained by the abundance of metals in biogenic particulate
matter having a planktonic composition (Buat-Ménard and Chesselet,
1979). If we assume that these data are correct, one interpretation is that metals are enriched in oceanic particles because of
conservation of the particulate input from the atmosphere. This
conservation may occur either because the input is permanently
preserved in the particulate form or, because the input (although
dissolvable in seawater) is processed by the biological cycle and
delivered to deep waters in the particulate form. Alternatively,
if the atmosphere is not a significant source to surface waters,
the observed enrichments in oceanic particles would reflect only
purely internal oceanic processes, as is the case for nutrients.
The answer can be given quantitatively from a flux balance for the
mixed layer between input via the atmosphere and removal by settling particles. Such a calculation will be presented in the next
section, where I will try to review the quantitative aspects of
air-to-sea and sea-to-air particulate transfer.

4. ASSESSMENT OF PARTICULATE FLUXES FROM AND TO THE AIR-SEA INTERFACE : GEOCHEMICAL IMPLICATIONS

4.1 Introduction

In this last section, I will describe some of the tools
available for calculating particle fluxes in the atmosphere. As
will be seen, direct measurements are scarce, and in many cases,
fluxes are calculated indirectly from atmospheric concentration
data. For some elements, the source functions are known with
enough certainty so that the right order-of-magnitude of net air-
to-sea and sea-to-air exchange rates can be obtained. Some insight into the variability of these fluxes with space and time
will be also given. Finally, when enough information is available,
geochemical implications of the calculated fluxes will be emphasized.

4.2 Basis for the Calculation of Atmospheric Particulate Fluxes

Sea salt flux from the ocean to the atmosphere : a world of
uncertainty. There is still a discrepancy of two orders-of-magnitude in the estimate of the global production of sea-salt particles (Eriksson, 1959, 1960 ; Blanchard, 1963 ; Petrenchuk, 1980 ;
Morelli, 1978). For reasons which are not yet clear to me, the
sea-salt source strength most commonly used in the literature is
the lowest, that of Eriksson : 1,200 Tg/yr. As pointed out by
Duce (1981), more accurate estimates are needed to properly assess the ocean source strength for global atmospheric cycles of
a number of elements. Taking sulfur as an example, Duce (1981)
indicated that the eleven estimates of the global cycle of this

element, published in the last 20 years, all use the sea-spray sulfate source strength of 44 Tg/yr derived from Eriksson's sea-salt
flux. That value could be 8 times higher if Blanchard's estimate
of 10,000 Tg/yr for sea-salt is correct. It is interesting to try
to understand why these two estimates are different because key
factors responsible for the discrepancy must be evaluated. Eriksson's and Blanchard's general approach is the same ; they have
derived global sea-salt production rates from sea-salt fallout estimates, assuming steady state between production and removal by
dry and wet deposition. With respect to dry fallout, the key factor was the dependance of sea-salt mass fallout as a function of
wind speed. Since the pioneering work of Woodcock (1953), the dependence of the sea-salt particle size distribution on wind speed
has been clearly shown. The concentration of the larger sea-salt
particles increases much more rapidly with wind speed than does
the total sea-salt concentration. Eriksson used an average 12 kt
wind over the entire ocean, while Blanchard used measurements of
the temporal and geographical variation of wind speeds over 5° to
20° latitude, longitude squares over the global ocean. With respect to wet removal by rain, the salt content of precipitation
used by Blanchard is higher than that assumed by Eriksson. These
general considerations show us some of the important parameters
which must be taken into account to assess properly aerosol deposition to the ocean surface. This whole topic is treated in detail
by Slinn in this volume.

Over the ocean, owing to inherent sampling difficulties,
direct measurements of dry deposition and wet deposition are scarce compared to aerosol particle concentration data. As a consequence, most often, indirect flux calculations have been based on
the latter type of data. It is, therefore, important to consider if
aerosol concentration data can provide the basis for reliable dry
and wet deposition estimates. With a few examples, I will present
different approaches to this problem. I must emphasize here that
measurements of rain concentrations and measurements or calculations of dry deposition fluxes only give us gross fallout values.
From this we have to factor out the "recycled" component, coming
from the ocean itself. We have seen in the first section that this
component can be identified in some cases. However, the recycled
fraction may be different in air filtration, dry deposition and
wet deposition samples, owing to the dependence of both deposition processes on particle size. In rain, a further complication
is to be expected for those elements or compounds present in the
atmosphere both in the vapor and particulate phases.

For all these reasons, large uncertainties exist in calculations of atmospheric-particle fluxes. Nevertheless, correct order-
of-magnitude estimates can be obtained, with some care, and these
allow reasonable geochemical conclusions to be made about the
global atmospheric cycles of many trace elements.

498 						P. BUAT-MÉNARD

<u>Dry deposition : Problems</u>. As recently shown by McDonald
al. (1982), dry deposition rates of sea-salt aerosol particles may
be satisfactorily estimated from measured concentrations and theoretical dry deposition velocity calculations, provided extreme
care is taken to obtain air samples that represent the true sea-
salt particulate mass and size distribution. Even giant particles
(radii >20 μm) must be efficiently collected. These results show
the strong wind speed dependence of the sea-salt deposition rate
(Table 6) : when the wind speed increases from 3.4 to 10 m s^{-1}, the
salt deposition rate increases 50-fold, compared to a 7-10 fold
increase in atmospheric salt concentration. This results, almost
uniquely, because large particles dominate the deposition, and the
mass concentration of such particles increases much faster with
increasing wind speed than that of smaller particles. Even at the
low wind speed of 3.4 m s^{-1}, 10 μm radius sea-salt particles which
constitute only 13% of the total sea-salt mass, yet contribute
70% of the deposition. Therefore, for any element which, in marine
aerosols, has a sizable sea-derived component, the dry deposition
flux may be mostly due to the recycled component of that flux.

I will recall here the example, relative to lead dry deposition data from Enewetak Atoll (Settle and Patterson, 1982 ; Duce,
1982 ; see Section 3.2), where the small percentage of Pb on the
largest particles (likely recycled) seems to be controlling the Pb
dry deposition. Obviously, in the lower marine atmosphere, direct
dry deposition measurements cannot provide any reliable estimate
of the net dry flux of submicron anthropogenic lead aerosol. On
the other hand, such measurements may provide the best data to
quantify the flux of lead from the sea surface to the atmosphere.
The best approach to calculate the net dry deposition of Pb to the
ocean therefore seems to be to consider only that fraction of Pb
not recycled (i.e. the submicron fraction) and use predicted dry
deposition velocities. This assumes that Pb recycled from the sea
surface exhibits the same particle size distribution as that of
sea-salt, which might not be true (see Section 3.3).

Dry deposition velocities for submicron aerosols over the
ocean surface are not yet known with great accuracy (Slinn, this
volume). In the 0.1 to 1 μm radius interval, there seems to be a
large range of uncertainty, from 0.02 cm s^{-1} to 1 cm s^{-1} . Currently
used cascade impactors do not discriminate well in this size range,
so that, in my opinion, predicted values can give only the correct
order of magnitude of the net flux. Such sampling devices can also
bias the true size distribution of atmospheric particles. For high
volume cascade impactors, it seems that particle "bounce" effects,
especially for dry particles such as aluminosilicate minerals, can
result in a size distribution shifted to smaller sizes. This seems
to be a serious problem for such particles not only in continental
air (Walsh et al., 1978) but also in the remote marine atmosphere
(Buat-Ménard et al., 1982) ; see Figure 17.

TABLE 6. A comparison of sea-salt concentrations and fluxes as a function of wind speed (adapted from McDonald et al., 1982)

Wind Speed $m\ s^{-1}$	Atmospheric salt concentration $\mu g\ m^{-3}$	Salt flux $ng\ cm^{-2}hr^{-1}$
3.4	2.7	8
6.5	14	170
10	18	410

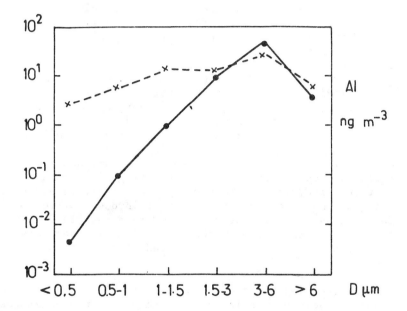

Figure 17. Aluminium concentration as a function of particle size in Enewetak air as found with a high-volume cascade impactor ----- (Duce, personal communication) and, as found by low volume bulk filtration and individual analysis of aluminosilicate particles by analytical scanning electron microscopy ——— (Buat-Ménard et al., 1982).

Wet deposition : Problems. We will consider first those elements or compounds that exist only in the particulate form in the
marine atmosphere. On a global scale, wet removal rates could be
assessed if we had representative rain concentration data and rain
precipitation statistics over the ocean. This is not yet the case.
At a given site, sampling over long time periods, integrating different meteorological regimes, is also required. Great care has
to be taken during collection to sample the entire rain duration
period. Indeed, sequential analysis of rains has shown that the
concentration of trace materials in rain is substantially greater
during the early period of rainfall, possibly caused by partial
evaporation of drops en route to the earth's surface, through air
not yet saturated, and/or by the clouds first use of the largest
condensation nuclei (Slinn, this volume).

In general, as described by Slinn in this volume, it is not
easy to predict accurately the wet removal rates of elements by
using washout factors (or scavenging ratios) and aerosol particle
concentration data from a few meters above the surface of the
ocean. An example of such a problem is given by recent results of
the mineralogical composition of aluminosilicate particles in
Tropical Pacific air and rain (Buat-Ménard et al., 1982). As shown
in Table 7, the mineralogy of these particles in the air sample
collected over a 3-week sampling period is different from that
from one, 1-hr rain sample collected during these 3 weeks. Whether this difference reflects different sampling time scales or
different source regions for the dust in air and rain requires
further investigation. It is interesting to note that at the same
sampling site, lead-isotope measurements (Settle and Patterson,
1982) do indicate that the lead in rain does not reflect directly
the lead aerosol content of the lower marine troposphere. It is
clear that data on the geochemistry of marine aerosols as a function of altitude are urgently needed if we want to solve such
problems.

Another important question is that of the scavenging efficiency of aerosol particles by cloud and rain droplets as a function of aerosol particle size. There are almost no data in the
remote marine atmosphere. Gatz (1977) and Duce et al. (1979) have
computed, for certain trace metals at urban and continental sites,
the washout factor or scavenging ratio W, where

$$W = \rho_a \, C_r \, / \, C_a \qquad\qquad (3)$$

C_r = concentration of metal in rain ($\mu g \ kg^{-1}$)

C_a = concentration of material in ground level air ($\mu g \ m^{-3}$)

ρ_a = density of air

Table 7. Mineralogical composition of insoluble particles in Enewetak air and rain (from Buat-Ménard et
al., 1982)

Mineral Phases	12/07 - 31/07 1979 air		17/07 1979 rain	
	n	%	n	%
Quartz	27	20	9	11
Plagioclase	1	1	5	6
Muscovite-Illite	18	14	25	31
Smectite	3	2	0	0
Vermiculite	1	1	0	0
Chlorite	2	2	7	9
Kaolinite	16	12	2	2
"Talc"	10	7	2	2
Fe	30	22	6	7
Ti	3	2	1	1
Cr	6	4	0	0
Not determined	17	13	25	31
Total number examined	134		81	

Values for W ranged from a few hundred to a few thousand.
Results from both studies indicate a trend for W to increase with
the mass-mean radius of the aerosol particles containing the element. In the remote marine atmosphere of Enewetak, this trend
does not seem to be observed. Ng and Patterson (1981) indicate
that W for Pb at Enewetak is slightly greater than for continental dust, despite the fact that the mass mean radius of Pb, about
0.2 μm, is one fourth that of dust, about 0.8 μm (Duce et al.,
1980). Data based on the analysis of individual aluminosilicate
particles by scanning electron microscopy (Buat-Ménard et al.,
1982) also show no trend for W at that site to increase with particle size. As indicated before, this may reflect that Pb and
aluminosilicate minerals in rain are not derived from the aerosol pool of the lower marine atmosphere. On the other hand, the
result may reflect that in relatively-clean remote areas (as opposed to polluted, urban areas) clouds use essentially all available particles as cloud condensation nuclei ; thereby, no particle-size dependence of W would be expected (see Slinn, this
volume). In any event, these observations show again the need for
additional studies including measurements of the distribution of
the aerosol with altitude. Thus, effects of size, particle chemis-

try and other factors make it difficult to predict W more accurately than to within a factor of ten.

A further complication arises from the presence of the recycled component of trace substances in marine rain. If the wet-removal efficiency for aerosols does effectively increase with increasing particle size, the relative contribution of the recycled component to elemental concentrations in rain might be up to 10 times larger than the recycled material's contribution to the marine aerosol. Again we will take lead as an example. Assume that 10% of the aerosol lead is recycled lead on large sea-salt particles (with a mean washout factor W = 2,000) and that 90% of aerosol lead is not recycled and is submicronic (with a mean washout factor of 200). It follows that the percentage of recycled lead in rain will be 55% of the total lead in rain. Such a possibility indicates that we must be cautious in interpreting total, wet-removal fluxes as representing a net input to the ocean. Also, we must be careful when considering the use of some tracers to derive deposition rates to the ocean surface. For example ^{210}Pb flux measurements may be to some extent biased by the existence of a recycled component ; if this isotope behaves like stable lead, an overestimation by a factor 2 in net flux assessment due to wet deposition could result.

Despite these pessimistic statements, washout factors are useful to assess the relative importance of wet and dry removal of marine aerosols. For small particles, it seems clear that (in areas of intense precipitation) dry deposition fluxes are almost negligible compared to wet deposition. On the other hand, dry and wet deposition of large particles over the ocean appear, on the average , to be of comparable importance (NAS, 1978 ; Slinn, this volume). From these considerations, and taking into account what we know about the marine versus non-marine components of elemental concentrations as a function of particle size, we can infer that atmospheric net fluxes from the continents to the oceans will be primarily due to wet deposition in the marine atmosphere. This is particularly true for the anomalously-enriched elements which are generally found on submicron sized aerosols (see Section 2.4). Dry deposition, on the other hand is certainly significant for sea-source aerosols, as well as for continental dust from soil erosion.

So far, we have only considered the case of conservative particulate elements in aerosols ; i.e., those which do not undergo appreciable gas-to-particle conversion. It is interesting to consider the removal of those elements or compounds that have both a vapor-phase component and a particle component in the marine atmosphere. Dry and wet removal rates of trace gases are the subject of other chapters in this book (see the chapters by

Peters,   and by Liss). Recent calculations indicate that for some
organic compounds and a trace element like Hg, the scavenging efficiency of the gas phase by rain is much smaller than the scavenging efficiency of the aerosol phase (Duce and Gagosian, 1982 ;
Fitzgerald et al., 1981). For example, the Hg content of marine
rain can be explained entirely by the scavenging of the particulate phase, despite the fact that particulate Hg is at least one
hundred times less abundant in the marine atmosphere than vapor-
phase Hg. Therefore, the impact of the atmospheric input of anthropogenic mercury on the marine chemistry of that element is
probably more dependent on the atmospheric cycle of aerosol-Hg
than on the atmospheric cycle of vapor-phase Hg, although the two
cycles may be interdependent.

All these considerations of the present state-of-the-art may
appear pessimistic. They show how difficult it is to make precise
assessments of particulate fluxes to and from the ocean. However,
with reasonable care and provided a good data base is used, our
present knowledge does permit first approximations of such fluxes
and their geochemical significance.

4.3 General Considerations of the Role of Air-Sea Particulate Exchange in Geochemical Cycling

The presence of natural and man-made substances in the marine atmosphere raises the question of how global geochemical cycles at the surface of the earth depend on atmospheric chemistry
and how these cycles are being perturbed by human's activities.

Depending on the substance being investigated, the nature of
what we need to know will vary. For example, present interest in
the global sulfur cycle arises, first, from the possible effect
of small-particle, excess S-aerosol on climate, and second, on its
impact on the acidity of rain. Human influence on the global sulfur aerosol-burden is mainly to be found in small particles mostly
derived from gas-to-particle conversion involving SO_2. But, the
detailed source strengths for SO_2 are not yet known. Data on both
the origin of vapor-phase sulfur compounds as well as on gas to
particle conversion rates are needed. With respect to air-sea particulate exchange, we need to know the amount of S-aerosol injected from the ocean surface to the atmosphere. That would appear
to be straightforward, if we assume that sea-source aerosol sulfur
is simply sea-salt sulfate. This may not be the case : there may
be some direct contribution of fine particle sulfate from the
ocean (Lawson and Winchester, 1979a,b). and vapor-phase sulfur compounds derived from biogenic sources and emitted from the ocean
may be important precursors for the small-particle sulfur (Bonsang et al., 1980). Thus, in a geochemical context, the assessment of the oceanic source function for S aerosol needs further

study. A recent compilation by Meszaros (1982) suggests strongly
that the atmospheric sulfur budget calculations must be revised
since the minimum sulfur emission from the ocean (sea-salt sulfur)
appears to be 4 times higher than the value of Eriksson (1959,1960),
which has been used by several workers.

The organic carbon cycle in marine aerosols resembles in
some way the aerosol particle sulfur cycle. Moreover, the interest
is also to assess correctly the net input of organic matter to
the oceans. The limited data available indicate that most of the
gross atmospheric flux to the ocean is due to wet removal of organic carbon (Duce and Duursma, 1977). This total-flux extimate
is higher than the present estimate of the quantity of dissolved
organic carbon transported by rivers, and could amount to as much
as 10% of the organic carbon produced in the ocean itself by primary productivity. We do not know yet how much of that gross flux
is recycled from the sea surface. As mentioned in section 2.7,
the use of carbon isotopes and biological source marker measurements in rain could help increase our knowledge.

The role of sea-salt particles as cloud condensation nuclei
and in the electrification of the atmosphere is well known. Still,
as pointed out by Blanchard many times, more accurate sea-salt
production rates are needed. In the context of global geochemical
cycles, the correct assessment of the contribution of "cyclic
salts" (atmospheric fallout of sea-salt over land) to river chemistry is also a major subject of debate. In a recent paper, Stallard and Edmond (1981) have criticized previous approaches to this
problem, mainly based on the assumption that all the chlorine in
rivers was cyclic in origin. They show the need of more quantitative assessments of sea-salt fluxes over land, in order to solve
the controversy. Also, as mentioned in Section 3.2, data on the
chemical mass balance of trace metals in rivers suggest that the
recycling of trace elements which are enriched in sea-derived aerosol particles, could contribute significantly to trace-metal
concentration in rivers (Martin and Meybeck, 1979 ; Li, 1981).
Most of the present data on trace-metal discharge by rivers are
based on measurements at the mouth, precisely where the contribution of the cyclic component is a maximum.

Very recently, interest has been growing in the phenomenon of
acidity of rainwater in the marine atmosphere. Data from "remote"
marine sites are rare but indicate that rain is usually more acidic (pH $\leq$ 5.6) than would be expected on the basis of atmospheric
CO_2-pure water equilibrium (Miller and Yoshinaga, 1981 ; Galloway
et al., 1982 ; Jickells et al., 1982 ; Pszenny et al., 1982).
Long-range transport of pollutants or marine biogenic reduced sulfur compounds are probably the more likely causes of this acidity.
Charlson and Rodhe (1982) have emphasized that, besides carbonic

acid,other naturally occuring acids can affect the pH of "natural rainwater" as can naturally occurring basic materials (such as NH_3 and $CaCO_3$). pH values may,however,vary considerably due to variability in scavenging efficiencies as well as geographical patchiness of the sulfur, nitrogen and water cycles. Pszenny et al. (1982) have shown that, at American Samoa, alkalinity due to sea-salt appears also to affect pH significantly when the salinity of the rain is greater than 0.01‰ or 3000 µg Na/kg (Figure 18). Much more work is needed to understand the factors controlling the pH and composition of natural rainwater. This is an important area for future research because of widespread concern regarding the acidification of rain. A large number of variables must be studied simultaneously a) to understand the relationships between the composition of rainwater, cloud water and the aerosol and trace gases that go into the cloud and b) to define clearly natural and perturbed conditions.

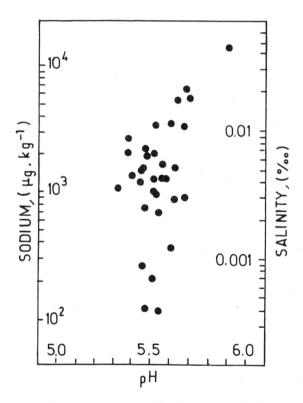

Figure 18. pH vs sodium concentration and salinity in Samoan rain. Adapted from Pszenny et al., 1982.

TABLE 8. The global atmospheric mass balance of arsenic,
unit of 10⁹ grams per year, adapted from Walsh et al.(1979)

	NORTHERN HEMISPHERE	SOUTHERN HEMISPHERE
1- Source strengths		
A- Natural sources (particulates d<5 μm)		
Ocean		
Bubble bursting	0.012	0.016
Gas exchange	0.05	0.06
Earth's crust		
Particle weathering	0.16	0.077
Direct volatilization	0.0005	0.0002
Volcanoes	3.5	3.5
Forest wild fires	0.11	0.05
Terrestrial Biosphere	0.17	0.09
Natural sources : total	4.0	3.8
B- Anthropogenic sources (particulates d<5μm)		
Copper production	12.	1.
Iron/steel production	3.8	0.4
Lead/zinc production	2.0	0.2
Arsenic agriculture	1.7	0.2
Other sources	2.1	0.2
Anthropogenic sources : total	21.6	2.0
Natural and anthropogenic sources : total	25.6	5.8
2- Removal +		
Over land	20.0	7.6
Over the ocean	2.2	0.56
Global removal	22.2	8.2

+ Calculated assuming the tropospheric burden of arsenic is
removed 40 times per year.

4.4 A Case Example : the Air-Sea Exchange of Trace Metals

Global fluxes of metals through the atmosphere. Among the
trace constituents of the marine aerosol, trace metals and especially heavy metals and metalloids have been extensively studied
in the last ten years, primarily because of concern about atmospheric pollution input to the ocean. Global fluxes of metals have
been estimated in different ways : consideration of emission
source strengths, removal rates by rain and dry deposition over
land and ocean, and consideration of concentration data in aerosols and mean residence times of aerosols in the troposphere. As
indicated previously, only first approximations are possible from
such estimates. The correct magnitude of such estimates can be
checked, however, if one looks for agreement among fluxes calculated in different ways.

An example of the state-of-the-art in this context is the
new global atmospheric cycle of arsenic (Walsh et al., 1979),
summarized in Table 8. Source strengths and removal rates have
been calculated separately for the Northern and the Southern Hemispheres. Based on the most reliable estimates, source strength
calculations indicate that 80% and 35% of the emissions are of
anthropogenic origin in the Northern and Southern Hemisphere respectively. The most significant anthropogenic sources appear to
be volatilization from metal processing and agricultural chemical
usage. The most important natural source may be volcanoes, which
also involves a volatilization process. The total arsenic emission
from all of these volatilization processes is found to be about
120 times greater than primary particle source strengths, for normal crustal weathering and sea-salt production. This estimate is
in agreement with the order-of-magnitude arsenic enrichments relative to crustal or bulk seawater composition in remote aerosol
particles. Walsh et al. suggest that arsenic concentrations are
predominantly industrial in the Northern Hemisphere and predominantly natural in the Southern Hemisphere.

Removal rates of arsenic by dry and wet deposition were calculated by Walsh et al. (1979) from representative global surface concentration data in three ways : a) assuming that tropospheric aerosols are removed 40 times per year, b) using total
washout factors for arsenic from continental areas and regional
precipitation volumes, and c) using the mean dry deposition velocity of arsenic over England, assuming that over land the removal
by rain is twice the removal by dry deposition. All three estimates have considerable uncertainties. Assumption a) implies an
average residence time of aerosols of 9 days. Since arsenic is
primarily present on submicron aerosols, this value is probably
correct within a factor 2, considering predicted residence times
for submicron radioactive aerosols (NAS, 1978 ; Lambert et al.,

Table 9. Source strengths (x 10^9 grams per year) for some trace metals on atmospheric particulate material according to Weisel (1981).

Element	"Natural Sources"						Anthropogenic	
	Ocean			Crust	Volcanoes	Vegetation	Fossil fuel	Other
	Mean	Min.	Max.					
Al	150	50	500	200,000	700	40	1,500	1,600
Cd	0.4	0.1	7	0.05	0.1	0.05	0.2	5
K	3,800	-	-	6,000	200	100	190	200
Mg	12,000	-	-	3,000	80	20	310	200
Mn	4	2	10	200	9	5	8	360
Pb	5	0.2	20	3	0.4	0.2	4	400
V	9	5	20	30	0.7	0.2	18	1.8
Zn	10	0.5	100	80	10	10	84	240

1982). Assumptions b) and c) are probably correct within a factor 3-5 (see earlier in this section). Nevertheless, the three esti- mates of arsenic removal rates are all within a factor 2 of each other and with the total arsenic source estimates. This mass ba- lance indicates that the ocean is a net sink for arsenic, arsenic inputs in the Northern Hemisphere being predominantly anthropo- genic. This geochemical conclusion is not in agreement with that reached by Lantzy and Mackenzie (1979). This latter work only considered data on arsenic in rain compiled by Wedepohl (1969). The large calculated output of arsenic from the atmosphere by precipitation could not be matched by known inputs, so that Lantzy and Mackenzie postulated a large source of arsenic by biological volatilization from the surface ocean. Although such a source has been postulated by Wood (1974), its strength is probably largely overestimated by Lantzy and Mackenzie for the following reasons : a) Rain data used by Lantzy and Mackenzie were found to be much higher than recent data on rain samples from remote marine areas (Andreae, 1980). Andreae's data are comparable to those predicted by Walsh et al. (1979) using washout factors. b) Chemical specia- tion studies of arsenic in rain by Andreae indicated that methy- lated forms of arsenic contributed a negligible fraction of the total concentration.

An idea of the order of magnitude of global fluxes through the atmosphere is given for some metals in Table 9. This table is based on the computations of Weisel (1981). Sea-to-air fluxes take into account enrichment of these elements in sea-source aerosols, as estimated from the results of the analysis of aerosol parti- cles artificially produced from bursting bubbles. From this ta- ble, one can conclude that, except for the alkali and alkaline earth metals, the ocean is a sink for trace metals. Two major types of input dominate : mineral dust from soil erosion and man- made inputs. However, the ocean could be a significant source of Cd to the atmosphere.

Metal fluxes from the atmosphere to the ocean : geographical variability. Per unit area of the ocean surface, the net atmosphe- ric input to the ocean will depend on the remoteness from conti- nental areas and on the initial load of continental air masses. If we assume that net fluxes are represented, Table 10 shows that the rate of input of metals decreases by many orders of magnitude from coastal areas to the remote regions of the Tropical North Pacific. Flux values for the North Sea and the Tropical North Pa- cific are based on direct measurements. The other values given in this table are based on mean atmospheric concentration data. The fluxes have been derived assuming either that tropospheric aero- sols are removed 40 times per year (Duce et al., 1976a), or that the total deposition velocity of continental aerosols in the ma- rine atmosphere is 1 cm s^{-1} (Buat-Ménard and Chesselet, 1979 ;

Table 10. Estimated mean fluxes of trace metals from the atmosphere to the sea-surface (ng cm^{-2}yr^{-1})

Element	New York Bight (1)	North Sea (2)	Western Mediterranean (3)	South Atlantic Bight (4)	Bermuda (5)	Tropical North Atlantic (6)	Tropical North Pacific (7)
Al	6,000	30,000	5,000	2,900	3,900	5,000	1,900
Sc	-	5	1	-	0.6	1.1	0.4
V	-	480	-	-	5	1.7	7
Cr	-	210	49	-	9	14	6
Mn	-	920	-	60	45	70	18
Fe	5,700	25,500	5,100	5,900	3,000	3,200	1,300
Co	-	39	3,5	-	1.2	2.7	0.6
Ni ::	-	260	-	390	3.	20	-
Cu ::	-	1,300	96	220	30	25	8
Zn ::	1,400	8,950	1,080	750	75	130	22
As ::	-	280	54	45	3	-	-
Se ::	-	22	48	-	3	14	10
Ag ::	-	-	3	-	-	0.9	0.7
Cd ::	30	43	13	9	4.5	5.	~1
Sb ::	-	58	48	-	1.0	3.5	0.24
Au ::	-	-	0.05	-	-	0.1	-
Hg ::	-	-	5	24	-	2.1	-
Pb ::	3,900	2,650	1,050	660	100	310	12
Th	-	4	1.2	-	-	0.9	0.9

Notes : " :: " denotes elements generally enriched in marine aerosols.

(1) Duce et al. (1976b); (2) Cambray et al. (1975) ; (5) Arnold et al. (1982) ;
(4) Windom (1981) ; (5) Duce et al. (1976a) ; (6) Buat-Ménard and Chesselet (1979) ;
(7) Duce (1982).

Arnold et al., 1982), based on the ^{210}Pb total deposition data
(Turekian et al., 1977 ; Turekian and Cochran, 1981). One can
note for the North Atlantic the relatively good agreement between the values calculated by Duce et al. from concentration data
obtained during 1973 and 1974 at Bermuda, and those calculated by
Buat-Ménard and Chesselet from atmospheric samples collected during 4 cruises in the Tropical North Atlantic in 1974 and 1975.

For coastal and semi-remote areas, it is interesting to compare the relative importance of atmospheric and river inputs (and
waste discharges), especially for those metals that are "enriched"
in marine aerosols and may therefore have a pollution origin. Such
a comparison is presented for some metals in Table 11.

Table 11. Ratio of atmospheric flux to riverine flux (dissolved) of some trace elements in coastal and semi-remote
oceanic areas.

ELEMENT	SOUTH ATLANTIC BIGHT (1)	NEW YORK BIGHT (2)	NORTH SEA (3)	WESTERN MEDITERRANEAN SEA (4)
Arsenic	2.1	1.0	1.7	–
Cadmium	2.7	3.1	1.1	–
Copper	1.9	–	1.9	–
Iron	5.8	6.4	1.7	–
Manganese	0.6	–	0.8	–
Mercury	22	–	2.1	0.8
Nickel	1.7	–	1.3	–
Lead	9.5	20	6.8	6.2
Zinc	2.3	3.1	1.9	0.8

(1) Windom (1981) ; (2) Duce et al. (1976b); (3) Cambray et al.
(1975) ; (4) Arnold et al. (1982).

From this table, it appears that the atmospheric input of
elements is the same order-of-magnitude as, if not greater than,
the estimated dissolved riverine input. But, such a comparison
must be considered with caution, since the riverine flux is for the
soluble fraction only. Essentially all suspended sediments in rivers are at the present time trapped in estuaries. Atmospheric
fluxes are for total particulates. However, except for crustal
elements, a large percentage of the trace metals on atmospheric
particles dissolve upon entering seawater. As we have seen in

Section 3.3, the fate of the dissolved flux is complex, since
recycling between dissolved phase and particulate mineral and organic phases will occur. Nevertheless, for elements such as Zn,
Cd, and Pb, the atmospheric flux will effectively compete with
the direct dissolved input from rivers and coastal discharges.
It appears, therefore, that the potential importance of atmospheric deposition to coastal and semi-remote marine pollution is
certain. The impact of atmospheric inputs will depend on vertical
and horizontal mixing processes (solar, tidal, wind driven) as
well as on biological and chemical recycling.

Atmospheric metal fluxes to the open ocean : geochemical
implications. If we assume that enriched elements in marine aerosols are predominantly of anthropogenic origin, it can be seen
from Table 10 that the North Atlantic waters are potentially more
influenced by pollution than is the Tropical North Pacific. Up to
now, with the exception of Pb (Settle and Patterson, 1982), almost
no data have been available for the Southern Hemisphere. We would
expect that atmospheric net fluxes of pollution source elements
should be, on the average, up to one order of magnitude lower in
the Southern than in the Northern Hemisphere. The question now is :
can we get an insight into the importance of the atmospheric input as a net source for these elements in the ocean ?

One approach is to search for an effect on the dissolved concentrations in sea-water. Up to now, this approach has worked only
for Pb. The reasons for this are found in the specificity of the Pb
cycle in the atmosphere and in the ocean, at the present time. Pb
has a very fast cycle in the ocean, since its residence time is
estimated to be about 2 years in surface waters and only about
100 years in the deep ocean (Settle and Patterson, 1980). At present, the burden of atmospheric lead (at least in the Northern
Hemisphere) is entirely dominated by industrial inputs, which have
increased dramatically in the last 200 years. The present lead
cycle in the ocean is,therefore,not at steady state with respect
to atmospheric inputs. Since the residence time of lead in the
ocean is about 10 times smaller than the turn-over time of the
deep waters, different vertical distributions of lead can be expected for one oceanic region to another, depending on the local
atmospheric source strength. This is illustrated by the data presented in Figure 19, from Settle and Patterson (1982). It can be
seen that lead concentrations in the surface waters are always
higher than in deep waters, which illustrates the importance of
the present atmospheric source strength. Also, the magnitude of
lead concentrations in the waters are related to the level of the
atmospheric input rate. Interestingly, lead concentrations in
North Atlantic surface waters are 10 times higher than in South
Tropical Pacific surface waters, whereas the atmospheric lead flux
is 100 times higher over the North Atlantic than over the Tropical

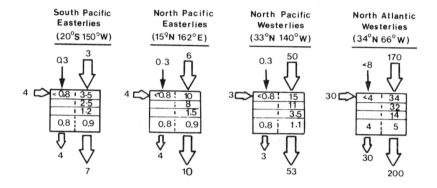

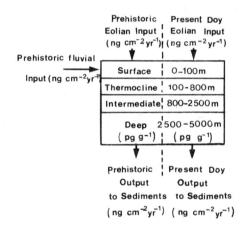

Figure 19. Relationships among prehistoric lead and present-day eolian inputs, vertical concentration profiles and sediment outputs of lead on a global scale. (Patterson, personal communication; data can be found in Settle and Patterson, 1982. Additional information is given in Fleagle, A.R. and C.C. Patterson, 1983: Vertical concentration profiles of lead in the Central Pacific at 15°N and 20°S. Earth Planet. Sc. Lett., in press).

South Pacific. Thus, when we look at dissolved lead concentrations, the effect of different atmospheric input rates seems to be
diluted. As suggested in the last section, the reason for this
dilution is probably that the transport of atmospheric lead in the
ocean does not result only from water mixing processes, but from
the involvement of this element in the particulate cycle of the
ocean (Buat-Ménard and Chesselet, 1979 ; Burnett and Patterson,
1980).

A second approach to the assessment of the importance of the
atmosphere as a source of metals in the ocean has been to compare
atmospheric inputs with the inputs of metals to the ocean from all
sources. Total inputs have been estimated on the basis of the
known elemental composition of deep-sea sediments and the measured
sediment accumulation rates. Such a comparison has been made for
the North Atlantic (Duce et al., 1976 ; NAS, 1978). This approach
has very serious caveats, primarily because it assumes that the
oceans are at steady state with respect to atmospheric inputs.
Clearly, this is not true. Atmospheric input rates, calculated at
present, reflect recent phenomena on the geological time scale.
This is the case not only for anthropogenic inputs, but also for
natural inputs such as eolian transport of mineral dust from soils
to the oceans. There is more and more evidence that this eolian
transport of mineral dust was much more important during the last
glaciation (Petit et al., 1981), between 30,000 and 13,000 years
before present. If we consider that the sedimentation rates of
deep-sea sediments are of the order of a few mm per millenium, only
data from the very surface of the sediments might be useful for a
comparison with dust inputs from the atmosphere. Moreover, bioturbation processes are known to homogenize sediments over a depth of
several cm (and therefore over a time-record of several thousand
years). Consequently such a comparison must be considered with a
great caution and, in my opinion, is valid only for truly conservative particulate fluxes from the atmosphere to the ocean : e.g.
clay minerals and quartz, which are essentially insoluble in seawater.

Another caveat is seen from consideration of the transport
velocity of particles in seawater. Most of the vertical particulate flux of terrigenous dust in the ocean is accelerated because
of the attachment of these small dust particles with fast sinking,
large biogenic particles which reach the ocean floor in a few
weeks at most. This would mean that atmospheric dust fluxes are
deposited within months in deep-sea sediments and undergo little
horizontal transport (perhaps less than 100 km). However, in some
oceanic areas, resuspension and relatively fast (a few cm s^{-1} ;
Biscaye and Ettreim, 1977) horizontal transport of recently-depo-
sited sediments does occur. This means that in such areas the
comparison between the atmospheric dust flux and the sedimented
flux is not valid. This has been shown to be likely the case in
the Tropical North Atlantic, west of the Mid-Atlantic Ridge, by
Biscaye et al. (1974), based on measurements of rubidium and

strontium isotopes in clay minerals.

In the previous Section, I have described what is the fate of atmospheric inputs of metals that dissolve upon entering seawater. The involvement of these elements in the biological cycle in surface waters and in deep waters proceeds down to the sediment-seawater interface. There, the organic particulate input is known to be primarily recycled back to the water column. As is the case for nutrients, many metals can be remobilized in the dissolved form (Lambert, 1981). They may, therefore, cycle one or several times in the ocean before being finally buried in sediments. At each cycle ($\sim$1,000 yrs), only a small fraction of the vertical particulate flux from surface waters would be trapped in sediments, as suggested by studies of bomb-produced metallic radionuclides (Labeyrie et al., 1976). This added to the fact that the present rate of the net atmospheric input of "enriched" metals may represent a recent phenomenon, precludes drawing any meaningful geochemical conclusion. One might infer, however, that if the present rate of the atmospheric input of enriched metals to the ocean is higher than the average clay-sedimentation rate for these elements, this is the case for Cd, Zn, Se, and Pb in the North Atlantic (NAS, 1978), there is evidence of the importance of the atmosphere as a source of these elements for the ocean. The restriction here is that deep-sea clays are not the only sink for trace elements in the ocean. Major sinks for nutrient elements are probably to be found in continental rise and shelf areas. As stated by Turekian (1977),"elements involved in the biological cycle of the ocean are sphinx-like in their silence about their origin and depositories."

I am personally convinced that the best available way to assess the importance of atmospheric deposition of trace metals as a source for open ocean waters is that described in the previous Section ; i.e., using a mass balance between input from the atmosphere and removal from the surface layers by sinking particles. I have used this approach for the North Atlantic (Buat-Ménard, 1979 ; Buat-Ménard and Chesselet, 1979). Figure 20 presents the relative contribution of F_a, the atmospheric flux to F_s, the flux associated with the settling of small and large particles from the surface layers. Since in this model we can assume a steady state between the atmospheric input (F_a) and the removal flux (F_s), the consideration of the ratio between these two fluxes provides an insight into the atmospheric source function to ocean waters. If we forget momentarily the inherent uncertainties in the calculated fluxes, a number of geochemical conclusions can be drawn from the comparison presented in this figure (see also figure 16).

It appears, first, that present atmospheric inputs to the North Atlantic represent, depending on the trace element being considered, between 5% and 100% of total inputs to surface waters. Thus,

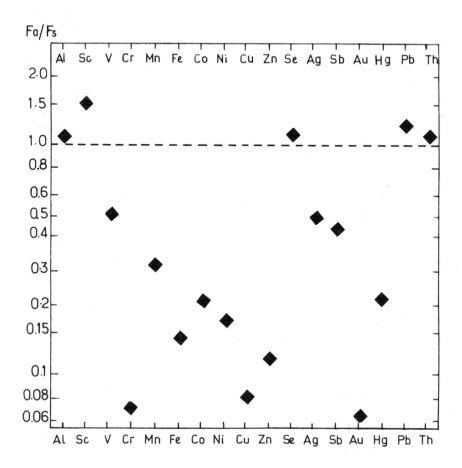

Figure 20. Ratio of flux from atmosphere, Fa, to downward
particulate flux from surface layers, Fs, in the Tropical
North Atlantic. Adapted from Buat-Ménard and Chesselet, 1979.

atmospheric deposition of trace metals to the ocean is a significant source in the North Atlantic. It appears that for several
elements (Al, Sc, Th, Pb, and Se), the atmospheric transport constitutes the major source to ocean surface waters. From what we
know about the origin of these elements in atmospheric aerosols,
the following conclusions can be drawn :a) Al, Sc and Th are associated with aluminosilicate minerals both in atmospheric and
oceanic particles (Buat-Ménard and Chesselet, 1979). Consequently
the vertical particulate flux of such particles in open North
Atlantic deep waters is primarily derived from an atmospheric
source. This conclusion has been recently extended by Chester
(1982) to the South Atlantic and probably applies to the Tropical
North Pacific (Duce, 1982). b) Lead in North Atlantic aerosols
has an anthropogenic origin. Therefore, the particulate flux of
lead leaving surface waters is also anthropogenic. This conclusion is in complete agreement with that drawn by Settle and Patterson (1980, 1982), based on dissolved lead vertical distributions. These authors have shown that this situation is probably
world-wide for this element. Further evidence of the recent atmospheric origin of deep water lead is given by lead-isotope data
both for dissolved lead (Patterson, personal communication) and
particulate lead (Petit and Buat-Ménard, in preparation). How
much of the anthropogenic Pb flux from the atmosphere is present
in deep-sea sediments cannot yet be assessed. c) Like lead, selenium is extremely enriched in marine aerosols. However the reasons for the Se enrichment in aerosols are not yet understood.
There is evidence, however, that atmospheric Se has a natural
origin, since over the open ocean its atmospheric concentrations
show very small differences between the Northern and the Southern
Hemispheres. Natural volatilization processes, either from volcanic activity (Buat-Ménard and Arnold, 1978) or from the ocean
surface (Mosher and Duce, 1981), could be the major sources. If
volcanic activity is the dominant source, Figure 20 would imply
that the atmospheric input of Se to the ocean is the major natural source of this element in ocean surface waters. On the other
hand,with a marine source, atmospheric input of Se to the ocean
would be simply a recycled flux.

Figure 20 indicates that for V, Cr, Mn, Fe, Co, Ni, Cu, Zn,
Ag, Sb, Au, and Hg, the vertical particulate flux in seawater is
higher than the atmospheric flux. Consequently, since these elements are enriched in oceanic particulates compared to their crustal distribution, the excess concentrations of these elements
in oceanic particulates reflect mostly their involvement in the
biological cycle. Among these elements, Ni, Cu, Zn, Ag, Sb, Cd,
Au, and Hg are enriched in marine aerosols, possibly because of
the presence of an anthropogenic component. Figure 20 indicates
that for these elements, the anthropogenic air-sea flux is small
relative to natural fluxes. However, one can speculate that if

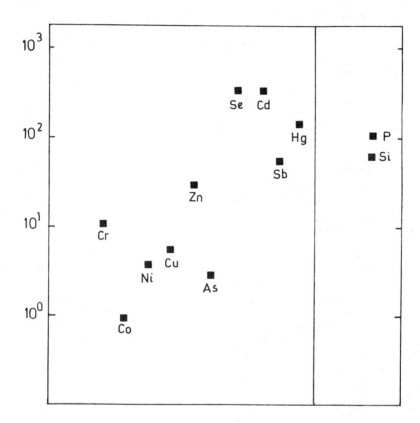

Figure 21. Ratio of the flux transported by settling particles from surface to deep waters in the Tropical North Atlantic, to the accumulation rate in deep-sea clays for that region.

Adapted from Buat-Ménard, 1979.

in the future, the atmospheric input was to be one order of magnitude higher than at present over the North Atlantic, then detectable effects would be recorded in oceanic suspended matter and
therefore would alter the oceanic cycle of such elements. Such
effects are nevertheless extremely difficult to predict at present. Indeed, changes in the rate of input from the atmosphere will
be biologically cycled at the same rate only if such changes do
not affect the biological cycle itself. For example, increased
levels of toxic metals in surface waters could reduce the biogeochemical cycling. However, if at the same time, increased nutrient availability was to occur, it would counteract the first
effect. Although we know that appreciable changes in nutrient
availability by large-scale changes in water circulation and trace element fluxes have occured in the past, the sedimentary record has not yet been able to tell us precisely how nature takes
care of such stresses.

    Another conclusion, which can be made for some of those ele-
ments naturally enriched in ocean particulates, is that they must
be remobilized at the water sediment interface and through water
circulation and upwelling, be reinjected in the dissolved form
into surface waters. Figure 21 shows that for Cu, Zn, Se, and
other elements, the vertical particulate flux to deep waters is,
as for nutrients, much higher than their accumulation rate in
deep-sea sediments. Since there is evidence that this vertical
flux is enriched in such elements throughout the entire water
column, and since the resuspended material from deep-sea sediments shows no enrichment for these elements (Buat-Ménard, 1979 ;
Lambert, 1981), it follows that a reflux from the sediments in
the dissolved form must take place. Evidence for this can clearly
be seen for Cu (Boyle et al., 1977) from the pattern of the vertical profile of the dissolved concentration of this element in
deep and bottom waters. It is, therefore, probable that such elements will recycle several times in the water column before being permanently buried in deep sediments. Depending on their residence time in the deep ocean, atmospheric inputs of these elements will accumulate in ocean waters for long periods of time
(a few hundred to a few thousand years) before being removed
from the ocean.

    The mass-balance model, which has been used here, requires
that the difference between the downward particulate flux of
trace metals in the ocean and the sedimentary flux should be
equal, at steady state, to the quantity of these elements re-
entering the surface reservoir in the dissolved form (Figure 15),
due to upwelling and diffusion. An agreement between the two
calculated fluxes would validate the conclusions drawn by using
this predictive model. There are at present few data available
for dissolved elements in the North Atlantic. When using those

from the North Pacific, for Cu, Ni, Cd, and Zn, the calculated
values of these two types of fluxes agree within a factor 5,
which is satisfactory for such first approximations. To me, the
geochemical implications of such conclusions urge carefully coor-
dinated field programs to estimate more accurately net atmospheric fluxes to the ocean, as well as vertical particulate fluxes
in the ocean, at mesoscale (coastal regions, enclosed seas) and
at global scale, in both Hemispheres. Data from the Southern
Hemisphere, where the atmospheric flux both natural and anthropogenic is minimum, should be obtained with high priority to assess natural fluxes in the ocean.

5. CONCLUSIONS

Using a few examples, I have tried to show how an accurate
assessment of particulate fluxes is the key to our understanding
many atmospheric and oceanic cycles and their alterations by humans. Critical uncertainties in the present estimations have been
pointed out, so that appropriate sampling strategies could be
adopted in future studies. I am convinced that the answers to remaining questions will be found only on the basis of interdisciplinary approaches, involving collaborative efforts between
atmospheric physicists and chemists together with physical, chemical, and biological oceanographers.

ACKNOWLEDGEMENTS

Many ideas and concepts in this paper have been developed
over the years in continuing discussions with my colleagues and
friends. I wish to express my gratitude to R. Chesselet who gui-
ded and inspired my work while I was a graduate student at Gif-
sur-Yvette. I have also benefited from the broad and different
insights of R.A. Duce, C.E. Lambert, G. Lambert, C. Patterson,
K.A. Rahn and K.K. Turekian ; I thank them for their help. I
would also like to thank J. Labeyrie and C. Lalou for the stimulating atmosphere which they have maintained throughout the years
in our laboratory. I am indebted to M. Spyridakis and S. Queran
for the final presentation of the manuscript.

REFERENCES

A.E., 1981: Atmospheric Environment: Symposium on Arctic Air
   Chemistry 15, 1345-1516.
Andreae, M.O., 1979: Arsenic speciation in seawater and
   interstitial waters: the influence of biological-chemical
   interactions on the chemistry of a trace element. Limnol.
   Oceanogr., 24, 440-452.
Andreae, M.O., 1980: Arsenic in rain and the atmospheric mass
   balance of arsenic. J. Geophys. Res., 85, 4513-4518.
Arnold, M., P. Buat-Ménard and R. Chesselet,1981: An estimate of
   the input of trace metals to the global atmosphere by
   volcanic activity. RVEAC Symp. IAMAP, Hamburg, 17-28
   August. Abstract
Arnold, M., A. Seghaier, D. Martin, P. Buat Ménard and R.
   Chesselet,1982. Géochimie de l'aérosol marin au-dessus de
   la Méditerranée occidentale. Workshop on Pollution of the
   Mediterranean, Cannes, December 2-4.
Atlas, E. and C.S. Giam, 1981: Global transport of organic
   pollutants:ambient concentration in the remote marine
   atmosphere. Science, 211, 163-165.
Bach, W., 1976: Global air pollution and climatic changes. Rev.
   Geophys. Space Phys., 14, 429-474.
Bacon, M.P. and R.F. Anderson,1982: Distribution of Th isotopes
   between dissolved and particulate forms in the deep sea. J.
   Geophys. Res., 87, 2045-2056.
Bacon, M.P., D.W. Spencer and P.O. Brewer, 1976:^{210}Pb/^{226}Ra and
   ^{210}Po/^{210}Pb disequilibrium in seawater and suspended
   particulate matter. Earth Planet. Sc. Lett., 32, 277-296.
Balistrieri, L., P.G. Brewer and J.W. Murray, 1981: Scavenging
   residence times of trace metals and surface chemistry of
   sinking particles in the deep ocean. Deep Sea Res., 28A,
   101-121.
Barbier, M., D. Tusseau, J.C. Marty and A. Saliot, 1981: Sterols
   in aerosols, surface microlayer and subsurface water in the
   North-Eastern Tropical Atlantic. Oceanologica Acta, 4,
   77-84.
Barger, W.R. and W.D. Garrett, 1970: Surface active organic
   material in the marine atmosphere. J. Geophys. Res., 75,
   4561-4566.
Barger, W.R. and W.D. Garrett, 1976: Surface active organic
   material in air over the Mediterranean and over the eastern
   Equatorial Pacific. J. Geophys. Res. 81, 3151-3157
Beauford, W.J., J. Barlser and A.R. Barringer, 1977: Release of
   particles containing metals from vegetation into the
   atmosphere. Science, 195, 571-573.
Belot, Y., C. Caput and D. Gauthier, 1982: Transfer of americium
   from seawater to atmosphere by bubble bursting. Atmos.
   Environ., 16, 1463-1466.
Berg, W.W. and J.W. Winchester, 1978: Aerosol chemistry of the

marine atmosphere. In:Chemical Oceanography, vol.7
(J.P.Riley and R.Chester, eds.) Academic Press, 173-231.

Bertine, K.K. and E.D. Goldberg, 1971: Fossil fuel combustion
and the major sedimentary cycle. Science, 173, 233-235.

Bidleman, T.F. and C.E. Olney, 1974: High volume collection of
atmospheric polychlorinated biphenyls. Bull. Envr. Contam.
Toxicol., 11, 442-450.

Biscaye, P.E., R. Chesselet and J.M. Prospero, 1974: Rb-Sr,
^{87}Sr/^{86}Sr isotope system as an index of provenance of
continental dust in the open Atlantic Ocean. J. Rech.
Atmosph., 8, 819-829.

Biscaye, P.E. and S.L. Ettreim, 1977: Suspended particulate
loads and transports in the nepheloid layer of the abyssal
Atlantic Ocean. Mar. Geol., 23, 155-172.

Bishop, J.K.B., M.R. Ketten and J.M. Edmond, 1978: The
chemistry, biology and vertical flux of particulate matter
from the upper 400m of the Cape Basin in the Southeast
Atlantic ocean. Deep Sea Res., 25, 1121-1161.

Blanchard, D.C., 1963: Electrification of the atmosphere by
particles from bubbles in the sea. In:Progress in
Oceanogr., vol. 1(M. Sears ed.), Pergamon press, 71-202.

Blanchard, D.C., 1975: Bubble scavenging and the water-to-air
transfer of organic material in the sea. Advances in
Chemistry Ser., 145, 360-387.

Blanchard, D.C. and A.H. Woodcock, 1980: The production,
concentration and vertical distribution of the sea-salt
aerosol. Ann. N.Y. Acad.Sc., 338, 330-347.

Bonsang, B., B.C. Nguyen, A. Gaudry and G. Lambert, 1980:
Sulfate enrichment in marine aerosols owing to biogenic
gaseous sulfur compounds. J. Geophys. Res., 85, 7410-7416.

Boyle, E.A., F.R. Sclater and J.M. Edmond, 1977: The
distribution of dissolved copper in the Pacific. Earth.
Planet. Sc. Lett., 37, 38-54.

Brewer, P.G., V. Nozaki, D.W. Spencer and A.P. Fleer, 1980:
Sediment trap experiments in the deep North Atlantic;
isotopic and elemental fluxes. J. Mar. Res., 38, 703-728.

Broecker, W.S., A. Kaufman and R.M. Trier, 1973: The residence
time of thorium in surface seawater and its implications
regarding the fate of reactive pollutants. Earth. Planet.
Sc. Lett., 20, 35-44.

Bruland, K.W., G.A. Knauer and J.H. Martin, 1978a: Zinc in
Northeast Pacific water. Nature, 271, 741-743.

Bruland, K.W., G.A. Knauer and J.H. Martin, 1978b: Cadmium in
Northeast Pacific waters. Limnol. Oceanogr., 23, 618-625.

Buat-Ménard, P., 1979: Influence de la retombée atmosphérique
sur la chimie des métaux en trace dans la matière en
suspension de l'Atlantique Nord. Thèse Dr. Sc. Paris,
434 pp.

Buat-Ménard, P. and M. Arnold, 1978: The heavy metal chemistry
of atmospheric particulate matter emitted by Mount Etna

volcano. Geophys. Res. Lett., 5, 245-248.

Buat-Ménard, P. and R. Chesselet, 1979: Variable influence of the atmospheric flux on the trace metal chemistry of oceanic suspended matter. Earth Planet. Sc. Lett., 42, 399-411.

Buat-Ménard, P., U. Ezat and A. Gaudichet, 1982: Size distribution and mineralogy of alumino-silicate dust particles in Tropical Pacific air and rain. 4th Intern. Conf. on Precipitation Scavenging, Dry Deposition and Resuspension, November 29-December 3, Santa Monica, Calif., Abstract.

Buat-Ménard, P., J. Morelli and R. Chesselet, 1974: Water-soluble elements in atmospheric particulate matter over Tropical and Equatorial Atlantic. J. Rech. Atmos., 8, 661-673.

Burnett, M.W. and C.C. Patterson, 1980: Perturbation of natural lead transport in nutrient calcium pathways of marine ecosystems by industrial lead.In:Isotope Marine Chemistry (E. Goldberg, Y. Horibe, and K. Saruhashi, eds.),Uchida Rokakuo Publishing Co., Tokyo, 413-438.

Cambray, R.S. and J.D. Eakins, 1980: Studies of Environmental Radioactivity in Cumbria. Part I: Concentration of Plutonium and Caesium-137 in Environmental Samples from West Cumbria and a possible Maritime Effect,U.K. A.E.A., Harwell Rept. 9807; 15 pp.

Cambray, R.S., D.F. Jefferies and G. Topping, 1975: An Estimate of the Input of Atmospheric Trace Elements into the North Sea and the Clyde Sea (72-73). U.K. A.E.A., Harwell Rept. 7733, 30 pp.

Charlson,R.J. and H. Rodhe, 1982: Factors controlling the acidity of natural rainwater. Nature, 295, 683-685.

Chesselet, R., 1969: Etude de la radioactivité artificielle du milieu marin par spectrométrie gamma. Rapport CEA-R-3698, 117 pp.

Chesselet, R., 1979: Modes of settling and organic input to the sediment seawater interface, a review. Coll. Intern. CNRS. n°293, 27-33.

Chesselet, R., M. Fontugne, P. Buat-Ménard, U. Ezat and C.E. Lambert, 1981: The origin of particulate organic carbon in the atmosphere as indicated by its stable carbon isotopic composition. Geophys. Res. Lett., 8, 345-348.

Chesselet, R., J. Morelli and P. Buat Ménard, 1972a: Some aspects of the geochemistry of marine aerosols. In:The Changing Chemistry of the Oceans, Nobel Symp. 20, Wiley, 93-114.

Chesselet, R., J. Morelli and P. Buat-Ménard, 1972b: Variations in ionic ratios between reference seawater and marine aerosols. J. Geophys. Res., 77, 5116-5131.

Chester, R. 1982: Particulate aluminium fluxes in the eastern Atlantic. Mar. Chem., 11, 1-16.

Cicerone, R.J., 1981: Halogens in the atmosphere. Rev. Geophys.
    Space Phys., 19, 123-139.
Cipriano, R.J. and D.C. Blanchard, 1981: Bubble and aerosol
    spectra produced by a laboratory "breaking wave". J.
    Geophys. Res., 86, 8055-8092.
Coantic, M., 1980: Mass transfer across the ocean-air interface:
    small scale hydrodynamic and aerodynamic mechanisms.
    Physicochemical Hydrodynamics, 1, 249-279.
Crozat, G., 1979: Sur l'émission d'un aérosol riche en potassium
    par la forêt tropicale. Tellus, 31, 52-57.
Dehairs, F., R. Chesselet and J. Jedwab, 1980: Discrete
    suspended particles of barite and the barium cycle in the
    open ocean. Earth Planet. Sc. Lett., 49, 528-550.
Duce, R.A., 1978: Speculations on the budget of particulate and
    vapor phase non-methane organic carbon in the global
    troposphere. Pure and Appl.Geophys., 116, 244-273.
Duce, R.A., 1981: Biogeochemical cycles and the air-sea exchange
    of aerosols. Prepared for the SCOPE Workshop on the
    Interaction of Biogeochemical Cycles., Orsundsbro, Sweden,
    20-30 May 1981. In press.
Duce, R.A., 1982: Sea Salt and Trace Element Transport Across
    the Sea-Air Interface. Abstract, Symposium Ocean/Atmosphere
    Material Exchange, Joint Oceanographic Assembly, Halifax,
    Nova Scotia Canada, 2-13 August 1982.
Duce, R.A. and E.K. Duursma, 1977: Inputs of organic matter to
    the ocean. Marine Chem., 5, 319-339.
Duce, R.A. and Gagosian, R.B., 1982: The input of atmospheric
    $n-C_{10}$ to $n-C_{30}$ alkanes to the ocean. J. Geophys. Res. Paper
    2C 0661, in press.
Duce, R.A. and E.J. Hoffman, 1976: Chemical fractionation at the
    air/sea interface. Ann. Rev. Earth Planet. Sc., 4, 187-228.
Duce, R.A., G.L. Hoffman and W.H. Zoller, 1975: Atmospheric
    trace metals at remote northern and southern hemisphere
    sites: Pollution or natural ? Science, 197, 551-557.
Duce, R.A., G.L. Hoffman, B.J. Ray, I.S. Fletcher, G.T. Wallace,
    J.L. Fasching, S.R. Piotrowicz, P.R. Walsh, E.J. Hoffman,
    J.M. Miller, and J.L. Heffter, 1976a: Trace metals in the
    marine atmosphere:Sources and Fluxes. In:Marine Pollutant
    Transfer (H. Windom and R. Duce, eds.), D.C. Heath and Co.,
    Lexington, Mass., 77-119.
Duce, R.A., J.G. Quinn, C.E. Olney, S.R. Piotrowicz, B.J. Ray
    and T.L. Wade, 1972, Enrichment of heavy metals and organic
    compounds in the surface microlayer of Narragansett Bay,
    Rhode Island. Science, 176, 161-163.
Duce, R.A., C.K. Unni, P.J. Harder, B.J. Ray, C.C. Patterson,
    D.M. Settle and W.F. Fitzgerald, 1979: Wet and dry
    deposition of trace metals and halogens in the marine
    environment. CACGP Symp. Boulder, Colorado, Abstract.
Duce, R.A., C.K. Unni, B.J. Ray, J.M. Prospero and J.T. Merrill,
    1980: Long-range atmospheric transport of soil dust from

Asia to the Tropical North Pacific: temporal variability. Science, 209, 1522-1524.

Duce, R.A., G.T. Wallace and B.J. Ray Jr., 1976b: Atmospheric Trace Metals over the New York Bight. NOAA Techn. Rept. ERL 361-MESA 4, 17pp.

Eichmann, R., G. Ketseridis, G. Schebeke, R. Jaenicke, J. Hahn, P. Warneck and C. Junge, 1980: N-Alkane studies in the troposphere. II: Gas and particulate concentrations in Indian Ocean air. Atmos. Environ., 14, 695-703.

Eichmann, R., P. Neuling, G. Ketseridis, J. Hahn, R. Jaenicke and C. Junge, 1979: N-Alkane studies in the troposphere. I- Gas and particulate concentrations in North-Atlantic air. Atmos. Environ., 13, 587-599.

Eriksson, E., 1959: The yearly circulation of chloride and sulfate in nature. Meteorological, geochemical, pedological implications. Part I. Tellus, 11, 317-403.

Eriksson, E., 1960: The yearly circulation of chloride and sulfate in nature, Part II. Tellus 12, 63-109.

Fitzgerald, W.F., G.A. Gill and A.D. Hewitt, 1981: Air-Sea exchange of mercury. In:Trace Metals in Seawater (C.S.Wong, J.D. Burton, E. Boyle , K. Bruland and E. Goldberg, eds.), NATO Symp. Series, Plenum Press, in press.

Fraizier, A., M. Masson and J.C. Guary, 1977: Recherches préliminaires sur le rôle des aérosols dans le transfert de certains radioéléments du milieu marin en milieu terrestre. J. Rech. Atmos., 11, 49-60.

Gagosian, R.B., E.T. Peltzer and O.C. Zafiriou, 1981: Atmospheric transport of continentally derived lipids to the Tropical North Pacific. Nature, 291, 312-314.

Gagosian, R.B., O.C. Zafiriou, E.T. Peltzer and J.B. Alford, 1982: Lipids in aerosols from the Tropical North Pacific: temporal variability. J. Geophys. Res., submitted.

Galloway, J.N., G.E. Likens, W.E. Keene and J.M. Miller, 1982: The composition of precipitation in remote areas of the world. J. Geophys. Res., in press.

Garrett, W.D. and R.A. Duce, 1980: Surface microlayer samplers. Chapter 25.In:Air-Sea Interaction,(F. Dobson, L. Hasse and R. Davis, eds.) Plenum Press, 471-490.

Gatz, D.F., 1977: Scavenging ratio measurements in Metromex. In: Precipitation Scavenging (1974), Conf. 741003, N.T.I.S., Springfield, Virginia, 71-87.

Gaudichet, A. and P. Buat-Ménard, 1982: Nature minéralogique et origine des particules atmospheriques insolubles du Pacifique Tropical Nord (Atoll d'Enewetak). Etude par microscopie électronique analytique en transmission. C.R. Acad. Sci., 294, 1241-1246.

Glaccum, R., 1978: The mineralogy and elemental composition of mineral aerosols over the Tropical North Atlantic: the influence of Saharan dust. MS thesis, University of Miami, 161 pp.

Glaccum, R.A. and J.M. Prospero, 1980: Saharan aerosols over
    the Tropical North-Atlantic, mineralogy. Marine Geol., 37,
    295-321.
Goldberg, E.D., 1976: Rock volatility and aerosol composition.
    Nature, 260, 128-129.
Graham, W.F. and R.A. Duce, 1979: Atmospheric pathways of the
    phosphorus cycle. Geochim. Cosmochim. Acta, 43, 1195-1208.
Graham, W.F. and R.A. Duce, 1981: Atmospheric input of
    phosphorus to remote tropical islands. Pacific Sc., 35,
    241-255.
Graham, W.F. and R.A. Duce, 1982: The atmospheric transport of
    phosphorus to the western North Atlantic. Atmos. Environ.,
    16, 1089-1097.
Graham, W.F., S.R. Piotrowicz and R.A. Duce, 1979: The sea as a
    source of atmospheric phosphorus. Marine Chem., 7, 325-342.
Granat, L., H. Rodhe and R.O. Halbert, 1976: The global sulfur
    cycle. In:Global Cycles, SCOPE Rept 7, Ecol. Bull.
    (Stockholm), 22, 89-134.
Gravenhorst, G., K.P. Muller and L. Schutz, 1981: Inorganic
    nitrogen compounds over the Atlantic. ROAC Symposium,
    IAMAP, Hamburg, 17-28 August, abstract.
Harvey, G.R. and W.G. Steinhauer, 1974: Atmospheric transport of
    polychlorobiphenyls to the North Atlantic. Atmos. Environ.
    8, 777-782.
Haulet, R., P. Zettwoog and J.C. Sabroux, 1977: Sulfur dioxide
    discharge from Mount Etna. Nature, 268, 715-717.
Hoang, C.T. and J. Servant, 1974: Exemple d'un apport
    continental de quelques métaux dans l'aérosol au-dessus de
    l'Atlantique Nord à la latitude de 40°N. J. Rech. Atmos.,
    8, 791-805.
Hoffman, G.L. and R.A. Duce, 1972: Consideration of the chemical
    fractionation of alkali and alkaline earth metals in the
    Hawaïan marine atmosphere. J. Geophys. Res., 77, 5161-5169.
Hoffman, E.J. and R.A. Duce, 1977a: Alkali and alkaline earth
    metal chemistry of marine aerosols generated in the
    laboratory with natural seawater. Atmos. Environ., 11,
    367-372.
Hoffman, E.J. and R.A. Duce, 1977b: Organic carbon in marine
    atmospheric particulate matter:concentration and particle
    size distribution. Geophys. Res. Lett., 4, 449-452.
Hoffman, E.J., G.L. Hoffman and R.A. Duce, 1974a: Chemical
    fractionation of alkali and alkaline earth metals in
    atmospheric particulate matter over the North Atlantic.
    J. Rech. Atmos., 8, 675-688.
Hoffman, G.L., R.A. Duce, P.R. Walsh, E.J. Hoffman, B.J. Ray and
    J.L. Fasching, 1974b: Residence times of some particulate
    trace metals in the oceanic surface microlayer:
    significance of atmospheric deposition. J.Rech. Atmos., 8,
    745-759.
Hoffman, E.J., G.L. Hoffman and R.A. Duce, 1980: Particle size

dependance of alkali and alkaline earth metal enrichment in marine aerosols from Bermuda. J. Geophys. Res., 85, 5499-5502.

Hunter, K.A., 1977: Chemistry of the sea surface microlayer. Ph.D. Thesis, University of East Anglia. 363 pp.

Hunter, K.A., 1980: Processes affecting particulate trace metals in the sea surface microlayer. Marine Chem., 9, 49-70.

Hunter, K.A. and P.S. Liss, 1976: Measurement of the solubility product of various metal ions carboxylates. J. Electroanal. Chem., 73, 347-358.

Hunter, K.A. and P.S. Liss, 1979: The surface charge of suspended particles in estuarine and coastal waters. Nature, 2821, 823-825.

Honjo, S., 1980: Material fluxes and modes of sedimentation in the mesopelagic and bathypelagic zones. J. Mar. Res., 38, 53-97.

Jaenicke, R. 1978: The role of organic material in atmospheric aerosols. Pure and Appl. Geophys., 116, 283-292.

Jickells, T.S., A.H. Knap, T.M. Church, J.N. Galloway and J.M. Miller, 1982: Acid rain in Bermuda. Nature, 297, 55-57.

Johnson, B.D. and R.C. Cooke, 1980: Organic particle and aggregate formation resulting from the dissolution of bubbles in seawater. Limnol. Oceanogr., 25, 653-661.

J.G.R:Journal of Geophysical Research, 1972: Working symposium on Sea-Air Chemistry. 77, 5059-5349.

J.R.A: Journal de Recherches Atmosphériques, 1974: International Symposium on the Chemistry of Sea/Air Particulate Exchange Processes. 8, 499-1002.

Jullien, D., 1982: L'interface air-mer: Composants organiques, budget et processus d'évolution. Thèse 3ème cycle, Université de Paris VI, 120 pp.

Labeyrie, L.D., H.D. Livingston and V.T. Bowen, 1976: Comparison of the distribution in marine sediments of the fall out derived nuclides ^{55}Fe and ^{239,240}Pu. In:Transuranium Nuclides in the Environment. I.A.E.A., Vienna, 121-137.

Lal, D., 1977: The organic microcosm of particles. Science, 198, 997-1009

Lal, D. 1980: Comments on some aspects of particulate transport in the oceans. Earth Planet. Sc. Lett., 49, 520-527.

Lambert, C.E., 1981: Le cycle interne du fer et du manganese et leurs interactions avec la matière organique dans l'océan. Ph.D. Thesis, Université de Picardie, 235 pp.

Lambert, C.E. and Jehanno, C., 1980-1981. La peau de la mer. Oceanis, 6, 153-165.

Lambert, C.E., C. Jehanno, J.S. Silverberg, J.C. Brun-Cottan and R. Chesselet, 1981: Log-normal distributions of suspended particles in the open ocean. J. Mar. Res., 39, 77-98.

Lambert, G., B. Ardouin and G. Polian, 1982: Volcanic output of long-lived radon daughters. J. Geophys. Res., in press.

Lantzy, R.J. and F.T. Mackenzie, 1979: Atmospheric trace metals:

global cycles and assessment of man's impact. Geochim.
Cosmochim. Acta, 43, 511-525.

Lawson, D.R. and J.W. Winchester, 1979a: Sulfur and trace
elements concentrations relationships in aerosols from the
South American Continent. Geophys. Res. Lett., 5, 195-198.

Lawson, D.R. and J.W. Winchester, 1979b: Sulfur, potassium and
phosphorus association in aerosols from South American
Tropical Rain Forests. J. Geophys. Res., 84, 3723-3727.

Li, Y.H., 1981: Geochemical cycles of elements and human
perturbation. Geochim. Cosmochim. Acta, 45, 2073-2084.

Lion, L.W. and J.O. Leckie, 1981: The biochemistry of the
air-sea interface. Ann. Rev. Earth Planet. Sc., 9, 449-486.

Liss, P.S., 1975: Chemistry of the sea surface microlayer.In:
Chemical Oceanography, vol.2, (J.P.Riley and G. Skirrow,
eds.) Academic Press. 193-240.

Lowman, F.G., T.R. Rice and F.A. Richards, 1971: Accumulation
and redistribution of radionuclides by marine organisms.
In: Radioactivity in the Marine Environment, N. A. S.,
Washington DC, 161-199.

MacIntyre, F., 1972: Flow patterns in breaking bubbles. J.
Geophys. Res., 77, 5211-5228.

MacIntyre, F., 1974a: Non-lipid-related possibilities for
chemical fractionation in bubble film caps. J. Rech. Atmos.
8, 515-527.

MacIntyre, F., 1974b: Chemical fractionation and sea-surface
microlayer processes. In:The Sea, volume 5 (E.D. Goldberg,
ed.), Wiley, 245-300.

Martin, J.M. and M. Meybeck, 1979: Elemental mass-balance of
material carried by major world rivers. Marine Chem., 7,
173-206.

Martin, J.M., A.J. Thomas and C. Jeandel, 1981: Transport
atmospherique des radionucleides artificiels de la mer vers
le continent. Oceanologica Acta, 4, 263-266.

Marty, J.C., A. Saliot, P. Buat-Ménard, R. Chesselet and K.
Hunter, 1979: Relationship between the lipid composition of
marine aerosols, the sea-surface microlayer and subsurface
water. J. Geophys. Res., 84, 5707-5716.

McCave, I.N., 1975: Vertical flux of particles in the ocean.
Science, 202, 429-431.

McDonald, R.L., C.K. Unni and R.A. Duce, 1982: Estimation of
atmospheric sea-salt dry deposition: wind speed and
particle size dependence. J. Geophys. Res., 87, 1246-1250.

Measures, C.I. and J.D. Burton, 1980: The vertical distribution
and oxidation states of dissolved selenium in the Northeast
Atlantic Ocean and their relationship to biological
processes. Earth Planet. Sc. Lett., 46, 385-396.

Meinert, D.C. and J.W. Winchester, 1977: Chemical relationships
in the North Atlantic marine aerosol. J. Geophys. Res., 82,
1778-1782.

Meszaros, E. 1982: On The atmospheric input of sulfur into the ocean. Tellus, 34, 277-282.

Miller, J.M. and A.M. Yoshinaga, 1981: The pH of Hawaiian precipitation: a preliminary report. Geophys. Res. Lett., 8, 779-782.

Morelli, J., 1978: Données sur le cycle atmosphérique des sels marins. J. Rech. Ocean., 3, 27-50.

Morelli, J., P. Buat-Ménard and R. Chesselet, 1974: Production experimentale d'aérosols à la surface de la mer. J. Rech. Atmos., 8, 961-986.

Mosher, B. and R.A. Duce, 1981: Vapor phase selenium in the marine atmosphere, ROAC Symp. IAMAP, Hamburg, 17-28 August, abstract.

Mroz, E.J. and W.H. Zoller, 1975: Composition of atmospheric particulate matter from the eruption of Heimaey, Iceland. Science, 190, 461-464.

N.A.S. 1978: The Tropospheric Transport of Pollutant and other Substances to the Ocean (J.M. Prospero, ed.), National Academy of Sciences, Washington DC, 243 pp.

Ng, A. and C. Patterson, 1981: Natural concentrations of lead in ancient Arctic and Antarctic ice. Geochim. Cosmochim. Acta, 45, 2109-2121.

Nozaki, Y., J. Thomson and K.K. Turekian, 1976: The distribution of ^{210}Pb and ^{210}Po in the surface waters of the Pacific Ocean. Earth. Planet. Sci. Lett., 22, 304-312.

Nozaki, Y. and S. Tsunogai, 1976: ^{210}Pb and ^{210}Po disequilibrium in western North Pacific. Earth Planet. Sc. Lett., 32, 313-321.

Osterberg, C., A.G. Carey and H. Curl, 1963: Acceleration of sinking rates of radionuclides in the ocean. Nature, 200, 1276-1277.

Pattenden, N.J., R.S. Cambray and M. Playford, 1981: Trace and major elements in the sea-surface microlayer. Geochim. Cosmochim. Acta, 45, 93-100.

Pattenden, N.J., R.S. Cambray, K. Playford, J.D. Eakins and E.M.R. Fisher, 1980: Atmospheric measurements on radionuclides previously discharged to sea. In: International Symp. on Impact of radionuclides releases into the marine environment, Vienna, 6-10 October 1980, I.A.E.A.

Pearcy, W.G., E.E. Krygier and H.H. Cutshall, 1977: Biological transport of ^{65}Zn into the deep-sea. Limnol. Oceanogr., 22, 846-855.

Petit, J.R., M. Briat and A. Royer, 1981: Ice aged aerosol content from East Antarctic ice core samples and past wind strength. Nature, 293, 391-394.

Petrenchuk, O.P., 1980: On the budget of sea salts and sulfur in the atmosphere. J. Geophys. Res., 85, 7439-7444.

Piotrowicz, S.R., R.A. Duce, J.L. Fasching and C.P. Weisel, 1979: Bursting bubbles and their effect on the sea to air transport of Fe, Cu and Zn. Marine Chem., 7, 307–324.

Piotrowicz, S.R., B.J. Ray, G.L. Hoffman and R.A. Duce, 1972: Trace metal enrichment in the sea–surface microlayer. J. Geophys. Res., 77, 5243–5254.

Prospero, J.M., 1981: Eolian transport to the world ocean. In: The Oceanic Lithosphere, The Sea, vol.7 (C. Emiliani, ed.), Wiley Interscience, 801–874.

Pszenny, A.A.P., F. MacIntyre and R.A. Duce, 1982: Sea salt and the acidity of marine rain on the windward coast of Samoa. Geophys. Res. Lett., 9, 751–754.

Quinn, J.G. and T.L. Wade, 1972: Lipid measurements in the marine atmosphere and sea surface microlayer. In: Baseline Studies of Pollutants in the Marine Environment (E.D. Goldberg, ed.), NSF, Washington DC, 633–663.

Rahn, K.A., 1976: The Chemical Composition of the Atmospheric Aerosol, Techn. Rep., Graduate School of Oceanography, University of Rhode Island, 265 pp.

Rahn, K.A., 1981: Relative importance of North America and Eurasia as sources of Arctic aerosol. Atmos. Environ., 15, 1447–1455.

Rancher, J. and M.A. Kritz, 1980: Diurnal fluctuations of Br and I in the tropical marine atmosphere. J. Geophys. Res., 85, 5581–5587.

Sackett, W.M., 1978: Suspended matter in seawater. In:Chemical Oceanography, vol.7 (J.P. Riley and R. Chester, eds.), Academic Press, 127–169.

Sanak, J., A. Gaudry and G. Lambert, 1981: Size distribution of ^{210}Pb aerosols over oceans. Geophys. Res. Lett., 8, 1067–1069.

Savoie, D.L. and J.M. Prospero,1982: Particle size distribution of nitrate and sulfate in the marine atmosphere.J. Geophys. Res., in press.

Schutz, L. and K.A. Rahn, 1982: Trace element concentrations in erodible soils. Atmos. Environ., 16, 171–176.

Sclater, F.R., E.A. Boyle and J.M. Edmond, 1976: On the marine geochemistry of nickel. Earth Planet. Sc. Lett., 31, 119–128.

Settle, D.M. and C.C. Patterson, 1980: Lead in Albacore:a guide to lead pollution in Americans. Science, 207, 1167–1176.

Settle, D.M. and C.C. Patterson, 1982: Magnitudes and sources of precipitations and dry deposition fluxes of industrial and natural leads to the north Pacific at Enewetak. J. Geophys. Res., in press.

Settle, D.M., C.C. Patterson, K.K.Turekian and J.K. Cochran, 1982: Lead precipitation fluxes at tropical oceanic sites determined from ^{210}Pb measurements. J. Geophys. Res., 87, 1239–1245.

Simoneit, B.R.T., 1977: Organic matter in aeolian dust over the Atlantic Ocean. Mar. Chem., 5, 443–464.

Simoneit, B.R.T., 1979: Biogenic lipids in aeolian particulates collected over the ocean. In:Carbonaceous particles in the atmosphere (T. Novakov, ed.), Berkeley Laboratory 9037, 233-244.

Simpson, H.J., 1977: Man and global nitrogen cycle: group report. Dahlem Konferenzen, Global Chemical Cycles and Their Alteration by Man, (W. Stumm, ed.), 253-274.

Slinn, S.A. and W.G.N. Slinn, 1980: Predictions for particle deposition on natural waters. Atmos. Environ., 14, 1013-1016.

Stallard, R.F. and J.M. Edmond, 1981: Geochemistry of the Amazon precipitation chemistry and marine contribution to the dissolved load at the time of peak discharge. J. Geophys. Res., 86, 9844-9858.

Turekian, K.K., 1977: The fate of metals in the oceans. Geochim. Cosmochim. Acta, 41, 1139-1144.

Turekian, K.K. and J.K. Cochran, 1981: ^{210}Pb in surface air at Enewetak and the Asian dust flux to the Pacific. Nature, 292, 522-524. Corrigenda, Nature, 294, 670.

Turekian, K.K., Y. Nozaki and L.K. Benninger, 1977: Geochemistry of atmospheric radon and radon products. Ann. Rev. Earth Planet. Sc., 5, 227-255.

Wallace, G.T.Jr. and R.A. Duce, 1978a: Transport of particulate organic matter by bubbles in marine waters. Limnol. Oceanogr., 23, 1155-1167.

Wallace, G.T. Jr. and R.A. Duce, 1978b: Open ocean transport of particulate trace metals by bubbles. Deep Sea Res., 25, 827-835.

Wallace, G.T.Jr., G.L. Hoffman and R.A. Duce, 1977: The influence of organic matter and atmospheric deposition on the particulate trace metal concentration of northwest Atlantic surface seawater. Marine Chem., 5, 143-190.

Walsh, P.R., R.A. Duce and J.L. Fasching, 1979: Considerations of the enrichment, sources and flux of arsenic in the troposphere. J. Geophys. Res., 84, 1719-1726.

Walsh, P.R., K.A. Rahn and R.A. Duce, 1978: Erroneous elemental mass-size fractions from a high volume cascade impactor. Atmosph. Environ., 12, 1493-1495.

Wedepohl, K.H., 1969: Handbook of geochemistry, vol.11, n°2.

Weisel, C.P., 1981: The atmospheric flux of elements from the ocean. Ph.D. Thesis, University of Rhode Island, 174 pp.

Weisel, C.P., R.A. Duce, J.L. Fasching and R.W. Heaton, 1982, Enrichment of potassium and magnesium on sea salt aerosols generated in situ by the BIMS, submitted to J. Geophys. Res..

Wilkniss, P.E., D.J. Bressan, R.A. Carr and R.E. Larson, 1974: Chemistry of marine aerosols and meteorological influences. J. Rech. Atmos., 8, 883-893.

Williams, P.M., 1967: Sea surface chemistry, organic carbon and organic and inorganic nitrogen and phosphorus in surface film and subsurface waters. Deep Sea Res., 14, 791–800

Windom, H.L., 1981: Comparison of atmospheric and riverine transport of trace elements to the continental shelf environment. In: River Inputs to Ocean Systems. UNEP and UNESCO, 360–369.

Winkler, P., 1975: Chemical analysis of Aitken particles (<0.2μm radius) over the Atlantic Ocean. Geophys. Res. Lett., 2, 45–48.

Wood, J.M., 1974: Biological cycles for toxic elements in the environment. Science, 183, 1049–1052.

Woodcock, A.H. 1953: Salt nuclei in marine air as a function of altitude and wind force. J. Meteor., 10, 362–371.

Zoller, W.H., E.S. Gladney and R.A. Duce, 1974: Atmospheric concentrations and sources of trace metals at the South Pole. Science, 183, 198–200.

APPENDIX A

List of Symbols

     In the first section of this list, some subscripts and other
notations are defined. Because of the great demand for symbols
from the many disciplines, we have sometimes been forced to use
multiple subscripts. In these, if no comma appears, the letters
are an abbreviation for a word or a single concept; if a comma
appears, two concepts are represented. Thus, C_{sn} is the concentration in snow; $F_{d,n}$ represents the normal component of the dry
deposition flux. At the end of this list is an indication of the
physical significance of most of the dimensionless groups used in
this book.

Subscripts

        a = air, atmospheric, aerodynamic
A,B,... = layer identification
        b = bulk
        c = critical
       cw = cloud water
        d = dry; dew point
        D = diffusion; drag
        E = evaporation (of water vapor)
       eq = equilibrium
        H = heat
        i = interface
       IM = impaction
       IN = interception
       lw = liquid water
        j = jth species; jth storm
        m = molecular; mass-mean; mixed; momentum
        M = momentum
        n = normal; neutral
        o = surface level; overall; initial (t = o)
        r = rain; reference level; ratio
        s = sea; sensible; surface
       sn = snow

    ss = steady state
    sat = saturated
    t = total
    v = vapor
    w = wet/water
    z = vertical component

Other Notations

$\vec{r}, t$ = position vector, time

$\bar{\xi}, \tilde{\xi}$ = time-average ξ

$\langle \xi \rangle$ = space-average ξ

ξ' = fluctuation of ξ about the average defined by the context

[ ] = concentration in moles per liter

Frequently Used Symbols

a = particle radius; Fourier amplitude; activity

A = area

b = Boltzmann constant, bulk

$\vec{c}$ = phase velocity of waves

$\vec{c}_g$ = group velocity

c_p, c_v = specific heat of specified gas at constant pressure, volume

C = mass concentration

C_a, C_s, C_r = mass concentrations in air, sea, and rain

C_D, C_H, C_E = bulk transfer coefficients for momentum (drag coefficient), heat, and water vapor

D = diameter, depth

$\mathcal{D}$ = molecular or Brownian diffusivity

e = water-vapor pressure

e_s = saturation pressure

E = magnitude of bulk, water-vapor flux (evaporation); collection efficiency; wave energy

$f = 2\Omega\sin\phi$ = Coriolis parameter; probability density function

$F = U-TS$ = Helmholtz free energy

$\vec{F}$ = force; flux

F_d, F_w = magnitude of (dry, wet) deposition flux

Fr = Froude number

g = gravitational acceleration

G = source (or gain) rate per unit volume; Gibbs function = U−TS+pV

h = height

H = U+pV = enthalpy; magnitude of bulk heat flux

H_m = height of mixed layer

$\mathcal{H}$ = Henry's law constant (conc. in air ÷ conc. in liquid)

$\mathcal{H}_*$ = effective Henry's law constant, accounting for reactions

$\hat{i},\hat{j},\hat{k}$ = unit vectors of cartesian coordinate system

I = interception parameter

k = wave number; thermal conductivity; reaction rate constant

$\vec{k}$ = wave number vector

k_a, k_r, k_s = transfer velocities (speeds) in air, rain, and the sea, when the flux is written in terms of the driving force (concentration difference) in a specific medium; e.g., $F = k_a(C_a - C_{a,i})$

K_a, K_r, K_s = transfer velocities when flux is written in terms of the overall driving force; e.g., $F = K_a(C_a - C_{a,eq})$, where $C_{a,eq} = \mathcal{H}C_{r,b}$

Kn = Knudsen number

K_z = zz-component of eddy diffusivity

K_C, K_H, K_M = vertical (zz-component) of eddy diffusivities for specified constituent, heat, and momentum

$\mathbb{K}$ = (second-order tensor) eddy diffusivity

ℓ = specific latent heat of vaporization; length scale; liters

$\hat{\ell} = \vec{k}/k$ = wave direction

ℓ,m,n = wave numbers

L = latent heat of vaporation; Monin−Obukov length; other length scale

$\mathcal{L}$ = sink (or loss) rate per unit volume

m = equilibrium mole-fraction distribution coefficient (e.g., $x_a = mx_r$)

m_r = mixing ratio

$\dot{m}''_v$ = local water-vapor mass flux

M = molecular weight

n = number of moles

$n(a)da$ = number of particles per unit volume and with radii a to $a + da$

$N(R)dR$ = same as preceding, but for hydrometers

N = Avogadro's number, Brunt-Väisälä frequency

p = pressure, precipitation rate

$\underline{P}$ = total precipitation during a specified time

Pe = Péclét number = Re Sc

q = specific humidity

$q*$ = saturation humidity

Q = heat

r = relative humidity; radial distance; resistance = k^{-1}

R = gas constant for a specific gas; raindrop radius; auto-correlation; resistance = K^{-1}; radiation heat flux

R = universal gas constant

Ra = Rayleigh number

Re = Reynolds number

Rf = flux Richardson number

Ri = gradient Richardson number

Ro = Rossby number

s = salinity (‰); specific entropy; Ostwald solubility coefficient = H^{-1}

s_r = scavenging ratio

S = entropy; Stokes number

S_* = critical Stokes number

Sc = Schmidt number

Sh = Sherwood number

t = time

T = temperature; total time

$T*$ = virtual temperature

T_d = dew-point temperature

T_e = equivalent temperature

T_w = wet-bulb temperature

u = specific internal energy; x component of velocity $\vec{v}$

$\bar{u}$ = x (or eastward) component of mean velocity

u_* = friction velocity

U = internal energy

$\vec{v}$ = (u,v,w) = velocity

v = y (northerly) component of $\vec{v}$; specific volume

v_d = dry deposition velocity (speed)

v_g = gravitational settling speed

$\vec{V}_g$ = geostrophic velocity

v_r = resuspension velocity (speed)

$\vec{v}_s$ = slip velocity

v_w = wet deposition velocity (speed)

v_∞ = approach or upstream velocity

V = volume

w = vertical component of velocity; specific work

W = work

$\dot{W}$ = rate of water vapor condensation

x = cartesian coordinate; mole fraction (x_a in air; x_r in rain)

y = cartesian coordinate

z = cartesian coordinate (vertical)

z_c = wave critical level

z_o = roughness height

z_+ = zu_*/ν = dimensionless height

α = enhancement factor representing chemical reactivity; coefficient of linear, thermal expansion; thermal diffusivity; angle between wind and isobars

β = coefficient of volume, thermal expansion; coefficient of absorption or wave attenuation; meridional variation of Coriolis force

γ = surface tension; ratio of heat capacities (c_p/c_v); temperature lapse rate

Γ_d, Γ_s = (dry, saturated) adiabatic (or isentropic) lapse rates

δ = thickness of viscous sublayer

ε = fraction; specific dissipation rate

θ = potential temperature; solar zenith angle

θ_e, θ_w = equivalent and wet-bulb potential temperature

$\kappa \approx 0.4$ = von Karman's constant

λ = wavelength; molecular mean-free path

μ = chemical potential; coefficient of dynamic viscosity

ν = coefficient of kinematic viscosity (= μ/ρ); stoichiometric coefficient; frequency

ρ = mass density; density of pure water

ρ_a, ρ_p = density of air, particle

$\rho_i, \rho_{sw}, \rho_{wv}$ = density of ice, sea water, and water vapor

σ = Stefan-Boltzmann constant; area; r.m.s. wave height; standard deviation

τ = time; period; shear stress on horizontal plane

$\vec{\tau}$ = stress vector

τ_d = residence time if only dry removal processes

τ_s = particle stopping or relaxation time

τ_w = residence time if only wet removal process

ϕ = latitude

Φ = gz = geopotential; wave spectrum

ψ = precipitation scavenging rate coefficient

Ψ = stability function

ω = angular frequency

Ω = angular velocity of the earth

Dimensionless Groups

Froude number = Fr = $v^2/(gL)$ or sometimes $v/\sqrt{gL}$. A measure of the ratio of the inertial acceleration of a fluid ($\partial\vec{v}/\partial t$ and/or $\vec{v}\cdot\nabla v$), as given by v^2/L, to the acceleration of gravity. For Fr>>1, gravitational effects are negligible (e.g., mechanical turbulence in the atmosphere); for Fr$\approx$1, gravity is important (e.g., gravity waves in water and air); for Fr<<1, gravity dominates. For a wave at the air-sea interface (of wave number k and amplitude A) Fr = A^2k^2, and then for Fr>1, the disruptive accelerations are larger than the restoring force; therefore, the wave will break.

Knudsen number = Kn = λ/L. The ratio of the mean-free path for molecules in a gas to a characteristic length-scale of the flow field (e.g., particle radius). For Kn<<1, the "object" is so large that the fluid can be treated as a continuum (the usual Navier-Stokes equations for fluids can then be used); for Kn>>1, then λ>>L, and in this "free-molecule flow regime", the continuum equations and the no-slip boundary condition fail (then requiring descriptions using some form of kinetic theory; e.g., with the Boltzmann equation). The regime with $0.1 \lesssim$ Kn $\lesssim 10$ is called the transition regime.

Péclét number = Pe = $vL/\mathcal{D}$. As sketched in Figure 8d in the chapter by Slinn (this volume), the Péclét number is the ratio of the speed of convection (v) to the "speed of diffusion" $(\mathcal{D}/L)$. Thus, for Pe>>1, convection dominates mass transfer; for Pe<<1 (e.g., no flow), diffusion dominates. It is noted that Pe = $vL/\mathcal{D}$ = $(vL/\nu)(\nu/\mathcal{D})$ = Re Sc.

Rayleigh number = Ra = $[\beta\Delta Tg\delta^2/\nu]/(\alpha/\delta)$. The ratio of the speed of heat transport by convection, w = $\beta\Delta Tg\delta^2/\nu$, to the speed of heat transport by conduction, α/δ, where α is the thermal diffusivity (e.g., in cm^2s^{-1}) and δ is a characteristic length scale of the heated layer. The estimate given for w is found from estimating a balance between buoyancy and viscous drag: $\nu w/\delta^2 \simeq \beta\Delta Tg$, where ν is the coefficient of viscosity and β is the coefficient of volume, thermal expansion. For Ra>>1, any convection (e.g., generated by random fluctuations) will dominate, and the fluid will spontaneously engage in "free convection". This is known as the Rayleigh instability, which is initiated at Ra $\simeq 10^3$ and maintained if Ra $\simeq 10^2$. For smaller Rayleigh numbers, conduction dominates.

Reynolds number = Re = vL/ν. As sketched in Figures 8a, b, and c in Slinn (this volume), the Reynolds number is the ratio of a fluid's characteristic transport speed, v, to the speed with which the "signal" (vorticity) diffuses into the fluid, "telling it there's a body present," ν/L, where ν is the kinematic coefficient of viscosity (e.g., in cm^2s^{-1}) and L is a characteristic size of the body. Alternatively, the Reynolds number can be viewed as the ratio of the inertial acceleration of the fluid ($\partial\vec{v}/\partial t + \vec{v}\cdot\nabla\vec{v} \sim v^2/L$) to the viscous force per unit mass ($\nu\nabla^2\vec{v} \sim v/L^2$). From this second view, it is easier to see that if the inertial acceleration is large compared to the viscous deceleration (Re>>1), then very large accelerations can occur; for a local Reynolds number (based on a local length scale) $\gtrsim 10^3$, turbulence results and is maintained if Re $\gtrsim 10^2$. For Re<<1 (e.g., Stokes' creeping flow about a sphere), viscous forces dominate the flow.

Richardson numbers, Rf and Ri. The flux Richardson number, Rf, is the ratio of the rate at which buoyancy forces ($\sim \rho'g$) do work on a fluid ($\sim g\overline{\rho'w'}$) to the rate of work by the shear stress $\sim\tau\partial\bar{u}/\partial z$. The gradient (or "original") Richardson number uses

K-theory to specify the buoyancy and momentum fluxes, $Ri = (gK_\rho \partial\tilde{\rho}/\partial z)/[\rho K_m (\partial\tilde{u}/\partial z)^2]$, and takes $K_\rho = K_m$. Since a thermally stable fluid, if displaced, will oscillate at the Brunt-Väisälä frequency $N = [(g/\rho)(\partial\tilde{\rho}/\partial z]^{1/2}$, the gradient Richardson number can also be written as $Ri = N^2/(\partial\tilde{u}/\partial z)^2$. In this form, the significance of the Richardson number is easiest to see: (i) for Ri large and negative, $N^2 < 0$, and the fluid is (convectively) unstable; (ii) for Ri large and positive, $N^2 > 0$, and an unsheared flow would tend to be stable (and to oscillate at frequency N if displaced). It might be expected that the Richardson number at which the switch-over would occur (from unstable to stable) would be $Ri = 0$; but in a turbulent shear flow, energy to drive the turbulence is obtained from the shear, and turbulence can be maintained (even against a stabilizing density gradient) up to $Ri \simeq 1/4$. Then, for $Ri \gtrsim 1/4$, the flow will be stable.

Rossby number $= Ro = v/(fL)$ is the ratio of the inertial (or advective) acceleration of a fluid ($\vec{v}\cdot\nabla\vec{v} \sim v^2/L$) to the Coriolis acceleration, vf, where $f = 2\Omega\sin\phi$, in which $\Omega = 2\pi/(24\ hr) = 0.76 \times 10^{-4}\ s^{-1}$ is the earth's angular velocity, and ϕ its latitude. When $Ro \gg 1$ (e.g., for a tornado, $v/(fL) \simeq 30\ m\ s^{-1}/(10^{-4}\ s^{-1})(300m) \simeq 10^3$) then the Coriolis effect can be neglected; when $Ro \ll 1$ (e.g., for a cyclonic storm, with $L \sim 10^3$ km) then the fluid's acceleration is negligible compared with the Coriolis "force" (which, when balanced by the pressure-gradient force, is called the geostrophic approximation).

Stokes number $= S = \tau_s/\tau_f$ is the ratio of the stopping (or velocity-relaxation) time of a particle in a fluid to the characteristic time scale of the flow (e.g., L/v). If $S \ll 1$ (e.g., a small particle), then the particle's relaxation time is so small compared to τ_f that the particle follows the fluid's motion (e.g., streamlines) essentially exactly. If $S \gg 1$ (e.g., a large particle), then the particle's motion is almost uninfluenced by variations in the fluid's motion.

Schmidt number $= Sc = \nu/\mathcal{D}$ is the ratio of diffusivities: for momentum (or better, for vorticity) to that for the particles (or trace-gas molecules). As sketched in Slinn's Figure 8d (this volume), near a body, $Sc = (\nu/L)/(\mathcal{D}/L)$ gives a measure of the relative thickness of the viscous and diffusion "boundary layers." For $Sc \simeq 1$ (as for low-molecular-weight gases), these thicknesses are almost the same; for $Sc \gg 1$ (e.g., for particles with radii $\sim 1\mu m$, $Sc \sim 10^6$), then the diffusion layer is very thin, density gradients are large, and the diffusive flux does not decrease so rapidly (with increasing particle size) as $\mathcal{D}^1$ (or Sc^{-1}) but only as $\mathcal{D}^{-2/3}$ (or $Sc^{-2/3}$), or even as $\mathcal{D}^{1/2}$ (or $Sc^{-1/2}$).

Sherwood number = Sh is the nondimensional mass flux. For example,
    for mass flux via molecular or Brownian diffusion with $\vec{F}$ =
    $-\mathcal{D}\nabla C$), then to obtain the Sherwood number, the flux is nondi-
    mensionalized with $\mathcal{D}C_0/L$, where C_0 is a reference concentra-
    tion (e.g., far from the collector), and L is a characteristic
    dimension of the collector (e.g., Peters, this volume, uses
    sphere diameter; Slinn, this volume, uses sphere radius--one
    of the least troublesome communication problems between chemists
    and physicists!).

APPENDIX B

Annotated Bibliography of Sampling

Meteorological and Oceanographic Measurements, and Sampling
Strategy

Given here are references useful to those interested in experiment design, including available instruments, techniques of
measurement, and the statistics of sampling, error analysis, and
spectral analysis. If it seems weighted towards the oceanic side,
this is because meteorological instruments and techniques are more
standardized. Meteorological measurements are organized into a
global weather analysis network by the World Meteorological
Organization, Geneva. The WMO publishes many handbooks on metero-
logical measurements, which are available from the national weather
bureaus.

Baker, D.J., 1981: Ocean instruments and experiment design. In
    Evolution of Physical Oceanography, B.A. Warren and C.Wunsch
    (eds.), MIT Press, 396-433.

This article gives a complete treatment of deep-sea physical
oceanographic instruments developed since publication of The Oceans
(see below), and goes into many of the advantages and pitfalls
associated with their use. It does not discuss remote sensing,
and leaves surface measurements to Dobson, Hasse and Davis (see
below).

Bernstein, R.L. (ed.), 1982: Seasat Special Issue I: Geophysical
    Evaluation. J. Geophys. Res. 87 (C5), 3175-3438.

This series of papers summarizes the state of the art in microwave
remote sensing of the sea surface attained by SEASAT I. It contains review articles on the radar altimeter, passive microwave
radiometer, radar scatterometer, and synthetic aperture radar.
Neither infrared sensing of sea surface temperature nor color
scanners are mentioned (they were not part of SEASAT).

Blackman, R.B. and J.W. Tukey, 1959: The Measurement of Power
      Spectra. Dover, Inc., New York, 190 pp.

This slim volume has long been the bible of those interested in
designing optimal sampling strategies and collecting data for
spectral analysis. The style is terse, the notation is curious
and at times hard to follow, but it is all there for the determined reader.

Dobson, F.W., L. Hasse and R. Davis (eds.), 1980: Air-Sea Interaction: Instruments and Methods. Plenum Press, 801 pp.

This book deals in considerable detail with both the instruments
and the measurement techniques in use today in the field of air-
sea interaction. Measurements in both air and sea are treated.
There is a chapter on sampling the surface microlayer, as well as
sections on velocity, temperature, salinity, pressure, and instrument platforms. Each section is written by an acknowledged expert.
Remote sensing is not included.

Gower, J.F.R. (ed.), 1981: Oceanography from Space. Plenum Press,
      Marine Science Series 13, 978 pp.

This book is the proceedings of an international symposium on the
subject, held in Venice in 1980. It is not truly a reference work,
but rather a collection of papers, as is the SEASAT issue edited
by Bernstein. The reader must therefore dig for information on
how to use remote sensing as a tool for investigating the ocean.
Sections are included on infrared measurements of sea surface temperature, ocean color scanners, and sea ice sensing, as well as on
the subjects covered by the Bernstein reference.

Jenkins, G.M. and D.B. Watts, 1969: Spectral Analysis and its
      Applications. Holden-Day, Inc., San Francisco, 525 pp.

This book includes and builds on the information contained in
Blackman and Tukey. It is less of a "handbook" and more of a
text, and delves into some of the underlying theory of sampling
and spectral analysis, as well as probability theory and statistics.
The only item important to this audience which is not covered is
the Cooley-Tukey Fast Fourier Transform.

Sverdrup, H.U., M.W. Johnson and R.H. Fleming, 1942: The Oceans:
      Their Physics, Chemistry and General Biology. Prentice-Hall,
      Englewood Cliffs, NJ, 1087 pp.

Long the only general source of information on the oceans, this
book remains the oceanographer's bible. Chapter 10: Observations
and Collections at Sea, is an excellent source of information on
the "classical" instruments and techniques.

Collection of Atmospheric Aerosols and Precipitation

The following give a comprehensive review of collection techniques as well as of possible sampling strategies in the marine atmosphere.

Air-Pollution, 1968: Vol. 2, Analysis, Monitoring and Surveying (A.C. Stern,ed.), Academic Press, 637 pp.

WMO, 1977: Proceedings of the World Meteorological Organization Technical Conference on Atmospheric Pollution Measurement Techniques. WMO Publication n° 460. Numerous papers on sampling and analytical techniques for a wide variety of trace substances.

N.A.S., 1978: The Tropospheric Transport of Pollutant and other Substances to the Ocean (J.M. Prospero, ed.), National Academy of Sciences, Washington D.C., Chapter 10: Techniques, 222-240.

More specific papers include:

Hoffman, E.J., Hoffman, G.L. and Duce, R.A., 1976: Contamination of Atmospheric Particulate Matter Collected at Remote Shipboard and Island Locations. National Bureau of Standards Special Publication 422: Accuracy in trace analysis: Sampling, Sample Handling, and Analysis, 377-388.

Galloway, J.N. and G.E. Likens, 1976: Calibration of Collection Procedures for Determination of Precipitation Chemistry. Water, Air and Soil Pollution, 6, 241-258.

Sampling Atmospheric Gases

Butcher, S.B. and R.J. Charlson, 1972: An Introduction to Air Chemistry, Academic Press, New York, 241 pp.

Chapters 3 and 4 provide an introduction to sampling and collection, treatment of data, and analytical methods useful for measurement of trace gases. The presentation is at a level that requires little prior knowledge.

Stern, A.C., editor, 1976: Air Pollution, Volume III: Measuring, Monitoring and Surveillance of Air Pollution, Third Edition, Academic Press, New York, 799 pp.

Part A of this volume is a compendium of sampling and analysis procedures written by experts in the field of air pollution analysis. The treatment includes chemical analysis of particulates as well as gaseous pollutants.

Warner, P.O., 1976: Analysis of Air Pollutants, John Wiley and
    Sons, New York, 329 pp.

This text attempts to present air pollutant chemical analysis in
an organized form for undergraduate science majors and beginning
graduate students. There is some detail on the identification of
particulate air pollutants, continuous automated methods, and
even step-by-step directions for the novice.

Sea-Surface Microlayer Sampling

    Different apparatus vary in their suitability for sampling
trace metals, organic substances, and microorganisms. A major
problem is that large volume (liter) samplers remove too much
bulkwater, while those that minimize this contamination yield
very small volumes (μls to mls). A sampler of variable capacity
that will lift off just the organic film without the underlying
water is needed. The following references and their cited literature supplement the references to sampling in this volume.

Hatcher, R.F. and B.C. Parker, 1974: Microbiological and chemical
    enrichment of freshwater-surface microlayers relative to the
    bulk-subsurface water. Can. J. Microbiol., 20, 1051-1057.

Screen, drum and tray samples all showed an enrichment over subsurface waters, but the results varied markedly between sampling
methods.

Wangersky, P.J., 1976: The surface film as a physical environment,
    Ann. Rev. Ecol. Syst., 7, 161-176.

A broad look at surface sampling and the inherent problems for
the different disciplines studying the air-sea exchange.

Sieburth, J. McN., 1979: Sea Microbes, Oxford University Press,
    New York, 78-80.

The major methods for microbiological sampling of the surface mi-
microlayer are reviewed and illustrated in the sampling section of
this introductory text on marine microorganisms.

Garrett, W.D. and R.A. Duce, 1980: Surface microlayer samplers.
    In: Air-Sea Interaction (F. Dobson, L. Hasse and R. Davis,
    eds.), Plenum Publishing Corp., New York, 471-490.

The problems in sampling and the in-situ characterization of sea
slicks and surface microlayers are discussed, with guidelines for

the selection of techniques.

Lion, L.W., and J.D. Leckie, 1981: The biogeochemistry of the air-
    sea interface, Ann. Rev. Ecol. Earth Planet. Sci., 9, 449-
    486.

An up-to-date review of data on metals and organic compounds
obtained by different sampling methods as well as an excellent
discussion on microbial ecology and transport at the air-sea
interface.

Sampling Dissolved and Particulate Matter in the Ocean

    The following provide basic information with respect to
"conventional" sampling techniques. Both are now somewhat out of
date, but still contain much useful information.

Riley, J.P., 1975: Analytical Chemistry of Sea Water. In: Chemical
    Oceanography, Vol. 3 (J.P. Riley and G. Skirrow, eds.),
    Academic Press, 193-514.

Grasshoff, K., 1976: Methods of Seawater Analysis, Verlag Chemie,
    Weinheim, 317 pp.

Papers dealing with specific aspects are listed below.

A review of the various types of sediment traps which have been
used in aquatic environments is given in:

Reynolds, C.S., S.W. Wiseman and W.D. Gartner, 1980: An annotated
    bibliography of aquatic sediment traps and trapping methods.
    Occasional publication No. 11, Freshwater Biological Asso-
    ciation, 54 pp.

The following three papers describe various approaches to the in-
situ filtration of large (> 10^3 l) samples in surface and deep
waters:

Krishnaswami, S., D. Lal and B.L.K. Somayajulu, 1976: Investigations
    of gram quantities of Atlantic and Pacific surface particu-
    lates, Earth Planet. Sci. Lett. 32, 403-419.

Krishnaswami, S., D. Lal, B.L.K. Somayajulu, R.F. Weiss and
    H. Craig, 1976: Large volume in situ filtration of deep
    Pacific waters: Mineralogical and Radioisotopes Studies.

Bishop, J.K.B. and J.M. Edmond, 1976: A new large volume filtra-
    tion system for the sampling of oceanic particulate matter,
    J. Mar. Res., 34, 181-198.

A novel technique for sampling waters for dissolved gas analysis and which appears to be successful (i.e. contaminant-free) for trace gases, such as the Freons, is described in:

Cline, J.D., H.B. Milburn and D.P. Wisegarver, 1982: A simple rosette-mounted syringe sampler for the collection of dissolved gases, Deep-Sea Res., 29, 1245-1250.

The ultra-clean techniques required for the sampling and analysis of trace metals and organics are detailed in the following two papers:

Bruland, K.W., R.P. Franks, G.A. Knauer and J.H. Martin, 1979: Sampling and analytical methods for the determination of copper, cadmium, zinc and nickel at the nanogram per liter level in seawater, Anal. Chim. Acta 105, 233-245.

Wong, C.S., W.J. Cretney, J. Piuze, P. Christensen and P.G. Berrang, 1977: Clean laboratory methods to achieve contaminat-free processing and determination of ultra-trace samples in marine environmental studies, Special publication No. 464, National Bureau of Standards, Washington, D.C., 249-258.

Sampling Methods for Bubbles and the Bubble-Produced Aerosol

Laboratory work with bubble-related phenomena requires techniques to produce bubbles of any given size one at a time. This can be done by using glass capillary tips of known size (Blanchard and Syzdek, 1977). For capillary tips stationary in the water, it is very difficult to produce a bubble less than about 300 μm diameter. But this can be done with a rotating tank (Blanchard and Syzdek, 1972), where the water in solid rotation shears the air as it emerges from a fixed capillary tip. There are times when it is necessary to age a bubble before it reaches the surface, i.e., provide a long travel distance through the water. Since it is often difficult to submerge a capillary tip to depths beyond a meter or so, a bubble aging tube (Blanchard, 1963; Blanchard and Syzdek, 1972) can effectively be used.

The production of a spectrum of bubbles to simulate the bubble spectrum in the sea has been done by allowing water to splash into water (Cipriano and Blanchard, 1981) and by letting oppositely-moving wave crests collide (Monahan and Zietlow, 1969).

When bubbles burst they produce both jet and film drops. Methods to detect these drops have been discussed by Blanchard (1963). The use of electrostatics, and gelatin and magnesium oxide-coated slides for drop sizing has been discussed in detail by Blanchard and Syzdek (1975). In the marine atmosphere, the jet and film drops comprise the sea-salt aerosol. The size distribution of the aerosol can be determined quite accurately by collection on small glass slides (Woodcock, 1972) or in a cascade impactor (Duce and Woodcock, 1971).

Methods to determine the concentrations of bacteria in jet and film drops have been developed by Blanchard and Syzdek (1970, 1982).

References are as follows:

Blanchard, D.C., 1963: The electrification of the atmosphere by particles from bubbles in the sea. Prog. Oceanog., 1, 71-202.

Blanchard, D.C.,and L.D. Syzdek, 1970: Mechanism for the water-to-air transfer and concentration of bacteria. Science , 170, 626-628.

Blanchard, D.C., and L.D. Syzdek, 1975: Concentration of bacteria in jet drops from bursting bubbles. J. Geophys. Res., 77, 5087-5099.

Blanchard, D.C., and L.D. Syzdek, 1975: Electrostatic collection of jet and film drops. Limnol. Oceanogr., 20, 762-774.

Blanchard, D.C., and L.D. Syzdek, 1977: Production of air bubbles of a specified size. Chem. Engr. Sci., 32, 1109-1112.

Blanchard, D.C., and L.D. Syzdek, 1982: Water-to-air transfer and enrichment of bacteria in drops from bursting bubbles. Appl. Environ. Microbiol., 43, 1001-1005.

Cipriano, R.J., and D.C. Blanchard, 1981: Bubble and aerosol spectra produced by a laboratory 'breaking wave'. J. Geophys. Res., 86, 8085-8092.

Duce, R.A., and A.H. Woodcock, 1971: Difference in chemical composition of atmospheric sea salt particles produced in the surf zone and on the open sea in Hawaii. Tellus, 23, 427-435.

Monahan, E.C., and C.R. Zietlow, 1969: Laboratory comparisons of freshwater and salt-water whitecaps. J. Geophys. Res., 74, 6961-6966.

Woodcock, A.H., 1972: Smaller salt particles in oceanic air and bubble behavior in the sea. J. Geophys. Res., 77, 5316-5321.

INDEX

Accomodation coefficient 223
Acidity 285
Acid rain 205
Adiabatic 30, 62
    lapse rate 8
        wet 32
        dry 8
    process 8
Advection 93
Aerobic processes 135
Aerosol particles
    interaction with trace
    gases 217
    log-area distributions 302
    number distributions 301
    size distributions 223, 301
    surface-area distributions 302
    volume distributions 301
Aerosols
    monodisperse 307, 336
    on climate, impact of 457
    polydisperse 307
    residence time of 507
    seasalt 498
    Se enrichment in 517
    total deposition velocity of
        509
Air
    friction velocity 106
    mass trajectories 465
Air-phase 189
    controlled process 198
    mass transfer coefficient 193
    resistance 190
Air-sea fluxes of water vapour
    269

Air-sea gas exchange 276
Air-sea gas transfer in the
        field 260
Air-sea interface, organic-
        material accumulation
        at the 486
Albedo 11, 14, 15
Algae 287
Aluminosilicate particles,
        mineralogical composi-
        tion of 500
Ammonia (NH$_3$) 177, 284
Anaerobic processes 138
Angular deviation (cross-isobar
        angle) 35
Anthropogenic sources, indicators
        of 470
    bomb-produced atmospheric
            radionuclides 470
    lead 470
        isotopes 470
    synthetic organics 470
Aqueous phase 189
Arsenic
    global atmospheric cycle 507
Atmospheric chemistry 181
Atmospheric flux 511
Atmospheric particulate organic
        carbon 474
Atmospheric stability 270

Bacteria 157
    as ice nuclei 443
Baroclinic instability 11, 18
Below-cloud scavenging, see
        Scavenging

Biogenic (sources) 503
Biological production 289
Biomass burning 287
Bioturbation 514
Bisulphite (HSO_3^-) 206
Boundary layer 322
    height of 39
    theory 251, 266, 277
Breeze
    land-sea 27
    sea 27
Brownian diffusion, see
            Diffusion
Brunt-Väisälä frequency 71
Bubble
Bubble
    background distribution 414
    clusters 417
    collection efficiency for
            bacteria 436
    flux distribution 420
    scavenging 435
    surface life 416
Bubbles 268, 479
    as gas exchangers 268
    density 269
    number distribution of 268
    size distribution of 268,
        408
    solution 269
    spectra 269
        laboratory simulations
            of 410

Capillary tips 416
Capillary wave suppression 272
Capillary waves 275
Carbon 280
Carbon-14
    bomb-produced 264
    natural 264
Carbon dioxide (CO_2) 175, 281
Carbon disulphide (CS_2) 286
Carbon monoxide (CO) 177
Carbon, stable isotopes 474
Carbon tetrachloride (CCL_4) 288
Carbonic anhydrase 258
Carbonyl sulphide (COS) 177, 286
Cascade impactors 498
Catalysis 258

Catspaws 98, 105
Chain propagation 185
Chain terminating 187
Chemical enhancement factor 255
Chemical reaction
    dissociation 206
    first order 202
    general order 203, 204
    heterogeneous 188
    instantaneous reversible 206
    reversible first order 204
    reversible general order 205
Chemical reactivity 242
Chemical transfer processes 254
Chlorine 287
Chloroform ($CHCL_3$) 288
Circulation 190
    time 212
    with chemical reaction 214
    with drops 211
Circulations
    secondary 27
    thermal 27
Climate 285
Clouds 173
    condensation nuclei 342
    condensed water content 340
    convective 341
    droplets 173
    drops 189
    street 32, 34
Coastal convergence 29
Collection
    by impaction 325
    by interception 325, 436
    diffusion 321
    efficiency 319, 324
    electrical influences 336
    from Brownian diffusion 324
    substrates 458
Collision efficiencies
    particle/raindrop 330
    particle/ice-crystal 330
Collision-reaction probability
        219
Combustion sources 287
Concentration
    bulk average 193
    gradients 191
Condensation 10, 249, 273

nuclei 173
Constant-flux
    conditions 264, 388
    layer 389
Contamination 458
Continuity equation 7, 382, 388
Continuity of stress 251
Convection 32
Coriolis acceleration 2
Coriolis parameter 58
Crapper ripples 247
Critical level 108
Crystal weathering 460
Crystal types
    ice needles 336
    powdered snow 336
    spatial dendrites 336
Cyclic salts 504

Deep-sea sediments 514
Deep waters 491
Density, potential 62
Deposition
    dry 428, 498
    to inland waters 307
    wet 280, 500
Deposition flux
    dry 303, 361
    wet 304
Deposition layer 308, 368
Deposition velocity 269, 387
    dry 361, 384
    for momentum 369
    particles mass
        -mean 395, 397
    wet 305
Depressions 24, 25, 26
Depth of no motion 56
Derivatives
    material 3
    partial 3
    substantive 3
Desert soils, composition of
        460
Dew point 249
Diel cycle 140, 147
Diffusion 68, 191
    Brownian 369
    cabbeling 69
    categories 48

Fickian 68
    models 48
    molecular 69, 193
    of heat 65
    salt fingering 69
    time 212
    turbulent 364
Diffusivities 194, 266, 277
    eddy 36, 37, 38
    molecular 37
Direct flux method 260
Dissociation 180
Dimethyl sulphide
        $[(CH_3)_2S]$ 177, 286
Disturbance 25
Double diffusive 69
Drag coefficient 43, 44, 80,
        270, 369
Drop distribution
    lognormal 337
Drops, terminal velocity 328
Drop size, mass mean 319
Dry deposition, see Deposition

Easterly waves 23
Eddy correlation 269
    techniques 263
Eddy diffusion 218
Eddy scale-length 39
Efficiency
    -collection 440
    -collision 440
    sticking or attachment 440
Effective film thickness 196
Electroneutrality 201
Electrostatic induction 441
Energy 14
    balance of the atmosphere 12
Enhancement 201
    coefficient 204
    factor 203
    of the transfer rate 201
Enrichment factor 460
    bacterial 434
Equation of state 4
Equations of motion 1
Equilibrium 190
    coefficients 206
    scavenging 230

Ethane (C_2H_6) 281
Ethene (C_2H_4) 284
Eucaryotes 125
Eulerian time 359
Evaporation 11, 22, 45, 249, 272

Fallout 339
Fatty acids 271
Fatty alcohols 271
Fetch 46
Fick's law 191
Film area 275
Film drop enrichment 441
Film-drop flux 44
Film droplets 479
Film-drop production
   lower-bound 420
   upper-bound 420
Film drops 417
Film-forming substances 271
Film model (see also Models)
   266, 277
   3-film 271
Film pressure 272, 274, 275
Films 270, 276
   at the sea surface 274
   interfacial 271
   laboratory studies of 270
   natural 270
   organic 275
   pollutant 270
   surface 190
Film theory 196
Flows
   baroclinic 54, 56
   barotrophic 54, 67
   Ekman 56
   geostrophic 54
   gyral 54
   Langmuir 94
   low-speed streaks 380
   mean wave-induced 109
   ocean wave-induced 109
   orographic 53
   Sverdrup 58
   thermohaline 65
   zonal 53
Flux 191
   chamber 260

Fluxes 263, 276
   diffusional 362
Formaldehyde (CH_2O) 178, 280, 281
Fossil fuels 281
Free radical chemistry in cloud
   droplets 234
Free radicals 175
Freon-11 (CCL_3F) 288
Freons 288
Freshwater ponds 160
Friction 2, 6, 7, 14, 35, 37
   velocity in the sea 44
Fronts 17, 25
   oceanic 76
   planetary 76
   polar 76
   plume 79
   shallow-sea 79
   shelf break 79
   upwelling 79
   western boundary current 79
Frössling correlation 324
Frössling equation 198
Froude number 71

Gas absorption 175, 202
Gas bubbles
   dissolution of 249
Gases
   cycling of 135, 138
   soluble 189
Gas-phase 179
Gas-to-particle conversion 476
Gas transfer 173, 241
   and bubbles 248
   resistance to 241
General order reactions 203, 204
Generation and loss 219
Geochemical cycles 283
Geochemical cycling 276
Geochemistry 276
GEOSECS 63
Geostrophic equilibrium 5, 6
Geostrophic shear 7
Geostrophic wind (current) 5, 44
Gradient techniques 269
Gravitational settling speed 304,
   319, 326
Gulf stream rings 72

Hadamard-Rybczynski circulation 211

Hadley Cell 17

Haloclines 84

Halogen cycles 478

Halostad 63

Health hazards 442

Heat transfer and gas exchange 249

Heavy metals 458

Henry's law 191
  constants 199, 242, 277, 284, 288

Heterogeneous loss constant 222

Heterogeneous processes 189, 217

Heterogeneous removal rate 220

Heterotrophic bacteria 130

Higbie equations for simple physical absorption 209

High pressure cells 25

Highs 6
  dynamic 26
  thermal 26

Homogeneous chemistry 174

Hydration rate 258
  constant 257

Hydrocarbons 474
  low molecular weight 280
  non methane 177

Hydrogen (H_2) 277

Hydrogen chloride (HCL) 199

Hydrogen peroxide (H_2O_2) 178

Hydrogen suphide (H_2S) 177, 285

Hydrolysis in the water 205

Hydrometeor 189, 320
  gas interaction 225
  life of 203
  size distribution 320

Hydroperoxide, free radical (HO_2) 186

Hydrophilic 271

Hydrophobic 271

Hydrostatic equation 4

Hydroxyl radicals (OH) 178, 283

Ice crystal types
  ice needles 335
  graupel particles 335
  plane dendrites 335

powder snow 335
rimed plates 335
spatial dendrites 335
stellar dendrites 335

Ice-freezing nuclei 342

In-cloud scavenging, see Scavenging

Inertial terms 3

Interface
  atmosphere-ocean 42
  liquid/gas 42

Interhemispheric transports 19, 22

Intertropical Convergence Zone (ITCZ) 17, 19, 22, 23

Inversions 26, 31, 38, 39

Iodine 289
  cycle 289

Ionization 200

Isentrophic surfaces 47

Island effect 30

Isobars 56

Isopiestic method 422

Isopycnals 56

Jet-drop flux 419

Jet droplets 479

Jet drops 415

Jet set 440

Jet streams 19
  subtropical 19

Katabatic winds 28

Kelvin-Helmholtz instability 246

Kinematic viscosity 41, 102

Kinetic energy 10, 25

Knudsen number 223

Kronig-Brink solution 213

K-theory 366

Lagrangian 312
  time 359

Laminar 40
  flow 36

Langmuir circulations 94, 150, 409

Latent heat 12

Layer
  Prandtl 39

surface or constant flux 39
Lead (Pb), ocean cycle 512
Legionnaires disease 442
Liquid phase 185
    chemical reactions 199
    controlled process 198
    resistance 200, 270
Local wind systems 27
Lognormal distribution 317
Lows 6

Marine aerosols 456
    enrichments in 459
    phosphorus in 464
Marine air 286
Marine atmosphere 173, 478
    acidity of rainwater in the
        504
    potassium in 461
    ^{210}Pb in 466
    ^{222}Rn in 466
Marine precipitation 285
Marine rain, Hg content of 503
Mass transfer 251
    coefficient 191
    penetration theories 378
    surface-renewal 378
Material derivative 3
Mechanical energy 18
Meridional cross section of the
        atmosphere 19
Meridional transport 22
Mesoscale oceanography 71
Mesoscale transport 48
Metals
    global atmospheric fluxes 507
Methane (CH_4) 175, 280
    -CO oxidation 184
Methyl chloride (CH_3CL) 287
Methyl iodide (CH_3I) 287, 289
Microbial films 154
Microbiological processes 281
Microlayer skimmed from the
        bubble 439
Microorganisms, planktonic 123
Mineral dust, eolian transport
        of 468
Mixed layer 393
    buoyancy 91
    conservation relations 92

depth 268
    efficiency 97
    entrainment 93
    forcing variables 87
    modelling the 85
    role in ocean dynamics and
        climate 97
    shear instabilities 92
    solar radiation 87
    time and space scales 85
    wind forcing 90
Mixing 68
    diabatic 85
    lateral 65
    length 39
    tidally-induced 82
    turbulent 69
Models
    film 251, 266, 277
    surface renewal 251
Molecular diffusivity 251
Monin-Obukhov length 43
Monin-Obukhov similarity 42
Monolayers 271, 272, 274
Monsoon 22

Nanoplankton 144
Neuston 148
Nitric acid (HNO_3) 178
Nitric oxide (NO) 175, 177, 283
Nitrite (NO_2) 283
Nitrogen 283, 476
Nitrogen dioxide (NO_2) 175,
        177, 283
    photolysis of 179
Nitrous acid (HNO_2) 187
Nitrous oxide (N_2O) 283

Obukhov length 270
Ocean currents 54
Ocean eddies 71
Oceanic mixed layer 84
Oceanography, large scale 53
Oil 276
    films 270, 271, 274
Oleyl alcohol ($C_{17}H_{33}CH_2OH$)
        272, 273
Organic aerosols 473
Organic compounds 458
Organic films 149, 268, 417

Organic matter 276
    dissolved 417
Organic size-fractions 143
Overall mass transfer coefficient
        196
Oxygen 285
    balance approach 260
    dissolved 261
Ozone (O_3) 177, 285
    formation 182
    steady stade for 184
    uptake 264

Parameterization 43, 48, 92
Partial derivative 3
Particle
    -gas interaction 219
    geometric mean of 337
    growth by water-vapour
        condensation 332, 338,
        346
    hygroscopic 342
    insoluble 347
    radius, geometric mean 337
    slip velocity 362
    soluble 347
    stopping time 326
    tropospheric residence time
        357, 399
Particulate matter
    biogenic cycle in the
        ocean 491
Pasquill-classes 48
Pathogen 442
Peclet number 212, 324
Perfect absorbers 205
Photochemical oxidation 285
Photochemical reactions 179
Photochemistry 186, 286
Photoinhibition 141, 162
Photolysis 283
Photooxidation 281
Photosynthesis 261
Physical absorption 208
Picoplankton 144
Planetary Boundary Layer (PBL)
        35
    height of 35
Plankton size-fractions 143

Pleuston 148
Polar front 17, 24, 25
    jet streams 17
Pollution
    vanadium (V) 470
    manganese (Mn) in
        aerosols 470
Potential energy 10, 18, 25
Potential flow 322, 325
    circular cylinder 329
    disk 329
    sphere 329
Potential temperature 30, 47
Prandtl layer, see Layer
Precipitation 22, 280
    cold-front 341
    efficiencies 349, 354
    orographic 341
    rate 334
    scavenging 173
        stochastic process 316
    warm-front 341
Pressure, warm-coupled 111
Primary producers 128, 158
Procaryote 123
Prodigiosin 440
Profile correlation techniques
        263
Propane (C_3H_8) 281
Propene (C_3H_6) 281
Protozoa 131, 133, 157
Pseudo-steady state 181
    approximation 181
Pycnocline 84
Pycnostad 63

Quantum yield 180

Radiation 11
    balance 285
    solar 11, 12, 13
    terrestrial 13
Radiocarbon 264
Radon deficiency method 264,
        267
Rain 285
Raindrops 189, 329
    internal circulation 332
    number distributions 334

Rainfall
    rate 318
    total 318
Random walk 325
Reaction chain 185
Redfield ratios 260
Reference elements 459
Relative velocity 190
Relaxation time 326
Removal, wet and dry, relative
        importance 396
Residence times 47, 175
    for water vapour 310
    in the troposphere 308
    of sulphur 310
    of the particle 221, 309
    stratospheric 308
    wet removal's contribution 312
Resistance 270
    interfacial 190
    law 44
    models 391
Respiration/decomposition 261
Resuspension 385
Retention-efficiency 330
Reynolds number 192, 322
Reynolds stress 370
Richardson number 36, 37, 43,
        70, 80, 83, 92, 93, 270
River chemistry 504
Riverine flux 511
Rossby deformation radius 71
Roughness 29
    hydrodynamic 42
    length 40

Salinity 271
Salt concentration
    variations in 423
    with altitude, decrease in 424
Salt-cycled through the
        atmosphere 421
Salt samples from island stations
        425
Salt inversion
    attempt to find 432
    explanation of 431
Saturation 10
    water vapour curve 9
Scavenging

below-cloud 227, 320, 338
by storms 338
    frontal 335
coefficient 228
convective storms 336
efficiencies 312, 350, 354, 358
    of aerosol particles by
            cloud and rain
            droplets 500
first-order 313
in-cloud 227, 320, 338
of metals in sea-water 487
nucleation 338, 347
process 313
rain 318, 319
rates 310, 319, 320, 358
    approximate 333
    first-order process 311
    n-averaged 337
    rain 320
        approximations 333
    snow 320
        approximations 333, 335
ratios 305, 309, 339,
        358, 500
    convective storms 344
    particle-size dependence
            344
    water vapour 352
reversible 230
snow 336
spray 387
stochastic process 313
total mass 336
Schmidt number 192, 251, 323,
        367
Sea salt
    aerosol 498
    concentration as a function
            of wind speed 422
    flux from the ocean to the
            atmosphere 496
    height variation of 430
    inversion 428
    particles 421
        size distribution of 426
Sea-source aerosols
    chemical composition of 479
    radioactive nuclides in 482
Sea surface 273

films 270
    material 274
    microlayer 272, 286
    roughness of 43
Sea water
    buoyancy 91
    physical properties of 62
Sediment accumulation rates 514
Settling particles 492
Sherwood number 191, 214, 322
Shipboard samples 425
Slicks 275, 276
Slope winds 28
Smooth flow, hydrodynamically
        42
Solar(or short-wave) energy 11
Solar flux 182
Solar radiation 147, 180
Solubility 243, 258, 268
    coefficient 306, 352
        Ostwald/Henry's law 386
Source identification 458
Sources
    anthropogenic 175
    natural 175
Stability
    categories 48
    functions 43
Stagnant liquid 193
Stefan flow 336, 363, 369
Stefan velocity 369
Sterol 487
Stokes flow 322, 326
Stokes number (or impaction
        parameter) 325, 327,
        329, 370
Storm efficiencies 354
    convective storms 356
Stratification
    conditionally unstable 34
    density 32
        stability of 37
    parameter 249
    ratio 79
    stable 31
    stability 32, 39
    unstable 31, 34
Stratosphere 18, 288
Substantive derivative 3
Subtropical highs 17, 19, 23

Subtropical jet streams 17
Sulphate ($SO_4^=$) 174, 188
    aerosol 188, 285
    excess 285
Sulphite ($SO_3^=$)
    oxidation of 215, 216, 231
    species 206, 207
Sulphur 285, 476
    compounds 503
    gases 260
    global budgets 285, 503
    volatile 285
Sulphur dioxide (SO_2) 174, 177
        188, 269, 286
    absorption 203
        flux 208
    conversion rates 189
    dissolved 206
Sulphuric acid (H_2SO_4) 188
Supersaturation 283
Surface-active compounds 271
Surface-active organic
            substances 486
Surface adsorption onto solid
            particles 487
Surface curvature effects 412
Surface films 190
Surface free energy 415
Surface layer, see Layer
Surface microlayer
    composition of the 481
    concentrations of
        S. marcescens 437
Surface-renewal theory 207,
        277
Surface seawater 491
    dissolved material 484
    particulate material 484
    trace metals in 487
Surface resistance 189
Surface tension 268
Surfactants
    dry 271, 272, 274
    soluble 273
    wet 272, 274

Temperature, potential 62
Terminal settling velocity 198
Terrestrial (or long-wave)
            radiation 11

Thermal wind 7, 11, 24, 29, 37
Thermoclines 84
    main 59
    permanent 65, 84
Thermodynamics, first law of 7
Thermophoresis 336
Thermostad 63
Tides 80
    baroclinic 82
    barotrophic 82
    internal 82
Time scale 181
    associated with the varia-
        tions of the air-phase
        concentrations 229, 230
    for mass transfer 229
Trace gas concentrations 175
Tracers 63
    and source markers 466
Trade winds 17, 19, 23
Trajectory analysis 47
Transfer coefficients 45
Transfer to a rough surface 253
Transfer velocities 241, 261,
            263, 264,266, 267, 269,
            272, 273, 276, 277,
            282, 289
    Schmidt number dependencies
        376
    Stokes number dependencies
        376
Transient tracers in the ocean
    (TTO) 65
Transport
    by convective motions 38
    Ekman 56, 59, 76
    equations 218
    geostrophic 59
    molecular 37, 40, 41
    resistances 46
    Sverdrup 58
    turbulent 37, 40, 41
    vertical 45
Trichloroethane (CH_3CCL_3) 177
Trichlorofluoromethane (CCL_3F)
    288
Tropical cyclones(hurricanes)
    23, 24
Tropical depressions 23
Tropical storms 23

Trophic structure 126
Troposphere 18, 288
Tropospheric residence times,
            particle-size
            dependent 398
T-S diagram 63
Turbulence 36, 39
    energy of 36
Turbulent bursts 379
Turbulent flow 36
Turnover time 309

Upwelling 283

Velocity
    ascension 397
    profile 41, 251
Vertical mixing 283
Viscous sublayer 40, 371
    thickness 44
Volatility 459
Volcanic aerosols 469
Vorticity 61, 94, 102, 111

Washout factor 500
Water 253
    boundary layer 251
    bubble interface 268
    droplets 173
    global balance 22
    property, distributions 63
    to-air transfer of bacteria
        434
    type 63
    vapour 8
Wave damping, capillary 276
Wavenumber 111
Wave slicks, internal 149
Waves 42
    and gas transfer 247
    breaking 98, 103, 249
        of small-wavelength 105
    capillaries 42, 100, 113, 247,
            272, 275
    coupled fluctuations in air
        pressure 107
    deep water 112
    dispersion relation 101, 112
    easterly 23
    energy density 112

generation and growth 106
groups 100
group speed 112
internal 71, 83
in the presence of a slick
         102
large-scale 80
lengths 100
momentum density 113
nonlinear interactions 107
phase speed 112
saturation 105
sea 100
sea surface 98
steepness 101
Stokes drift 102
swell 100
tidal 100
tsunamis 100
Westward intensification 59
Wet deposition, see Deposition
Wet flux 339
Whitecap coverage 413
Wind
    and gas transfer 245
    geostrophic, shear of 37
    katabatic 28
    mountain slope 28
    shear 37
    stress 37
    tunnels 243
        circular 245
        linear 245
    valley 28
Windrows 94